LEONARD LO

RANDOM POINT PROCESSES

SIMÉON DENIS POISSON 1781–1840)
(Photo. Bibl. nat. Paris)

Siméon Denis Poisson was born June 21, 1781, in Pithiviers, France, and died April 25, 1840, in Paris. At first he studied medicine but later abandoned it in favor of mathematics. In 1798, at the age of 17, he entered the École Polytechnique where he attracted the attention of J. L. Lagrange, P. S. Laplace, and J. B. J. Fourier. At the age of 18 he wrote a memoir on finite differences that was recommended for publication by A. M. Legendre. He became a deputy professor at the École Polytechnique in 1802 and a full professor in 1806.

At least 300 scientific publications are attributed to Poisson. The distribution bearing his name appears first in his expositions on the application of probability to the administration of justice. His most famous publication on this topic, *Recherches sur la Probabilité des Jugements*, was written in 1837.

RANDOM POINT PROCESSES

DONALD L. SNYDER

Washington University
St. Louis, Missouri

A Wiley-Interscience Publication

JOHN WILEY & SONS, New York · London · Sydney · Toronto

Library of Congress Cataloging in Publication Data:

Snyder, Donald Lee.
Random point processes.

"A Wiley-Interscience publication."
Includes bibliographical references and indexes.
1. Point processes. I. Title.

QA274.42.S69 519.2'3 75-11556

ISBN 0-471-81021-5

Printed in the United States of America

10 9 8 7 6 5 4 3 2 1

PREFACE

It was Jerome R. Cox, Jr., founder and director of Washington University's Biomedical Computer Laboratory, who first interested me in the aspects of random point processes that form the theme of this book. At the time I joined BCL in 1969, there was already an active project dealing with the computer processing of radioactive-tracer data obtained daily in clinical and research laboratories in the Mallinckrodt Institute of Radiology of the Washington University School of Medicine. There was then, as now, a sizable group of enthusiastic participants in this BCL effort: Carol S. Coble, Rexford L. Hill, Kenneth B. Larson, Joanne Markham, Nizar A. Mullani, and Maxine L. Rockoff were all members of this group. BCL's effort was conducted in collaboration with E. James Potchen and Michel M. Ter-Pogossian, directors of the Divisions of Nuclear Medicine and Radiation Sciences, respectively, and John O. Eichling, Michael E. Phelps, and Roger H. Secher-Walker. My thoughts on point processes have been influenced to some extent by all these people who became my colleagues in 1969. But it was Jerome R. Cox, Jr. who suggested my using the statistical estimation and decision theory I had previously applied only to communication and control problems for the analysis of radioactive-tracer data. Thus, the theme of this book is the application of concepts and techniques now used routinely by communication and control theorists to point-process models.

A detailed preview of the book may be found in Section 1.3 of Chapter 1, so I will not elaborate further on its contents here. Basically, what I have done is develop the Poisson process in detail in Chapter 2. Then in subsequent chapters, a hierarchy of increasingly more complex point-process models are derived from the Poisson process by making a series of modifications of its properties. Each modification is introduced by illustrating its usefulness in a physical application. The final model in Chapter 7 is, I think, one of substantial generality and potential usefulness. Although such matters are difficult to judge, I feel the procedure I have followed of gradually building up to this final model, rather than starting with it in Chapter 2 and examining numerous special cases, will help the reader appreciate its deficiencies as well as its potentialities.

I have used early drafts of this book for a one-semester graduate course in the Electrical Engineering Department at Washington University. Students in the course have had a wide range of interests and backgrounds.

Most have been advanced-degree candidates in either the Electrical Engineering or System Science and Mathematics Departments. Several were enrolled in the Biomedical Engineering Program, which spans several departments in the engineering school. An elementary course in probability and random processes is a prerequisite for the course; the level and material required is surveyed in Section 1.4 of Chapter 1. In addition, students in the course have had previous exposure to statistical estimation and decision theory, but I have incorporated this material in the text, particularly in Chapter 2, so it is not a prerequisite. While my own students in this course have been in engineering disciplines, I have attempted to write the book for a broader audience. I feel statisticians, physicists, operations analysists, and others will also find it useful.

Several people have contributed to the preparation of the text, and it is a pleasure to acknowledge them. Dr. Jerome R. Cox, Jr., not only launched me on the project by his suggestions, but he also continually encouraged and supported me while I pursued it. I am grateful to Professor Russell R. Pfeiffer, Chairman of the Electrical Engineering Department, for his support during the preparation of the manuscript. I thank Philip M. Fishman for his many constructive suggestions. Two students, Marco V. Vaca and Richard C. Setzer, worked the exercises that appear at the end of Chapters 2 through 6. Philip M. Fishman, Donald J. Halski, Joseph L. Hibey, and Marco V. Vaca all read what I thought was the final manuscript and made several valuable suggestions for its improvement.

I thank the Bibliothèque Nationale, Paris, for permission to reproduce their fine portrait of S. Poisson. I also thank the Institute of Electrical and Electronic Engineers, New York, for permission to reproduce in Chapter 6 figures from my articles, "Statistical Analysis of Dynamic Tracer-Data," and, "Filtering and Detection for Doubly Stochastic Poisson Processes," which appear in the January, 1973 *Transactions on Biomedical Engineering* and the January, 1972 *Transactions on Information Theory*, respectively; the Society of Nuclear Medicine for permission to reproduce in Chapter 6 figures from my chapter in *Computer Processing of Dynamic Images From an Anger Scintillation Camera*, published by the Society in 1974; and the M. I. T. Lincoln Laboratory for permission to reproduce in Chapters 1, 4, and 5 figures from Technical Note 1969–18, "Preliminary Analysis of ELF Noise," written by Dr. James E. Evans in 1969.

The preparation of the manuscript was done under partial support of the National Institutes of Health under Research Grant Number RR00396 from the Division of Research Resources and by the National Science Foundation under Research Grant Number GK-32239.

DONALD L. SNYDER

St. Louis, Missouri
June, 1974

CONTENTS

1 Point and Counting Processes: Introduction, Preliminaries 1

1.1 Introduction, 2
1.2 Counting Processes, 7
1.3 Organization of the Book, 9
1.4 Mathematical Preliminaries, 13

Probability Spaces, 13
Random Variables, 14
Stochastic Processes, 24
Sequences and Convergence Concepts, 30

2 The Poisson Process 37

2.1 Introduction, 38
2.2 Conditions for the Poisson Counting Process, 41
2.3 Time Statistics for the Poisson Counting Process, 56

Sample Function Density, 62
Conditional Distribution of Occurrence Times, 65

2.4 Parameter Estimation for Observed Poisson Processes, 70

Maximum Likelihood Estimation, 75
Parameter Estimation Accuracy, 79

2.5 Hypothesis Testing for Observed Poisson Processes, 87

Binary Hypothesis Testing, 88
Binary Hypothesis Testing Performance, 94
M-ary Hypothesis Testing, 103

References, 106
Problems, 108

3 Marked Point Processes: The Compound Poisson Process 127

3.1 Introduction, 128
3.2 Statistics of a Compound Poisson Process, 130

Characteristic Functional for a Compound Poisson Process, 130
First- and Second-Order Statistics for a Compound Poisson Process, 135
3.3 Representation of a Compound Poisson Processes, 138

Applications to Statistical Inference, 145

References, 157
Problems, 158

4 Filtered Poisson Processes **163**

4.1 Filtered Poisson Processes Generated by a Superposition of Point Responses, 164

Statistics for a Filtered Poisson Process, 168
A Central Limit Theorem, 178
4.2 Poisson Driven Markov Processes, 185

Calculus for Poisson Driven Markov Processes, 192
Statistics for Poisson Driven Markov Processes, 210

References, 222
Problems, 223

5 Self-Exciting Point Processes **235**

5.1 Introduction, 236
5.2 General Self-Exciting Point Processes, 237

Counting Statistics for Self-Exciting Point Processes, 240
Time Statistics for Self-Exciting Point Processes, 244
A Limit Theorem for Sums of Self-Exciting Point Processes, 253
5.3 Self-Exciting Point Processes with Limited Memory, 256

Characterization of m-Memory Processes for $m=0$, 257
Characterization of m-Memory Processes for $m>0$, 264

References, 272
Problems, 273

6 **Doubly Stochastic Poisson Processes** 283

6.1 Introduction, 284
6.2 Counting Statistics for Doubly Stochastic Poisson Processes, 286

Counting Statistics by the Method of Conditioning, 287
Counting Statistics by the Method of Self-Exciting Point Processes, 292
Limit Theorems for Counting Statistics, 295

6.3 Time Statistics for Doubly Stochastic Poisson Processes, 300

Time Statistics by the Method of Conditioning, 301
Time Statistics by the Method of Self-Exciting Point Processes, 309

6.4 Hypothesis Testing for Doubly Stochastic Poisson Processes, 310
6.5 Filtering for Doubly Stochastic Poisson Processes, 314

Linear Minimum Mean Square Error Filtering, 317
Nonlinear Minimum Mean Square Error Filtering, 330
Suboptimal Nonlinear Filtering, 352
Nonlinear Filtering Performance for Gaussian Information Processes, 371

6.6 Multichannel Doubly Stochastic Poisson Processes, 383
6.7 Smoothing for Doubly Stochastic Poisson Processes, 392

Fixed-Point Smoothing, 399
Fixed-Interval Sweep Smoothing, 401
Fixed-Lag Smoothing, 402

References, 403
Problems, 405

7 Marked Point Processes 433

7.1 Introduction, 434
7.2 Marked Point Processes Having a Denumerable Mark Space, 436

Sample-Function Density: Denumerable Mark Space, 443
Doubly Stochastic Marked Point Processes: Denumerable Mark Space, 449
Filtering for Marked Point Processes, 455

7.3 Marked Point Processes Having a Nondenumerable Mark Space, 466

Sample Function Density:Nondenumerable Mark Space, 469
Doubly Stochastic Marked Point Processes: Nondenumerable Mark Space, 471

References, 476

Author Index, 477

Subject Index, 479

Examples Index, 483

POINT AND COUNTING PROCESSES: INTRODUCTION, PRELIMINARIES

CHAPTER ONE

1.1 INTRODUCTION

A random point process is a mathematical model for a physical phenomenon characterized by highly localized events distributed randomly in a continuum. Each localized event is represented in the model by an idealized point to be conceived of as identifying the position of the event in the continuum. If $\mathfrak{X}$ denotes the continuum space, then a realization of a random point process on $\mathfrak{X}$ is a set of points having coordinates in $\mathfrak{X}$. The number and variety of phenomena for which this type of stochastic process provides a reasonable mathematical model is surprisingly large. Here are ten examples.

Example 1.1.1 (Electron Emissions). Suppose the heating element in a vacuum tube is energized at time t_0 and that electron emissions from the cathode are observed for the subsequent T seconds. Here, the space $\mathfrak{X}$ is an interval $\{t: t_0 \leqslant t \leqslant t_0 + T\}$ of the real line. Each element t of $\mathfrak{X}$ represents a time during the observation period. A realization of the random point process on $\mathfrak{X}$ is a sequence of times $\{t_1, t_2, \ldots, t_N\}$, where $t_0 \leqslant t_1 < t_2 < \cdots < t_N \leqslant t_0 + T$. Here, t_i represents the emission time of the ith electron and N the total number of emitted electrons during the observation period.

Example 1.1.2 (Electrical Activity in Nerves). A microelectrode inserted into a nerve fiber can detect electrical activity in the form of randomly occurring "spike discharges." Suppose the electrode is inserted at time t_0 and that discharges are observed for the following T seconds. The space $\mathfrak{X}$ is that of Example 1.1.1. Here, t_i represents the occurrence time of the ith discharge and N the total number of discharges during the observation interval. An example of spontaneous electrical activity observed in the auditory nerve of a cat is shown in Figure 1.1.

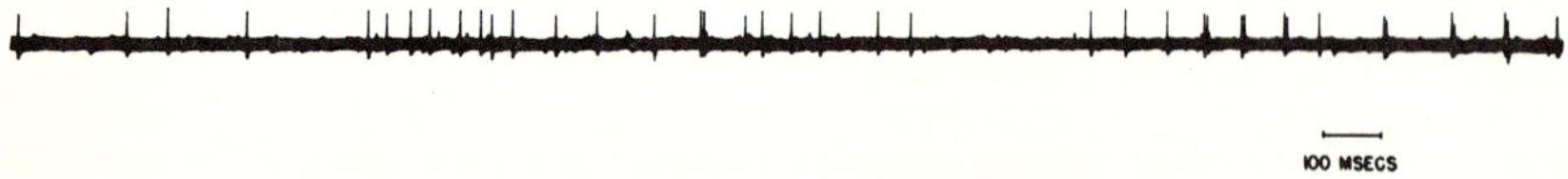

Figure 1.1. Spontaneous electrical activity recorded in the auditory nerve of a cat.

Example 1.1.3 (Radioactivity). Gamma or annihilation photons generated in radioactive decay are emitted at a rate that depends on the quantity of radioactive material present. Emissions in a T-second interval starting at time t_0 lead to the space $\mathfrak{X}$ of Example 1.1.1. The times t_i are photon emission times, and N is the total number of photons emitted during the interval $[t_0, t_0 + T]$.

Example 1.1.4 (Lightning Discharges). Lightning discharges occurring in the earth's atmosphere produce interference in a wideband radio receiver operating at low frequencies. This type of interference is called atmospheric noise and is significant at frequencies below about thirty kilohertz. Suppose the receiver is energized at time t_0 and discharges are observed for the following T seconds. The space $\mathfrak{X}$ is that of Example 1.1.1. Each t_i corresponds to the instant a discharge is observed. Examples of atmospheric noise observed in Malta at the output of a loop antenna are shown in Figure 1.2.

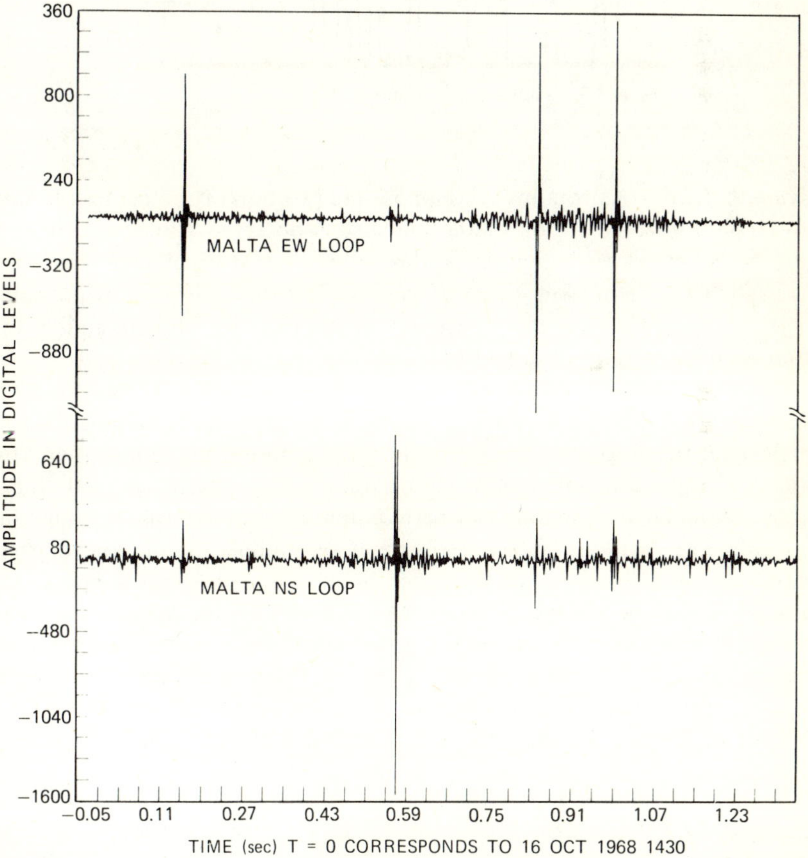

Figure 1.2. Low-frequency atmospheric-noise recorded in Malta from a loop antenna (having East–West and North–South orientation in the upper and lower traces, respectively). From J. Evans [3].

Example 1.1.5 (Photoelectron Emissions). An optical field incident on a detector such as a photomultiplier or avalanche photodiode results in the emission at the detector output of photoelectrons intermixed with extraneous, internally generated thermoelectrons. Emissions in a T-second interval starting at t_0 lead to the space $\mathfrak{X}$ of Example 1.1.1. The times t_i are electron emission times. An example of a pulse process observed at the output of a photomultiplier is shown in Figure 1.3. These data were obtained by F. Davidson using coherent laser light. An individual pulse is of about 30 nanoseconds duration, and the pulses occur at a rate of about 2×10^6 per second.

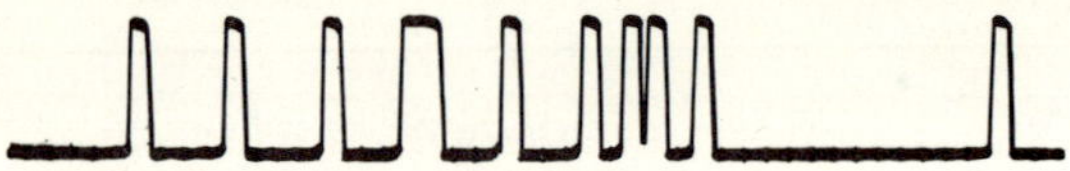

Figure 1.3. Photomultiplier pulses.

Example 1.1.6 (Astronomy). Suppose the position of the center of gravity of stars in some region of the universe is observed. Here, the space $\mathfrak{X}$ can be taken as a region in a three-dimensional Euclidean space. A realization of the random point process on $\mathfrak{X}$ is the collection of positions $\{\vec{r}_1, \vec{r}_2, \ldots, \vec{r}_N\}$, where $\vec{r}_i$ represents the position of the ith star and N the total number of stars observed in the region.

Example 1.1.7 (Scintigraphy in Nuclear Medicine). A technique used in nuclear medicine to visualize internal organs is to introduce into them (by injection, ingestion, inhalation, etc.) a radioactive pharmaceutical tracer whose radioactive emissions can be monitored by an external scintillation detector. The positions of observed gamma or annihilation photons depend on the spatial distribution of the tracer. Suppose the positions are observed as light flashes in a large, two-dimensional scintillation crystal. Then the space $\mathfrak{X}$ can be taken as a region in a two-dimensional Euclidean space. A realization of the random point process on $\mathfrak{X}$ is the collection $\{\vec{r}_1, \vec{r}_2, \ldots, \vec{r}_N\}$, where $\vec{r}_i$ is the position of the ith light flash and N the total number of flashes. An example of a scintigraph of human lungs is shown in Figure 1.4.

Example 1.1.8 (Electron Microscopy). In electron microscopy, a flood of electrons illuminates a microscopic object to be imaged. The electrons are collected in a scintillation crystal and their positions observed. The space $\mathfrak{X}$ can be taken as a region of a two-dimensional Euclidean space

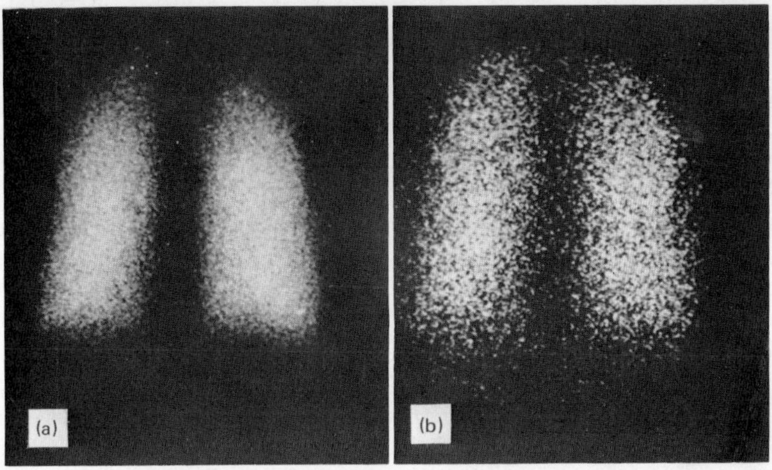

Figure 1.4. Scintigraph of the human lungs following inhalation of xenon-133 at (a) high concentration and (b) low concentration.

corresponding to the crystal surface. A realization of the stochastic point process on $\mathfrak{X}$ is the collection $\{\vec{r}_1, \vec{r}_2, \ldots, \vec{r}_N\}$ of positions of the electron impacts in the crystal and N is the total number.

Example 1.1.9 (Dynamic Studies in Nuclear Medicine). A technique used in nuclear medicine to determine parameters relating to dynamic physiological phenomena such as blood-flow rate, metabolic-exchange rates, and lung ventilation is to introduce a radioactive pharmaceutical tracer into the organ or region of interest and then to monitor the entire time course of radioactivity as the pharmaceutical is removed from the field of view of the detector as a result of the phenomenon of interest. The space $\mathfrak{X}$ can be taken as the Cartesian product of the time interval $[t_0, t_0 + T]$ corresponding to the observation interval and a region of a two-dimensional Euclidean space corresponding to the surface of a scintillation crystal used to detect the radioactivity. A realization of the stochastic point process on $\mathfrak{X}$ is the collection $\{(t_1, \vec{r}_1), (t_2, \vec{r}_2), \ldots, (t_N, \vec{r}_N)\}$ of times and locations of detected photons, where t_i and $\vec{r}_i$ are the time and location of the ith photon and N is the total number.

Example 1.1.10 (Seismic Events). Seismic events such as earthquakes can be observed on a worldwide basis with an array of seismometers. Suppose that seismic events are observed for T seconds starting at time t_0 and the epicenter of each such event in the earth is determined from the observations. Here the space $\mathfrak{X}$ can be taken as the Cartesian product of the time interval $[t_0, t_0 + T]$ and a region of three-dimensional Euclidean space corresponding to the earth. A realization of the stochastic point process on $\mathfrak{X}$ is the collection $\{(t_1, \vec{r}_1), (t_2, \vec{r}_2), \ldots, (t_N, \vec{r}_N)\}$ of times and locations of the seismic events.

An observer of point phenomena is often not directly interested in the point realizations themselves but, rather, in what can be inferred from them about some underlying mechanism that influences where the points occur. For instance, in Example 1.1.2, an observer of spike discharges in an auditory nerve may be interested in the transformation that takes place between an acoustic pressure signal he applies at the outer ear and the electrical activity he measures. In Example 1.1.3, an observer of radioactivity may want to know the half-life of the isotope he monitors. In Example 1.1.5, photoelectron emissions in optical communication and radar systems are influenced by analog or digital signals that modulate the optical field incident on a photodetector in the receiver; there is a desire to recover these modulating signals from the photodetector's output. In Example 1.1.8, an electron microscopist wants structural information about the object he illuminates with electrons. Finally, in Example 1.1.9, a radiologist using radioactive tracers in dynamic-function studies wants to determine physiological parameters from the time course of radioactivity that he observes.

Thus underlying mechanisms of interest are often associated with observed point phenomena. As a result, there is a need for a systematic theory of information processing for such observations. The formulation of this theory requires mathematical models for point phenomena; a wide variety of models are developed in the text. These models are then used with statistical estimation theory and statistical hypothesis testing to make inferences about underlying mechanisms that influence observed point phenomena.

The space $\mathfrak{X}$ in the above examples is seen to represent time (Examples 1.1.1–1.1.5), a geometrical space (Examples 1.1.6–1.1.8), or a combination of these (Examples 1.1.9–1.1.10). In a general setting of the theory, the space $\mathfrak{X}$ can be quite abstract. However, this generality will not be needed for our purposes, and for the first six chapters of the book we will take $\mathfrak{X}$ to be a semiinfinite real line representing time. In Chapter 7, we examine multidimensional point processes, and $\mathfrak{X}$ will then be taken as a Euclidean space or as a Euclidean space with an additional dimension representing time.

There are two particular features of a stochastic point process for which statistical characterizations are normally sought. The first is interpoint spacings, with most emphasis placed on nearest-neighbor spacing. Thus, for instance, the statistics of intervals between spike discharges in Example 1.1.2 or photoelectron emissions in Example 1.1.5 are desired. These statistics are termed *time statistics* in this book. Time statistics are characterized for a wide variety of models in Chapters 2, 5, 6, and 7.

The second feature of a point process for which statistical characterizations are sought is the number of points falling in subregions of the space where points occur. For instance, with reference to Example 1.1.3, the statistics of the total number of photons emitted during $[t_0, t_0+T]$ may be desired, or, with reference to Example 1.1.10, the statistics of the number of seismic events in the years 1940 to 1970 in New Zealand may be desired (see D. Vere-Jones[21] for an interesting discussion of this latter topic). These statistics are termed *counting statistics* in this book. Counting statistics are characterized for a wide variety of models in Chapters 2, 5, 6, and 7. A particularly useful concept in discussing counting statistics is that of a *counting process*. This is defined in the following section.

1.2 COUNTING PROCESSES

It is often of interest in practice to count numbers of points in subsets of the space $\mathfrak{X}$ on which a point process is defined. A *counting process* is introduced for this purpose and can be associated with every point process. The idea is as follows. We assume that a realization, call it ω, of a stochastic point process on $\mathfrak{X}$ is a denumerable point set of $\mathfrak{X}$. This means that ω can be enumerated as $\omega = \{X_1, X_2, \ldots\}$, where each X_i denotes the coordinate of a point in $\mathfrak{X}$. Now, let A be a subset of $\mathfrak{X}$, and denote by $N(A;\omega)$ the number of points in ω that lie in A. Formally,

$$N(A;\omega) = \sum_i 1_A(X_i) \tag{1.1}$$

where each X_i is the coordinate of a point in ω and where $1_A(X)$ denotes the characteristic function for the set A; that is,

$$1_A(X) = \begin{cases} 1, & X \in A \\ 0, & X \notin A. \end{cases} \tag{1.2}$$

When viewed as a function of A and ω, $N(A;\omega)$ defines a nonnegative, integer-valued random process on $\mathfrak{X}$. A process constructed this way is called a counting process. It is customary to suppress the dependency of $N(A;\omega)$ on ω and write the number of points in A simply as $N(A)$. We adhere to this custom in the remainder of the book.

As an example, suppose $\mathfrak{X} = \{t; t \geq t_0\}$ corresponding to a point process in time for times after an initial time t_0. The stochastic point process on $\mathfrak{X}$ can be interpreted as modeling electron emission times for $t \geq t_0$, lightning discharges for $t \geq t_0$, photoelectrons for $t \geq t_0$, and so forth. If A is an interval of time, $N(A)$ is the number of events having epochs in the interval. For the set $A=\{t: s \leqslant t<u\}$ where $t_0 \leqslant s$, we introduce the notation

$$N(\{t: s \leqslant t<u\}) \triangleq N_{s,u}. \tag{1.3}$$

Thus, $N_{s,u}$ is the number of points in the interval from s to u possibly including a point at s but not at u. We further introduce a stochastic process $\{N_t; t \geqslant t_0\}$ according to the definition

$$N_t \triangleq N_{t_0} + N_{t_0,t} \tag{1.4}$$

where N_{t_0} is a nonnegative integer that most often will simply be set to zero. Viewed as a function of t, $\{N_t; t \geqslant t_0, N_{t_0}=0\}$ has a unit jump at the epoch of each point as shown for a typical sample function in Figure 1.5; N is continuous from the left at each of its jumps.

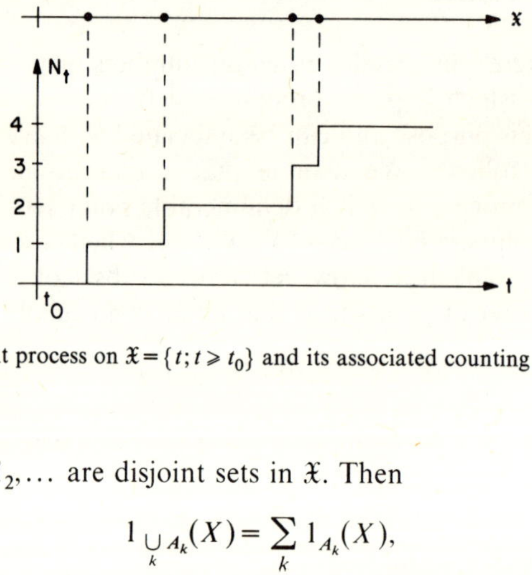

Figure 1.5. A point process on $\mathfrak{X}=\{t; t \geqslant t_0\}$ and its associated counting process $\{N_t; t \geqslant t_0\}$.

Suppose $A_1, A_2, \ldots$ are disjoint sets in $\mathfrak{X}$. Then

$$1_{\bigcup_k A_k}(X) = \sum_k 1_{A_k}(X),$$

and we have from (1.1) that

$$N\left(\bigcup_k A_k\right) = \sum_i \sum_k 1_{A_k}(X_i) = \sum_k N(A_k). \tag{1.5}$$

Thus, $N(A)$ is an additive set function on $\mathfrak{X}$; the number of points in a union of disjoint sets in $\mathfrak{X}$ is the sum of the number in each set. As an example, suppose $\mathfrak{X}=\{t; t \geqslant t_0\}$. If $s \in [t_0, t)$, then $[t_0, t) = [t_0, s) \cup [s, t)$ and $[t_0, s) \cap [s, t) = \phi$. Consequently, by the additivity property of N we have

$$N_{s,t} = N_t - N_s \tag{1.6}$$

Thus, $N_{s,t}$ is the random variable denoting the increment of the counting process N on the interval $[s,t)$.

1.3 ORGANIZATION OF THE BOOK

The organization of the text material is indicated in Figure 1.6. The *Poisson process* in time is the basic building block. We begin with its definition and an investigation of its properties in Chapter 2. Qualitative conditions a point process must possess to be a Poisson process are given. These facilitate making judgments about the appropriateness of a Poisson model in particular applications. The time and counting statistics for a Poisson process are developed, and, on the basis of these statistics, procedures are outlined for numerically generating a Poisson process for use in simulation or Monte Carlo studies of situations too complex for direct analysis. The important practical problem of estimating parameters that influence the rate at which points occur in a Poisson process is discussed. Major emphasis in the chapter is placed on maximum likelihood estimation, which is developed in detail.

Compound Poisson processes are studied in Chapter 3. These are obtained from the Poisson process by associating an auxiliary random variable, called a *mark*, with each point occurrence. Successive marks are selected independently from a common distribution. A compound Poisson process might be used, for instance, as a first approximation model for the lightning discharges of Example 1.1.4. Here, the Poisson process would model the occurrence times of the discharges and the marks would indicate the energy of each discharge. In this way, the amplitude as well as the location of the pulses in Figure 1.2 can be modeled. A compound Poisson process is shown to be decomposable into a linear superposition of Poisson processes. Decompositions of this type have a traditional role in the theory of random processes, particularly independent-increment processes. We use the decomposition to specify procedures for estimating unknown parameters that influence an observed, compound Poisson process.

Point processes often have a direct influence on systems or devices of interest. The classic example is shot noise, which occurs in connection with the electron emissions of Example 1.1.1 and which, for example, limits the signal detectability in radio and radar recievers. Another example is lightning-induced atmospheric noise, which also limits signal detectability in radio receivers; here, the energy marks described above also play a role in characterizing the noise. Thus, in Chapter 4, we study *filtered Poisson processes* and *Poisson-driven Markov processes*. These processes are the response of linear and nonlinear dynamical systems to the marked Poisson

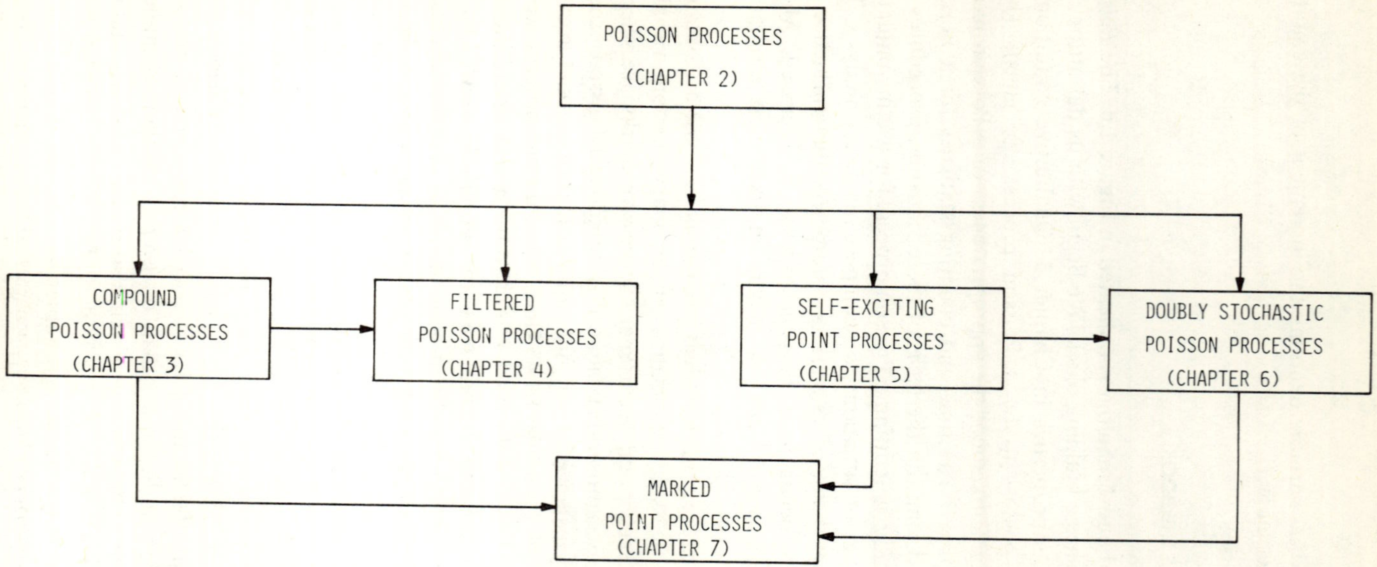

Figure 1.6. Organization of book.

processes of Chapter 3. There are two major sections in the chapter. The response of linear systems is considered in the first. Here, the response is a superposition of the individual responses to each marked point. Statistics for such processes are developed in detail, including the historic Campbell's theorem. Applications to the modeling of shot noise and photoelectron noise are given. The second major section of Chapter 4 deals with the modeling of responses that are not a superposition of separate responses to each point. The main mathematical tool is the theory of stochastic differential and integral equations, which we develop in some detail but not with the generality or rigor that may be found in the more mathematical texts by A. Skorokhod [19] and I. Gihman and A. Skorohod [6]. The use of stochastic differential and integral equations for describing the response of linear and nonlinear systems driven by a continuous random process, rather than a point process, is by now nearly standard in the engineering and applied sciences literature; this approach is developed in the texts by A. Jazwinski [8], A. Sage and J. Melsa [18], and E. Wong [22]. Our treatment in Chapter 4 for the response to a point process parallels these developments. While a familiarity with these alternative discussions is, of course, helpful, it is not needed for understanding Chapter 4.

A distinguishing property of the Poisson process is that it evolves in time without aftereffects. This means that the occurrence times and number of points realized before some time have no influence on the occurrence times and number of points subsequent to that time. While this property is often met to a reasonable degree for many point phenomena encountered in practice, it is not a good hypothesis universally. An example where it is not a good hypothesis occurs in connection with the nerve discharges described in Example 1.1.2. It is observed experimentally that subsequent to each discharge there is a brief refractory period during which additional discharges do not occur. A similar effect can be observed in scintillation crystals used in the dynamic studies in nuclear medicine described in Example 1.1.9. The restriction that points occur without aftereffects is removed in Chapter 5 where we define and study *self-exciting point processes*. For these processes, the future evolution of points can be influenced by all past occurrences of points. In many applications, the evolution does not depend on the entire past but only a particular feature of the past such as the most recent occurrence time or the total number of past occurrences. The most notable of these processes because of their wide use are *renewal processes* and *Markov birth processes*. Self-exciting point processes with limited memory are characterized in Chapter 5.

Doubly stochastic Poisson processes are obtained as a generalization of the Poisson process when the rate at which points occur is allowed to be a random process; a doubly stochastic Poisson process is conditionally a Poisson process given the stochastic rate process. Doubly stochastic Poisson processes are reasonable models for the photoelectron emissions of Example 1.1.5 when the incident optical field is incoherent Gaussian light as emitted, for example, by a light-emitting diode or an astronomical source. The photoelectron generation rate in an optical communication system employing coherent laser light as a carrier is also stochastic, because of intentional modulation at the transmitter by information-bearing signals and because of unintentional modulation by fading processes in atmospheric propagation. In Chapter 6, we develop algorithms for processing a sequence of observed point occurrences in order to estimate random signals that influence the underlying and unobservable rate at which the observed points occur. Two approaches are considered. In the first, the estimate is constrained to be a linear function of the observed point process. A realization of the estimator as a Kalman–Bucy filter is given. This obviates the need for spectral factorization, and the estimator can be readily implemented numerically. The linearity constraint is removed in the second approach, which is remarkably parallel to the estimation theory initiated by R. Stratonovich [20] and H. Kushner [9] for observations of a continuous random process rather than a point process. Familiarity with the Stratonovich–Kushner theory provides a useful background for Chapter 6 but is not needed for its understanding.

The Poisson process of Chapter 2, compound Poisson process of Chapter 3, self-exciting point process of Chapter 5, and the doubly-stochastic Poisson process of Chapter 6 are all subsumed as special cases of the *marked point process* of Chapter 7. These are processes having an ancillary variable associated with each point. Point processes in multidimensional spaces are marked point processes in which marks indicate a spatial coordinate; jump processes, such as Markov jump processes, are marked point processes in which marks indicate state transitions or jump amplitudes. In the general marked point process, the future evolution of both points and marks can be influenced by all previous points and marks and, furthermore, by random signals. Thus, marked point processes are self-exciting and doubly stochastic. General principles are developed in Chapter 7 for estimating underlying parameters and random signals that influence an observed, marked point process.

Major sections are numbered in each chapter; for instance, Section 3.4 is the fourth section of Chapter 3. Examples and problems are numbered by sections; for instance, Example 3.4.2 and Problem 3.4.2 are the second example and problem of Section 3.4. References and problems appear at the end of each chapter. Indices to cited authors and to examples are at the end of the text.

1.4 MATHEMATICAL PRELIMINARIES

In this section, we survey certain elements of the theory of probability and stochastic processes that are used frequently in the substantive chapters which follow. The survey is intended mainly as a reference and not as source material for the uninitiated. This section also serves to establish the notation to be used subsequently. More detailed treatments of the material summarized here are readily available; in particular, we mention the elementary texts by E. Parzen [15, 16], A. Papoulis [14], and J. Melsa and A. Sage [11] and the more advanced texts by I. Gikman and A. Skorokhod [7], J. Doob [2], J. Neveu [13], and M. Loève [10].

Probability Spaces

A basic concept in the theory of probability and stochastic processes is that of a *probability space*. A probability space is comprised of three quantities: a sample space Ω, a collection $\mathcal{F}$ of certain subsets of Ω called a sigma field of events, and a probability measure $\mathcal{P}$. These quantities are defined as follows.

Sample Space. An element ω of Ω is a possible outcome of a random experiment. In the simplest situations, Ω consists of a denumerable collection of elements $\omega_1, \omega_2 \ldots$. Examples include the random experiment of tossing a die, for which Ω consists of six elements with ω_i corresponding to the ith face showing on the toss, and the random experiment of tossing a coin until a head shows, for which Ω consists of a denumerably infinite number of elements with ω_i corresponding to a head on the ith toss. Nondenumerable sample spaces are common. Examples arise with the random experiment of breaking a unit stick, for which $\Omega = \{\omega | 0 \leqslant \omega \leqslant 1\}$, and with the random experiment of failures of a lightbulb placed in service at time t_0, for which $\Omega = \{\omega | \omega \geqslant t_0\}$. Nondenumerable sample spaces predominate in later chapters.

More complicated sample spaces are possible. These include function spaces in which, for instance, each ω corresponds to a continuous function of t for $0 \leqslant t \leqslant 1$. A situation of more interest here occurs when each ω corresponds to a denumerable point set in some Euclidean space. While we will not do so, it is possible to develop a theory of stochastic point processes starting with such a space; see J. Moyal [12], P. Fishman [4], and P. Fishman and D. Snyder [5].

Sigma Field of Events. A subset of Ω is called an event. An example is the subset $A = \{\omega | 0.25 \leqslant \omega \leqslant 0.75\}$ of the sample space $\Omega = \{\omega | 0 \leqslant \omega \leqslant 1\}$ associated with the random experiment of breaking a unit stick; in words, A is the event that the break occurs in the central half of the stick. The sigma field of events $\mathcal{F}$ is a particular collection of events. For denumerable sample spaces, $\mathcal{F}$ may consist of all possible subsets of Ω, but this is not necessary. For nondenumerable sample spaces, $\mathcal{F}$ cannot consist of all possible subsets of Ω but, rather, only of certain subsets called *measurable events*. What is required of $\mathcal{F}$ are the following three properties:

1. $\mathcal{F}$ is not empty;
2. if an event A is in $\mathcal{F}$, so is its complement $A^c = \{\omega | \omega \in \Omega, \omega \notin A\}$;
3. if $A_1, A_2, \ldots$ are events in $\mathcal{F}$, so is their union $\cup_i A_i = \{\omega | \omega \in A_i$ for some $i\}$.

It follows immediately from these properties that Ω and its complement $\emptyset = \Omega^c$ are sets in $\mathcal{F}$; in this context, Ω and $\emptyset$ are called the sure and null events, respectively. Moreover, if $A_1, A_2, \ldots$ are events in $\mathcal{F}$, so is their intersection $\cap_i A_i = \{\omega | \omega \in A_i$ for all $i\}$.

Probability Measure. The probability measure $\mathcal{P}$ is a function that maps each set in $\mathcal{F}$ into the unit interval $[0, 1]$. This function is required to have the following properties:

1. $0 \leqslant \mathcal{P}(A) \leqslant 1$ for every set $A \in \mathcal{F}$;
2. $\mathcal{P}(\Omega) = 1$;
3. if $A_1, A_2, \ldots$ are disjoint sets in $\mathcal{F}$, then $\mathcal{P}(\cup_i A_i) = \Sigma_i \mathcal{P}(A_i)$.

So far as the theory is concerned, these are the only restrictions on $\mathcal{P}$; in other words, any function on $\mathcal{F}$ satisfying (1)–(3) is an admissible probability measure. However, it is only by a judicious selection of $\mathcal{P}$ that the mathematical model can be made useful in predicting occurrences in an actual physical application. An event A is said to occur *with probability one* or *almost surely* if $\mathcal{P}(A) = 1$.

Thus, a probability space, denoted by the triple $(\Omega, \mathcal{F}, \mathcal{P})$, consists of a sample space Ω, a sigma field $\mathcal{F}$ of subsets of Ω, and a probability measure $\mathcal{P}$.

Random Variables

Certain real-valued functions defined on a probability space are called *random variables*. To specify which functions, let $(\Omega, \mathcal{F}, \mathcal{P})$ be a probability space, and denote the real line by R. A function $x(\omega)$ with domain Ω (that is, $\omega \in \Omega$) and range R [that is, $x(\omega) \in R$ for each ω] is a random variable if for every $X \in R$, the set $\{\omega | x(\omega) \leqslant X\}$ is in the sigma field of events $\mathcal{F}$. The restriction to functions of this type ensures that the probability $\mathcal{P}(\{\omega | x(\omega) \leqslant X\})$ is well defined for all $X \in R$. Moreover, if B is any subset of R obtainable by forming countable unions, intersections, and complements of subsets of R of the form $\{r | r \leqslant X\}$, then $\mathcal{P}(\{\omega | x(\omega)$

$\in B\}$) is well defined. The function

$$P_x(X) = \mathcal{P}(\{\omega | x(\omega) \leqslant X\})$$

maps the real line R into the unit interval $[0, 1]$ and is called the *probability distribution function* for the random variable $x(\omega)$. We note that $P_x(X)$ is a nondecreasing function of X with $P_x(-\infty) = 0$ and $P_x(+\infty) = 1$.

A particular form for $P_x(X)$ is often postulated in applying the theory. In these cases, the probability space only plays a subsidiary role, and the explicit dependence of $x(\omega)$ on ω is normally suppressed. Thus, throughout most of the book, we simply refer to a random variable x having a given probability distribution function $P_x(X)$. We also use the notation $\Pr(B)$ for the probability a condition B on the random variable holds; that is $\Pr(B) = \mathcal{P}(\{\omega | B \text{ holds}\})$. Thus, $\Pr(x \leqslant X)$ is another way we write the probability distribution function $P_x(X)$ of x. It is readily verified that $\Pr(X_1 < x \leqslant X_2) = P_x(X_2) - P_x(X_1)$ for $X_1 < X_2$.

A random variable $x(\omega)$ is called *discrete* if it has only a countable number of possible values in R. Suppose the set of values is $\{X_1, X_2, \ldots\}$, where $X_i \in R$ for each i. Then

$$P_x(X) = \sum_i 1_{B(X)}(X_i) \mathcal{P}(\{\omega | x(\omega) = X_i\}), \tag{1.7}$$

where $B(X) = \{r | r \in R, r \leqslant X\}$ is the set of real numbers less than or equal to X, and where

$$1_{B(X)}(X_i) = \begin{cases} 1, & X_i \in B(X) \\ 0, & X_i \notin B(X). \end{cases}$$

It follows from (1.7) that the probability distribution function for a discrete random variable is piecewise constant with jump discontinuities at $X_1, X_2, \ldots$; the amplitude of the discontinuity at X_i is $\mathcal{P}(\{\omega | x(\omega) = X_i\}) = \Pr(x = X_i)$.

A random variable $x(\omega)$ is called *continuous* if there exists a function $p_x(X)$, called the *probability density function* for $x(\omega)$, such that

$$\Pr(x \in B) = \int_B p_x(X)\, dX$$

holds for any subset B of R for which $P(\{\omega | x(\omega) \in B\})$ is defined. The following relations between the probability distribution and density functions for $x(\omega)$ apply whenever the density exists:

$$\text{(a)} \qquad P_x(X) = \int_{-\infty}^{X} p_x(\alpha)\, d\alpha$$

and

$$\text{(b)} \qquad p_x(X) = \frac{dP_x(X)}{dX}.$$

Both discrete and continuous random variables will be encountered in our studies of stochastic point processes. The most prominent discrete variable, which arises repeatedly, is the number of points falling in some specified set; the values this random variable may take are nonnegative integers. The time interval between successive points of a point process evolving in time will always be a continuous random variable in the point processes we study.

The extension to more than one random variable defined on the same probability space is straightforward. Thus, if $x_1(\omega)$, $x_2(\omega),\ldots,$ and $x_n(\omega)$ are n random variables defined on $(\Omega, \mathcal{F}, \mathcal{P})$ the vector

$$x(\omega)=[x_1(\omega)x_2(\omega)\ldots x_n(\omega)]'$$

defines a mapping from Ω to R^n, where R^n denotes n-dimensional Euclidean space. This mapping is such that the set $\{\omega|x_1(\omega)\leqslant X_1,\ldots,x_n(\omega)\leqslant X_n\}$ is in $\mathcal{F}$ for every choice of $\mathbf{X}=[X_1\; X_2\cdots X_n]'\in R^n$. Consequently, the probability of this set is well defined, and we let

$$P_{x_1,x_2,\ldots,x_n}(X_1,X_2,\ldots,X_n)=P(\{\omega|x_1(\omega)\leqslant X_1,\ldots,x_n(\omega)\leqslant X_n\}). \quad (1.8)$$

This function is called the *joint probability distribution* for $x_1(\omega)$, $x_2(\omega),\ldots,x_n(\omega)$. For conciseness in notation, we also use $P_{\mathbf{x}}(\mathbf{X})$ and even $P_{\mathbf{x}}^{(n)}(\mathbf{X})$ for this function, the latter especially when there is possible ambiguity in the number n of variables involved. The random variables $x_1(\omega)$, $x_2(\omega),\ldots,x_n(\omega)$ are said to be statistically independent if their joint distribution function is the product of their separate distribution functions; that is, if

$$P_{\mathbf{x}}(\mathbf{X})=\prod_{i=1}^{n} P_{x_i}(X_i)$$

for all $\mathbf{X}\in R^n$. If there is a function

$$p_{\mathbf{x}}(\mathbf{X})=p_{x_1,x_2,\ldots,x_n}(X_1,X_2,\ldots,X_n)$$

such that

$$\Pr(\mathbf{x}\in B)=\int\cdots\int_B p_{\mathbf{x}}(\mathbf{X})\,dX_1\,dX_2\cdots dX_n$$

holds for an arbitrary set B in R^n for which the probability $\mathcal{P}(\{\omega|\mathbf{x}(\omega)\in B\})$ is defined, then the variables $x_1(\omega)$, $x_2(\omega),\ldots,x_n(\omega)$ are jointly continuous. The function $p_{\mathbf{x}}(\mathbf{X})$ is called the *joint probability density* for $x_1(\omega)$, $x_2(\omega),\ldots,x_n(\omega)$.

Suppose $\mathbf{g}(\cdot)$ is a function mapping R^n into R^m; a point $\mathbf{X}$ in R^n is mapped into the point $\mathbf{Y}=\mathbf{g}(\mathbf{X})$ in R^m. If $\mathbf{x}(\omega)$ is an R^n-valued random variable on $(\Omega, \mathcal{F}, \mathcal{P})$, then $\mathbf{y}(\omega)=\mathbf{g}(\mathbf{x}(\omega))$ is a composite function mapping Ω into R^m, but $\mathbf{y}(\omega)$ is not necessarily a random variable. $\mathbf{y}(\omega)$ is an R^m-valued random variable on $(\Omega, \mathcal{F}, \mathcal{P})$, defined by $\mathbf{x}(\omega)$ and $\mathbf{g}(\cdot)$, provided $\mathbf{g}(\cdot)$ is such that $\{\omega|y_1(\omega)\leqslant Y_1,\ldots,y_m(\omega)\leqslant Y_m\}$ is in $\mathcal{F}$ for all $\mathbf{Y}\in R^m$. Continuous functions mapping R^n into R^m satisfy this condition.

Suppose now that $\mathbf{g}(\cdot)$ is a one-to-one, invertible mapping of R^n into R^n such that both $\mathbf{g}(\mathbf{X})$ and its inverse function $\mathbf{g}^{-1}(\mathbf{Y})$ have continuous partial derivatives with respect to the elements of $\mathbf{X}$ and $\mathbf{Y}$, respectively. Then, $\mathbf{g}(\cdot)$ is a continuous function, and $\mathbf{y}(\omega)=\mathbf{g}(\mathbf{x}(\omega))$ is an R^n-valued random variable. Furthermore, if $\mathbf{x}(\omega)$ is a continuous random variable with joint probability density $p_{\mathbf{x}}(\mathbf{X})$, then $\mathbf{y}(\omega)$ is also a continuous random variable with joint density

$$p_{\mathbf{y}}(\mathbf{Y})=\left|\det\left(\frac{\partial \mathbf{g}^{-1}(\mathbf{Y})}{\partial \mathbf{Y}}\right)\right| p_{\mathbf{x}}(\mathbf{g}^{-1}(\mathbf{Y})), \tag{1.9}$$

where $\det(\mathbf{M})$ denotes the determinant of the matrix $\mathbf{M}$, and where $\partial \mathbf{g}^{-1}(\mathbf{Y})/\partial \mathbf{Y}$ denotes the Jacobian matrix of $\mathbf{g}^{-1}(\mathbf{Y})$; the *i,j-element* is $\partial g_j^{-1}(\mathbf{Y})/\partial Y_i$. As an application of (1.9), suppose $x_1(\omega), x_2(\omega), \ldots,$ and $x_n(\omega)$ are statistically independent, normally distributed random variables, $p_{x_i}(X_i) = (2\pi\sigma_i^2)^{-1/2}\exp(-X_i^2/2\sigma_i^2)$. Then

$$p_{\mathbf{x}}(\mathbf{X})=(2\pi)^{-n/2}[\det(\Lambda)]^{-1/2}\exp(-\tfrac{1}{2}\mathbf{X}'\Lambda^{-1}\mathbf{X}), \tag{1.10}$$

where

$$\Lambda=\operatorname{diag}(\sigma_i^2)=\begin{bmatrix} \sigma_1^2 & & & 0 \\ & \sigma_2^2 & & \\ 0 & & \ddots & \\ & & & \sigma_n^2 \end{bmatrix}.$$

Let $\mathbf{Y}=\mathbf{g}(\mathbf{X})=\mathbf{M}\mathbf{X}+\boldsymbol{\mu}$, where $\mathbf{M}$ is a nonsingular matrix. Then, $\mathbf{g}^{-1}(\mathbf{Y}) = \mathbf{M}^{-1}(\mathbf{Y}-\boldsymbol{\mu})$, and the probability density function for $\mathbf{y}(\omega)=\mathbf{g}(\mathbf{x}(\omega))$ is given by

$$p_{\mathbf{y}}(\mathbf{Y})=(2\pi)^{-n/2}[\det(\boldsymbol{\Sigma})]^{-1/2}\exp[-\tfrac{1}{2}(\mathbf{Y}-\boldsymbol{\mu})'\boldsymbol{\Sigma}^{-1}(\mathbf{Y}-\boldsymbol{\mu})], \tag{1.11}$$

where $\boldsymbol{\Sigma}=\mathbf{M}\Lambda\mathbf{M}'$. Random variables $y_1(\omega), y_2(\omega), \ldots, y_m(\omega)$ having this density are called jointly normal; we sometimes abbreviate $p_{\mathbf{y}}(\mathbf{Y})$ in (1.11) by $N(\boldsymbol{\mu},\boldsymbol{\Sigma})$.

The *expectation* or *mean value* of a random variable (r.v.) $\mathbf{x}(\omega)$ is denoted by $E(\mathbf{x})$ and defined by

$$\begin{aligned} E(\mathbf{x}) &= \int_{R^n} \mathbf{X}\,dP_{\mathbf{x}}(\mathbf{X}) \\ &= \begin{cases} \sum_i \mathbf{X}_i \Pr(\mathbf{x}=\mathbf{X}_i), & \mathbf{x} \text{ a discrete r.v.} \\ \int_{R^n} \mathbf{X} p_{\mathbf{x}}(\mathbf{X})\, d^n\mathbf{X}, & \mathbf{x} \text{ a continuous r.v.} \end{cases} \end{aligned} \tag{1.12}$$

Suppose $\mathbf{y}(\omega)=\mathbf{g}(\mathbf{x}(\omega))$ is a random variable defined by $\mathbf{x}(\omega)$ and the function $\mathbf{g}(\cdot)$, which maps R^n into R^m. Then (1.12) also defines the expectation of $\mathbf{y}(\omega)$ according to

$$E(\mathbf{y})=\int_{R^m}\mathbf{Y}\,dP_{\mathbf{y}}(\mathbf{Y}), \tag{1.13}$$

where $P_{\mathbf{y}}(\mathbf{Y})$ is the probability distribution function for $\mathbf{y}(\omega)$. This expression for the expectation of $\mathbf{y}(\omega)$ is often inconvenient because of the need to determine $P_{\mathbf{y}}(\mathbf{Y})$ before the integral (1.13) can be evaluated. This complication can be avoided because it can be shown that an alternative expression for the expectation is

$$\begin{aligned} E(\mathbf{y}) &= \int_{R^n}\mathbf{g}(\mathbf{X})\,dP_{\mathbf{x}}(\mathbf{X}) \\ &= \begin{cases} \sum_i \mathbf{g}(\mathbf{X}_i)\Pr(\mathbf{x}=\mathbf{X}_i), & \mathbf{x} \text{ a discrete r.v.} \\ \int_{R^n}\mathbf{g}(\mathbf{X})p_{\mathbf{x}}(\mathbf{X})\,d^n\mathbf{X}, & \mathbf{x} \text{ a continuous r.v.} \end{cases} \end{aligned} \tag{1.14}$$

Special cases of (1.14) occur so frequently in later chapters that it is convenient to introduce additional terminology for them. Suppose, first, that

$$g(\mathbf{x}(\omega))=\mathbf{e}_i'\{\mathbf{x}(\omega)\mathbf{x}'(\omega)-E(\mathbf{x})E(\mathbf{x}')\}\mathbf{e}_j,$$

where $\mathbf{e}_i$ is an n-dimensional unit vector with a one in row i and zeros elsewhere. The expectation of $g(\mathbf{x}(\omega))$ is called the *covariance* of $x_i(\omega)$ and $x_j(\omega)$ when $i\neq j$ and the *variance* of $x_i(\omega)$ when $i=j$. We denote $E[g(\mathbf{x}(\omega))]$ by λ_{ij}. The matrix $\mathbf{\Lambda}=(\lambda_{ij})$ is termed the *covariance matrix* of $\mathbf{x}(\omega)$. These second moments occur naturally in estimation problems where the estimator is constrained to be a linear transformation of the observable data. As an example, suppose $x_1(\omega)$, $x_2(\omega),\ldots,$and $x_{n-1}(\omega)$ model observed data, and on the basis of an observation of these quantities it is desired to estimate $x_n(\omega)$, which cannot be observed directly. It is a straightforward calculation to verify that if the estimate $\hat{x}_n(\omega)$ is constrained to have the linear form

$$\hat{x}_n(\omega)=a+\mathbf{b}'\mathbf{x}^{(n-1)}(\omega) \tag{1.15}$$

for some choice of a and $\mathbf{b}$, where $\mathbf{x}^{(n-1)}=[x_1 \quad x_2\cdots x_{n-1}]'$, then the selection of a and $\mathbf{b}$ minimizing $E[(x_n-\hat{x}_n)^2]$ is given by

$$a=E(x_n)-\mathbf{b}'E[\mathbf{x}^{(n-1)}]$$

$$\mathbf{b}=\begin{bmatrix} \lambda_{11} & \lambda_{12} & \cdots & \lambda_{1,n-1} \\ \lambda_{21} & \lambda_{22} & \cdots & \lambda_{2,n-1} \\ \vdots & & & \\ \lambda_{n-1,1} & \cdots & & \lambda_{n-1,n-1} \end{bmatrix}^{-1} \begin{bmatrix} \lambda_{1n} \\ \lambda_{2n} \\ \\ \lambda_{n-1,n} \end{bmatrix}.$$

In Chapter 6, we examine a version of this linearly constrained estimation problem in which the data $x_1, x_2, \ldots, x_{n-1}$ are derived from observations of a point process and x_n is an underlying variable influencing the point process.

Complex valued random variables will occasionally occur in our subsequent studies. The real and imaginary parts of these are random variables defined on the same probability space as above. A frequently occurring example in Chapters 2 through 7 is associated with the complex exponential function $g(\mathbf{X})=\exp[j\langle\mathbf{v},\mathbf{X}\rangle]$, where $j=\sqrt{-1}$, $\mathbf{v}\in R^n$, $\mathbf{X}\in R^n$, and $\langle\mathbf{v},\mathbf{X}\rangle=\sum_{i=1}^n v_i X_i$. As $\operatorname{Re}[g(\mathbf{X})]=\cos(\langle\mathbf{v},\mathbf{X}\rangle)$ and $\operatorname{Im}[g(\mathbf{X})]=\sin(\langle\mathbf{v},\mathbf{X}\rangle)$ are continuous mappings from R^n to R, the random variables $y_r(\omega)=\cos(\langle\mathbf{v},\mathbf{x}(\omega)\rangle)$ and $y_i(\omega)=\sin(\langle\mathbf{v},\mathbf{x}(\omega)\rangle)$ are well defined. Thus, we can define the complex random variable $y(\omega)$ by

$$y(\omega)=y_r(\omega)+jy_i(\omega)=\exp[j\langle\mathbf{v},\mathbf{x}(\omega)\rangle].$$

The expectation of $y(\omega)$ plays an important role in later chapters; we denote it by $M_{\mathbf{x}}(j\mathbf{v})$. Thus,

$$\begin{aligned} M_{\mathbf{x}}(j\mathbf{v}) &= E\{\exp[j\langle\mathbf{v},\mathbf{x}\rangle]\} \\ &= \int_{R^n}\exp[j\langle\mathbf{v},\mathbf{X}\rangle]\,dP_{\mathbf{x}}(\mathbf{X}) \\ &= \begin{cases} \sum_i \exp[j\langle\mathbf{v},\mathbf{X}_i\rangle]\Pr(\mathbf{x}=\mathbf{X}_i), & \mathbf{x}\text{ discrete} \\ \int_{R^n}\exp[j\langle\mathbf{v},\mathbf{X}\rangle]\,p_{\mathbf{x}}(\mathbf{X})\,d^n\mathbf{X}, & \mathbf{x}\text{ continuous.} \end{cases} \end{aligned} \tag{1.16}$$

The function $M_{\mathbf{x}}(j\mathbf{v})$ is called the *characteristic function* of $\mathbf{x}(\omega)$. There is a one-to-one correspondence between the characteristic function and probability distribution function for $\mathbf{x}(\omega)$ because they form a Fourier trans-

Table 1.1. Some Discrete Probability Laws and Their Characteristic Functions

Discrete Probability Law	$\Pr(x=n)$	Characteristic Function $M_x(jv)$	Mean $E(x)$	Variance $E(x^2)-E^2(x)$
Poisson $\lambda>0$ $n=0,1,2,\ldots$	$\frac{\lambda^n}{n!}\exp(-\lambda)$	$\exp[\lambda(e^{jv}-1)]$	λ	λ
Negative binomial $m=1,2,\ldots$ $n=0,1,\ldots$ $0\leqslant p\leqslant 1$	$\frac{(m+n-1)!}{(m-1)!n!}p^m(1-p)^n$	$p^m[1-(1-p)\exp(jv)]^{-m}$	$mp^{-1}(1-p)$	$mp^{-2}(1-p)$
Bose – Einstein $\lambda>0$ $n=0,1,\ldots$	$\lambda^n(1+\lambda)^{-n-1}$	$[1+\lambda-\lambda\exp(jv)]^{-1}$	λ	λ

Table 1.2. Some Continuous Probability Laws and Their Characteristic Functions

Continuous Probability Law	Probability Density $p_x(X)$	Characteristic Function $M_x(jv)$	Mean $E(x)$	Variance $E(x^2)-E^2(x)$
Exponential $\lambda > 0$ $X \geqslant 0$	$\lambda \exp(-\lambda X)$	$\lambda(\lambda - jv)^{-1}$	λ^{-1}	λ^{-1}
Gamma $\alpha > 0$ $\beta > 0$ $X \geqslant 0$	$\frac{\alpha^{\beta+1}}{\Gamma(\beta+1)} X^{\beta} \exp(-\alpha X)$	$\alpha^{\beta+1}(\alpha - jv)^{-\beta-1}$	$\alpha^{-1}(\beta+1)$	$\alpha^{-2}(\beta+1)$
Uniform $\alpha < X < \beta$	$(\beta-\alpha)^{-1}$	$\frac{\exp(jv\beta)-\exp(jv\alpha)}{jv(\beta-\alpha)}$	$\frac{1}{2}(\alpha+\beta)$	$\frac{1}{12}(\beta-\alpha)^2$
Normal $-\infty < \mu < \infty$ $\sigma^2 > 0$ $-\infty < X < \infty$	$(2\pi\sigma^2)^{-1/2}\exp[-(X-\mu)^2/2\sigma^2]$	$\exp[-\frac{1}{2}\sigma^2 v^2 + jv\mu]$	μ	σ^2

form pair. For instance, if $\mathbf{x}$ is a continuous random variable, then

$$p_{\mathbf{x}}(\mathbf{X})=(2\pi)^{-n}\int_{R^n}\exp[-j\langle\mathbf{v},\mathbf{X}\rangle]M_{\mathbf{x}}(j\mathbf{v})\,d^n\mathbf{v}. \tag{1.17}$$

Characteristic functions for some distributions encountered in this book are given in Tables 1.1 and 1.2 for discrete and continuous random variables, respectively.

In later chapters, we will often be interested in information observed data provides about random quantities that cannot be measured directly. The concepts of *conditional probability* and *conditional expectation* play a central role in this issue. We now give an heuristic overview of these concepts; for a precise treatment, we refer to J. Doob [2, p. 15].

If $\mathbf{x}(\omega)$ and $\mathbf{y}(\omega)$ are discrete random variables defined on a common probability space and having values $\mathbf{X}_1, \mathbf{X}_2, \ldots$ in R^n and $\mathbf{Y}_1, \mathbf{Y}_2, \ldots$ in R^m, respectively, then the conditional probability that $\mathbf{x}(\omega)=\mathbf{X}_i$ given that $\mathbf{y}(\omega)=\mathbf{Y}_j$ is defined for all $\mathbf{Y}_j$ such that $\Pr(\mathbf{y}=\mathbf{Y}_j)>0$ by

$$\Pr(\mathbf{x}=\mathbf{X}_i|\mathbf{y}=\mathbf{Y}_j)=\frac{\Pr(\mathbf{x}=\mathbf{X}_i,\mathbf{y}=\mathbf{Y}_j)}{\Pr(\mathbf{y}=\mathbf{Y}_j)}.$$

The conditional probability distribution function for $\mathbf{x}(\omega)$ given $\mathbf{y}(\omega)=\mathbf{Y}_j$ is then defined to be

$$P_{\mathbf{x}|\mathbf{y}}(\mathbf{X}|\mathbf{Y}_j)=\sum_i 1_{B(\mathbf{X})}(\mathbf{X}_i)\Pr(\mathbf{x}=\mathbf{X}_i|\mathbf{y}=\mathbf{Y}_j), \tag{1.18}$$

provided $\Pr(\mathbf{y}=\mathbf{Y}_j)>0$, which is the counterpart of (1.7) given that $\mathbf{y}(\omega)=\mathbf{Y}_j$. The conditional expectation of $\mathbf{x}(\omega)$ given that $\mathbf{y}(\omega)=\mathbf{Y}_j$ is defined for all $\mathbf{Y}_j$ such that $\Pr(\mathbf{y}=\mathbf{Y}_j)>0$ by

$$E(\mathbf{x}|\mathbf{y}=\mathbf{Y}_i)=\int_{R^n}\mathbf{X}\,dP_{\mathbf{x}|\mathbf{y}}(\mathbf{X}|\mathbf{Y}_j)=\sum_i \mathbf{X}_i P_{\mathbf{x}|\mathbf{y}}(\mathbf{X}_i|\mathbf{Y}_j). \tag{1.19}$$

The situation for continuous random variables is analogous. If $\mathbf{x}(\omega)$ and $\mathbf{y}(\omega)$ are jointly continuous random variables defined on a common probability space and having values in R^n and R^m, respectively, then the conditional probability density for $\mathbf{x}(\omega)$, given that $\mathbf{y}(\omega)=\mathbf{Y}$, is defined for all $\mathbf{Y}$ such that $p_{\mathbf{y}}(\mathbf{Y})>0$ by

$$p_{\mathbf{x}|\mathbf{y}}(\mathbf{X}|\mathbf{Y})=\frac{p_{\mathbf{x},\mathbf{y}}(\mathbf{X},\mathbf{Y})}{p_{\mathbf{y}}(\mathbf{Y})}, \tag{1.20}$$

and the conditional probability distribution function is

$$P_{\mathbf{x}|\mathbf{y}}(\mathbf{X}|\mathbf{Y}) = \Pr(\mathbf{x} \leqslant \mathbf{X}|\mathbf{y} = \mathbf{Y}) = \int_{R^n} 1_{B(\mathbf{X})}(\xi) p_{\mathbf{x}|\mathbf{y}}(\xi|\mathbf{Y}) d^n\xi. \qquad (1.21)^\dagger$$

The event $\{\omega|\mathbf{y}(\omega) = \mathbf{Y}\}$ associated with these expressions is to be interpreted in the sense of the limit of the event $\{\omega| Y_i - \Delta Y_i < y_i(\omega) \leqslant Y_i; i = 1,2,\dots,m\}$ as $\max_{1 \leqslant i \leqslant m} \Delta Y_i$ tends to zero. The conditional expectation of $\mathbf{x}(\omega)$ given that $\mathbf{y}(\omega) = \mathbf{Y}$ is defined for all $\mathbf{Y}$ such that $p_{\mathbf{y}}(\mathbf{Y}) > 0$ by

$$E(\mathbf{x}|\mathbf{y} = \mathbf{Y}) = \int_{R^n} \mathbf{X}\, d\mathbf{P}_{\mathbf{x}|\mathbf{y}}(\mathbf{X}|\mathbf{Y}) = \int_{R^n} \mathbf{X} p_{\mathbf{x}|\mathbf{y}}(\mathbf{X}|\mathbf{Y})\, d^n\mathbf{X}. \qquad (1.22)$$

The conditional expectation of $\mathbf{x}(\omega)$ given $\mathbf{y}(\omega) = \mathbf{Y}$ defines an R^n-valued function of $\mathbf{Y}$, $\mathbf{g}(\mathbf{Y}) = E(\mathbf{x}|\mathbf{y} = \mathbf{Y})$. Hence, $\mathbf{g}(\mathbf{y}(\omega)) = E(\mathbf{x}|\mathbf{y}(\omega))$ is an R^n-valued random variable. A property of conditional expectation that will be useful in later chapters is that the expectation of this random variable is the expectation of $\mathbf{x}(\omega)$; that is,

$$E(\mathbf{x}) = E[E(\mathbf{x}|\mathbf{y}(\omega))] = \int_{R^m} E(\mathbf{x}|\mathbf{y} = \mathbf{Y})\, dP_{\mathbf{y}}(\mathbf{Y}). \qquad (1.23)$$

This *iterated expectation* formula is frequently useful for evaluating the expectation of expressions involving more than one random variable. As an application motivated by Example 1.1.4, suppose the number n of lightning strokes in a day is a discrete random variable with given probabilities $\Pr(n = i)$, $i = 0,1,2,\dots,$and that the energy released in the ith stroke z_i is a continuous random variable. Assume the energies released in successive strokes are statistically independent and all are distributed as a random variable z having a probability density function $p_z(Z)$. The total energy released in a day is $x = \Sigma_{j=1}^{n} z_j$. The expected energy released in a day can easily be evaluated using the iterated expectation formula $E(x) = E[E(x|n)]$. We see that

$$E(x|n = i) = E(z_1 + z_2 + \dots + z_i) = iE(z).$$

Hence, $E(x) = E[nE(z)] = E(n)E(z)$. The total expected energy released per day is the product of the expected number of strokes per day and the average energy released per stroke.

There is another property of conditional expectation that will be useful in later chapters, particularly in Chapters 5, 6, and 7 where we encounter estimation problems having a minimum mean-square error fidelity criterion. To isolate this property, consider again the estimation problem associated with (1.15). The variables $x_1(\omega)$, $x_2(\omega)$,..., and $x_{n-1}(\omega)$ model

†The vector inequality $\mathbf{x} \leqslant \mathbf{X}$ means $x_i \leqslant X_i$ for $1 \leqslant i \leqslant n$.

observed data, and on the basis of these data, it is desired to estimate $x_n(\omega)$, which cannot be observed directly. Suppose now that the estimate $\hat{x}_n(\omega)$ is not constrained to have the linear form (1.15). Rather, what we seek is that function $\hat{x}_n(\omega)=g(x_1(\omega),x_2(\omega),\ldots,x_{n-1}(\omega))$ of the data which minimizes $E[(x_n-\hat{x}_n)^2]$. By iterating the expectation, we note after some manipulation that

$$E\left[(x_n-\hat{x}_n)^2\right]=E\left\{E(x_n^2|x_1,x_2,\ldots,x_{n-1})+[\hat{x}_n-E(x_n|x_1,x_2,\ldots,x_{n-1})]^2 - E^2(x_n|x_1,x_2,\ldots,x_{n-1})\right\}.$$

Only the second term on the right side depends on the selection $\hat{x}_n(\omega)$, and, for any selection, the contribution of this term to $E[(x_n-\hat{x}_n)^2]$ is nonnegative. Consequently, the choice

$$\hat{x}_n(\omega)=E[x_n|x_1(\omega),x_2(\omega),\ldots,x_{n-1}(\omega)] \tag{1.24}$$

minimizes $E[(x_n-\hat{x}_n)^2]$ over all other functions of the observables. The conditional mean, therefore, has an important property for optimization problems: it defines the structure of minimum mean square-error estimators. In Chapters 5, 6, and 7, we study versions of this estimation problem in which x_1, $x_2,\ldots$, and x_{n-1} are derived from observations of a point process and x_n is an underlying random quantity influencing where the points occur.

Stochastic Processes

A stochastic process $\{\mathbf{x}_t(\omega);t\in T\}$ is a family of random variables all defined on the same probability space $(\Omega,\mathcal{F},\mathcal{P})$ and indexed by a real parameter t that takes values in a parameter set T called the *index set* of the process. The indexing parameter t represents time in nearly all our considerations. The process $\{x_t(\omega);t\in T\}$ is called a *discrete-time* process if the index set T is a countable set $t_1,t_2,\ldots$, and it is called a *continuous-time* process if T is an interval. Almost all processes we consider in subsequent chapters are continuous-time processes with $T=\{t|t\geqslant t_0\}$. For fixed $t\in T$, $x_t(\omega)$ as a function of ω is an R^n-valued random variable. For each fixed $\omega\in\Omega$, $x_t(\omega)$ as a function of t is an R^n-valued time function called a *sample function*, *realization*, or *path* of the process. The set of all these time functions is called the *ensemble* or *sample-function space* of the process.

Example. Let each $\omega\in\Omega$ be a finitely denumerable point set in the interval $[0,1]$. Therefore, each ω can be enumerated in the form $\{t_1(\omega),t_2(\omega),\ldots,t_n(\omega)\}$, where $0\leqslant t_1(\omega)\leqslant t_2(\omega)\leqslant\cdots\leqslant t_n(\omega)\leqslant 1$ are the coordinates of the points and $n<\infty$. Define $\{N_t(\omega); 0\leqslant t<1\}$ by

$$N_t(\omega)=\begin{cases} 0, & 0\leqslant t\leqslant t_1(\omega)\\ 1, & t_1(\omega)<t\leqslant t_2(\omega)\\ \vdots & \\ n-1, & t_{n-1}(\omega)<t\leqslant t_n(\omega)\\ n, & t_n(\omega)<t<1. \end{cases}$$

Then, for each fixed $t\in[0,1]$, $N_t(\omega)$ is an integer-valued, discrete random variable. For each fixed $\omega\in\Omega$, $N_0(\omega)=0$ and $N_t(\omega)$ is a piecewise-constant, left-continuous function of t with unit positive jumps at $t_1(\omega)$, $t_2(\omega),\ldots,$ and $t_n(\omega)$. The sample-function space consists of all functions of this type that are initially zero and have a finite number of unit jumps. Defined in this way, $\{N_t(\omega); 0\leqslant t\leqslant 1\}$ is a counting process.

A stochastic process $\{\mathbf{x}_t(\omega); t\in T\}$ is said to be *completely characterized statistically* if the joint distribution function

$$P_{\mathbf{x}_{t1},\mathbf{x}_{t2},\ldots,\mathbf{x}_{tk}}(\mathbf{X}_1,\mathbf{X}_2,\ldots,\mathbf{X}_k)=\Pr(\mathbf{x}_{t_1}\leqslant\mathbf{X}_1,\,\mathbf{x}_{t_2}\leqslant\mathbf{X}_2,\ldots,\mathbf{x}_{t_k}\leqslant\mathbf{X}_k)$$

for the random variables $\mathbf{x}_{t_1}(\omega)$, $\mathbf{x}_{t_2}(\omega),\ldots,$ and $\mathbf{x}_{t_k}(\omega)$ is known for any finite collections $\{t_1,t_2,\ldots,t_k\}$ and $\{\mathbf{X}_1,\mathbf{X}_2,\ldots,\mathbf{X}_k\}$, where $t_i\in T$ and $\mathbf{X}_i\in R^n$ for $1\leqslant i\leqslant k$. The following are remarks on the complete statistical characterization of some particular stochastic processes we shall encounter in later chapters.

Discrete-Time Markov Processes. Let the index set T of the process $\{\mathbf{x}_t(\omega); t\in T\}$ be a countable set $\{t_1,t_2,\ldots\}$ that is ordered $t_1<t_2<\cdots$. This process is by definition a discrete-time Markov process if for any $k\geqslant 2$, $t_k\in T$, there holds

$$\Pr(\mathbf{x}_{t_k}\leqslant\mathbf{X}_k|\mathbf{x}_{t_1}\leqslant\mathbf{X}_1,\ldots,\mathbf{x}_{t_{k-1}}\leqslant\mathbf{X}_{k-1})=\Pr(\mathbf{x}_{t_k}\leqslant\mathbf{X}_k|\mathbf{x}_{t_{k-1}}\leqslant\mathbf{X}_{k-1}). \quad (1.25)$$

This condition can be interpreted qualitatively in two useful ways. The first is that for a process to be a Markov process, it must be amnesic to the extent that given the "past" $\mathbf{x}_{t_1}, \mathbf{x}_{t_2},\ldots,\mathbf{x}_{t_{k-1}}$, the "present" $\mathbf{x}_{t_k}$ depends on only the most recent past $\mathbf{x}_{t_{k-1}}$. For the second interpretation, note that if (1.25) holds, so does

$$\begin{aligned}\Pr(&\mathbf{x}_{t_1}\leqslant\mathbf{X}_1,\ldots,\mathbf{x}_{t_{j-1}}\leqslant\mathbf{X}_{j-1},\mathbf{x}_{t_{j+1}}\leqslant\mathbf{X}_{j+1},\ldots,\mathbf{x}_{t_k}\leqslant\mathbf{X}_k|\mathbf{x}_{t_j}\leqslant\mathbf{X}_j)\\ &=\Pr(\mathbf{x}_{t_1}\leqslant\mathbf{X}_1,\ldots,\mathbf{x}_{t_{j-1}}\leqslant\mathbf{X}_{j-1}|\mathbf{x}_{t_j}\leqslant\mathbf{X}_j)\Pr(\mathbf{x}_{t_{j+1}}\leqslant\mathbf{X}_{j+1},\ldots,\mathbf{x}_{t_k}\leqslant\mathbf{X}_k|\mathbf{x}_{t_j}\leqslant\mathbf{X}_j).\end{aligned} \quad (1.26)$$

Thus, a Markov process has the property that its "past" $\mathbf{x}_{t_1},\ldots,\mathbf{x}_{t_{j-1}}$ and "future" $\mathbf{x}_{t_{j+1}},\ldots,\mathbf{x}_{t_k}$ are independent given the "present" $\mathbf{x}_{t_j}$.

A discrete-time Markov process can be completely characterized statistically by only two quantities, the distribution for $\mathbf{x}_{t_1}$ and the conditional distribution for $\mathbf{x}_{t_k}$ given $\mathbf{x}_{t_{k-1}}$, for $k \geqslant 2$. This follows by first noting that in general

$$\Pr(\mathbf{x}_{t_1} \leqslant \mathbf{X}_1,\ldots,\mathbf{x}_{t_k} \leqslant X_k)$$

$$= \Pr(\mathbf{x}_{t_1} \leqslant \mathbf{X}_1) \prod_{i=2}^{k} \Pr(\mathbf{x}_{t_i} \leqslant \mathbf{X}_i | \mathbf{x}_{t_1} \leqslant \mathbf{X}_1,\ldots,\mathbf{x}_{t_{i-1}} \leqslant \mathbf{X}_{i-1}).$$

Thus, if $\{\mathbf{x}_t(\omega);\ t \in T\}$ is a Markov process, (1.25) implies

$$\Pr(\mathbf{x}_{t_1} \leqslant X_1,\ldots,\mathbf{x}_{t_k} \leqslant \mathbf{X}_k) = \Pr(\mathbf{x}_{t_1} \leqslant X_1) \prod_{i=2}^{k} \Pr(\mathbf{x}_{t_i} \leqslant \mathbf{X}_i | \mathbf{x}_{t_{i-1}} \leqslant \mathbf{X}_{i-1}). \quad (1.27)$$

The quantity $\Pr(\mathbf{x}_{t_i} \leqslant X_i | \mathbf{x}_{t_{i-1}} \leqslant X_{i-1})$ is called the *transition distribution function* and is the principal quantity characterizing the time evolution of $\{\mathbf{x}_t(\omega); t \in T\}$.

We may also employ the joint probability density $p_{\mathbf{x}_{t1},\mathbf{x}_{t2},\ldots,\mathbf{x}_{tk}}(\mathbf{X}_1,\mathbf{X}_2,\ldots,\mathbf{X}_k)$ for $\mathbf{x}_{t_1}(\omega),\mathbf{x}_{t_2}(\omega),\ldots,$ and $\mathbf{x}_{t_k}(\omega)$ when it exists. For a Markov process,

$$p_{\mathbf{x}_{t1},\mathbf{x}_{t2},\ldots,\mathbf{x}_{tk}}(\mathbf{X}_1,\mathbf{X}_2,\ldots,\mathbf{X}_k) = p_{\mathbf{x}_{t1}}(\mathbf{X}_1) \prod_{i=2}^{k} p_{\mathbf{x}_{ti}|\mathbf{x}_{t,i-1}}(\mathbf{X}_i|\mathbf{X}_{i-1}). \quad (1.28)$$

The conditional probability density function $p_{\mathbf{x}_{ti}|\mathbf{x}_{t,i-1}}(\mathbf{X}_i|\mathbf{X}_{i-1})$ of $\mathbf{x}_{ti}(\omega)$ given that $\mathbf{x}_{t_{i-1}}(\omega) = \mathbf{X}_{i-1}$ is called the *transition density function*.

Continuous-time Markov processes have similar properties and characterizations. These are defined and discussed in detail in Chapter 4 where they are developed as models for the response of physical systems to a point process excitation and in Chapter 6 where they model random phenomena influencing an observed point process.

We refer to I. Gikhman and A. Skorokhod [7, Chapters 7 and 8] and to A. Bharucha–Reid [1] for further discussion of Markov processes and their analytic properties.

Independent and Orthogonal Increment Processes. A random process $\{\mathbf{x}_t(\omega); t \in T\}$ is called an *independent increment* process if the k random variables $\mathbf{x}_{t_1}(\omega)$, $\mathbf{x}_{t_2}(\omega) - \mathbf{x}_{t_1}(\omega),\ldots,\mathbf{x}_{t_k}(\omega) - \mathbf{x}_{t_{k-1}}(\omega)$ are statistically independent for any finite collection of times $t_1 < t_2 < \ldots < t_k$, where $t_i \in T$ for

$1 \leqslant i \leqslant k$. If these variables are pairwise uncorrelated, $\{\mathbf{x}_t(\omega); t \in T\}$ is called a process with *uncorrelated increments* or, if in addition the increment variables have zero expectation, *orthogonal increments*.

A stochastic process $\{\mathbf{x}_t(\omega); t \in T\}$ with independent increments is completely characterized statistically by its first-order distribution [that is, the distribution for $\mathbf{x}_t(\omega)$, $t \in T$] and by its increment distribution [that is, the distribution for $\mathbf{x}_t(\omega) - \mathbf{x}_s(\omega)$, $s < t$, $s \in T$, $t \in T$]. This can be verified as follows. Let t_i, $1 \leqslant i \leqslant k$, be an ordered set of times in T, and denote the joint characteristic function for $\mathbf{x}_{t_1}(\omega)$, $\mathbf{x}_{t_2}(\omega), \ldots,$ and $\mathbf{x}_{t_k}(\omega)$ by

$$M_{\mathbf{x}_{t1},\mathbf{x}_{t2},\ldots,\mathbf{x}_{tk}}(j\mathbf{v}_1, j\mathbf{v}_2, \ldots, j\mathbf{v}_k) = E\left\{\exp\left[j\sum_{i=1}^{k}\langle \mathbf{v}_i, \mathbf{x}_{t_i}\rangle\right]\right\}.$$

Now,

$$\sum_{i=1}^{k}\langle \mathbf{v}_i, \mathbf{x}_{t_i}\rangle = \left\langle \sum_{m=1}^{k} \mathbf{v}_m, \mathbf{x}_{t_1}\right\rangle + \sum_{i=2}^{k}\left\langle \sum_{m=i}^{k} \mathbf{v}_m, \mathbf{x}_{t_i} - \mathbf{x}_{t_{i-1}}\right\rangle.$$

Hence, if $\{\mathbf{x}_t(\omega); t \in T\}$ has independent increments,

$$M_{\mathbf{x}_{t1},\mathbf{x}_{t2},\ldots,\mathbf{x}_{tk}}(j\mathbf{v}_1, j\mathbf{v}_2, \ldots, j\mathbf{v}_k) = M_{\mathbf{x}_{t1}}\left(j\sum_{m=1}^{k}\mathbf{v}_m\right)\prod_{i=2}^{k} M_{\mathbf{x}_{ti}-\mathbf{x}_{t,i-1}}\left(j\sum_{m=i}^{k}\mathbf{v}_m\right). \tag{1.29}$$

The joint characteristic function is, therefore, determined by the characteristic functions for $\mathbf{x}_{t_1}$ and $\mathbf{x}_{t_i} - \mathbf{x}_{t_{i-1}}$. As characteristic functions and distributions have a one-to-one relationship, it follows that the joint distribution of $\mathbf{x}_{t_1}(\omega), \mathbf{x}_{t_2}(\omega), \ldots,$ and $\mathbf{x}_{t_k}(\omega)$ is determined by the first-order and increment distribution for $\{\mathbf{x}_t(\omega); t \in T\}$.

Independent increment processes are basic building blocks for a wide variety of stochastic processes used as models for physical phenomena. In later chapters, we will encounter two particularly important independent increment processes, the *Poisson counting process* and the *Wiener process*. A fundamental role is played by these processes because all other independent increment processes are a linear combination of them. The Poisson counting process is defined and studied in detail in Chapter 2; it is an independent increment process with discontinuous sample functions. The Wiener process appears in Chapters 4, 6, and 7. It is an independent increment process with continuous sample functions. By definition,

$\{x_t(\omega); t \geq t_0\}$ is a Wiener process if

a. $\{x_t(\omega); t \geq t_0\}$ has independent increments;
b. $x_{t_0}(\omega) = 0$ with probability one;
c. the increment $x_t(\omega) - x_s(\omega)$, where $t_0 \leq s \leq t$, is normally distributed with mean zero and variance $(t-s)\alpha$, where α is a positive parameter.

Property (b) means that the event $A = \{\omega | x_{t_0}(\omega) \neq 0\}$ has zero probability measure in the underlying probability space $(\Omega, \mathcal{F}, \mathcal{P})$; that is, $A \in \mathcal{F}$ and $\mathcal{P}(A) = 0$. According to property (c), the distribution of the increment depends only on the difference $t-s$ and not t and s separately. A process with this property is said to possess *stationary increments*. $x_t(\omega)$ is normally distributed with zero mean and a variance $(t-t_0)\alpha$ that increases linearly with time. The constant α determines how rapid this increase is. The quantity α^{-1} is a characteristic time for the process, which accumulates one (rms) unit of drift in $t - t_0 = \alpha^{-1}$ units of time. A Wiener process is said to be *normalized* or *standardized* if $\alpha = 1$. An R^n-valued, standard Wiener process is a vector of n independent, standard Wiener processes. By using the expression in Table 1.2 for the characteristic function of a normal random variable, and by using (1.29), we see for a Wiener process that

$$M_{x_{t1}, x_{t2}, \ldots, x_{tk}}(jv_1, jv_2, \ldots, jv_k)$$

$$= \prod_{i=1}^{k} \exp\left[-\tfrac{1}{2}\alpha(v_i + v_{i+1} + \cdots + v_k)^2 (t_i - t_{i-1})\right]$$

$$= \exp\left[-\tfrac{1}{2}\mathbf{v}'\mathbf{\Lambda}\mathbf{v}\right], \tag{1.30}$$

where $\mathbf{v} = [v_1 \quad v_2 \ldots v_k]'$ and $\mathbf{\Lambda}$ is a $k \times k$ matrix with *i,j-element* λ_{ij} given by

$$\lambda_{ij} = \begin{cases} \alpha(t_i - t_0), & i \leq j \\ \alpha(t_j - t_0), & i > j \end{cases}$$

$$= \alpha \min(t_i - t_0, t_j - t_0).$$

Hence,

$$P_{x_{t1}, x_{t2}, \ldots, x_{tk}}(X_1, X_2, \ldots, X_k) = (2\pi)^{-k/2}[\det(\mathbf{\Lambda})]^{-1/2} \exp\left[-\tfrac{1}{2}\mathbf{X}'\mathbf{\Lambda}^{-1}\mathbf{X}\right], \tag{1.31}$$

where $\mathbf{X}=[X_1 \quad X_2 \ldots X_k]$. Thus, we conclude from (1.11) that for a Wiener process $\{x_t(\omega); t \geqslant t_0\}$, the random variables $x_{t_1}(\omega), x_{t_2}(\omega), \ldots, x_{t_k}(\omega)$ for $t_0 < t_1 < t_2 < \cdots < t_k$ are jointly normal with mean zero and covariance matrix $\boldsymbol{\Lambda}$. While the sample functions of a Wiener process are continuous with probability one, they are nondifferentiable. For a proof of this assertion as well as further properties of the Wiener process, we refer to J. Doob [2, Chapter 3].

Gaussian Processes. A stochastic process $\{x_t(\omega): t \in T\}$ is called a *Gaussian* or *normal* process if for any finite collection $(t_1, t_2, \ldots, t_k) \subset T$, the random variables $x_{t_1}(\omega), x_{t_2}(\omega), \ldots, x_{t_k}(\omega)$ are jointly normal $N(\boldsymbol{\mu}, \boldsymbol{\Sigma})$, where the ith element of the mean vector $\boldsymbol{\mu}$ is $m_x(t_i) = E(x_{t_i})$ and where the i,jth element of the covariance matrix $\boldsymbol{\Sigma}$ is $K_x(t_i, t_j) = E(x_{t_i} x_{t_j}) - E(x_{t_i})E(x_{t_j})$. The functions $m_x(t)$ and $K_x(t,u)$ are called the *mean-value function* and *covariance function* of the process. It is seen that these first and second moments completely characterize a Gaussian process statistically because knowing them is equivalent to knowing the joint distributions. From (1.31), a Wiener process is a Gaussian process with mean-value function $m_x(t)=0$ and covariance function $K_x(t,u) = \alpha \min(t,u)$.

Wide-Sense Stationary Processes. A stochastic process $\{x_t(\omega); -\infty < t < \infty\}$ is said to be *wide-sense stationary* if its mean-value function $m_x(t) = E(x_t)$ is a constant independent of t and its covariance function $K_x(t,u) = E(x_t x_u) - E(x_t)E(x_u)$ is a function only of $t-u$ and not t and u separately. For a wide-sense stationary process, $K_x(t,u) = K_x(t-u, 0)$. It is common practice to write $K_x(t-u,0)$ as $K_x(\tau)$, a function of a single variable $\tau = t-u$. The power-density spectrum for a wide-sense stationary process is the Fourier transform of its covariance function,

$$S_x(f) = \int_{-\infty}^{\infty} K_x(\tau) \exp(-j2\pi f \tau)\, d\tau. \tag{1.32}$$

Conversely,

$$K_x(\tau) = \int_{-\infty}^{\infty} S_x(f) \exp(j2\pi f \tau)\, df. \tag{1.33}$$

While the power-density spectrum introduces nothing new statistically, it has important physical interpretations. Its total area

$$K_x(0) = \int_{-\infty}^{\infty} S_x(f)\, df$$

is the variance of x_t for any t. By virtue of the relation

$$K_x(0)= \lim_{T\to\infty} (2T)^{-1}\int_{-T}^{T} x^2(t)\,dt,$$

which holds for an ergodic process, this variance is often interpreted as the total power in x. The integral $\int_{f_1}^{f_2} S_x(f)\,df$ is interpreted as the power in x associated with components in the frequency band $[f_1,f_2]$. This may be motivated using the relationship that exists between the power-density spectra of the processes at the input and output of a linear time-invariant filter. Thus, if $y_t=\int_{-\infty}^{\infty} h(t-u)x_u\,du$, it can be shown that

$$S_y(f)=|H(f)|^2 S_x(f),$$

where

$$H(f)=\int_{-\infty}^{\infty} h(t)\exp(-j2\pi ft)\,dt \tag{1.34}$$

is the frequency response function of the filter. Setting $H(f)=1$ for $f_1 \leqslant f \leqslant f_2$ and $H(f)=0$ otherwise shows that

$$K_y(0)=\int_{-\infty}^{\infty} S_y(f)\,df=\int_{f_1}^{f_2} S_x(f)\,df.$$

Hence, the total power in y is the power in those components of x in the range $[f_1,f_2]$. It is intuitively clear that the power in y, namely

$$\int_{-\infty}^{\infty} |H(f)|^2 S_x(f)\,df,$$

will differ by only a small amount from $N_0\int_{-\infty}^{\infty}|H(f)|^2\,df$ if $S_x(f)\cong N_0$ over all frequencies where $H(f)$ has significant nonzero values. In this case, it is common to assume that x is a *white noise process*. The power-density spectrum for a white noise process is a constant for all frequencies; equivalently, its covariance function is an impulse function. White noise processes do not exist physically or mathematically. Nevertheless, they provide a convenient manipulative tool. We do not use the notions of white noise and power-density spectra in most of the text; the main exception is in Chapter 4, where we discuss shot noise. E. Wong [22, Chapter 3] may be consulted for further discussion of wide-sense stationary processes.

Sequences and Convergence Concepts

A *sequence* of random variables $x_1(\omega), x_2(\omega), \ldots$ is a particular form of stochastic process in which the index set is the nonnegative integers

$\{1,2,\ldots\}$. A recurring problem in the analytical investigation of stochastic processes is the resolution of what transpires in the limit as the indexing parameter tends to infinity. For instance, this is the central issue in the study of the continuity, differentiability, and integrability of processes. We now briefly review convergence concepts for sequences of random variables. A thorough treatment of this topic is given by M. Loève [10, Chapter 2] and I. Gikhman and A. Skorokhod [7, Chapter 2].

Convergence Everywhere. Let the sequence of random variables $x_1(\omega), x_2(\omega), \ldots$ and the random variable $x(\omega)$ all be defined on the same probability space $(\Omega, \mathcal{F}, \mathcal{P})$. This sequence is said to converge *everywhere* to $x(\omega)$ if for every $\omega \in \Omega$, there holds

$$\lim_{n\to\infty} x_n(\omega) = x(\omega). \tag{1.35}$$

The probability measure $\mathcal{P}$ plays no role in this form of convergence. The following forms do reflect $\mathcal{P}$ in various ways. These alternatives are of greater importance in the theory of probability and stochastic processes.

Convergence Almost Everywhere. The sequence $x_1(\omega), x_2(\omega), \ldots$ is said to converge *almost everywhere*, *almost surely*, or *with probability one* to $x(\omega)$ if the set of ω for which (1.35) fails to hold has zero probability measure. Thus, the sequence may not converge for every $\omega \in \Omega$, but the event that it does not occurs with zero probability. We occasionally use the notation $x_n \xrightarrow{\text{a.s.}} x$ to indicate that $x_n(\omega)$ converges almost surely to $x(\omega)$.

Convergence in Probability. The sequence $x_1(\omega), x_2(\omega), \ldots$ is said to *converge in probability* to $x(\omega)$ if for every $\epsilon > 0$,

$$\lim_{n\to\infty} \Pr(|x_n - x| > \epsilon) = 0. \tag{1.36}$$

If the sequence $x_1(\omega), x_2(\omega), \ldots$ converges almost everywhere to $x(\omega)$, it also converges in probability to $x(\omega)$, but the converse need not hold. We sometimes use $x_n \xrightarrow{p} x$ or $p\lim_{n\to\infty} x_n = x$ to indicate that $x_n(\omega)$ converges in probability to x.

Convergence in rth Mean. The sequence $x_1(\omega), x_2(\omega), \ldots$ is said to *converge in the rth mean*, for $r > 0$, if

$$\lim_{n\to\infty} E(|x_n - x|^r) = 0. \tag{1.37}$$

The most important special case is $r = 2$. Convergence in the rth mean for $r = 2$ is also called *quadratic-mean* and *mean-square* convergence. If a sequence converges in the rth mean, it also converges in the r'th mean where $0<r'\leqslant r$. Moreover, it also converges in probability. This last assertion can be easily demonstrated for $r=2$ by use of the Tchebycheff inequality, which says for any random variable y that

$$E(y^2)=\int_{-\infty}^{\infty} Y^2 dP_y(Y) \geqslant \epsilon^2 \int_{|Y|>\epsilon} dP_y(Y)=\epsilon^2 \Pr(|y|>\epsilon). \quad (1.38)$$

Thus,

$$\Pr(|x_n - x|>\epsilon) \leqslant \epsilon^{-2} E(|x_n - x|^2),$$

which shows that if x_n converges in the quadratic mean to x, it also converges in probability to x.

Convergence in Distribution. The sequence $x_1(\omega), x_2(\omega), \ldots$ is said to converge in *distribution* to $x(\omega)$ if

$$\lim_{n\to\infty} P_{x_n}(X)=P_x(X) \quad (1.39)$$

for $-\infty<X<\infty$. Convergence in distribution is the weakest form of convergence as it is implied by convergence in probability, which, in turn, is implied by both almost everywhere and rth mean convergence. We will encounter convergence in distribution in two contexts. The first occurs in Chapter 4 where we study the response of linear systems to a point process excitation. We show that under certain conditions the response tends to have a normal distribution as the rate at which input points occur tends to infinity. The second context arises in Chapter 5 where we study the effect of pooling the points of many point processes. We show that under certain conditions the resulting pooled point process tends to have the Poisson distribution as the number of component processes tends to infinity.

There are several instances in the book where we require conditions for the validity of the relation

$$\lim_{n\to\infty} E(x_n)=E\Big(\lim_{n\to\infty} x_n\Big)=E(x), \quad (1.40)$$

in which $x_1(\omega), x_2(\omega), \ldots$ is a sequence of random variables converging to a limit $x(\omega)$ in one of the above senses. These conditions validate the operation of interchanging the order of taking a limit and expectation. The simplest assertion to establish is that if $x_n(\omega)$ converges in the mean-square sense to $x(\omega)$, then (1.40) holds. This is an immediate consequence of the inequality

$$|E(x_n - x)| \leqslant E(|x_n - x|). \quad (1.41)$$

As convergence in the rth mean for $r=2$ implies convergence for $r=1$, we see using (1.41) that

$$\lim_{n\to\infty} |E(x_n - x)| = \lim_{n\to\infty} |E(x_n) - E(x)| = 0.$$

provided x_n converges to x in the quadratic mean. The conditions given in the following theorem for the validity of (1.40) will be used often in the book.

Theorem 1.4.1 (Generalized Bounded Convergence). Let $\{g_n\}$ be a sequence of nonnegative random variables converging with probability one to a random variable g. Assume for each n that $E(g_n) < \infty$ and that $E(g) < \infty$. Let $\{f_n\}$ be a sequence of random variables converging with probability one to a random variable f. Assume for each n that $E(|f_n|) < \infty$ and that $|f_n| \leqslant g_n$ with probability one. If

$$E(g) = E\left(\lim_{n\to\infty} g_n\right) = \lim_{n\to\infty} E(g_n),$$

then

$$E(f) = E\left(\lim_{n\to\infty} f_n\right) = \lim_{n\to\infty} E(f_n) \tag{1.42}$$

and

$$E(f|y) = E\left(\lim_{n\to\infty} f_n | y\right) = \lim_{n\to\infty} E(f_n|y), \tag{1.43}$$

where y is a random variable.

Proof. If $g_n = g$ for all n, then (1.42) follows from the bounded convergence theorem proved, for example, by H. Roydan [17, p. 88], and (1.43) follows from a bounded convergence theorem for conditional expectations proved by J. Doob [2, p. 23]. If $g_n \neq g$ for all n, then (1.42) follows from a generalized bounded convergence theorem proved by H. Roydan [17, p. 89]. We slightly modify Doob's procedure to prove (1.43) when $g_n \neq g$ for all n. Thus, define

$$\hat{f}_n = \sup_{j \geqslant n} |f_j - f|.$$

Then, with probability one, $\hat{f}_1 \geqslant \hat{f}_2 \geqslant \cdots \geqslant 0$, $\lim_{n\to\infty} \hat{f}_n = 0$, and $\hat{f}_n \leqslant \hat{g}_n$, where we define

$$\hat{g}_n = g + \sup_{j \geqslant n} g_j.$$

We note that $E(\hat{g}_n) < \infty$, $\hat{g}_n \leqslant \hat{g}_1$, and $\lim_{n\to\infty} \hat{g}_n = 2g$ with probability one.

Consequently, by the bounded convergence theorem, there holds

$$E(2g) = E\left(\lim_{n\to\infty} \hat{g}_n\right) = \lim_{n\to\infty} E(\hat{g}_n).$$

We conclude from the generalized bounded convergence theorem (1.42) that

$$0 = E\left(\lim_{n\to\infty} \hat{f}_n\right) = \lim_{n\to\infty} E\left(\hat{f}_n\right).$$

Now observe that

$$\begin{aligned} 0 \leqslant |E(f|y) - E(f_n|y)| &= |E(f - f_n|y)| \\ &\leqslant E(|f - f_n||y) \\ &\leqslant E\left(\hat{f}_n|y\right). \end{aligned}$$

Thus, to establish assertion (1.43), it is sufficient to prove that $\lim_{n\to\infty} E(\hat{f}_n|y) = 0$ with probability one. As $\hat{f}_1 \geqslant \hat{f}_2 \geqslant \cdots \geqslant 0$, we have that

$$E\left(\hat{f}_1|y\right) \geqslant E\left(\hat{f}_2|y\right) \geqslant \cdots \geqslant 0$$

with probability one. Thus, with probability one, the sequence $\{E(\hat{f}_n|y)\}$ converges to a nonnegative random variable. Denote the limit by w; then, $w = \lim_{n\to\infty} E(\hat{f}_n|y)$. Moreover,

$$E(w) \leqslant E\left[E\left(\hat{f}_n|y\right)\right] = E\left(\hat{f}_n\right).$$

The right side tends to zero as n tends to infinity. Hence $E(w) = 0$ and, as w is almost surely nonnegative, there holds $w = 0$ with probability one. This proves assertion (1.43), thereby establishing Theorem 1.4.1.

REFERENCES

1. A. T. Bharucha-Reid, *Elements of the Theory of Markov Processes and Their Applications*, McGraw-Hill, New York, 1960.
2. J. L. Doob, *Stochastic Processes*, Wiley, New York, 1953.
3. J. Evans, "Preliminary Analysis of ELF Noise," Tech. Note 1969-18, M.I.T. Lincoln Laboratory, Lexington, Mass., March 1969.
4. P. M. Fishman, "Statistical Inference for Space-Time Point Processes," D.Sc. Thesis, Sever Institute, Washington University, St. Louis, Mo., June 1974.

5. P. M. Fishman and D. L. Snyder, "Estimation of a Random Space-Time Field That Influences an Observed Space-Time Point Process," in Proceedings of the Eleventh Annual Allerton Conference on Circuit and System Theory, University of Illinois, Urbana, October 1973.

6. I. I. Gihman and A. V. Skorokhod, *Stochastic Differential Equations*, Springer-Verlag, New York, 1972.

7. I. I. Gikhman and A. V. Skorokhod, *Introduction to the Theory of Random Processes*, Saunders, Philadelphia, 1965.

8. A. H. Jazwinski, *Stochastic Processes and Filtering Theory*, Academic, New York, 1970.

9. H. J. Kushner, "On the Differential Equations Satisfied by Conditional Probability Densities of Markov Processes with Applications," *J. SIAM Control*, Ser. A, **2**, No. 1, 106–119 (1964).

10. M. Loève, *Probability Theory*, Van Nostrand, New York, 1963.

11. J. L. Melsa and A. P. Sage, *An Introduction to Probability and Stochastic Processes*, Prentice-Hall, Englewood Cliffs, N.J., 1973.

12. J. E. Moyal, "The General Theory of Stochastic Population Processes," *Acta Math.*, **108**, 1–31 (1962).

13. J. Neveu, *Mathematical Foundations of the Calculus of Probability*, Holden-Day, San Francisco, 1965.

14. A. Papoulis, *Probability, Random Variables, and Stochastic Processes*, McGraw-Hill, New York, 1965.

15. E. Parzen, *Modern Probability Theory and Its Applications*, Wiley, New York, 1960.

16. E. Parzen, *Stochastic Processes*, Holden-Day, San Francisco, 1962.

17. H. L. Roydan, *Real Analysis*, Macmillan, London, 1968.

18. A. P. Sage and J. L. Melsa, *Estimation Theory with Applications to Communication and Control*, McGraw-Hill, New York, 1971.

19. A. V. Skorokhod, *Studies in the Theory of Random Processes*, Addison-Wesley, Reading, Mass., 1965.

20. R. L. Stratonovich, "On the Theory of Optimal Nonlinear Filtering of Random Functions," *Theory Probability Appl. (USSR)* (English transl.), **4**, 223–225 (1959).

21. D. Vere-Jones, "Stochastic Models for Earthquake Occurrence," *J. Roy. Stat. Soc. B*, **32**, No. 1, 1–62 (1970).

22. E. Wong, *Stochastic Processes in Information and Dynamical Systems*, McGraw-Hill, New York, 1971.

THE POISSON PROCESS

CHAPTER TWO

2.1 INTRODUCTION

The Poisson process is the simplest process associated with counting random numbers of points. We study it in detail in this chapter when the space where the points occur is a one-dimensional, semiinfinite, real line. While there is no mathematical reason to do so, we refer to this space exclusively as "time" because temporal phenomena seem to predominate in applications. The study of the Poisson process in only one dimension is no real restriction. All its novel features are exhibited, and the extension to spaces of higher dimension is conceptually straightforward.

Its simplicity notwithstanding, the Poisson counting process is important in at least three respects. First, it is found to be an accurate model in many and varied applications. One may well wonder why. A partial explanation is provided in this chapter in the form of some relatively weak qualitative conditions under which a point process is a Poisson process. Further explanation is provided in Chapter 5 when we discuss the pooling of points from numerous subsidiary point processes. A second reason that the Poisson counting process is important is that it is a natural introductory topic in the investigation of more complicated counting processes. This will be evident when we get to Chapters 5 and 6 where the conditions we impose here for the Poisson process are either removed entirely or relaxed substantially. A third reason the Poisson counting process is important is that it is an essential building block for other processes having sample paths strikingly different from those of a counting process. These include discontinuous processes having nonidentical, nonunit jumps as well as the most general independent-increment processes.

A Poisson counting process for times $t \geqslant t_0$ is by definition a counting process $\{N_t; t \geqslant t_0\}$ with the following three properties:

i. $\Pr[N_{t_0}=0]=1$;

ii. for $t_0 \leqslant s < t$, the increment $N_{s,t} = N_t - N_s$ is Poisson distributed with parameter $\Lambda_t - \Lambda_s$,

$$\Pr[N_{s,t}=n]=(n!)^{-1}(\Lambda_t-\Lambda_s)^n \exp[-(\Lambda_t-\Lambda_s)] \tag{2.1}$$

for $n=0,1,2,\ldots$, where Λ_t is a nonnegative, nondecreasing function of t;

iii. $\{N_t; t \geqslant t_0\}$ has independent increments.

Property (iii) is the distinguishing property. It means that for a Poisson counting process, the number of points in nonoverlapping intervals are statistically independent no matter how large or small the invervals and no matter how close or distant they may be; more precisely, if $[t_i, u_i)$ for $i=1,2,\ldots,k$ are k disjoint intervals on $[t_0, \infty)$, then

$$\Pr[N_{t_1,u_1}=n_1, N_{t_2,u_2}=n_2,\ldots,N_{t_k,u_k}=n_k]=\prod_{i=1}^{k}\Pr[N_{t_i,u_i}=n_i] \tag{2.2}$$

for $k=1,2,\ldots$. Moreover, property (iii), along with properties (i) and (ii), provides a complete statistical characterization for the Poisson counting process in the sense that the joint counting probability

$$\Pr[N_{t_1}=n_1, N_{t_2}=n_2,\ldots,N_{t_k}=n_k] \tag{2.3}$$

can be determined for any collection of times $t_1<t_2<\cdots<t_k$ in $[t_0,\infty)$ and any collection of nonnegative integers $n_1,n_2,\ldots,n_k$, where $k=1,2,\ldots$. To see this, note by property (i) that $N_{t_1}=N_{t_0}+N_{t_0,t_1}=N_{t_0,t_1}$. Hence,

$$\begin{aligned}
&\Pr[N_{t_1}=n_1, N_{t_2}=n_2,\ldots,N_{t_k}=n_k]\\
&\quad=\Pr[N_{t_0,t_1}=n_1, N_{t_1,t_2}=n_2-n_1,\ldots,N_{t_{k-1},t_k}=n_k-n_{k-1}]\\
&\quad=\prod_{i=1}^{k}\Pr[N_{t_{i-1},t_i}=n_i-n_{i-1}]
\end{aligned} \tag{2.4}$$

where the last equality follows from the independent increment property (iii). Thus, the joint probability (2.3) can be expressed in terms of the increment probabilities given in property (ii). It is evident that the independent increment property (iii) plays a crucial role in characterizing the Poisson process. The effort to eliminate this property will lead us to the point processes of Chapter 5 that do not possess it.

The function Λ in property (ii) is termed the *parameter function* of the Poisson counting process. According to property (ii), the only analytic restrictions required in general on this function are that it be nonnegative and nondecreasing. In particular, there is no requirement that Λ be continuous; even if continuous, Λ need not be differentiable. It is of interest, therefore, to investigate the influence of additional restrictions on this function. If $\Lambda_t-\Lambda_s$ is finite, then points do not occur with certainty on the interval $[s,t)$. That this is so follows from the equality $\Pr[N_{s,t} \geqslant 1] = 1 - \exp[-(\Lambda_t - \Lambda_s)]$ derived from (2.1); $\Lambda_t - \Lambda_s < \infty$ implies that $\Pr[N_{s,t} \geqslant 1] < 1$. But, also, $\Lambda_t - \Lambda_s$ finite implies that an infinite number of points do not occur on $[s,t)$ because from (2.1), $\Pr[N_{s,t} = n]$ tends to zero as n tends to infinity for $\Lambda_t - \Lambda_s < \infty$. A further examination of (2.1) indicates that the points of a Poisson process can occur at predetermined times where the parameter function has a jump discontinuity, and the number of points occurring at such times has the Poisson distribution with parameter equal to the size of the jump. As Λ_t is a finite-valued, nondecreasing function of t, the collection of these predetermined times in $[s,t)$ is denumerable and the sum of the jump sizes is finite. A. Khinchin [22] shows that $N_{s,t}$ can be decomposed into two statistically independent components $N_{s,t}^{(d)}$ and $N_{s,t}^{(c)}$, which he terms singular and nonsingular, respectively. $N_{s,t}^{(d)}$ is the contribution to $N_{s,t}$ of points occurring only at the predetermined times in $[s,t)$ where Λ is discontinuous, and $N_{s,t}^{(c)} = N_{s,t} - N_{s,t}^{(d)}$ is the contribution to $N_{s,t}$ of points occurring at the times in $[s,t)$ where Λ is continuous. If the parameter function is continuous for all $t \geqslant t_0$, points do not occur at predetermined times, and there holds $\lim_{\delta \downarrow 0} \Pr[N_{t,t+\delta} = 0] = 1$ for all $t \geqslant t_0$; in other words, $\{N_t; t \geqslant t_0\}$ is stochastically continuous. Finally, let us note for a Poisson process that the derivative of the parameter function, if it exists, is the instantaneous average rate that points occur. This follows because the expected number of points on the interval $[s,t)$ has the evaluation

$$E(N_{s,t}) = \sum_{n=0}^{\infty} n \Pr[N_{s,t} = n] = \Lambda_t - \Lambda_s \tag{2.5}$$

for a Poisson counting process. Consequently, the existence of the limit

$$\lim_{\delta \downarrow 0} \delta^{-1}(\Lambda_{t+\delta} - \Lambda_t) \tag{2.6}$$

implies that the point process has an instantaneous average rate at time t. Thus, if Λ_t is an absolutely continuous function of t so that it can be expressed as $\Lambda_t = \int_{t_0}^{t} \lambda_\sigma \, d\sigma$ for all $t \geqslant t_0$, where λ_t is a nonnegative function of t for $t \geqslant t_0$, then the limit in (2.6) exists and equals λ_t. We term this function the *intensity function* of the process; at any time $t \geqslant t_0$, the intensity function λ_t is the instantaneous average rate that points occur.

Our aim in this chapter is to study the Poisson process in detail. We begin in Section 2.2 by giving some sufficient conditions for a counting process to be a Poisson process. The notions of *orderliness* and *evolution without aftereffects* are introduced here. In Section 2.3, we examine the statistics of the point locations and interpoint spacings for a Poisson process. One of the more important quantities introduced is the *sample function density*. The sample function density can be interpreted very roughly as the probability of obtaining a particular realization of the Poisson counting process on a specified observation interval. It therefore plays a central role in solving estimation and decision problems for observed Poisson processes, as discussed in Sections 2.4 and 2.5. The sample function density will be generalized in later chapters for other counting processes. A surprising aspect of these generalizations is that the form of the sample function density is essentially always the same as for the Poisson counting process.

2.2 CONDITIONS FOR THE POISSON COUNTING PROCESS

We now give some qualitative conditions for an arbitrary counting process to be a Poisson counting process. An explicit statement of such conditions is important because alternative counting processes can be obtained by making obvious modifications of them. Furthermore, whether or not a Poisson counting process is a reasonable model in a given application can be judged in part by the degree to which the qualitative conditions are met.

Two attributes a point process may possess will be prominent in specifying these qualitative conditions. The first is *orderliness*, which is defined as follows.

Orderliness. A counting process $\{N_t; t \geqslant t_0\}$, and the underlying point process as well, is called orderly at time $t \geqslant t_0$ if for any given $\epsilon > 0$, there exists a $\delta \equiv \delta(t, \epsilon) > 0$ such that

$$\Pr[N_{t,t+\delta'} > 1] \leqslant \epsilon \Pr[N_{t,t+\delta'} = 1] \tag{2.7}$$

for all $\delta' \in (0, \delta)$. A point process is orderly on an interval of time if it is orderly for each time in the interval, and it is uniformly orderly on the interval if a single quantity $\delta \equiv \delta(\epsilon)$ can be chosen independently of t so (2.7) holds simultaneously for all times in the interval.

Were it not for the possibility that $\Pr[N_{t,t+\delta} = 1]$ may be zero, this somewhat complicated definition of orderliness could be written more compactly as

$$\lim_{\delta \downarrow 0} \frac{\Pr[N_{t,t+\delta} > 1]}{\Pr[N_{t,t+\delta} = 1]} = 0.$$

Orderliness according to (2.7) is sometimes referred to as *Khinchin orderliness* in honor of A. Khinchin [22], who first recognized its consequence. Other varieties of orderliness have been proposed, not all of which are equivalent; see either D. Daley [9] or D. Daley and D. Vere-Jones [10] for a thorough summary of these. For an orderly point process, the probability of there being more than one point in an interval can be made an arbitrary small fraction of there being one point provided the interval is sufficiently small. The qualitative interpretation is that points do not occur simultaneously, although the fact that points are distinct does not guarantee orderliness (see Problem 2.2.1). Orderliness is frequently a valid assumption for modeling physical phenomena, but it is not universally so. For instance, it is inappropriate for modeling job arrivals in a computer queue when jobs arrive in batches. Similarly, an optical field incident on a photoelectron converter can induce the emission of a random number of photoelectrons, and the assumption of orderliness for the photoelectron process may be inappropriate. As we have already seen in Section 2.1, points can occur simultaneously in a Poisson process where the parameter function has step discontinuities. Thus, in general, a Poisson counting process need not be orderly everywhere; it is orderly at those times t where Λ_t is continuous.

The second attribute a point process may possess and which we introduce now is that of *evolution without aftereffects*:

Evolution Without Aftereffects. A point process on $[t_0, \infty)$ is said to evolve without aftereffects if for any $t \geqslant t_0$, the realization of points during the interval $[t, \infty)$ does not depend in any way on the sequence of events that have transpired in the interval $[t_0, t)$. This expresses the independence of the past and future of the point process. The notion is defined mathematically as follows. Let P be any event that can be associated with the random variables $\{N_\sigma; t_0 \leqslant \sigma < t\}$. Thus, P can be any of the following events: the number of points on $[t_0, t)$ is n; the number of points on $[t_0, t)$ is n and they are located at times $W_1, W_2, \ldots, W_n$; the first point to occur in $[t_0, t)$ is at W_1; etc. Similarly, let F be any event that can be associated with the random variables $\{N_\sigma; \sigma \geqslant t\}$. If t denotes the present time, then P and F are arbitrary events in the past and future of the point process, respectively. A point process evolves without aftereffects if the conditional probability of F given P equals the unconditional probability of F for all $t \geqslant t_0$.

A process with independent increments evolves without aftereffects; that the converse also holds is obvious. The hypothesis that a point process evolves without aftereffects can fail in applications when the occurrence of a point causes either the generation or elimination of subsequent points. An example of a phenomenon in which the occurrence of an event leads directly to the generation of subsequent events arises in a lightning stroke in which a discharge is identified as an event; there is first a predischarge, called a pilot stroke, which is followed by a series of small energy discharges, called leaders, and then a large energy discharge, called a return stroke. Thus, there is a clustering of events in which the occurrence of the first indicates with high probability that more will follow. An example of a phenomenon in which the occurrence of a point leads directly to the elimination of subsequent points occurs in connection with practical gamma-ray detectors; an incident gamma photon can cause the detector to become inoperative for some brief period of time, called a dead time, during which subsequent gamma photons are not detected. It is evident that a Poisson process evolves without aftereffects.

We now establish two theorems that give qualitative conditions for an arbitrary counting process to be a Poisson counting process. The first is important because of its generality; it contains the weaker conditions. We find, however, that these conditions cannot be so easily generalized as those of the second theorem. Consequently, the remainder of the book is based on conditions and techniques that most closely resemble those of the second theorem. The first theorem is motivated by the work of A. Khinchin [22]. The proof we give is largely due to I. Gikhman and A. Skorokhod [13, p. 261].

Theorem 2.2.1 (Conditions for a Poisson Process). Let $\{N_t; t \geqslant t_0\}$ be the counting process associated with a point process on $[t_0, \infty)$. Suppose that:

a. the point process is uniformly orderly on $[t_0, t)$ for $t \geqslant t_0$;
b. the point process evolves without aftereffects;
c. points do not occur at preassigned times;
d. there is no finite interval in $[t_0, \infty)$ where points occur with certainty;
e. $\Pr[N_{t_0} = 0] = 1$.

Then $\{N_t; t \geqslant t_0\}$ is a Poisson counting process with a continuous parameter function.

We establish four lemmas before proving the theorem.

Lemma 2.2.1. Under hypotheses (b), (c), (d), and (e) of Theorem 2.2.1, there exists a nonnegative, nondecreasing, continuous function Λ_t, defined for all $t \geqslant t_0$, such that

$$\Pr[N_{s,t}=0]=\exp[-(\Lambda_t-\Lambda_s)] \tag{2.8}$$

for $t_0 \leqslant s \leqslant t$. Furthermore, $0<\Pr[N_{s,t}=0] \leqslant 1$.

Proof. Define $P_0(t)=\Pr[N_t=0]$ for $t \geqslant t_0$. We have $P_0(t_0)=1$ by (e). As $N_{t_2}=0$ implies $N_{t_1}=0$ for $t_2 \geqslant t_1$, there holds $P_0(t_2) \leqslant P_0(t_1)$ for $t_2 \geqslant t_1$. It follows that $P_0(t)$ is a nonincreasing function of t. According to (d), $1 \geqslant P_0(t)=1-\Pr[N_t \geqslant 1]>0$. By virtue of these properties of $P_0(t)$, we can define $\Lambda_t=-\ln P_0(t)$ for all $t \geqslant t_0$, where $\Lambda_{t_0}=0$, $0<\Lambda_t<\infty$, and Λ_t is a nondecreasing function of t. Let us next examine the increment $N_{s,t}=N_t-N_s$ for $t_0 \leqslant s \leqslant t$. We have

$$\begin{aligned}\Pr[N_t=0]&=\Pr[N_s=0, N_t-N_s=0]\\&=\Pr[N_t-N_s=0|N_s=0]\Pr[N_s=0].\end{aligned}$$

Consequently, using (b) and the definition of Λ_t, we have (2.8). The inequality $\Pr[N_{s,t}=0]>0$ is implied by (d). According to (c), $\lim_{\delta\downarrow 0}\Pr[N_{t,t+\delta}=0]=1$ for all $t \geqslant t_0$. Therefore, $\lim_{\delta\downarrow 0}(\Lambda_{t+\delta}-\Lambda_t)=0$ for all $t \geqslant t_0$ so that Λ_t is a continuous function of t for $t \geqslant t_0$. The proof of Lemma 2.2.1 is complete.

The uniform orderliness hypothesis (a) of Theorem 2.2.1 is not used in Lemma 2.2.1. We now add it to the hypotheses of Lemma 2.2.1 to show that by selecting a sufficiently small interval $[\alpha,\beta)$ both $\Pr[N_{\alpha,\beta}=1]-(\Lambda_\beta-\Lambda_\alpha)$ and $\Pr[N_{\alpha,\beta}>1]$ can be made an arbitrarily small fraction of $\Lambda_\beta-\Lambda_\alpha$. We use the property that a continuous function, namely Λ, is uniformly continuous on a closed, bounded interval (see T. Apostol [2, p. 75]).

Lemma 2.2.2. Suppose hypotheses (a)–(e) of Theorem 2.2.1 hold and that $[s,t]$, $t_0 \leqslant s \leqslant t$, is a bounded interval. Then for all $\epsilon>0$, there exists a $\delta \equiv \delta(\epsilon)>0$ such that for any interval $[\alpha,\beta)$ in $[s,t]$ with $\beta-\alpha<\delta$ there holds

$$|\Pr[N_{\alpha,\beta}=1]-(\Lambda_\beta-\Lambda_\alpha)| \leqslant \epsilon(\Lambda_\beta-\Lambda_\alpha) \tag{2.9}$$

and

$$\Pr[N_{\alpha,\beta}>1] \leqslant \epsilon(\Lambda_\beta-\Lambda_\alpha). \tag{2.10}$$

Proof. Both assertions (2.9) and (2.10) hold trivially if $\Lambda_\beta - \Lambda_\alpha = 0$ because this implies $\Pr[N_{\alpha,\beta} = 0] = 1$ from Lemma 2.2.1. Thus, assume $\Lambda_\beta - \Lambda_\alpha > 0$. Then

$$\begin{aligned}\Pr[N_{\alpha,\beta} \geqslant 1] &= \Pr[N_{\alpha,\beta} = 1] + \Pr[N_{\alpha,\beta} > 1] \\ &= 1 - \exp[-(\Lambda_\beta - \Lambda_\alpha)] \\ &= (\Lambda_\beta - \Lambda_\alpha) + o(\Lambda_\beta - \Lambda_\alpha)\end{aligned}$$

where $(\Lambda_\beta - \Lambda_\alpha)^{-1} o(\Lambda_\beta - \Lambda_\alpha)$ tends to zero as $\beta - \alpha$ tends to zero. Hence, we have the upper bound

$$\Pr[N_{\alpha,\beta} = 1] \leqslant (\Lambda_\beta - \Lambda_\alpha) + o(\Lambda_\beta - \Lambda_\alpha) \tag{2.11}$$

We next use the uniform orderliness hypothesis (a) to get a lower bound. Thus, choose $\delta > 0$ so that both

$$\Pr[N_{\alpha,\beta} > 1] \leqslant \left(\frac{\epsilon}{2}\right) \Pr[N_{\alpha,\beta} = 1] \tag{2.12}$$

and

$$|(\Lambda_\beta - \Lambda_\alpha)^{-1} o(\Lambda_\beta - \Lambda_\alpha)| \leqslant \left(\frac{\epsilon}{2}\right) \tag{2.13}$$

for all intervals $[\alpha, \beta)$ in $[s, t]$ satisfying $\beta - \alpha < \delta$. Such a δ exists for arbitrary $\epsilon > o$ by (a) and the uniform continuity of Λ on $[s, t]$. Then, by adding $\Pr[N_{\alpha,\beta} = 1]$ to both sides of (2.12), we obtain

$$\left(1 + \left(\frac{\epsilon}{2}\right)\right) \Pr[N_{\alpha,\beta} = 1] \geqslant (\Lambda_\beta - \Lambda_\alpha) + o(\Lambda_\beta - \Lambda_\alpha). \tag{2.14}$$

Upon combining (2.11) and (2.13), we obtain

$$(\Lambda_\beta - \Lambda_\alpha)^{-1} \Pr[N_{\alpha,\beta} = 1] - 1 \leqslant \left(\frac{\epsilon}{2}\right) < \epsilon. \tag{2.15}$$

Similarly, by combining (2.13) and (2.14), we obtain

$$(\Lambda_\beta - \Lambda_\alpha)^{-1} \Pr[N_{\alpha,\beta} = 1] - 1 \geqslant -\epsilon \left(1 + \left(\frac{\epsilon}{2}\right)\right)^{-1} > -\epsilon. \tag{2.16}$$

The upper and lower bounds (2.15) and (2.16) establish (2.9). To obtain (2.10), note first that

$$\begin{aligned}\Pr[N_{\alpha,\beta} > 1] &= 1 - \Pr[N_{\alpha,\beta} = 0] - \Pr[N_{\alpha,\beta} = 1] \\ &= o(\Lambda_\beta - \Lambda_\alpha) - \{\Pr[N_{\alpha,\beta} = 1] - (\Lambda_\beta - \Lambda_\alpha)\}.\end{aligned} \tag{2.17}$$

Equation (2.10) is an immediate consequence of applying the inequalities (2.12) and (2.13) to (2.17). This ends the proof of Lemma 2.2.2.

Lemma 2.2.3 (I. Gikhman and A. Skorokhod). Suppose hypotheses (a)–(e) of Theorem 2.2.1 hold. Then, for an arbitrary bounded interval $[s,t]$ in $[t_0,\infty)$,

$$\Pr[N_{s,t}>n]\leqslant\{\Pr[N_{s,t}>0]\}^{n+1}. \tag{2.18}$$

Proof. Consider a sequence of partitions $s=t_{K0}\leqslant t_{K1}\leqslant t_{K2}\leqslant\cdots\leqslant t_{KK}=t$ of $[s,t]$, for $K=1,2,\ldots,$ such that

$$\lim_{K\to\infty}\max_{0\leqslant k\leqslant K-1}|t_{K,k+1}-t_{Kk}|=0.$$

Define

$$\xi_{Kk}=\begin{cases}1, & \text{for } N_{t_{Kk},t_{K,k+1}}\geqslant 1\\ 0, & \text{for } N_{t_{Kk},t_{K,k+1}}=0\end{cases} \tag{2.19}$$

for $0\leqslant k\leqslant K-1$. Note for $0<\delta<1$ that

$$\Pr\left[N_{s,t}-\sum_{k=0}^{K-1}\xi_{Kk}>\delta\right]=\Pr\left[\sum_{k=0}^{K-1}(N_{t_{Kk},t_{K,k+1}}-\xi_{Kk})>\delta\right]$$

$$\leqslant\sum_{k=0}^{K-1}\Pr[N_{t_{Kk},t_{K,k+1}}>1].$$

According to Lemma 2.2.2, we can select K sufficiently large that

$$\Pr[N_{t_{Kk},t_{K,k+1}}>1]\leqslant\epsilon(\Lambda_{t_{K,k+1}}-\Lambda_{t_{Kk}}).$$

Hence, for K sufficiently large

$$\Pr\left[N_{s,t}-\sum_{k=0}^{K-1}\xi_{Kk}>\delta\right]\leqslant\epsilon(\Lambda_t-\Lambda_s).$$

As $\Lambda_t-\Lambda_s<\infty$ by Lemma 2.2.1 and ϵ can be selected to be arbitrarily small, it follows that $\sum_{k=0}^{K-1}\xi_{Kk}$ converges to $N_{s,t}$ in probability. We now use this result to establish (2.18). By using the evolution without aftereffects hypothesis (b), we obtain

$$\Pr\left[\sum_{k=0}^{K-1}\xi_{Kk}>n\right]=\Pr[\xi_{K0}=1]\Pr\left[\sum_{k=1}^{K-1}\xi_{Kk}>n-1\right]$$

$$+\sum_{i=1}^{K-1}\Pr\left[\sum_{k=0}^{i-1}\xi_{Kk}=0,\xi_{Ki}=1\right]\Pr\left[\sum_{k=i+1}^{K-1}\xi_{Kk}>n-1\right]$$

$$\leqslant\left\{\Pr[\xi_{K0}=1]+\sum_{i=1}^{K-1}\Pr\left[\sum_{k=0}^{i-1}\xi_{Kk}=0,\xi_{Ki}=1\right]\right\}$$

$$\times\ \Pr\left[\sum_{k=0}^{K-1}\xi_{Kk}>n-1\right]$$

$$=\Pr\left[\sum_{k=0}^{K-1}\xi_{Kk}>0\right]\Pr\left[\sum_{k=0}^{K-1}\xi_{Kk}>n-1\right]$$

Thus, in the limit as $K\to\infty$,

$$\Pr[N_{s,t}>n]\leqslant\Pr[N_{s,t}>0]\Pr[N_{s,t}>n-1]. \tag{2.20}$$

Recursive application of this equation starting with $n=1$ yields (2.18), and Lemma 2.2.3 is proven.

Lemma 2.2.4 (I. Gikhman and A. Skorokhod). Under hypotheses (a)–(e) of Theorem 2.2.1, $N_{s,t}$ has finite expectation for any bounded interval $[s,t]$, $t_0\leqslant s\leqslant t$.

Proof. Straightforward manipulation using $\Pr[N_{s,t}=n]=\Pr[N_{s,t}>n-1]-\Pr[N_{s,t}>n]$ shows

$$E[N_{s,t}]=\sum_{n=0}^{\infty}\Pr[N_{s,t}>n].$$

Consequently, using Lemma 2.2.3 and the result $\Pr[N_{s,t}>0]<1$ obtained from Lemma 2.2.1, we have

$$E[N_{s,t}]\leqslant\sum_{n=0}^{\infty}\{\Pr[N_{s,t}>0]\}^{n+1}$$

$$=\Pr[N_{s,t}>0]\{\Pr[N_{s,t}=0]\}^{-1}.$$

It follows that $E[N_{s,t}] < \infty$ because $\Pr[N_{s,t}=0]>0$. This ends the proof of Lemma 2.2.4. We now prove Theorem 2.2.1.

Proof. We need to establish that the increment $N_{s,t} = N_t - N_s$ has the Poisson distribution. The assertion of the theorem then follows because $\{N_t; t \geqslant t_0\}$ has independent increments as an immediate consequence of hypothesis (b). Consider a sequence of partitions $s = t_{K0} \leqslant t_{K1} \leqslant t_{K2} \leqslant \cdots \leqslant t_{KK} = t$ of $[s,t]$, for $K = 1,2,\ldots$, such that

$$\lim_{K\to\infty} \max_{1\leqslant k\leqslant K} |t_{Kk} - t_{K,k-1}| = 0.$$

Using hypothesis (b), we have for any $K \geqslant 1$ that

$$\Pr[N_{s,t}=n] = \sum_{k=0}^{K-1}\sum_{i=1}^{n} \Pr[N_{s,t_{Kk}}=0]\Pr[N_{t_{Kk},t_{K,k+1}}=i] \times \Pr[N_{t_{K,k+1},t}=n-i]. \tag{2.21}$$

Thus, by eliminating all terms containing $i \geqslant 2$, we obtain the lower bound

$$\Pr[N_{s,t}=n] \geqslant \sum_{k=0}^{K-1} \Pr[N_{s,t_{Kk}}=0]\Pr[N_{t_{Kk},t_{K,k+1}}=1]\Pr[N_{t_{K,k+1},t}=n-1].$$

This inequality can be further extended by using the relation

$$\Pr[N_{t_{Kk},t_{K,k+1}}=1] = 1 - \Pr[N_{t_{Kk},t_{K,k+1}}=0] - \Pr[N_{t_{Kk},t_{K,k+1}}>1]. \tag{2.22}$$

Neglecting the last term in (2.22), we have

$$\Pr[N_{s,t}=n] \geqslant -\sum_{k=0}^{K-1} \Pr[N_{t_{K,k+1},t}=n-1]\left\{\Pr[N_{s,t_{K,k+1}}=0] - \Pr[N_{s,t_{Kk}}=0]\right\}. \tag{2.23}$$

To get an upper bound from (2.21), note that

$$\begin{aligned}\sum_{i=2}^{n} \Pr[N_{t_{Kk},t_{K,k+1}}=i]\Pr[N_{t_{K,k+1},t}=n-i] &\leqslant \sum_{i=2}^{n} \Pr[N_{t_{Kk},t_{K,k+1}}=i] \\ &\leqslant \sum_{i=2}^{\infty} \Pr[N_{t_{Kk},t_{K,k+1}}=i] = \Pr[N_{t_{Kk},t_{K,k+1}}>1].\end{aligned}$$

Furthermore, according to the uniform orderliness hypothesis (a), for all $\epsilon > 0$, it is possible to select K sufficiently large that

$$\sum_{k=0}^{K-1} \Pr[N_{t_{Kk},t_{K,k+1}} > 1] \leqslant \epsilon \sum_{k=0}^{K-1} \Pr[N_{t_{Kk},t_{K,k+1}} = 1]$$

$$\leqslant \epsilon \sum_{k=0}^{K-1} E[N_{t_{Kk},t_{K,k+1}}] = \epsilon E[N_{s,t}].$$

By using this result to bound all terms in (2.21) containing $i \geqslant 2$, using the orderliness hypothesis (a) to bound the last term in (2.22), and collecting terms, we obtain

$$0 \leqslant \Pr[N_{s,t} = n] + S_K(s,t) \leqslant \epsilon[E(N_{s,t}) + S_K(s,t)]$$

$$\leqslant \epsilon[E(N_{s,t}) + 1] \tag{2.24}$$

where $S_K(s,t)$ denotes the sum appearing on the right side of (2.23), and where we have used the easily verified property that $|S_K(s,t)| \leqslant \Pr[N_{s,t} = 0]$ for all $K \geqslant 1$. Since the coefficient of ϵ in (2.24) is finite, we then deduce in the limit as $\epsilon \to 0$ and $K \to \infty$ that

$$\Pr[N_{s,t} = n] = -\int_s^t \Pr[N_{\tau,t} = n-1]\, d_\tau \Pr[N_{s,\tau} = 0]$$

$$-\int_s^t \Pr[N_{\tau,t} = n-1] \exp[-(\Lambda_\tau - \Lambda_s)]\, d\Lambda_\tau. \tag{2.25}$$

This Riemann–Stieltjes integral exists by virtue of the continuity of Λ_τ as a function of τ for $s \leqslant \tau \leqslant t$ (see T. Apostol [2, p. 211]). The integral can be evaluated recursively to demonstrate that the increment $N_{s,t}$ is Poisson distributed with parameter $\Lambda_t - \Lambda_s$. Thus, for $n = 1$,

$$\Pr[N_{s,t} = 1] = \int_s^t \exp[-(\Lambda_t - \Lambda_\tau)] \exp[-(\Lambda_\tau - \Lambda_s)]\, d\Lambda_\tau$$

$$= (\Lambda_t - \Lambda_s) \exp[-(\Lambda_t - \Lambda_s)].$$

Reinserting this expression into (2.25), we obtain for $n = 2$ that

$$\Pr[N_{s,t} = 2] = \tfrac{1}{2}(\Lambda_t - \Lambda_s)^2 \exp[-(\Lambda_t - \Lambda_s)].$$

Repeating this procedure, we obtain for $n \geqslant 0$ that

$$\Pr[N_{s,t}=n]=(n!)^{-1}(\Lambda_t-\Lambda_s)^n \exp[-(\Lambda_t-\Lambda_s)]. \tag{2.26}$$

This shows that the increment is Poisson distributed, and the proof of Theorem 2.2.1 is complete.

Remarks. The function Λ introduced somewhat arbitrarily in Lemma 2.2.1 emerges with the proof of Theorem 2.2.1 as the parameter of the Poisson distribution of $N_{s,t}$; according to (2.5), $\Lambda_t-\Lambda_s$ is the expected number of points in the interval $[s,t)$. The relation (2.18) was used in Lemma 2.2.4 to establish the finiteness of the first moment of $N_{s,t}$. I. Gikhman and A. Skorokhod [13, p. 256] use (2.18) to show that the rth moment of $N_{s,t}$ is finite for all $r>0$ (see Problem 2.2.5). The most general Poisson counting process on $[t_0, \infty)$ is obtained from the Poisson process of Theorem 2.2.1 by adding a nondecreasing, nonnegative, piecewise-constant function to the continuous function Λ_t.

We now state and prove a second theorem giving sufficient conditions for a counting process to be a Poisson counting process. In contrast to Theorem 2.2.1, the conditions do not imply the most general Poisson process. However, we shall see in later chapters, particularly Chapter 5, that they can be modified easily to give a wide class of non-Poisson models useful in practice.

We need the notion of *conditional orderliness*, which is somewhat stronger than orderliness.

Conditional Orderliness. Denote by P an arbitrary event determined by the random variables $\{N_\sigma; t_0 \leqslant \sigma < t\}$. A counting process $\{N_t; t \geqslant t_0\}$, and also the underlying point process, is conditionally orderly at time $t \geqslant t_0$ if for any P and any $\epsilon>0$, there exists a $\delta \equiv \delta(t,\epsilon)>0$ such that $\Pr[N_{t,t+\delta'} > 1|P] \leqslant \epsilon \Pr[N_{t,t+\delta'}=1|P]$ for all $\delta' \in (0,\delta)$. A point process will be called conditionally orederly if it is conditionally orderly for all $t \geqslant t_0$.

A point process which is conditionally orderly at time t is also unconditionally orderly at time t, but the converse need not hold. By virtue of its independent increments, a Poisson process with a continuous parameter function is conditionally orderly.

Theorem 2.2.2 (Conditions for a Poisson Process). Let $\{N_t; t \geqslant t_0\}$ be the counting process associated with a point process on $[t_0, \infty)$. Suppose that:

a. the point process is conditionally orderly;

b. for all $t \geqslant t_0$ and for an arbitrary event P associated with the random variables $\{N_\sigma; t_0 \leqslant \sigma < t\}$, the limit of $\delta^{-1}\Pr[N_{t,t+\delta}=1|P]$ exists as δ tends to zero, and the limit is a finite, integrable function of t alone. Denote the limiting function by λ_t; thus,

$$\lambda_t \triangleq \lim_{\delta \downarrow 0} \delta^{-1}\Pr[N_{t,t+\delta}=1|P], \tag{2.27}$$

and $\int_s^t \lambda_\sigma \, d\sigma$ exists and is finite for all finite intervals $[s,t)$, $t_0 \leqslant s \leqslant t$;

c. $\Pr[N_{t_0}=0]=1$.

Then $\{N_t; t \geqslant t_0\}$ is a Poisson counting process with an absolutely continuous parameter function $\Lambda_t \triangleq \int_{t_0}^t \lambda_\sigma \, d\sigma$.

Hypothesis (b) implies not only that the limit exists, but also that $\{N_t; t \geqslant t_0\}$ evolves *infinitesimally* without aftereffects. In proving the theorem, we will demonstrate that the process also evolves (globally) without aftereffects.

Proof. We need to show both that the increment $N_{s,t}=N_t-N_s$ is Poisson distributed with parameter $\int_s^t \lambda_\sigma \, d\sigma$ and that $\{N_t; t \geqslant t_0\}$ has independnet increments. The objective of the first part of the proof is to derive a differential equation for $\Pr[N_{s,t}=n]$ and then to demonstrate that the solution is the Poisson distribution. For $n \geqslant 1$, the event $\{N_{s,t+\delta}=n\}$ can occur in $n+1$ mutually exclusive ways; namely,

1.	$N_{s,t}=n$	and	$N_{t,t+\delta}=0$
2.	$N_{s,t}=n-1$	and	$N_{t,t+\delta}=1$
⋮			
$n+1$.	$N_{s,t}=0$	and	$N_{t,t+\delta}=n$.

Consequently,

$$\begin{aligned}\Pr[N_{s,t+\delta}=n] &= \sum_{k=0}^{n} \Pr[N_{s,t}=n-k, N_{t,t+\delta}=k] \\ &= \Pr[N_{t,t+\delta}=0|N_{s,t}=n]\Pr[N_{s,t}=n] \\ &\quad + \Pr[N_{t,t+\delta}=1|N_{s,t}=n-1]\Pr[N_{s,t}=n-1] \\ &\quad + \sum_{k=2}^{n} \Pr[N_{t,t+\delta}=k|N_{s,t}=n-k]\Pr[N_{s,t}=n-k] \end{aligned} \tag{2.28}$$

The last two terms are absent for $n=0$, and the last term is absent for $n=1$. Let $\epsilon>0$ be given and select $\delta>0$ so that both

$$\Pr[N_{t,t+\delta}>1|N_{s,t}=n-k]\leqslant\epsilon\Pr[N_{t,t+\delta}=1|N_{s,t}=n-k] \tag{2.29}$$

and

$$-\epsilon\leqslant\delta^{-1}\Pr[N_{t,t+\delta}=1|N_{s,t}=n]-\lambda_t\leqslant\epsilon. \tag{2.30}$$

Such a δ exists by hypotheses (a) and (b). Inequalities (2.29) and (2.30) can be used to upper bound the last term in (2.28) according to

$$\sum_{k=2}^{n}\Pr[N_{t,t+\delta}=k|N_{s,t}=n-k]\Pr[N_{s,t}=n-k]$$

$$\leqslant\sum_{k=2}^{n}\Pr[N_{t,t+\delta}>1|N_{s,t}=n-k]\Pr[N_{s,t}=n-k]$$

$$\leqslant\epsilon\sum_{k=2}^{n}\Pr[N_{t,t+\delta}=1|N_{s,t}=n-k]\Pr[N_{s,t}=n-k]$$

$$\leqslant\epsilon\Pr[N_{t,t+\delta}=1]\leqslant\epsilon\lambda_t\delta+\epsilon^2\delta.$$

From the relation

$$\Pr[N_{t,t+\delta}=0|N_{s,t}=n]=1-\Pr[N_{t,t+\delta}=1|N_{s,t}=n]$$
$$-\Pr[N_{t,t+\delta}>1|N_{s,t}=n]$$

and some manipulations involving (2.29) and (2.30), we obtain

$$1-(1+\epsilon)(\lambda_t\delta+\epsilon\delta)\leqslant\Pr[N_{t,t+\delta}=0|N_{s,t}=n]\leqslant1-\lambda_t\delta+\epsilon\delta. \tag{2.31}$$

Upon substituting these inequalities into (2.28) and rearranging, we have

$$-\epsilon(2+\lambda_t+\epsilon)\leqslant\delta^{-1}\{\Pr[N_{s,t+\delta}=n]-\Pr[N_{s,t}=n]\}$$
$$+\lambda_t\Pr[N_{s,t}=n]-\lambda_t\Pr[N_{s,t}=n-1]$$
$$\leqslant\epsilon(2+\lambda_t+\epsilon).$$

By virtue of the arbitrariness of ϵ and the finiteness of λ_t, we conclude that the derivative of the counting probability $\Pr[N_{s,t}=n]$ with respect to t exists and satisfies

$$\frac{d\Pr[N_{s,t}=n]}{dt}=-\lambda_t\Pr[N_{s,t}=n]+\lambda_t\Pr[N_{s,t}=n-1] \tag{2.32}$$

for $n \geqslant 1$ and $t_0 \leqslant s \leqslant t$. Following a similar argument, we also conclude that

$$\frac{d\Pr[N_{s,t}=0]}{dt} = -\lambda_t \Pr[N_{s,t}=0] \tag{2.33}$$

for $t_0 \leqslant s \leqslant t$. It is evident from (2.29), (2.30), and (2.31) that the initial conditions for these equations are given by

$$\lim_{t \downarrow s} \Pr[N_{s,t}=n] = \delta_{0,n}$$

for $n=0,1,2,\ldots,$ where $\delta_{0,n}$ denotes the Kronecker delta function. As outlined below, these differential-difference equations can be solved sequentially to demonstrate that $N_{s,t}$ is Poisson distributed with parameter $\int_s^t \lambda_\sigma \, d\sigma$; that is,

$$\Pr[N_{s,t}=n] = (n!)^{-1}\left(\int_s^t \lambda_\sigma \, d\sigma\right)^n \exp\left[-\left(\int_s^t \lambda_\sigma \, d\sigma\right)\right] \tag{2.34}$$

for $n=0,1,2,\ldots$ and $t_0 \leqslant s \leqslant t$.

We show next that $\{N_t; t \geqslant t_0\}$ has independent increments. For this, the salient observation is that the argument used above to derive (2.34) is unchanged if $\Pr[N_{s,t}=n]$ is replaced throughout by $\Pr[N_{s,t}=n|P]$, where P is any event determined by the random variables $\{N_\sigma; t_0 \leqslant \sigma < s\}$. In this way, we deduce that

$$\Pr[N_{s,t}=n|P] = (n!)^{-1}\left(\int_s^t \lambda_\sigma \, d\sigma\right)^n \exp\left[-\int_s^t \lambda_\sigma \, d\sigma\right]$$

and, hence, that $\Pr[N_{s,t}=n|P] = \Pr[N_{s,t}=n]$. In other words, the number of points occurring in the interval $[s,t)$ is independent of any events that transpire in the interval $[t_0,s)$. That $\{N_t; t \geqslant t_0\}$ has independent increments is an immediate consequence of this, and Theorem 2.2.2 is thereby established.

The sequential solution of the differential-difference equations, (2.32) and (2.33), for $\Pr[N_{s,t}=n]$ can be conveniently accomplished as follows. For s fixed, define

$$f_t(n) \triangleq \exp\left[\int_s^t \lambda_\sigma \, d\sigma\right] \Pr[N_{s,t}=n].$$

Then, it is easily concluded that f satisfies the following differential-difference equations and boundary conditions:

$$\frac{df_t(0)}{dt}=0 \qquad \text{with } f_s(0)=1$$

and

$$\frac{df_t(n)}{dt}=\lambda_t f_t(n-1) \qquad \text{with } f_s(n)=0 \text{ for } n\geqslant 1.$$

Thus, $f_t(0)=1$ for $t\geqslant s$, and $f_t(n)=\int_s^t\lambda_\sigma f_\sigma(n-1)\,d\sigma$ for $t\geqslant s$ and $n\geqslant 1$. The equality

$$f_t(n)=(n!)^{-1}\left(\int_s^t\lambda_\sigma\,d\sigma\right)^n, \qquad n=0,1,2,\ldots$$

can be shown straightforwardly by induction for the expression clearly holds when $n=0$, and

$$\int_s^t\lambda_\sigma(n!)^{-1}\left(\int_s^\sigma\lambda_\zeta\,d\zeta\right)^n d\sigma=\int_s^t(n!)^{-1}[\Lambda_\sigma-\Lambda_s]^n\,d\Lambda_\sigma$$

where $\Lambda_\sigma\triangleq\int_{t_0}^\sigma\lambda_{\sigma'}\,d\sigma'$.

The function λ_t defined as the limit in hypothesis (b) emerges according to (2.5) as the instantaneous average rate that points occur at time t. We have termed this rate the *intensity function*. When the intensity λ_t is a constant independent of time, the corresponding Poisson counting process is said to be *homogeneous*. In this case, $\int_s^t\lambda_\sigma\,d\sigma$ is proportional to $t-s$, which implies that the counting statistics on the interval $[s,t)$ are the same as those on $[s+\tau,t+\tau)$ for all τ such that $t_0\leqslant s+\tau$. It further follows that a homogeneous Poisson counting process has stationary independent increments. Whenever λ_t is not a constant, the corresponding Poisson counting process is said to be *inhomogeneous*.

Let us note from (2.34) that

$$\Pr[N_t-N_s>0]=1-\exp\left[-\int_s^t\lambda_\sigma\,d\sigma\right]$$

and that for $t_0\leqslant t<s$,

$$\Pr[N_s-N_t>0]=1-\exp\left[-\int_t^s\lambda_\sigma\,d\sigma\right].$$

Consequently, for each $t>t_0$, N_s approaches N_t as s approaches t from either the right or the left in the sense that $\Pr[|N_t-N_s|>0]$ approaches zero regardless of how s approaches t; for this reason $\{N_t;t\geqslant t_0\}$ is said to be stochastically continuous for each $t>t_0$. Similarly, for $t_0\leqslant t$,

$$\Pr[N_t-N_{t_0}>0]=1-\exp\left[-\int_{t_0}^{t}\lambda_\sigma\,d\sigma\right],$$

and we conclude that $\{N_t;t\geqslant t_0\}$ is stochastically continuous from the right at t_0. It is of interest to note that while an inhomogeneous Poisson process is stochastically continuous at each point in an interval, the probability is in general less than one that it is continuous simultaneously throughout the interval. This obtains because

$$\Pr\{N_\sigma=N_s \text{ for all } \sigma\in[s,t)\}=\exp\left[-\int_{s}^{t}\lambda_\sigma\,d\sigma\right],$$

which is less than one whenever the intensity function is nonzero on all or part of the interval $[s,t)$.

The characteristic function

$$M_{s,t}(jv)\triangleq E[\exp(jvN_{s,t})],$$

for the number of points on $[s,t)$ for a Poisson process can be determined by straightforward calculation using (2.1). Thus,

$$\begin{aligned}M_{s,t}(jv)&=\exp[-(\Lambda_t-\Lambda_s)]\sum_{n=0}^{\infty}(n!)^{-1}(\Lambda_t-\Lambda_s)^n\exp[jvn]\\&=\exp[-(\Lambda_t-\Lambda_s)]\sum_{n=0}^{\infty}(n!)^{-1}\{(\Lambda_t-\Lambda_s)\exp[jv]\}^n\\&=\exp[(e^{jv}-1)(\Lambda_t-\Lambda_s)].\end{aligned}\tag{2.35}$$

From this and the moment generating property of characteristic functions, we obtain

$$E[N_{s,t}]=M_{s,t}^{(1)}(j0)=\Lambda_t-\Lambda_s\tag{2.36}$$

and

$$E[N_{s,t}^2]=M_{s,t}^{(2)}(j0)=(\Lambda_t-\Lambda_s)+(\Lambda_t-\Lambda_s)^2,\tag{2.37}$$

where $M_{s,t}^{(k)}(j0)$ denotes the kth derivative of $M_{s,t}(jv)$ with respect to jv with $v=0$. A frequently used property that is evident from these relations is that the variance, $\text{var}[N_{s,t}]=E[N_{s,t}^2]-E^2[N_{s,t}]$, of the number of points on the interval $[s,t)$ equals the expected number of points, $E[N_{s,t}]$, for a Poisson process. This property is generalized in Problem 2.2.3.

The condition $\Pr[N_{t_0}=0]=1$ in the definition of a Poisson process can be easily modifed to model counters that are not reset to zero at the initial time t_0. If N_{t_0} is a nonnegative, integer-valued, random variable having value k with probability $\Pr[N_{t_0}=k]$, then

$$\Pr[N_t=n]=\sum_{k=0}^{n}\Pr[N_{t_0,t}=n-k]\Pr[N_{t_0}=k] \tag{2.38}$$

because by definition $N_t=N_{t_0}+N_{t_0,t}$ and $\{N_t;t\geqslant t_0\}$ has independent increments.

Summary. Two theorems providing conditions for a point process to be a Poisson process have been established in this section. The notion of a point process evolving without aftereffects is the prominent hypothesis of each theorem because this property distinguishes the Poisson process from all other point processes. It expresses the complete absence of any influence by points occurring before some time on points occurring after that time. There are additional ways to arrive at the Poisson process from simple starting assumptions. Three constructive techniques are developed in Problems 2.2.2, 2.3.11, and 2.3.12. Additional techniques are summarized by F. Haight [14, Chapter 2].

2.3 TIME STATISTICS FOR THE POISSON COUNTING PROCESSES

Two aspects of a point process are of interest. The first may be termed *interval characteristics* and relates to the number of points occurring in arbitrary intervals of time. We have investigated the interval, or counting, statistics for the Poisson process. The second may be termed the *location characteristics* and relates to the point locations and interpoint spacings. The development of the location, or time, statistics for the Poisson process is the topic of this section.

Two interrelated time sequences are of interest and indicated in Figure 2.1. The sequence $\{w_n\}$ is called the *waiting time* or *occurrence time* sequence. The former terminology seems preferred in the queueing-theory literature where points correspond to service times. We will mostly use the latter terminology. The sequence $\{t_n\}$ is called the *interarrival time* sequence; t_n is the nth interarrival time and is the random time between the $(n-1)$st and nth occurrence times. We also refer to t_n as the forward occurrence time from w_{n-1}.

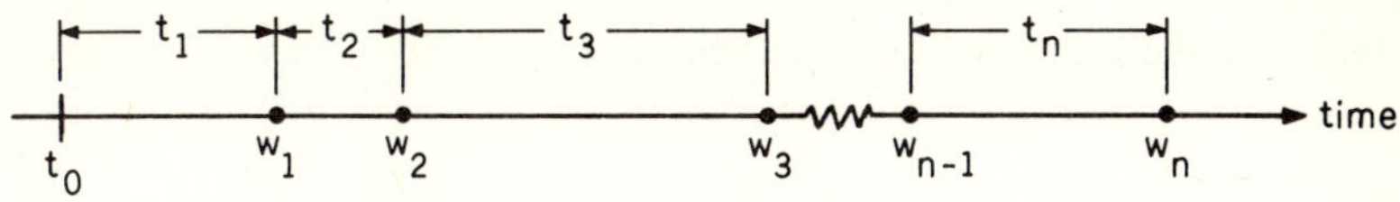

Figure 2.1. Waiting and interarrival time sequences.

Throughout this section, we assume that the conditions of Theorem 2.2.2 hold. Thus, $\{N_t; t \geqslant t_0\}$ is a Poisson counting process possessing an intensity function λ_t for all $t \geqslant t_0$. The parameter function for the process is $\Lambda_t = \int_{t_0}^{t} \lambda_\sigma \, d\sigma$ for $t \geqslant t_0$.

Our main objective is to derive an expression for the *sample function density* of a Poisson process. This important quantity is then applied in subsequent sections to problems of statistical inference for observed Poisson processes. In the course of deriving the sample function density, we also obtain several other important statistical properties of the occurrence and interarrival times for a Poisson process.

We begin with the occurrence times. Denote by $p_{\mathbf{w}}^{(n)}(\mathbf{W})$ the joint probability density for the first n occurrence times $\mathbf{w} = (w_1, w_2, \ldots, w_n)$. For an inhomogeneous Poisson counting process with intensity λ_t on $[t_0, t)$, we have that

$$p_{\mathbf{w}}^{(n)}(\mathbf{W}) = \begin{cases} \left(\prod_{i=1}^{n} \lambda_{W_i} \right) \exp\left(-\int_{t_0}^{W_n} \lambda_\sigma \, d\sigma \right), & t_0 \leqslant W_1 \leqslant W_2 \leqslant \cdots \leqslant W_n \\ 0, & \text{otherwise.} \end{cases} \tag{2.39}$$

To establish this expression for the *joint occurrence density*, consider the partitioning of time into the disjoint intervals shown in Figure 2.2. We now introduce the counting statistics of the Poisson process by noting the identity of the event $\{w_i \in [W_i, W_i + \Delta W_i);\ i = 1, 2, \ldots, n\}$ with the event

$$\{N_{t_0, W_1} = 0, N_{W_1, W_1 + \Delta W_1} = 1, N_{W_1 + \Delta W_1, W_2} = 0,$$
$$N_{W_2, W_2 + \Delta W_2} = 1, \ldots, N_{W_n, W_n + \Delta W_n} = 1\}.$$

By using the independence of the increments of N and the counting

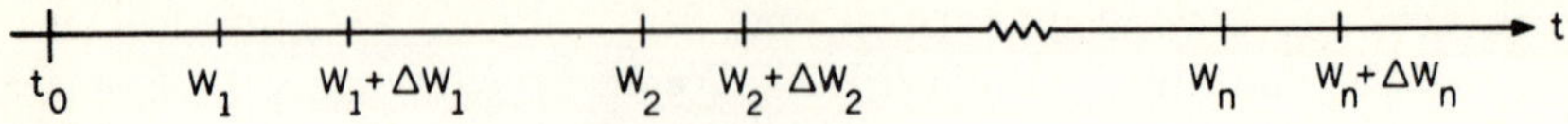

Figure 2.2. Partitioning time into disjoint intervals.

statistics of these increments, we then have

$$\Pr\{w_i \in [W_i, W_i + \Delta W_i); i = 1, 2, \ldots, n\}$$
$$= \left(\prod_{i=1}^{n} \int_{W_i}^{W_i + \Delta W_i} \lambda_\sigma \, d\sigma \right) \exp\left(-\int_{t_0}^{W_n + \Delta W_n} \lambda_\sigma \, d\sigma \right).$$

Consequently, the joint occurrence density,

$$p_{\mathbf{w}}^{(n)}(\mathbf{W}) \triangleq \lim_{\max \Delta W_i \to 0} \left(\prod_{i=1}^{n} \Delta W_i \right)^{-1} \Pr\{w_i \in [W_i, W_i + \Delta W_i); i = 1, 2, \ldots, n\},$$

has the form asserted in (2.39).

The joint occurrence density for a homogeneous Poisson process with constant intensity λ is obtained from (2.39) as

$$p_{\mathbf{w}}^{(n)}(\mathbf{W}) = \begin{cases} \lambda^n \exp[-\lambda(W_n - t_0)], & t_0 \leqslant W_1 \leqslant W_2 \leqslant \cdots \leqslant W_n \\ 0, & \text{otherwise.} \end{cases} \quad (2.40)$$

As an application of (2.39), let us demonstrate that the occurrence time sequence $w_1, w_2, \ldots, w_n$ for an inhomogeneous Poisson process is a Markov sequence with the transition density

$$p_{w_n | w_{n-1}}(W_n | W_{n-1}) = \lambda_{W_n} \exp\left[-(\Lambda_{W_n} - \Lambda_{W_{n-1}}) \right], \quad (2.41)$$

where $\Lambda_t = \int_{t_0}^{t} \lambda_\sigma \, d\sigma$. Using (2.39) and the definition of conditional probability densities, we have

$$p_{w_n | w_{n-1}, \ldots, w_1}(W_n | W_{n-1}, \ldots, W_1) = \frac{p_{\mathbf{w}}^{(n)}(\mathbf{W})}{p_{\mathbf{w}}^{(n-1)}(\mathbf{W})} = \lambda_{W_n} \exp\left(-\int_{W_{n-1}}^{W_n} \lambda_\sigma \, d\sigma \right)$$

Thus, the conditional density of w_n given that $w_i = W_i$, for $i = 1, 2, \ldots, n-1$, is not a function of W_i, for $i=1,2,\ldots,n-2$, and we easily deduce that

$$p_{w_n|w_{n-1}}(W_n|W_{n-1})=\int\cdots\int p_{w_n|w_{n-1},\ldots,w_1}(W_n|W_{n-1},\ldots,W_1)$$

$$\times p_{w_{n-2},\ldots,w_1|w_{n-1}}(W_{n-2},\ldots,W_1|W_{n-1})\,dW_{n-2}\cdots dW_1$$

$$=p_{w_n|w_{n-1},\ldots,w_1}(W_n|W_{n-1},\ldots,W_1), \tag{2.42}$$

which shows that the occurrence time sequence for an inhomogeneous Poisson counting process is a Markov sequence with the transition density given in (2.41).

We next examine a quantity termed the *forward-occurrence density*. This is the conditional probability density for the nth interarrival time, t_n, given the $n-1$ occurrence times $w_1, w_2, \ldots, w_{n-1}$, for $n=2,3,\ldots$. We also term the unconditional density for t_1 a forward-occurrence density. If points correspond to failures of some equipment, the forward-occurrence density is useful for determining the conditional probability the equipment will be operative some period of time after the last failure given records of previous failure times. We will see that if such failures follow a Poisson process, it is unnecessary to record more than the most recent failure time for this determination.

The relation $w_n=w_{n-1}+t_n$ and the Markovian nature of the occurrence-time sequence imply

$$p_{t_n|w_{n-1},\ldots,w_1}(T|W_{n-1},\ldots,W_1)=p_{w_n|w_{n-1},\ldots,w_1}(W_{n-1}+T|W_{n-1},\ldots,W_1)$$

$$=p_{w_n|w_{n-1}}(W_{n-1}+T|W_{n-1}) \tag{2.43}$$

for $n=2,3,\ldots,$ and $T\geqslant 0$. We conclude from this and (2.41) that the forward-occurrence density for a Poisson process is given by

$$p_{t_n|w_{n-1},\ldots,w_1}(T|W_{n-1},\ldots,W_1)=\lambda_{W_{n-1}+T}\exp\left[-(\Lambda_{W_{n-1}+T}-\Lambda_{W_{n-1}})\right] \tag{2.44}$$

for $n=2,3,\ldots,$ and $T\geqslant 0$. This expression also holds for $n=1$ by defining $w_0=W_0=t_0$. According to (2.43), the identical expression is the conditional density of t_n given only that $w_{n-1}=W_{n-1}$.

As an example suppose points correspond to times errors are made in decoding a message and that these errors occur as a homogeneous Poisson process with a rate of λ errors per second. Assume errors are detected at times $W_1, W_2, \ldots, W_{n-1}$. As $\Lambda_t=\lambda(t-t_0)$ in this instance, we have from (2.44) that the conditional probability density for the time to the nth error is given by

$$p_{t_n|w_{n-1},\ldots,w_1}(T|W_{n-1},\ldots,W_1)=\lambda\exp(-\lambda T) \tag{2.45}$$

for $n=1,2,\ldots,$ and $T\geqslant 0$. Thus, t_n is exponentially distributed with parameter λ regardless of the number of previous errors and when they occurred. The mean time from the $(n-1)$st error to the nth error is $E(t_n)=\lambda^{-1}$. The conditional probability that no error is made for a period of T seconds following the $(n-1)$st error is given by

$$\begin{aligned}\Pr[t_n>T|w_1=W_1,\ldots,w_{n-1}=W_{n-1}]&=\int_T^{\infty}\lambda\exp(-\lambda\sigma)\,d\sigma\\&=\exp(-\lambda T).\end{aligned} \tag{2.46}$$

It is easy to see from (2.45) that the interarrival times for a homogeneous Poisson process with intensity λ are identically distributed with the common distribution being exponential with parameter λ. We will show, further, that the interarrivals are independent. To prove this, we first examine the probability that t_n exceeds T given that $t_i=T_i$ for $i=1,2,\ldots,n-1$. Thus,

$$\begin{aligned}&\Pr[t_n>T|t_1=T_1,t_2=T_2,\ldots,t_{n-1}=T_{n-1}]\\&\qquad=\Pr[w_n>t_0+T_1+\cdots+T_{n-1}+T|w_1\\&\qquad\qquad=t_0+T_1,\ldots,w_{n-1}=t_0+T_1+\cdots+T_{n-1}]\\&\qquad=\Pr[w_n>t_0+T_1+\cdots+T_{n-1}+T|w_{n-1}=t_0+T_1+\cdots+T_{n-1}],\end{aligned}$$

where the last equality follows because the waiting time sequence is a Markov sequence. By differentiating with respect to T and using (2.41), we obtain

$$\begin{aligned}&p_{t_n|t_{n-1},\ldots,t_1}(T|T_{n-1},\ldots,T_1)\\&\quad=\lambda_{t_0+T_1+\cdots+T_{n-1}+T}\exp\left[-(\Lambda_{t_0+T_1+\cdots+T_{n-1}+T}-\Lambda_{t_0+T_1+\cdots+T_{n-1}})\right].\end{aligned} \tag{2.47}$$

An immediate consequence of this is that interarrivals are independent for a homogeneous Poisson process. This result is such an important part of the folklore of Poisson processes that we state it as a theorem.

Theorem 2.3.1. For a homogeneous Poisson counting process with intensity λ, the interarrival times $t_1,t_2,\ldots,t_n$ are independent and identically distributed with the common distribution being exponential with parameter λ.

Proof. Write the joint probability density for the interarrivals, $p_{\mathbf{t}}(\mathbf{T})$, in the factored form

$$p_{\mathbf{t}}(\mathbf{T}) = p_{t_n|t_{n-1},\ldots,t_1}(T_n|T_{n-1},\ldots,T_1)$$
$$\times p_{t_{n-1}|t_{n-2},\ldots,t_1}(T_{n-1}|T_{n-2},\ldots,T_1)\cdots p_{t_1}(T_1)$$

and use (2.47) to obtain

$$p_{\mathbf{t}}(\mathbf{T}) = \lambda^n \exp\left(-\lambda \sum_{i=1}^{n} T_i\right), \qquad T_i \geqslant 0, \quad i = 1,\ldots,n.$$

The conclusion of Theorem 2.3.1 follows from this.

It is clear from (2.47) that interarrivals for an inhomogeneous Poisson process are not independent.

Occurrence times for a homogeneous Poisson process are gamma distributed. In particular, if the intensity is λ, the nth occurrence time has the following probability density:

$$p_{w_n}(W) = \begin{cases} \dfrac{(\lambda W)^{n-1}}{(n-1)!}\lambda \exp(-\lambda W), & W \geqslant 0 \\ 0, \text{ otherwise.} \end{cases} \tag{2.48}$$

That this is so is an immediate consequence of Theorem 2.3.1 and the fact that $w_n = t_1 + t_2 + \cdots + t_n$ because the density for w_n must then be the n-fold convolution of the exponential density with itself. It follows from this that the mean and variance for w_n are (n/λ) and (n/λ^2), respectively. Somewhat more can be said about the distribution of w_n for n large because w_n is the sum of independent, identically distributed random variables with finite variance (if $\lambda \neq 0$). Specifically, the central limit theorem implies that for n large, the distribution function for w_n is approximately that of a normal random variable with mean n/λ and variance n/λ^2. See Problem 2.3.4 for an application of this asymptotic approximation.

The converse to Theorem 2.3.1 is established in Problem 2.3.12. In particular, it is shown that if $\{t_i\}$ is a sequence of independent, identically distributed, exponential random variables, and if a point process is constructed by assigning these successively as interarrival times, then the resulting point process is a homogeneous Poisson process with an intensity equal to the parameter of the exponential distribution. This construction is useful in practice for simulating Poisson processes in Monte Carlo studies of situations too complicated for direct analytical investigation; this includes both homogeneous and inhomogeneous Poisson processes. The procedure is outlined below.

Simulation of Homogeneous Poisson Processes. Algorithms for the generation of pseudorandom variables independently and uniformly distributed on $[0,1]$ are given by B. Jansson [20] and A. Ralston and H. Wilf [31]. Suppose $\{u_i\}$ is a sequence of such variables. Then a sequence of independent, exponentially distributed variables can be obtained from these by the transformation $t_i = -\lambda^{-1}\ln(u_i)$. It is easily verified that under this transformation, the probability density for t_i is $\lambda\exp(-\lambda T)$ for $T \geqslant 0$. A homogeneous Poisson process with intensity λ can be simulated by: generating u_1, transforming to obtain t_1, then assigning t_1 as the time from t_0 to the first occurrence time; generating u_2, transforming to obtain t_2, then assigning t_2 as the time from the first to second occurrence time; and so forth *ad infinitum*.

Simulation of Counts for Inhomogeneous Poisson Process. Suppose it is desired to generate a realization of a random variable $N_{s,t}$ that is Poisson distributed with parameter $\int_s^t \lambda_\sigma \, d\sigma$. Define $\{M_t; t \geqslant 0\}$ to be a homogeneous Poisson process with unit intensity. Then M_τ is Poisson distributed with parameter τ. M_τ and $N_{s,t}$ have the same distribution for $\tau = \int_s^t \lambda_\sigma \, d\sigma$. A sequence of independent, unit-parameter, exponentially distributed variables can be used to obtain a realization of $\{M_t; t \geqslant 0\}$, as described above. The number of points in $[0,\tau)$ is a realization of $N_{s,t}$.

Simulation of an Inhomogeneous Poisson Process. A homogeneous Poisson process can be transformed into an inhomogeneous Poisson process, and vice versa, by a rescaling of the time coordinate. The rescaling required to achieve an inhomogeneous process with a specified intensity function is developed in Problem 2.3.5. In this way, an inhomogeneous Poisson process can be simulated by first simulating a homogeneous process as outlined above and then rescaling time.

Sample Function Density

The sample function density is one of the more important quantities we shall encounter. It will play a central role in Sections 2.4 and 2.5, where we study problems of statistical inference for observed Poisson processes.

Moreover, the sample function density for several non-Poisson processes will be developed in later chapters, and its role in statistical inference will be the same. A not easily anticipated result is that for a very broad class of non-Poisson counting processes developed in Chapters 5–7, the form of the sample function density is essentially identical to that of a Poisson process. Consequently, the interpretation of the sample function density developed here, as well as the techniques indicated for its use, will carry over unchanged in later chapters.

The quantity $p[\{N_\sigma; t_0 \leqslant \sigma < t\}]$ defined below will be called the *sample-function density* for the counting process N on the interval $[t_0, t)$:

$$p[\{N_\sigma; t_0 \leqslant \sigma < t\}] \triangleq \begin{cases} \Pr[N_t = 0], & N_t = 0 \\ p_{\mathbf{w}}(\mathbf{W}, N_t = n), & N_t = n \geqslant 1, \end{cases} \tag{2.49}$$

where

$$p_{\mathbf{w}}(\mathbf{W}, N_t = n) \triangleq \Pr[N_t = n | w_1 = W_1, w_2 = W_2, \ldots, w_n = W_n] p_{\mathbf{w}}^{(n)}(\mathbf{W}).$$

Roughly speaking, the sample-function density determines the probability of obtaining a particular realization of the point process on $[t_0, t)$ with $N_t = n$ points located, for $n \geqslant 1$, at times $w_1 = W_1$, $w_2 = W_2, \ldots, w_n = W_n$. This interpretation follows from the approximation

$$p[\{N_\sigma; t_0 \leqslant \sigma < t\}]\left(\prod_{i=1}^{n} \Delta W_i\right)$$
$$\simeq \Pr[N_t = n, w_i \in [W_i, W_i + \Delta W_i); i = 1, 2, \ldots, n].$$

The sample-function density for a Poisson process is given in the following theorem.

Theorem 2.3.2 (Sample-Function Density for a Poisson Process). Let $\{N_t; t \geqslant t_0\}$ be an inhomogeneous Poisson counting process with intensity function λ_t for $t \geqslant t_0$. Then the sample-function density for N defined in (2.49) has the evaluation

$$p[\{N_\sigma; t_0 \leqslant \sigma < t\}] = \begin{cases} \exp\left(-\int_{t_0}^{t} \lambda_\sigma \, d\sigma\right), & N_t = 0 \\ \left(\prod_{i=1}^{n} \lambda_{W_i}\right) \exp\left(-\int_{t_0}^{t} \lambda_\sigma \, d\sigma\right), & N_t = n \geqslant 1. \end{cases} \tag{2.50}$$

Proof. Equation (2.50) follows immediately from the definition of the sample-function density in (2.49), the joint occurrence density in (2.39), and the relation

$$\Pr[N_t = n | w_1 = W_1, w_2 = W_2, \ldots, w_n = W_n] = \Pr[N_t - N_{W_n+} = 0]$$
$$= \exp\left(-\int_{W_n}^{t} \lambda_\sigma \, d\sigma\right).$$

We find it convenient to incorporate into (2.50) a particular sample function of the counting process on $[t_0, t)$, namely, the sample function for which $N_t = n$ and for which the n waiting times occur at the instants W_1, $W_2, \ldots, W_n$ as shown in Figure 2.3. For this sample function, (2.50) can be alternatively written as

$$p[\{N_\sigma; t_0 \leqslant \sigma < t\}] = \exp\left(-\int_{t_0}^{t} \lambda_\sigma \, d\sigma + \int_{t_0}^{t} \ln \lambda_\sigma \, dN_\sigma\right), \tag{2.51}$$

where the second integral can be interpreted as a Riemann–Stieltjes integral with the evaluation

$$\int_{t_0}^{t} \ln \lambda_\sigma \, dN_\sigma = \begin{cases} 0, & N_t = 0 \\ \sum_{i=1}^{N_t} \ln \lambda_{w_i}, & N_t \geqslant 1. \end{cases}$$

We term integrals of this type *counting integrals*. They occur often in later chapters and are developed more carefully in Chapter 4.

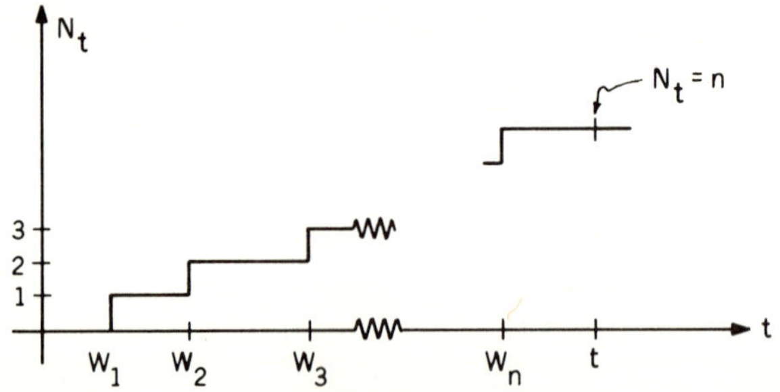

Figure 2.3. A particular sample function of N.

As a function of the random process $\{N_t; t \geqslant t_0\}$, the sample-function density may also be viewed as a random variable for each fixed t. The reader may thus wish to verify that the kth moment of the sample-function density for a Poisson process is given by (see Problem 4.2.5)

$$E(p^k[\{N_\sigma; t_0 \leqslant \sigma < t\}]) = \exp\left\{\int_{t_0}^{t}[\lambda_\sigma^{k+1} - (k+1)\lambda_\sigma]\,d\sigma\right\}. \tag{2.52}$$

We now give several applications where the sample-function density proves to be useful.

Conditional Distribution of Occurrence Times

The first application we will make of the sample function density is to derive a useful additional property of the occurrence times. Under the condition $N_T = n$, where $n \geqslant 1$, the n occurrence times $w_1, w_2, \dots, w_n$ for an inhomogeneous Poisson counting process with intensity λ_t have the same distribution as the order statistics of n independent, identically distributed random variables $u_1, u_2, \dots, u_n$ with the common probability density being

$$p_{u_i}(U) = \begin{cases} \lambda_U\left(\int_{t_0}^{T}\lambda_\sigma\,d\sigma\right)^{-1}, & t_0 \leqslant U \leqslant T \\ 0, & \text{otherwise.} \end{cases} \tag{2.53}$$

By "order statistics" we mean the following. The u_i, being distributed throughout the interval $[t_0, T]$, are unordered. In terms of the u_i we define n random variables ordered in magnitude; for instance, the first and last order variables are $\min(u_1, u_2, \dots, u_n)$ and $\max(u_1, u_2, \dots, u_n)$, respectively. The assertion is that these new variables have the same distribution as the occurrence times given that $N_T = n$.

To verify this assertion, let us note first that by using the expression for the sample-function density in (2.50) and the Poisson distribution for N_T, we can obtain

$$p_{\mathbf{w}}(\mathbf{W}|N_T = n) = \frac{p_{\mathbf{w}}(\mathbf{W}, N_T = n)}{\Pr[N_T = n]} = (n!)\prod_{i=1}^{n}\left[\lambda_{W_i}\left(\int_{t_0}^{T}\lambda_\sigma\,d\sigma\right)^{-1}\right] \tag{2.54}$$

for $t_0 \leqslant W_1 \leqslant \cdots \leqslant W_n \leqslant T$. Secondly, the joint probability density for the unordered variables $u_1, u_2, \dots, u_n$ is

$$p_{\mathbf{u}}(\mathbf{U}) = \prod_{i=1}^{n}\left[\lambda_{U_i}\left(\int_{t_0}^{t}\lambda_\sigma\,d\sigma\right)^{-1}\right]$$

for $t_0 \leqslant U_i \leqslant T$ and $i = 1, 2, \ldots, n$. Let $o_1, o_2, \ldots, o_n$ denote the ordered random variables associated with $u_1, u_2, \ldots, u_n$ as shown in Figure 2.4. There are $n!$ permutations of the unordered variables that result in the same ordered variables, and these correspond to mutually exclusive events in the realization of the unordered variables. Since probabilities for mutually exclusive events add, we have the following joint density for the ordered variables:

$$p_{\mathbf{o}}(\mathbf{O}) = (n!) \prod_{i=1}^{n} \left[\lambda_{O_i} \left(\int_{t_0}^{t} \lambda_\sigma \, d\sigma \right)^{-1} \right]$$

for $t_0 \leqslant 0_1 \leqslant \cdots \leqslant 0_n \leqslant T$. Comparison of this with (2.54) establishes the assertion.

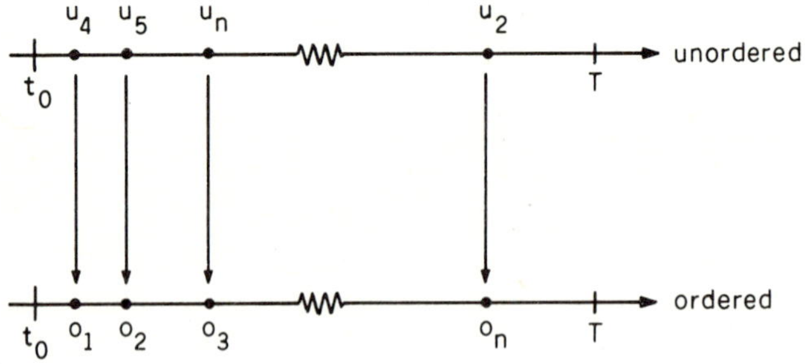

Figure 2.4. A realization of the unordered variables $u_1, \ldots, u_n$ and the corresponding ordered variables $o_1, \ldots, o_n$.

Examination of (2.53) shows that under the condition that $N_T = n$, the n occurrence times $w_1, w_2, \ldots, w_n$ for a homogeneous Poisson counting process with intensity λ have the same distribution as the order statistics of n independent, uniformly distributed random variables on $[t_0, T]$ *regardless* of the particular value of λ. This fact provides the means to test whether or not a set of points observed experimentally is reasonably modeled as a homogeneous Poisson process. A simple test based on the central limit theorem is developed in Problem 2.3.4. E. Parzen [29, p. 141] describes some more powerful tests.

As an example demonstrating the usefulness of these conditional properties of the occurrence times, let us determine the *characteristic functional* for an inhomogeneous Poisson counting process. A characteristic functional is a generalization of the idea of a characteristic function; it is important because it provides a complete statistical description of its associated random process. The characteristic functional associated with the random variables $\{N_\sigma; t_0 \leqslant \sigma < T\}$ is defined by the expectation

$$\phi_N(jv) \triangleq E\left[\exp\left(j\int_{t_0}^{T} v_\sigma \, dN_\sigma\right)\right], \tag{2.55}$$

where v is an arbitrary real-valued function. The integral can be interpreted as a Riemann–Stiltjes integral having the evaluation

$$\int_{t_0}^{T} v_\sigma \, dN_\sigma = \begin{cases} 0, & N_T = 0 \\ \sum_{i=1}^{N_T} v_{w_i}, & N_T \geqslant 1. \end{cases}$$

The expectation in (2.55) is straightforward to evaluate using the properties of conditional expectation and (2.53). Thus,

$$\begin{aligned} \phi_N(jv) &= \sum_{n=0}^{\infty} \Pr[N_T = n] E\left[\exp\left(j\int_{t_0}^{T} v_\sigma \, dN_\sigma\right) \middle| N_T = n\right] \\ &= \Pr[N_T = 0] + \sum_{n=1}^{\infty} \Pr[N_T = n] E\left[\exp\left(j\sum_{i=1}^{N_T} v_{w_i}\right) \middle| N_T = n\right]. \end{aligned} \tag{2.56}$$

The following observations are useful for evaluating the conditional expectation in this expression. Suppose $N_T = n$ points occur in $[t_0, T)$. The occurrence times $w_1, w_2, \ldots, w_n$ represent one labeling of these n points; this labeling is ordered. Suppose we relabel the points by assigning the labels $u_1, u_2, \ldots, u_n$ in a completely random manner; this labeling is unordered. We note for this relabeling that the sum appearing in the conditional expectation remains unchanged; that is,

$$\sum_{i=1}^{n} v_{w_i} = \sum_{i=1}^{n} v_{u_i}.$$

Consequently, the expectation in (2.56) can be evaluated by relabeling the occurrence times in this fashion and using the fact for a Poisson process that, as unordered variables, the points are independent and identically distributed with the common probability density being that in (2.53).

Doing this, we obtain

$$E\left[\exp\left(j\sum_{i=1}^{N_T} v_{w_i}\right)\Big| N_T = n\right] = E\left[\exp\left(j\sum_{i=1}^{N_T} v_{u_i}\right)\Big| N_T = n\right]$$

$$= \left[\left(\int_{t_0}^{T} \lambda_\sigma \, d\sigma\right)^{-1} \int_{t_0}^{T} \lambda_\sigma e^{jv_\sigma} \, d\sigma\right]^n. \quad (2.57)$$

Substitution of this expression into (2.56) and use of the Poisson distribution for N_T then shows that the characteristic functional for an inhomogeneous Poisson process on the interval $[t_0, T)$ is given by

$$\phi_N(jv) = \exp\left[\int_{t_0}^{T} \lambda_\sigma (e^{jv_\sigma} - 1) \, d\sigma\right]. \quad (2.58)$$

The characteristic function can be obtained by specialization of the characteristic functional. For illustration of this, take the arbitrary function v in (2.55) to have the form

$$v_\sigma = \begin{cases} 0, & t_0 \leqslant \sigma < s \\ \alpha, & s \leqslant \sigma < t \\ 0, & t \leqslant \sigma < T. \end{cases}$$

Then $\int_{t_0}^{T} v_\sigma \, dN_\sigma = \alpha N_{s,t}$, and the characteristic functional becomes the characteristic function for the increment $N_{s,t} = N_t - N_s$. Equation (2.58) becomes

$$\phi_N(jv) = \exp\left[\int_{s}^{t} \lambda_\sigma \, d\sigma (e^{j\alpha} - 1)\right].$$

Comparison of this expression with (2.35) shows for this choice of v that $\phi_N(jv) = M_{s,t}(j\alpha)$.

As a further illustration, suppose v has the piecewise constant form shown in Figure 2.5 where the intervals $[t_i, u_i)$, $i = 1, 2, \ldots, k$, are mutually disjoint on $[t_0, T)$. For this choice of v,

$$\int_{t_0}^{T} v_\sigma \, dN_\sigma = \sum_{i=1}^{k} \alpha_i N_{t_i, u_i}$$

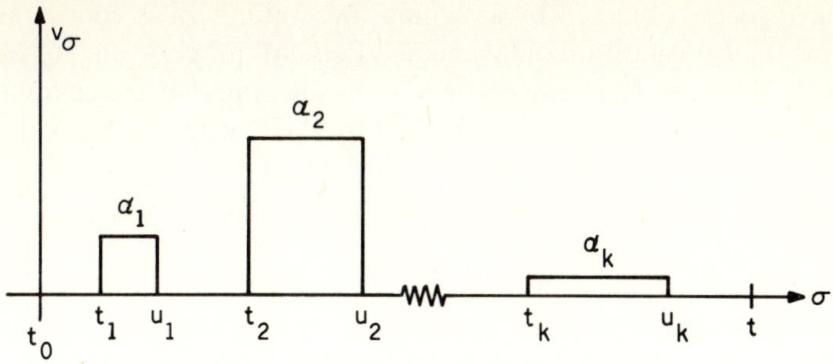

Figure 2.5. A particular choice of v for establishing increment statistics.

and the characteristic functional is the joint characteristic function for the k increments N_{t_i,u_i}, $i=1, 2,\ldots,k$. Equation (2.58) becomes

$$\phi_N(jv)=\prod_{i=1}^{k}\exp\left[\int_{t_i}^{u_i}\lambda_\sigma\,d\sigma\,(e^{j\alpha_i}-1)\right],$$

which demonstrates that N has independent, Poisson distributed increments, as we already knew. These two specializations of the characteristic functional of (2.55) show that it provides a complete statistical characterization of an inhomogeneous Poisson counting process since we have been able to deduce from it the increment distribution of N and the fact that N has independent increments.

We have shown that under the condition $N_T=n$, the n occurrence times $w_1, w_2,\ldots,w_n$ for an inhomogeneous Poisson process with intensity λ_t have the same distribution as the order statistics of n independent, identically distributed random variables with the common probability density given in (2.53). The converse is also true and sometimes useful in the simulation of Poisson processes. Specifically, a point process constructed on $[t_0, T)$ in the following manner is an inhomogeneous Poisson process with intensity λ_t. First select a nonnegative integer N according to the Poisson distribution

$$\Pr[N=n]=(n!)^{-1}\left(\int_{t_0}^{T}\lambda_\sigma\,d\sigma\right)^n\exp\left(-\int_{t_0}^{T}\lambda_\sigma\,d\sigma\right).$$

This can be accomplished using the procedure outlined above for simulating counts for an inhomogeneous Poisson process. Secondly, select N numbers $u_1, u_2,\ldots,u_N$ independently over the interval $[t_0, T)$ each with probability density

$$p_{u_i}(U)=\lambda_U\left(\int_{t_0}^{T}\lambda_\sigma\,d\sigma\right)^{-1}, \qquad t_0\leqslant U\leqslant T.$$

The point process obtained by assigning these numbers as coordinates for points in $[t_0, T)$ is an inhomogeneous Poisson process on $[t_0, T)$ with intensity λ_t. This assertion can be proven by showing that the characteristic functional for the point process specified by the construction has the form in (2.58). That this is indeed the form is evident starting from (2.56) and paralleling the derivation of (2.58); see Problem 2.3.11.

Summary. In this section, we have developed the statistical properties of the occurrence and interarrival times for a Poisson counting process. The important results can be summarized as follows:

a. the joint occurrence time density is given for an inhomogeneous process in (2.39) and for a homogeneous process in (2.40);
b. the occurrence times form a Markov sequence;
c. interrarrival times for a homogeneous process are independent and identically distributed exponential random variables;
d. occurrence times for a homogeneous process are gamma distributed;
e. given the number of points in an interval is n, the n occurrence times have the same distribution as the order statistics for n independent and identically distributed random variables with the common distribution being that in (2.53).

One of the most important quantities developed in the section is the sample-function density, which completely characterizes the statistical properties of paths of the counting process. The sample-function density will be used in the next two sections, where we study problems of statistical inference for observed Poisson processes. Generalizations of the sample function density for other counting processes will appear throughout the book. The results of this section are used in Problems 2.3.11 and 2.3.12 to develop constructive derivations of the Poisson process, which complement the derivations of Section 2.2.

2.4 PARAMETER ESTIMATION FOR OBSERVED POISSON PROCESSES

The desire to know parameters that cannot be directly observed often arises in practice. In the typical situation, data influenced by the parameters are collected and analyzed to estimate them. Two issues arise. The first is that of designing the analysis to be performed and the second that of predicting the accuracy of the resulting estimate. The purpose of this section is to address these issues when the collected data are reasonably modeled as a Poisson process. In later chapters, we will reencounter these issues when the data are modeled by more complicated point processes and their generalizations.

It is not our intention to give a complete theory of statistical parameter estimation here, since it is extensive and readily available in numerous textbooks, including those by L. J. Savage [35], M. Fisz [12], and H. L. Van Trees [40]. The monograph by D. R. Cox and P. A. W. Lewis [7] contains specialized estimation techniques for the Poisson process. We will restrict our attention in this section to maximum likelihood estimation for the Poisson process.

In general terms, the problem of statistical estimation for inhomogeneous Poisson counting processes can be stated as follows. Let $\{N_t; t \geqslant t_0\}$ be an inhomogeneous Poisson counting process with intensity $\lambda_t(\mathbf{X})$, where λ is a known function of t and $\mathbf{X}$, and where $\mathbf{X}$ is a vector of unknown parameters having values in the space $\mathfrak{X}$ of possible parameter values. Let $D_{t_0,T}$ denote data available from observations of N on the interval $[t_0, T)$; for example, $D_{t_0,T}$ can be simply the number of counts observed on the entire interval, the number of counts observed on several subintervals of $[t_0, T)$, or the entire counting path $\{N_\sigma; t_0 \leqslant \sigma < T\}$. Let $P(D_{t_0,T}|\mathbf{X})$ denote the statistics of $D_{t_0,T}$, where we have explicitly indicated in this notation that these statistics depend on the parameters $\mathbf{X}$ of the intensity function. The problem is as follows: Knowing $P(D_{t_0,T}|\mathbf{X})$ for every $\mathbf{X} \in \mathfrak{X}$, and having observed a particular set of data, guess the value of the parameters $\mathbf{X}$. This means that some function $\hat{\mathbf{X}} = \hat{\mathbf{X}}(D_{t_0,T})$ of $D_{t_0,T}$ is to be determined; we refer to $\hat{\mathbf{X}}$ as an estimate of $\mathbf{X}$.

Some specific situations where parameter estimation problems of the type described arise are given in the following examples.

Example 2.4.1 (Radioactive Decay). As shown by R. D. Evans [11, p. 470] the emission of gamma or annihilation photons by a radioactive source can be modeled to a close approximation as an inhomogeneous Poisson process with an intensity of the form $X_1 \exp(-X_2 t)$ for $t \geqslant 0$, where $X_1 > 0$ depends on the quantity of the source material and $X_2 > 0$ is the reciprocal of the mean life of the source. The problem of estimating X_1 and X_2 from observed radiation for various sources is a problem of standing interest in nuclear physics, nuclear medicine, geochronology, and numerous other disciplines.

Example 2.4.2 (Nuclear Medicine). The medical use of radioactive tracers has developed rapidly in recent years because such tracers provide relatively noninvasive diagnostic procedures for clinical medicine and a basic research tool in biochemical and physiological research. Radiotracers are employed in both static and dynamic studies. In static studies, the desire is to obtain an image of the spatial distribution of the tracer. Dynamic studies are designed to obtain quantitative information about parameters relating to time-dependent physiological phenomena such as blood flow rate, metabolic exchange rates, and lung ventilation [37,41,42].

Dynamic radiotracer data are typically collected with some type of scintillation detector and have the form of a series of discrete events each of which is associated with the emission of a light pulse in the crystal of the detector. Each light pulse originates, in turn, because of the absorption in the crystal of a gamma or annihilation photon emanating from the radiotracer. To a first approximation the series of light pulses can be modeled as an inhomogeneous Poisson process. As discussed by J. A. Jacques [19] and C. W. Sheppard [36], a form of intensity function that is frequently assumed is

$$\lambda_t(\mathbf{X}) = X_0 + \sum_{i=1}^{n} X_i \exp(-X_{n+i}t) \tag{2.59}$$

for $t \geqslant 0$ where $X_i > 0$ for $i=0, 1,\ldots,2n$. In practice, n is typically between one and five. This intensity reflects physiological transport phenomena; for instance, the intensity decreases with time because, in part, of the radioactive source being removed from the field of view of the detector by blood flow. Thus, the $2n+1$ parameters $X_0,\ldots,X_{2n}$ are of interest in both clinical and physiological studies.

Example 2.4.3 (Optical Detection). The stream of photoelectrons produced when coherent light is focused on a photosensitive surface has been shown by L. Mandel [25,26] to be modeled to a good approximation by an inhomogeneous Poisson process. This model is also reviewed by S. Karp, E. L. O'Neill, and R. M. Gagliardi [21]. Suppose the incident light can be characterized classically by the scalar electric field

$$s(t,\vec{r}) = \sqrt{2}\,\mathrm{Re}[S(t,\vec{r})\exp(j2\pi\nu t)] \tag{2.60}$$

at time t and position $\vec{r}$, where ν is the optical frequency, $S(t,\vec{r})$ is a complex space-time envelope that is slowly varying in comparison to ν, and $\mathrm{Re}(\cdot)$ denotes the real part of $(\cdot)$. The light intensity at time t and position $\vec{r}$ is $|S(t,\vec{r})|^2$. The rate (or intensity) at which electrons emanate from the photodetector is related to the light intensity incident on its sensitive surface according to

$$\lambda_t = \left(\frac{\eta}{h\nu}\right)\int_{\mathcal{Q}} |S(t,\vec{r})|^2 d^2\vec{r} + \lambda_0 \tag{2.61}$$

where η is the quantum efficiency of the detector (the probability that a quantum of light energy is converted into an electron), h is Planck's constant, the integral is over the sensitive surface $\mathcal{Q}$ of the detector, and λ_0 is the rate of generation of extraneous electrons in the detector and accounts for dark current. Parameter estimation problems arise in a number of ways with the use of photodetectors. For simplicity in describing these, we will assume that spatial variations of the field on $\mathcal{Q}$ are negligible and write the parameterized rate function as $\lambda_t(\mathbf{X}) = \alpha|S(t,\mathbf{X})|^2 + \lambda_0$, where α is a constant. Some special cases of interest in optical communication and radar systems are

1. *Pulse Amplitude Modulation.* Suppose the intensity of a pulsed light source is adjusted in accordance with a parameter X. The electron generation rate at the output of a photodetector used to observe the light is then $\lambda_t(X) = \alpha X|S(t)|^2 + \lambda_0$. The problem of estimating X from observations of the photodetector output has been studied by I. Bar-David [5].
2. *Pulse-Position Modulation, Optical Range Finding.* Suppose the timing of a pulse from a light source is adjusted in accordance with a parameter X. The electron generation rate at the output of a photodetector used to observe the light is then $\lambda_t(X) = \alpha|S(t-X)|^2 + \lambda_0$. The problem of estimating X from observations of the photodetector output has been studied by I. Bar-David [5] and E. Hoversten and D. Snyder [18].
3. *Pulse-Frequency Modulation, Optical Range-Rate Finding.* In an effort to measure the velocity of an object, the intensity of a light beam directed toward it is modulated sinusoidally. The light reflected has all its frequencies shifted because of the Doppler effect; the frequency of the modulation is shifted by an amount proportional to the modulation frequency and the range-rate of the object. The electron generation rate at the output of a photodetector used to observe the reflected light is then of the form

$$\lambda_t(X) = \alpha\{1 + m\cos[2\pi(f_m + X)t]\} + \lambda_0 \tag{2.62}$$

where α and m are constants ($\alpha > 0, |m| < 1$), f_m is the modulation frequency, and X is the Doppler shift. The problem of estimating X has been studied by C. Helstrom [16], who attributes it originally to N. McEvoy.

Example 2.4.4 (Auditory Electrophysiology). A common procedure used in auditory electrophysiology is to insert a microelectrode into an exposed auditory nerve fiber and observe the electrical activity in the nerve in response to an acoustic pressure stimulus applied to the outer ear. An example of the electrical signals obtained this way is shown in Figure 1.1; the narrow pulses are called spike discharges. W. Siebert [38, 39] argues that for a restricted class of stimuli—sinusoidal tones of limited duration, amplitude, and frequency—the spike discharge sequence can be modeled to a first approximation as an inhomogeneous Poisson process with an intensity function of the form

$$\lambda_t(\mathbf{X}) = X_1 \exp[X_2 \cos(2\pi f t + X_3)], \tag{2.63}$$

where f is the frequency of the applied stimulus and the parameters X_1, X_2, and X_3 reflect the physiological mechanisms involved in converting the pressure stimulus into electrical nerve activity. W. Littlefield [24] has conducted extensive measurements to estimate X_1, X_2, and X_3 in order to make inferences about the physiological mechanisms involved.

A form of data that are frequently collected for observed point processes is obtained by partitioning an observation interval $[t_0, T)$ into a set of disjoint subintervals $[t_0, t_1)$, $[t_1, t_2), \ldots, [t_{k-1}, t_k = T)$ and then collecting counts in each subinterval. In practice, the subintervals are often not of equal duration so that roughly the same number of counts are obtained in each. For instance, the interval durations may be increased systematically for the exponentially decaying intensities encountered in Examples 2.4.1 and 2.4.2 above. We refer to data of this type as *histogram data*. Thus, histogram data is the set of k subinterval counts n_1, $n_2, \ldots, n_k$, where $n_i = N_{t_{i-1}, t_i}$ is the observed number of points in the interval $[t_{i-1}, t_i)$.

A *count-rate histogram*, formed from histogram data, is a kind of estimate of the intensity function that is frequently used. It is constructed from the points

$$\left(\frac{n_i}{t_i - t_{i-1}}, \frac{t_i + t_{i-1}}{2}; \qquad i = 1, 2, \ldots, k \right). \tag{2.64}$$

An example of such a count-rate histogram is shown in Figure 2.6. The data shown were obtained by monitoring radioactivity over the brain following rapid injection of oxygen-15 labeled water into the internal carotid artery. For a discussion of this procedure, see Ter-Pogossian et al. [37]. The 780-second observation interval is divided into subintervals that increase systematically from 0.1 to 20 seconds. Counts were collected in the subintervals and are plotted as histogram points connected by straight lines. An advantage of this type of estimate of the intensity function is that it is nonparametric; no assumptions need to be made about the form of the intensity function. In what follows, however, we assume that the form of the intensity is known to within a set of parameters to be determined. The form of the intensity may be known in practice either by having a good understanding of the physical mechanisms giving rise to the observed point process or by having constructed a sufficient number of count-rate histograms to postulate it.

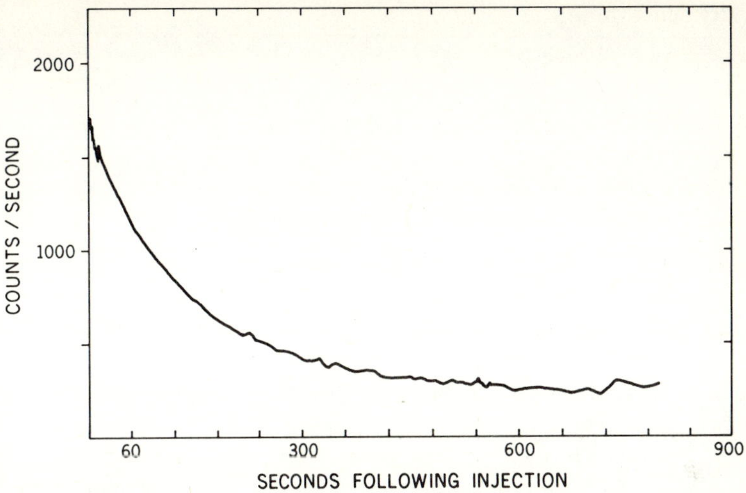

Figure 2.6. Count-rate histogram of radioactivity following injection of oxygen-15 labeled water.

Maximum Likelihood Estimation

Maximum likelihood estimation can be used when the form of the intensity is known. The maximum likelihood estimate $\hat{\mathbf{X}}_{\mathrm{ML}}=\hat{\mathbf{X}}_{\mathrm{ML}}\,(n_1,n_2,\ldots,n_k)$ of $\mathbf{X}$ in terms of the observed subinterval counts n_1, $n_2,\ldots,n_k$ is by definition the value of $\mathbf{X}$ that maximizes the probability of having observed n_1, $n_2,\ldots,n_k$. This maximization is, of course, subject to the constraint that $\hat{\mathbf{X}}_{\mathrm{ML}}\in\mathfrak{X}$; that is, the estimate must lie in the space of possible parameter values. The probability of observing n_1, $n_2,\ldots,n_k$ for a given parameter vector $\mathbf{X}$ can be easily evaluated for an inhomogeneous Poisson process by using (2.1) and (2.2). Thus,

$$\Pr\left[N_{t_{i-1},t_i}=n_i;\, i=1,\ldots,k|\mathbf{X}\right]$$

$$=\prod_{i=1}^{k}\left\{[(n_i)!]^{-1}\left(\int_{t_{i-1}}^{t_i}\lambda_\sigma(\mathbf{X})\,d\sigma\right)^{n_i}\exp\left[-\int_{t_{i-1}}^{t_i}\lambda_\sigma(\mathbf{X})\,d\sigma\right]\right\}. \tag{2.65}$$

The value of $\mathbf{X}$ that maximizes this probability also maximizes its logarithm because the logarithm is a monotonically increasing function of its argument. Consequently, the maximum likelihood estimate of $\mathbf{X}$ in terms of the observed counts n_1, $n_2,\ldots,n_k$ maximizes the *likelihood function*,

$l(\mathbf{X})$, defined by

$$l(\mathbf{X}) = -\int_{t_0}^{T} \lambda_\sigma(\mathbf{X})\,d\sigma + \sum_{i=1}^{k} n_i \ln\left(\int_{t_{i-1}}^{t_i} \lambda_\sigma(\mathbf{X})\,d\sigma\right), \tag{2.66}$$

where this maximization is subject to the constraint $\mathbf{X} \in \mathfrak{X}$.

Example 2.4.5 (Scaled Inhomogeneous Poisson Process). Suppose $\{N_t; t \geqslant t_0\}$ is an inhomogeneous Poisson process with a parametrized intensity function $\lambda_t(X) = X\mu_t$ for $t \geqslant t_0$, where μ_t is a known function of t and $X > 0$ is an unknown scale factor. An observation of N on the interval $[t_0, T)$ is made in order to determine X. Suppose the observations are in the form of histogram data. Thus, the observations result in k subinterval counts $n_1, n_2, \ldots, n_k$ from which X can be estimated; here, n_i is the number of points observed to occur in the subinterval $[t_{i-1}, t_i)$. The likelihood function defined in (2.66) becomes

$$l(X) = -X\int_{t_0}^{T} \mu_\sigma\,d\sigma + N_T \ln X + \sum_{i=1}^{k} n_i \ln\left(\int_{t_{i-1}}^{t_i} \mu_\sigma\,d\sigma\right) \tag{2.67}$$

where $N_T = n_1 + n_2 + \cdots + n_k$ is the total number of observed points. The maximum-likelihood estimate $\hat{X}_{\mathrm{ML}}$ of X in terms of n_1, $n_2, \ldots,$ and n_k maximizes this expression as a function of X. Therefore, it is easily seen that

$$\hat{X}_{\mathrm{ML}} = N_T\left(\int_{t_0}^{T} \mu_\sigma\,d\sigma\right)^{-1}. \tag{2.68}$$

We observe for this example that only the total number of points is needed to form the maximum-likelihood estimate and not the separate subinterval counts; in other words, N_T is a *sufficient statistic* for estimating X. If $\mu_t \equiv 1$, then N is homogeneous with an unknown intensity X. Thus, from (2.68), the maximum-likelihood estimate of the intensity of a homogeneous Poisson process is the observed time-average count rate $N_T/(T - t_0)$.

In situations more general than Example 2.4.5 and, indeed, in almost all applications, it is not possible to obtain an analytic solution for the value of $\mathbf{X}$ maximizing the likelihood function (2.66). Thus, the maximum-likelihood estimate of $\mathbf{X}$ usually has to be obtained numerically. The numerical problem is to find the value of $\mathbf{X}$ in $\mathfrak{X}$ that maximizes $l(\mathbf{X})$ given $n_1, n_2, \ldots, n_k$. A number of algorithms are described in the literature for implementing this search [1,28]. The efficiency and accuracy of several algorithms (steepest descent, conjugate gradient, variable metric, Newton–Raphson, gradient with respect to a non-Euclidean norm, and weighted least squares) are compared by J. Markham, D. Snyder, and J. Cox [27] for exponentially decreasing intensities in the form of (2.59).

Histogram data suppresses any useful information about the parameters that may be contained in the occurrence times of the observed point process. For this reason, our main focus in later chapters is on the use of the occurrence times rather than count-rate histograms derived from them. This implies a greater data gathering requirement in practice, but it makes possible the extraction of all available information about the parameters from the time series of detected events. Therefore, we also consider as observed data on the interval $[t_0, T)$ an entire counting path $\{N_\sigma; t_0 \leqslant \sigma < T\}$. We refer to such data as *count-record* data. Count-record data are, of course, entirely equivalent to a record $\{N_T; w_1, w_2, \ldots, w_{N_T}\}$ of the number of points and their occurrence times in $[t_0, T)$. It is clear that count-record data contain more information about the parameters than histogram data because histogram data can be obtained from count-record data but not vice versa.

Let $p[\{N_\sigma; t_0 \leqslant \sigma < T\}|\mathbf{X}]$ denote the sample function density for the Poisson process $\{N_t; t \geqslant t_0\}$ when the intensity function $\lambda_t(\mathbf{X})$ is parametrized by $\mathbf{X}$. Thus, from (2.51)

$$p[\{N_\sigma; t_0 \leqslant \sigma < T\}|\mathbf{X}] = \exp\left[-\int_{t_0}^{T} \lambda_\sigma(\mathbf{X})\, d\sigma + \int_{t_0}^{T} \ln \lambda_\sigma(\mathbf{X})\, dN_\sigma\right]. \quad (2.69)$$

From the interpretation of the sample function density developed in Section 2.3, we see that the value of $\mathbf{X}$ which maximizes (2.69) for a particular observed count record is the value of $\mathbf{X}$ which maximizes the probability of observing that record. This value of $\mathbf{X}$ is by definition the maximum likelihood estimate $\hat{\mathbf{X}}_{\text{ML}} = \hat{\mathbf{X}}_{\text{ML}}(\{N_\sigma; t_0 \leqslant \sigma < T\})$ of $\mathbf{X}$. Consequently, the maximum likelihood estimate of $\mathbf{X}$ maximizes the sample function density (2.69) or, equivalently, its logarithm, the likelihood function

$$l(\mathbf{X}) = -\int_{t_0}^{T} \lambda_\sigma(\mathbf{X})\, d\sigma + \int_{t_0}^{T} \ln \lambda_\sigma(\mathbf{X})\, dN_\sigma. \quad (2.70)$$

This maximization is subject to the constraint that $\mathbf{X}$ must be in the allowed parameter set $\mathfrak{X}$. As with histogram data, the maximization of the likelihood function will often need to be accomplished by a numerical search.

Example 2.4.6 (Pulse Position Modulation, Optical Range Finding). Suppose in an optical pulse-position modulation system or optical ranging system that electrons emanate from the photodetector at a rate $\lambda_t(X) = \mu(t-X)+\lambda_0$, where $\mu(t)$ is the photoelectron generation rate and λ_0 is the thermoelectron generation rate. It is desired to estimate the delay X on the basis of electron emissions observed at the output of the photodetector during the interval $[t_0, T)$. This interval may be assumed to be sufficiently large to capture all the signal energy for all possible delays. The likelihood function (2.70) becomes

$$l(X) = -\int_{t_0}^{T} [\mu(\sigma-X)+\lambda_0]\,d\sigma + \int_{t_0}^{T} \ln[\mu(\sigma-X)+\lambda_0]\,dN_\sigma. \quad (2.71)$$

The first term on the right can be disregarded since it will not depend on X when the observation interval is sufficiently large, which we have already assumed. The search for $\hat{X}_{\mathrm{ML}}$, which yields a maximum for $l(X)$, can then be mechanized by processing the photodetector output with a filter having an impulse response $h(t) = \ln[\mu(-t)+\lambda_0]$. According to (2.71), the output of the filter achieves a maximum at time $\hat{X}_{\mathrm{ML}}$.

Example 2.4.7 (Auditory Electrophysiology). Suppose spike discharges of the form shown in Figure 1.1 and discussed in Example 2.4.4 are observed with a microelectrode in an auditory nerve fiber during the interval $[0, T)$. Assume that a sinusoidal acoustic stimulation is applied and that the discharge sequence can be modeled as an inhomogeneous Poisson process with the parametrized intensity function given in (2.63). It is convenient to assume that the observation interval contains an integral number of cycles of the stimulation. The desire is to estimate X_1, X_2, and X_3 from the observed spike-discharge sequence. After some manipulation, the likelihood function (2.70) becomes

$$l(\mathbf{X}) = -X_1 I_0(X_2)T + N_T \ln X_1 + X_2 \xi \cos(X_3+\phi) \quad (2.72)$$

where $I_0(\cdot)$ is a modified Bessel function of the first kind of zero order and where we define $\xi^2 = A_c^2 + A_s^2$ and $\phi = \tan^{-1}(A_s/A_c)$ in which A_c and A_s depend on the number N_T and occurrence times $w_1, w_2, \ldots, w_{N_T}$ of the observed spikes according to

$$A_c = \int_0^T \cos(2\pi f\sigma)\,dN_\sigma = \sum_{i=1}^{N_T} \cos(2\pi f w_i)$$

and

$$A_s = \int_0^T \sin(2\pi f\sigma)\,dN_\sigma = \sum_{i=1}^{N_T} \sin(2\pi f w_i). \quad (2.73)$$

It is evident by inspection of (2.72) that

$$\hat{X}_{3,\mathrm{ML}} = -\phi = -\tan^{-1}\left(\frac{A_s}{A_c}\right). \tag{2.74}$$

By setting the derivatives of $l(\mathbf{X})$ with respect to X_1 and X_2 equal to zero, we find that

$$\hat{X}_{1,\mathrm{ML}} = N_T\left[TI_0\left(\hat{X}_{2,\mathrm{ML}}\right)\right]^{-1} \tag{2.75}$$

and that $\hat{X}_{2,\mathrm{ML}}$ is the unique value of X_2 satisfying the transcendental equation

$$\frac{I_1(X_2)}{I_0(X_2)} = N_T^{-1}\xi = N_T^{-1}\left[A_c^2 + A_s^2\right]^{1/2}, \tag{2.76}$$

where $I_1(X_2) = dI_0(X_2)/dX_2$.

Parameter Estimation Accuracy

The maximum likelihood estimate $\hat{\mathbf{X}}_{\mathrm{ML}}$ of $\mathbf{X}$ in terms of a particular record of observed data is the value of $\mathbf{X}$ that maximizes the probability of observing that record. Usually, $\hat{\mathbf{X}}_{\mathrm{ML}}$ will not be identically equal to the actual and unobservable value of $\mathbf{X}$; call this value $\mathbf{X}_a$. Thus, in any given experiment, there is a nonzero error $\tilde{\mathbf{X}} = \hat{\mathbf{X}}_{\mathrm{ML}} - \mathbf{X}_a$. We now seek a characterization of this error so that judgements about the accuracy of the maximum likelihood estimate can be made. This characterization will be on the ensemble of errors that result with repetitions of the experiment to collect data and generate $\hat{\mathbf{X}}_{\mathrm{ML}}$, that is, with $\tilde{\mathbf{X}}$ viewed as a random variable. The most complete characterization of $\tilde{\mathbf{X}}$ in this sense is its probability distribution. Unfortunately, the distribution of $\tilde{\mathbf{X}}$ is usually difficult, if not impossible, to determine analytically. Therefore, we are forced to seek less complete characterizations such as the moments of $\tilde{\mathbf{X}}$. The first and second moments are usually sought. The first moment

$$\mathbf{b}(\mathbf{X}) \triangleq E\left[\hat{\mathbf{X}}_{\mathrm{ML}} - \mathbf{X} | \mathbf{X}\right] \tag{2.77}$$

is termed the *bias* of the maximum likelihood estimate, and the second moment

$$\boldsymbol{\Sigma}(\mathbf{X}) \triangleq E\left[\left(\hat{\mathbf{X}}_{\mathrm{ML}} - \mathbf{X}\right)\left(\hat{\mathbf{X}}_{\mathrm{ML}} - \mathbf{X}\right)' | \mathbf{X}\right] \tag{2.78†}$$

the *mean square-error matrix*. The mean square-error matrix is also termed

†The prime denotes *transpose*.

the error covariance matrix when the bias is zero. The expectations in (2.77) and (2.78) are with respect to the distribution of the observed data used to generate the estimate (e.g., histogram data, count-record data) for an arbitrary but fixed value of $\mathbf{X} \in \mathfrak{X}$. The estimate is said to be *unbiased* if $\mathbf{b}(\mathbf{X}) = \mathbf{0}$ for all $\mathbf{X} \in \mathfrak{X}$.

Even the first two moments, (2.77) and (2.78), of the error are generally difficult to determine. Thus, rather than approach their evaluation directly, it is convenient first to derive a lower bound on the mean square-error matrix for an arbitrary estimate of **X** in terms of the available data. The mean square-error matrix for the maximum likelihood estimate can then be compared to this bound. Under favorable circumstances the maximum likelihood estimate actually achieves the lower bound so that its mean square-error matrix is the smallest possible using any other conceivable estimate with the same data.

The lower bound on the mean square-error matrix for an arbitrary estimate $\mathbf{X}^*$ of **X** is given in the following theorem for both histogram and count-record data. For histogram data, the observations are of the counts N_{t_{i-1},t_i} occurring in k subintervals $[t_{i-1}, t_i)$, $i = 1, 2, \ldots, k$, of the observation interval $[t_0, T)$, where $t_k = T$. For count-record data, the observations are of the random variables $\{N_\sigma; t_0 \leqslant \sigma < T\}$ or, equivalently, $\{N_T; w_1, w_2, \ldots, w_{N_t}\}$. The lower bound matrix we give for these Poisson distributed observations is a special case of a more general bound called the Cramér–Rao bound. Consult C. R. Rao [32, p. 265] and H. L. Van Trees [40, p. 79] for a discussion of the bound in a general setting.

Theorem 2.4.1 (Cramér–Rao Lower Bound). Let $\{N_t; t \geqslant t_0\}$ be an inhomogeneous Poisson process with a parametrized intensity function $\lambda_t(\mathbf{X})$ for $t \geqslant t_0$. Let $\mathbf{X}^*$ denote an arbitrary estimate of **X** based on histogram or count-record observations of N on the interval $[t_0, T)$. Then, the mean square-error matrix $\mathbf{\Sigma}(\mathbf{X}) = E[(\mathbf{X}^* - \mathbf{X})(\mathbf{X}^* - \mathbf{X})' | \mathbf{X}]$ satisfies

$$\mathbf{\Sigma}(\mathbf{X}) \geqslant \mathbf{b}(\mathbf{X})\mathbf{b}'(\mathbf{X}) + \left[\mathbf{I} + \frac{\partial \mathbf{b}(\mathbf{X})}{\partial \mathbf{X}}\right] \mathbf{F}^{-1}(\mathbf{X}) \left[\mathbf{I} + \frac{\partial \mathbf{b}(\mathbf{X})}{\partial \mathbf{X}}\right]' \quad (2.79)\dagger$$

where $\mathbf{b}(\mathbf{X}) = E[\mathbf{X}^* - \mathbf{X} | \mathbf{X}]$ is the bias of $\mathbf{X}^*$, $\partial \mathbf{b}(\mathbf{X}) / \partial \mathbf{X}$ is the Jacobian matrix of the bias [the i, j-element is $\partial b_j(\mathbf{X}) / \partial X_i$], **I** denotes the identity matrix, and $\mathbf{F}(\mathbf{X})$ is the Fisher information matrix given for histogram data

†The matrix inequality is in the sense that the left side minus the right side of (2.79) is a nonnegative definite matrix.

by

$$\mathbf{F}(\mathbf{X}) = \sum_{i=1}^{k} \left[\int_{t_{i-1}}^{t_i} \lambda_\sigma(\mathbf{X})\, d\sigma \right]^{-1} \left[\int_{t_{i-1}}^{t_i} \frac{\partial \lambda_\sigma(\mathbf{X})}{\partial \mathbf{X}}\, d\sigma \right] \left[\int_{t_{i-1}}^{t_i} \frac{\partial \lambda_\sigma(\mathbf{X})}{\partial \mathbf{X}}\, d\sigma \right]' \tag{2.80}$$

and for count-record data by

$$\mathbf{F}(\mathbf{X}) = \int_{t_0}^{T} \lambda_\sigma^{-1}(\mathbf{X}) \left[\frac{\partial \lambda_\sigma(\mathbf{X})}{\partial \mathbf{X}} \right] \left[\frac{\partial \lambda_\sigma(\mathbf{X})}{\partial \mathbf{X}} \right]' d\sigma. \tag{2.81}$$

In (2.80) and (2.81), $\partial \lambda_\sigma(\mathbf{X})/\partial \mathbf{X}$ denotes the gradient vector of $\lambda_\sigma(\mathbf{X})$ with respect to $\mathbf{X}$ [the ith element is $\partial \lambda_\sigma(\mathbf{X})/\partial X_i$]. Equality holds in (2.79) if and only if the estimate $\mathbf{X}^*$ satisfies

$$\mathbf{X}^* = \mathbf{X} + \mathbf{b}(\mathbf{X}) + \left[\mathbf{I} + \frac{\partial \mathbf{b}(\mathbf{X})}{\partial \mathbf{X}} \right] \mathbf{F}^{-1}(\mathbf{X}) \left[\frac{\partial l(\mathbf{X})}{\partial \mathbf{X}} \right] \tag{2.82}$$

for all $\mathbf{X}$.

Proof. Define

$$\mathbf{z}(\mathbf{X}) = \begin{bmatrix} \mathbf{X}^* - \mathbf{X} - \mathbf{b}(\mathbf{X}) \\ \dfrac{\partial l(\mathbf{X})}{\partial \mathbf{X}} \end{bmatrix}$$

where $l(\mathbf{X})$ is the likelihood function defined in (2.66) for histogram data and in (2.70) for count-record data. The bound (2.79) will follow from the expectation $E[\mathbf{z}(\mathbf{X})\mathbf{z}'(\mathbf{X})|\mathbf{X}]$. To evaluate this expectation, we need first to evaluate the expectations

$$E\left\{ [\mathbf{X}^* - \mathbf{X} - \mathbf{b}(\mathbf{X})] \left[\frac{\partial l(\mathbf{X})}{\partial \mathbf{X}} \right]' \Big| \mathbf{X} \right\} \quad \text{and} \quad E\left\{ \left[\frac{\partial l(\mathbf{X})}{\partial \mathbf{X}} \right] \left[\frac{\partial l(\mathbf{X})}{\partial \mathbf{X}} \right]' \Big| \mathbf{X} \right\}$$

for both histogram and count-record data. For brevity, denote the counting probability in (2.65) by $P(n_1, n_2, \ldots, n_k|\mathbf{X})$. Then, for histogram data, we have that

$$E\{[\mathbf{X}^* - \mathbf{X} - \mathbf{b}(\mathbf{X})]|\mathbf{X}\}$$

$$= \sum_{n_1, n_2, \ldots, n_k} [\mathbf{X}^*(n_1, n_2, \ldots, n_k) - \mathbf{X} - \mathbf{b}(\mathbf{X})] P(n_1, n_2, \ldots, n_k|\mathbf{X})$$

$$= \mathbf{0}. \tag{2.83}$$

Therefore, by taking the gradient of (2.83) with respect to $\mathbf{X}$ and using $\partial P / \partial \mathbf{X} = P \partial l / \partial \mathbf{X}$, we have for histogram data that

$$E\left\{ [\mathbf{X}^* - \mathbf{X} - \mathbf{b}(\mathbf{X})] \left[\frac{\partial l(\mathbf{X})}{\partial \mathbf{X}} \right]' |\mathbf{X} \right\} = \mathbf{I} + \frac{\partial \mathbf{b}(\mathbf{X})}{\partial \mathbf{X}}. \tag{2.84}$$

The analog to (2.83) for count-record data is

$$E\{[\mathbf{X}^* - \mathbf{X} - \mathbf{b}(\mathbf{X})]|\mathbf{X}\} = [\mathbf{X}^*(0) - \mathbf{X} - \mathbf{b}(\mathbf{X})] \Pr(N_T = 0|\mathbf{X})$$

$$+ \sum_{n=1}^{\infty} \int \cdots \int [\mathbf{X}^*(n; W_1, \ldots, W_n)$$

$$- \mathbf{X} - \mathbf{b}(\mathbf{X})] p_{\mathbf{w}}(\mathbf{W}, N_T = n) \, dW_1 \cdots dW_n$$

$$= \mathbf{0} \tag{2.85}$$

where the integrals are over the range $t_0 \leqslant W_1 \leqslant \cdots \leqslant W_n \leqslant T$ and where $p_w(W, N_T = n)$ is defined in (2.49). The gradient of this expression with respect to $\mathbf{X}$ again yields (2.84) for count-record data. From (2.66), we have

$$\frac{\partial l(\mathbf{X})}{\partial \mathbf{X}} = \sum_{i=1}^{k} \left[\int_{t_{i-1}}^{t_i} \frac{\partial \lambda_\sigma(\mathbf{X})}{\partial \mathbf{X}} \, d\sigma \right] \left\{ -1 + N_i \left[\int_{t_{i-1}}^{t_i} \lambda_\sigma(\mathbf{X}) \, d\sigma \right]^{-1} \right\}$$

where we define $N_i = N_{t_{i-1}, t_i}$. By using

$$E(N_i N_j|\mathbf{X}) = \begin{cases} \left[\int_{t_{i-1}}^{t_i} \lambda_\sigma(\mathbf{X}) \, d\sigma \right] \left[\int_{t_{j-1}}^{t_j} \lambda_\sigma(\mathbf{X}) \, d\sigma \right], & \text{for } i \neq j \\ \int_{t_{i-1}}^{t_i} \lambda_\sigma(\mathbf{X}) \, d\sigma + \left[\int_{t_{i-1}}^{t_i} \lambda_\sigma(\mathbf{X}) \, d\sigma \right]^2, & \text{for } i = j \end{cases}$$

we see after some manipulation that

$$E\left\{\left[\frac{\partial l(\mathbf{X})}{\partial \mathbf{X}}\right]\left[\frac{\partial l(\mathbf{X})}{\partial \mathbf{X}}\right]' \middle| \mathbf{X}\right\} = \mathbf{F}(\mathbf{X}), \tag{2.86}$$

where $\mathbf{F}(\mathbf{X})$ is the Fisher information matrix for histogram data, as given in (2.80). For count-record data, this same procedure yields (2.86) with $\mathbf{F}(\mathbf{X})$ as given in (2.81). By collecting these results together, we have for both histogram and count-record data that

$$\Lambda(\mathbf{X}) \triangleq E[\mathbf{z}(\mathbf{X})\mathbf{z}'(\mathbf{X})|\mathbf{X}] = \begin{bmatrix} [\boldsymbol{\Sigma}(\mathbf{X}) - \mathbf{b}(\mathbf{X})\mathbf{b}'(\mathbf{X})] & \left[\mathbf{I} + \dfrac{\partial \mathbf{b}(\mathbf{X})}{\partial \mathbf{X}}\right] \\ \left[\mathbf{I} + \dfrac{\partial \mathbf{b}(\mathbf{X})}{\partial \mathbf{X}}\right]' & \mathbf{F}(\mathbf{X}) \end{bmatrix}$$

It is easy to see for an arbitrary choice of $\mathbf{v}$ that $\mathbf{v}'\Lambda(\mathbf{X})\mathbf{v} = E[(\mathbf{v}'\mathbf{z}(\mathbf{X}))^2|\mathbf{X}] \geqslant 0$. This inequality holds in particular for the choice

$$\mathbf{v} = \begin{bmatrix} \mathbf{v}_1 \\ -\mathbf{F}^{-1}(\mathbf{X})\left[\mathbf{I} + \dfrac{\partial \mathbf{b}(\mathbf{X})}{\partial \mathbf{X}}\right]' \mathbf{v}_1 \end{bmatrix} \tag{2.87}$$

where $\mathbf{v}_1$ is arbitrary. For this choice, we conclude that

$$\mathbf{v}_1'\boldsymbol{\Sigma}(\mathbf{X})\mathbf{v}_1 \geqslant \mathbf{v}_1'\left\{\mathbf{b}(\mathbf{X})\mathbf{b}'(\mathbf{X}) + \left[\mathbf{I} + \frac{\partial \mathbf{b}(\mathbf{X})}{\partial \mathbf{X}}\right]\mathbf{F}^{-1}(\mathbf{X})\left[\mathbf{I} + \frac{\partial \mathbf{b}(\mathbf{X})}{\partial \mathbf{X}}\right]'\right\}\mathbf{v}_1.$$

This establishes (2.79) by virtue of the arbitrariness of $\mathbf{v}_1$. Furthermore, the arbitrariness of $\mathbf{v}_1$ also implies that

$$E\left[(\mathbf{v}_1'\mathbf{z}(\mathbf{X}))^2|\mathbf{X}\right]$$

$$= E\left[\left(\mathbf{v}_1'\left\{\mathbf{X}^* - \mathbf{X} - \mathbf{b}(\mathbf{X}) - \left[\mathbf{I} + \frac{\partial \mathbf{b}(\mathbf{X})}{\partial \mathbf{X}}\right]\mathbf{F}^{-1}(\mathbf{X})\left[\frac{\partial l(\mathbf{X})}{\partial \mathbf{X}}\right]\right\}\right)^2 \middle| \mathbf{X}\right]$$

will be zero if and only if (2.82) holds for all $\mathbf{X}\in\mathfrak{X}$. This completes the proof of Theorem 2.4.1.

Some important conclusions can be drawn from the assertions of Theorem 2.4.1. These hold for both histogram and count-record data.

a. Let $\mathbf{e}_i$ denote a unit vector with ith element equal to one and all other elements equal to zero. Then for arbitrary $\mathbf{X}^*$ there holds

$$E\left[(X_i^* - X_i)^2|\mathbf{X}\right] = \mathbf{e}_i'\boldsymbol{\Sigma}(\mathbf{X})\mathbf{e}_i \geqslant b_i^2(\mathbf{X}) + \Delta_{ii}$$

where Δ_{ii} denotes the (ii)-element of the matrix $[\mathbf{I}+\partial\mathbf{b}(\mathbf{X})/\partial\mathbf{X}]\mathbf{F}^{-1}(\mathbf{X})$ $[\mathbf{I}+\partial\mathbf{b}(\mathbf{X})/\partial\mathbf{X}]'$. This shows that the mean square error for any unbiased estimate of the parameter X_i is not less than a precomputable quantity that is defined independently of the method by which the estimate is selected.

b. An estimate having a mean square-error matrix satisfying (2.79) with equality is said to be efficient. Suppose there exists an efficient estimate $\mathbf{X}^*$. Then, in terms of this estimate, the maximum likelihood estimate satisfies $\hat{\mathbf{X}}_{\mathrm{ML}}+\mathbf{b}(\hat{\mathbf{X}}_{\mathrm{ML}})=\mathbf{X}^*$, where $\mathbf{b}(\hat{X}_{\mathrm{ML}})$ is the bias $\mathbf{b}(\mathbf{X})$ of $\mathbf{X}^*$ evaluated at $\mathbf{X}=\hat{\mathbf{X}}_{\mathrm{ML}}$. This follows immediately from (2.82) and the property that the gradient of the likelihood function $l(\mathbf{X})$ is zero for $\mathbf{X}=\hat{\mathbf{X}}_{\mathrm{ML}}$. Thus, the estimate $\hat{\mathbf{X}}_{\mathrm{ML}}+\mathbf{b}(\hat{\mathbf{X}}_{\mathrm{ML}})$ is efficient. It follows that the maximum likelihood estimate $\hat{\mathbf{X}}_{\mathrm{ML}}$ will be efficient only if the bias of $\mathbf{X}^*$ is zero at $\mathbf{X}=\hat{\mathbf{X}}_{\mathrm{ML}}$. The most important situation occurs when there exists an unbiased, efficient estimate for, then, the maximum likelihood estimate is also unbiased and efficient.

c. Suppose $\mathbf{X}^*$ is an unbiased estimate of $\mathbf{X}$. Then from (2.79), the mean square-error matrix for $\mathbf{X}^*$ satisfies $\boldsymbol{\Sigma}(\mathbf{X})\geqslant\mathbf{F}^{-1}(\mathbf{X})$. Consider now that the experiment of collecting data to estimate $\mathbf{X}$ is repeated independently M times. Denote the estimate of $\mathbf{X}$ based on these M trials by $\mathbf{X}_M^*$. Then the mean square-error matrix for $\mathbf{X}_M^*$ satisfies $\boldsymbol{\Sigma}_M(\mathbf{X})\geqslant(1/M)$ $\mathbf{F}^{-1}(\mathbf{X})$; for a derivation see Problem 2.4.4. The estimate $\mathbf{X}_M^*$ is said to be asymptotically efficient if $[M\boldsymbol{\Sigma}_M(\mathbf{X})-\mathbf{F}^{-1}(\mathbf{X})]$ tends to zero as M tends to infinity for all $\mathbf{X}\in\mathfrak{X}$. It is said to be asymptotically unbiased if $\mathbf{b}(\mathbf{X})$ tends to zero as M tends to infinity for all $\mathbf{X}\in\mathfrak{X}$. The maximum likelihood estimate is both asymptotically efficient and asymptotically unbiased under relatively weak conditions (consult S. Wilks [43, p. 363]).

Example 2.4.8 (Scaled Inhomogeneous Poisson Process). Consider again the model of Example 2.4.5 in which histogram data are collected in order to estimate the parameter X of the intensity $\lambda_t(X)=X\mu_t$. It is easily seen from the expression for $\hat{X}_{\mathrm{ML}}$ in (2.68) that $E[\hat{X}_{\mathrm{ML}}|X]=X$, so $\hat{X}_{\mathrm{ML}}$ is an unbiased estimate of X in terms of histogram data. We see from (2.67) that

$$\frac{\partial l(X)}{\partial X}=-\int_{t_0}^{T}\mu_\sigma\,d\sigma+X^{-1}N_T \tag{2.88}$$

and from (2.80) that

$$F(X)=X^{-1}\int_{t_0}^{T}\mu_\sigma\,d\sigma. \tag{2.89}$$

Thus,

$$\hat{X}_{\mathrm{ML}}=X+F^{-1}(X)\frac{\partial l(X)}{\partial X}$$

and the condition (2.82) is satisfied by the maximum likelihood estimate. The maximum likelihood estimate is both unbiased and efficient. As (2.79) is satisfied with equality, we have

$$E\left[\left(\hat{X}_{\mathrm{ML}}-X\right)^2\middle|X\right]=F^{-1}(X)=X\left(\int_{t_0}^{T}\mu_\sigma\,d\sigma\right)^{-1}. \tag{2.90}$$

This mean square error in estimating X is the smallest that can be achieved with any conceivable estimate using histogram data.

Example 2.4.9 (Pulse Position Modulation, Optical Range Finding). Assume as in Example 2.4.6 that $\lambda_t(X)=\mu(t-X)+\lambda_0$, where $\mu(t)$ is the photoelectron rate and λ_0 is the thermoelectron rate. According to Example 2.4.3, $\mu(t)=\alpha|S(t)|^2$, where $|S(t)|$ is the envelope of the optical field incident on the photodetector and $\alpha>0$ is a constant assumed to be known. An estimate of X satisfying (2.82) cannot be found, so an efficient estimate of X cannot be found. From (2.79) and (2.81), the mean-square error in estimating X using any unbiased estimate X^* satisfies the inequality

$$E\left[(X^*-X)^2|X\right]\geqslant\left[\int_{t_0}^{T}\frac{\dot{\mu}^2(\sigma-X)}{\mu(\sigma-X)+\lambda_0}\,d\sigma\right]^{-1}$$

$$=\left[\int_{t_0}^{T}\frac{\dot{\mu}^2(t)}{\mu(t)+\lambda_0}\,dt\right]^{-1} \tag{2.91}$$

where $\dot{\mu}(t)=d\mu(t)/dt$. In writing the last equality, we have assumed that the observation interval $[t_0, T)$ is sufficiently large to capture all the signal energy for all possible values of X. Equation (2.91) summarizes all that can be said about the performance of X^* from the Cramér–Rao inequality when the efficiency of X^* cannot be established. The inequality provides only a lower bound and no assurance that it can be approached. The implication of remark (c) above is that with repeated illuminations of the reflector, the maximum likelihood estimate will tend to become unbiased and have a mean-square error given by the right side of (2.91) divided by the number of illuminations. The lower bound in (2.91) can be given a qualitative interpretation in the limits of small and large thermoelectron rate λ_0. For this purpose, let $a(t)$ be a pulse in $[t_0, T)$ and $A(f)$ its Fourier transform. Define

$$B_a^2=\frac{\int_{t_0}^{T}\dot{a}^2(t)\,dt}{\int_{t_0}^{T}a^2(t)\,dt}=\frac{\int_{-\infty}^{\infty}|f|^2|A(f)|^2\,df}{\int_{-\infty}^{\infty}|A(f)|^2\,df},$$

where the second equality follows from Parseval's theorem. Then, B_a can be interpreted qualitatively as the rms bandwidth of $a(t)$. Also, let

$$E_a=\int_{t_0}^{T}a^2(t)\,dt$$

denote the energy of $a(t)$. In the limit $\lambda_0 \gg \mu(t)$, and for M repeated illuminations of the reflector, we have for the maximum likelihood estimate that

$$E\left[\left(\hat{X}_{\mathrm{ML}}-X\right)^2|X\right]\approx\lambda_0(MB_\mu E_\mu)^{-1}. \tag{2.92}$$

Thus, in the limit of high thermoelectron rate, the mean square error for $\hat{X}_{\mathrm{ML}}$ is given approximately by the noise-to-signal ratio (λ_0/E_μ) reduced by a factor MB_μ dependent on the number of illuminations M and the rms bandwidth B_μ of the square of the envelope $|S(t)|$ of the incident optical field. In the limit $\lambda_0 \ll \mu(t)$, and for M repeated illuminations of the reflector, we have

$$E\left[\left(\hat{X}_{\mathrm{ML}}-X\right)^2|X\right]\approx(4MB_{|S|}E_{|S|})^{-1}. \tag{2.93}$$

Thus, in the limit of low thermoelectron rate, the mean square-error for $\hat{X}_{\mathrm{ML}}$ is dependent on the rms bandwidth and energy of the envelope $|S(t)|$.

Summary. In this section, we have studied procedures for estimating unknown parameters of the intensity of an observed Poisson process for both histogram and count-record data. The maximum likelihood estimate of the parameters was emphasized. The difficult issue of parameter estimation performance was addressed using the Cramér–Rao inequality. This gives a precomputable lower bound on the mean square-error matrix for an arbitrary estimate in terms of its bias and the Fisher information matrix. An efficient estimate is one that achieves this lower bound. Our motivation for emphasizing the maximum likelihood estimate is based on the property that it is unbiased and efficient if such an estimate exists. More generally, the maximum likelihood estimate is asymptotically unbiased and efficient.

2.5 HYPOTHESIS TESTING FOR OBSERVED POISSON PROCESSES

This section is devoted to the basic principles of deciding between alternative mechanisms that govern an observed point process. A simple example arising in the context of optical communication and radar systems motivates the terminology we shall use. Referring to Figure 2.7, suppose in a communication system that the output of a binary information source is transmitted to a remote receiver by switching a light beam directed at him on or off in correspondence with the state of the information source. After some preliminary optical filtering to suppress the effects of background radiation, such as scattered sunlight, and to focus the light beam onto a photosensitive surface, the receiver observes the stream of electrons emanating from the photodetector. The rate at which electrons emanate will depend on the state of the information source. If the source is such that the light beam is on, the rate, according to (2.61), will be of the form $\mu_t + \lambda_0$, where μ_t is the photoelectron rate and λ_0 the thermoelectron rate. We term this condition "hypothesis one," and denote it by H_1. The rate is simply λ_0 when the light beam is off, a condition we term "hypothesis zero," H_0.

The receiver's task is to decide the state of the information source on the basis of his observations of the photodetector output. These observations may be either in the form of histogram or count-record data, both defined in Section 2.4. Regardless of the procedure the receiver adopts, there will always remain the possibility of incorrect decisions, of which there are two types. The first is to assert the light beam is on when it is off. We will term this error a *false alarm*. This nomenclature is common in radar practice where the presence of μ_t indicates the presence of a reflector. The second form of error is to assert the light beam is off when it is on. We will term

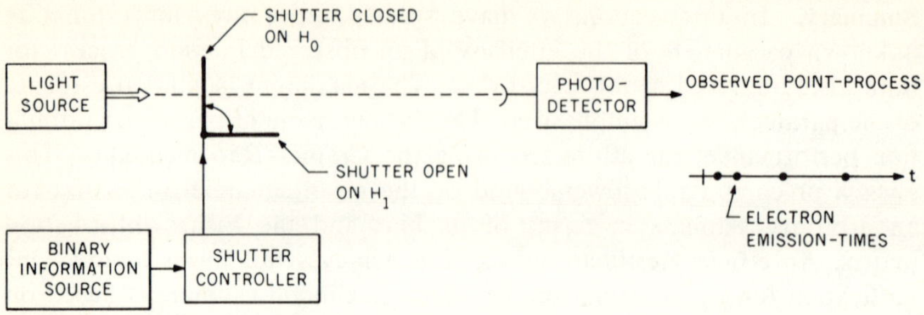

Figure 2.7. Optical communication system for transmitting the output of a binary source.

this a *miss*. Statisticians call a false alarm and a miss Type-I and Type-II errors, respectively. The strategy the receiver adopts for making decisions will usually reflect his view of the importance of not making false alarms and misses.

The receiver in this communication system must decide between two alternative mechanisms that govern the observed point process. We say that this is a *binary hypothesis testing* problem. The situation when there are M alternatives is termed an *M-ary hypothesis testing* problem. We first examine the binary hypothesis testing problem in some generality. The extension to M alternatives is conceptually straightforward and is discussed at the end of the section.

Binary Hypothesis Testing

The general binary hypothesis testing problem motivated by the above example is the following. An experiment is performed in which point-process data are collected during an observation interval $[t_0, T)$. These data are influenced by one of two underlying mechanisms called hypotheses H_0 and H_1. The problem is to decide which hypothesis given the observation. To retain some generality, let us denote the observed data by $D_{t_0,T}$. For example, $D_{t_0,T}$ can be histogram or count-record data.

There are four possible outcomes associated with performing the experiment of collecting the data $D_{t_0,T}$ and subsequently making a decision about the underlying influence, H_0 or H_1. These are:

1. Decide H_0 when H_0 is the influence,
2. Decide H_0 when H_1 is the influence,
3. Decide H_1 when H_0 is the influence,
4. Decide H_1 when H_1 is the influence.

Outcomes 2 and 3 are incorrect decisions, a miss and false alarm, respectively. Outcomes 1 and 4 are correct decisions. The decision strategy that the observer adopts will depend on the relative importance he attaches to these outcomes. With the *Bayes' criterion*, this importance is identified in a negative fashion as a cost incurred with each possible outcome, and the decision strategy is designed to minimize the average cost per decision that would accrue in a large number of successive and independent repetitions of the experiment. The costs for each outcome can be listed conveniently as elements of a cost matrix **C**,

$$\mathbf{C}=\begin{bmatrix} C_{00} & C_{01} \\ C_{10} & C_{11} \end{bmatrix},$$

where C_{ij} is the cost of choosing hypothesis H_i when hypothesis H_j is the actual influence. The relative costs of a miss and false alarm are $C_{01}-C_{11}$ and $C_{10}-C_{00}$, respectively; these are assumed to be positive.

We now seek an expression for the average cost per decision. For this, we require an expression for the *conditional average cost* of deciding in favor of each hypothesis given $D_{t_0,T}$ and, also, we need to introduce the notion of a *decision indicator function*, which defines the decision strategy. The decision indicator function will subsequently be selected to minimize the average cost per decision.

Denote by $\Pr(H_0|D_{t_0,T})$ and $\Pr(H_1|D_{t_0,T})$ the conditional probabilities of H_0 and H_1 given the observations $D_{t_0,T}$. These probabilities will play a critical role in defining the decision procedure. We define the conditional average cost $C(H_0|D_{t_0,T})$ of deciding in favor of H_0 given $D_{t_0,T}$ by

$$C(H_0|D_{t_0,T})=C_{00}\Pr(H_0|D_{t_0,T})+C_{01}\Pr(H_1|D_{t_0,T}). \tag{2.94}$$

The conditional average cost $C(H_0|D_{t_0,T})$ depends on the data $D_{t_0,T}$. It is the average cost per decision for a particular realization of the data if H_0 is always selected when that realization occurs. By averaging across all possible realizations, we obtain

$$C(H_0)=E\left[C(H_0|D_{t_0,T})\right]=C_{00}\Pr(H_0)+C_{01}\Pr(H_1).$$

That is, the expectation of $C(H_0|D_{t_0,T})$ is the average cost per decision if H_0 is always selected regardless of the observations. Similarly, we define the conditional average cost $C(H_1|D_{t_0,T})$ of deciding in favor of H_1 given

$D_{t_0,T}$ by

$$C(H_1|D_{t_0,T}) = C_{10}\Pr(H_0|D_{t_0,T}) + C_{11}\Pr(H_1|D_{t_0,T}). \tag{2.95}$$

The receiver must develop a strategy for deciding upon either H_0 or H_1 for every possible realization of $D_{t_0,T}$. His strategy can be summarized in terms of decision indicator functions, $I_{H_0}(D_{t_0,T})$ and $I_{H_1}(D_{t_0,T})$, defined by

$$I_{H_0}(D_{t_0,T}) = \begin{cases} 1, & \text{if the decision for } D_{t_0,T} \text{ is } H_0 \\ 0, & \text{otherwise} \end{cases} \tag{2.96}$$

and

$$I_{H_1}(D_{t_0,T}) = \begin{cases} 1, & \text{if the decision for } D_{t_0,T} \text{ is } H_1 \\ 0, & \text{otherwise.} \end{cases} \tag{2.97}$$

These functions of $D_{t_0,T}$ can be designed by the receiver. They map the space of all possible realizations of $D_{t_0,T}$ into the numbers zero and one corresponding to H_0 and H_1, respectively. From the definitions, we have that the product of $I_{H_0}(D_{t_0,T})$ and $I_{H_1}(D_{t_0,T})$ is zero and the sum one regardless of $D_{t_0,T}$. The expectations $E[I_{H_0}(D_{t_0,T})]$ and $E[I_{H_1}(D_{t_0,T})]$ are the probabilities the receiver will decide in favor of H_0 and H_1, respectively. The conditional expectations $E[I_{H_0}(D_{t_0,T})|H_1]$ and $E[I_{H_1}(D_{t_0,T})|H_0]$ are of more importance as these are the probabilities of a miss and false alarm, respectively.

By combining $C(H_0|D_{t_0,T})$ and $C(H_1|D_{t_0,T})$ with $I_{H_0}(D_{t_0,T})$ and $I_{H_1}(D_{t_0,T})$, we have that the conditional average cost per decision for a particular realization $D_{t_0,T}$ of the observations is given by

$$C(D_{t_0,T}) = I_{H_0}(D_{t_0,T})C(H_0|D_{t_0,T}) + I_{H_1}(D_{t_0,T})C(H_1|D_{t_0,T}). \tag{2.98}$$

It follows that the average cost per decision $\overline{C}$ is the expectation of $C(D_{t_0,T})$,

$$\begin{aligned}\overline{C} &= E\left[C(D_{t_0,T})\right] \\ &= E\left[I_{H_0}(D_{t_0,T})C(H_0|D_{t_0,T}) + I_{H_1}(D_{t_0,T})C(H_1|D_{t_0,T})\right].\end{aligned} \tag{2.99}$$

The receiver naturally attempts to minimize $\overline{C}$. It is evident that this can be accomplished by minimizing $C(D_{t_0,T})$ for each realization of $D_{t_0,T}$. In particular, the functions $I_{H_0}(D_{t_0,T})$ and $I_{H_1}(D_{t_0,T})$ should be one and zero, respectively, whenever $D_{t_0,T}$ is such that $C(H_0|D_{t_0,T})$ is smaller than $C(H_1|D_{t_0,T})$; otherwise, $I_{H_0}(D_{t_0,T})$ should be zero and $I_{H_1}(D_{t_0,T})$ one. Note that we have defined $I_{H_1}(D_{t_0,T})=1$ whenever $D_{t_0,T}$ is such that $C(H_0|D_{t_0,T}) = C(H_1|D_{t_0,T})$; the choice between defining $I_{H_0}(D_{t_0,T})=1$ or $I_{H_1}(D_{t_0,T})=1$ in this instance is arbitrary as neither $C(D_{t_0,T})$ nor $\overline{C}$ is affected by the choice. We see, therefore that to minimize $\overline{C}$, the receiver should simply make the decision given $D_{t_0,T}$ which on the average costs the least. We combine these conditions into a single expression by writing the strategy minimizing the average cost per decision as

$$C(H_0|D_{t_0,T}) \underset{H_0}{\overset{H_1}{\gtrless}} C(H_1|D_{t_0,T}). \tag{2.100}$$

Equivalently, by introducing (2.94) and (2.95) and performing some minor manipulations, the rule (2.100) is

$$\Pr(H_1|D_{t_0,T}) \underset{H_0}{\overset{H_1}{\gtrless}} \frac{C_{10}-C_{00}}{C_{01}-C_{11}} \Pr(H_0|D_{t_0,T}). \tag{2.101}$$

If for a particular realization of $D_{t_0,T}$ either of the conditional probabilities $\Pr(H_1|D_{t_0,T})$ or $\Pr(H_0|D_{t_0,T})$ is zero while the other is not, the decision to be made is obvious. In applications, the decision will rarely, if ever, be so unambiguous, and this unlikely condition is termed singular. Disregarding the singular situation, the receiver strategy that minimizes his average cost is summarized by

$$\Lambda(D_{t_0,T}) \triangleq \frac{\Pr(H_1|D_{t_0,T})}{\Pr(H_0|D_{t_0,T})} \underset{h_0}{\overset{H_1}{\gtrless}} \frac{C_{10}-C_{00}}{C_{01}-C_{11}} \tag{2.102}$$

The function $\Lambda(D_{t_0,T})$ is called the *likelihood ratio* and (2.102) a *likelihood-ratio test*. The likelihood-ratio test consists of evaluating the likelihood ratio for the particular realization of observed data and then comparing the result to a precomputable constant defined as the ratio of the relative costs of a false alarm and miss. This constant is called the *threshold* of the test.

If the relative costs of the two types of error are equal,

$$C_{10}-C_{00}=C_{01}-C_{11},$$

the hypothesis selected is the one having the greater conditional probability given the observations $D_{t_0,T}$. This procedure of selecting the most likely hypothesis given the data is termed *maximum likelihood detection* in communication practice where it arises in a natural way with the desire to minimize the total probability of error in deciding between H_0 and H_1. To see this, let the cost matrix be of the form

$$\mathbf{C}=\begin{bmatrix} C_c & C_e \\ C_e & C_c \end{bmatrix}. \tag{2.103}$$

That is, assume equal costs, C_c and C_e, for correct and erroneous decisions, respectively. The relative costs of a miss and false alarm both equal $C_e - C_c$, which is assumed positive. By combining (2.94), (2.95), (2.98), and (2.99), we observe for this choice of cost matrix that the average cost per decision becomes

$$\begin{aligned}\overline{C} &= E\Big[I_{H_0}(D_{t_0,T})\big\{ C_c + (C_e - C_c)\Pr(H_1|D_{t_0,T})\big\} \\ &\qquad + I_{H_1}(D_{t_0,T})\big\{ C_c + (C_e - C_c)\Pr(H_0|D_{t_0,T})\big\}\Big] \\ &= C_c + (C_e - C_c)E\Big[I_{H_0}(D_{t_0,T})\Pr(H_1|D_{t_0,T}) \\ &\qquad + I_{H_1}(D_{t_0,T})\Pr(H_0|D_{t_0,T})\Big]. \end{aligned} \tag{2.104}$$

However,

$$\begin{aligned} E\Big[I_{H_0}(D_{t_0,T})\Pr(H_1|D_{t_0,T})\Big] &= E\Big[I_{H_0}(D_{t_0,T})|H_1\Big]\Pr(H_1) \\ &= P_M \Pr(H_1), \end{aligned} \tag{2.105}$$

where P_M is the miss probability and $\Pr(H_1)$ is the prior probability that H_1 occurs. Similarly,

$$\begin{aligned} E\Big[I_{H_1}(D_{t_0,T})\Pr(H_0|D_{t_0,T})\Big] &= E\Big[I_{H_1}(D_{t_0,T})|H_0\Big]\Pr(H_0) \\ &= P_F \Pr(H_0), \end{aligned} \tag{2.106}$$

where P_F is the false-alarm probability and $\Pr(H_0)$ is the prior probability that H_0 occurs. Thus,

$$\overline{C} = C_c + (C_e - C_c)\Pr(\text{error}), \tag{2.107}$$

in which Pr(error) is the total probability of error,

$$\Pr(\text{error}) = P_M \Pr(H_1) + P_F \Pr(H_0). \tag{2.108}$$

Therefore, minimizing $\overline{C}$ for the cost matrix (2.103) is equivalent to minimizing Pr(error). We conclude that maximum likelihood detection minimizes the total probability of error.

The likelihood ratio $\Lambda(D_{t_0,T})$ defined in (2.102) will depend on the form of the data. Two forms are of particular interest with observations of a point process, histogram and count-record data. For histogram data, let $P(n_1, n_2, \ldots, n_k | H_0)$ and $P(n_1, n_2, \ldots, n_k | H_1)$ be the joint probabilities on hypotheses zero and one that the number of points N_{t_{i-1},t_1} in the interval $[t_{i-1}, t_i)$ is n_i for $i = 1, 2, \ldots, k$. In terms of these joint counting probabilities, the likelihood ratio becomes

$$\begin{aligned}\Lambda(n_1, n_2, \ldots, n_k) &= \frac{\Pr(H_1 | n_1, n_2, \ldots, n_k)}{\Pr(H_0 | n_1, n_2, \ldots, n_k)} \\ &= \frac{P(n_1, n_2, \ldots, n_k | H_1)\Pr(H_1)}{P(n_1, n_2, \ldots, n_k | H_0)\Pr(H_0)},\end{aligned} \tag{2.109}$$

where $\Pr(H_0)$ and $\Pr(H_1)$ are the prior probabilities of hypotheses zero and one, respectively. For count-record data, denote the sample-function densities on each hypothesis by $p[\{N_\sigma; t_0 \leqslant \sigma < T\} | H_0]$ and $p[\{N_\sigma; t_0 \leqslant \sigma < T\} | H_1]$. The likelihood ratio becomes

$$\begin{aligned}\Lambda(\{N_\sigma; t_0 \leqslant \sigma < T\}) &= \frac{\Pr(H_1 | N_\sigma; t_0 \leqslant \sigma < T)}{\Pr(H_0 | N_\sigma; t_0 \leqslant \sigma < T)} \\ &= \frac{p[\{N_\sigma; t_0 \leqslant \sigma < T\} | H_1]\Pr(H_1)}{p[\{N_\sigma; t_0 \leqslant \sigma < T\} | H_0]\Pr(H_0)}.\end{aligned} \tag{2.110}$$

As the prior probabilities in (2.109) and (2.110) do not depend on the observations, they can be incorporated into the threshold constant of the likelihood-ratio test (2.102).

These expressions for the likelihood ratio do not require that the observed point process be a Poisson process for their validity. The practical impediment in their use for non-Poisson processes is that explicit formulas for the joint counting probability and sample-function density are often not available. Nevertheless, (2.109) and (2.110) will be quite important in later chapters when we study hypothesis testing for non-Poisson point processes. Their use for the Poisson process is straightforward and outlined in the following two examples.

Example 2.5.1 (Binary Hypothesis Testing for a Poisson Process: Histogram Data). Suppose $\{N_t; t \geqslant t_0\}$ is a Poisson counting process with intensity $\lambda_t^{(0)}$ on hypothesis zero and $\lambda_t^{(1)}$ on hypothesis one. Let P_0 and $P_1 = 1 - P_0$ be the prior probabilities that the intensity is $\lambda_t^{(0)}$ and $\lambda_t^{(1)}$, respectively. Histogram data are collected during the interval $[t_0, T)$ to decide which intensity governs the point process. The strategy that minimizes the average cost per decision is found by combining (2.1), (2.2), and (2.109) to evaluate the likelihood ratio and then incorporating this into the likelihood ratio test (2.102). After taking the logarithm of both sides of (2.102), we find that the receiver's best strategy is defined by

$$-[\Lambda_T^{(1)} - \Lambda_T^{(0)}] + \sum_{i=1}^{k} n_i \ln\left[\frac{\Lambda_{t_i}^{(1)} - \Lambda_{t_{i-1}}^{(1)}}{\Lambda_{t_i}^{(0)} - \Lambda_{t_{i-1}}^{(0)}}\right] \underset{H_0}{\overset{H_1}{\gtrless}} \ln\left[\frac{P_0(C_{10} - C_{00})}{P_1(C_{01} - C_{11})}\right], \tag{2.111}$$

where $\Lambda_t^{(j)} = \int_{t_0}^{T} \lambda_\sigma^{(j)}\, d\sigma$ for $j = 0, 1$.

Example 2.5.2 (Binary Hypothesis Testing for a Poisson Process: Count-Record Data). Suppose $\{N_t; t \geqslant t_0\}$ is the Poisson process of Example 2.5.1. Assume now that count-record data are collected during the interval $[t_0, T)$ to decide which intensity governs the point process. The strategy that minimizes the average cost per decision is found by incorporating the sample function density (2.51) into (2.110) and then applying the likelihood ratio test (2.102). After taking the logarithm of both sides of (2.102), we find that the receiver's best strategy is defined by

$$-\int_{t_0}^{T} [\lambda_\sigma^{(1)} - \lambda_\sigma^{(0)}]\, d\sigma + \int_{t_0}^{T} \ln\left[\frac{\lambda_\sigma^{(1)}}{\lambda_\sigma^{(0)}}\right] dN_\sigma \underset{H_0}{\overset{H_1}{\gtrless}} \ln\left[\frac{P_0(C_{10} - C_{00})}{P_1(C_{01} - C_{11})}\right]. \tag{2.112}$$

This test can be implemented as shown schematically Figure 2.8. The controlled samplers generate impulse samples of $\ln \lambda_t^{(j)}$ for $j = 0, 1$ at the observed occurrence times.

Binary Hypothesis Testing Performance

The likelihood-ratio test, (2.102), and the likelihood-ratio expressions, (2.111) and (2.112), specify the processing that should be done on histogram and count-record data to decide between two alternative mechanisms that govern an observed Poisson process. We now want to examine the performance to be expected using this processing. For the

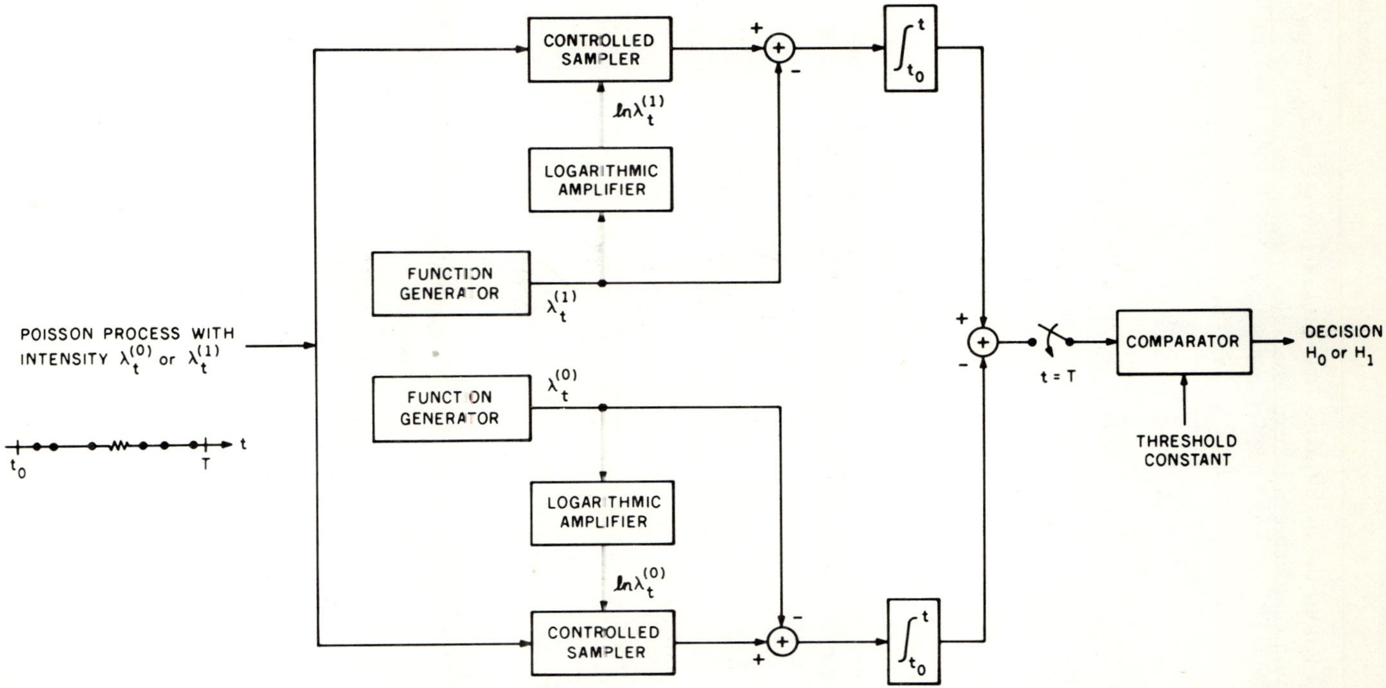

Figure 2.8. Optimum test for count-record observations of a Poisson process.

Bayes' criterion, the performance measure of interest is the average cost per decision $\overline{C}$, which can be expressed as

$$\overline{C}= C_{00}(1-P_F)\Pr(H_0)+C_{11}(1-P_M)\Pr(H_1)$$
$$+C_{10}P_F\Pr(H_0)+C_{01}P_M\Pr(H_1), \tag{2.113}$$

where P_M and P_F are the miss and false-alarm probabilities, and $\Pr(H_0)$ and $\Pr(H_1)$ are the prior probabilities of H_0 and H_1. For the minimum probability of error criterion, the performance measure of interest is given by

$$\Pr(\text{error})=P_F\Pr(H_0)+P_M\Pr(H_1). \tag{2.114}$$

Thus we see for both criteria that the quantities required for evaluating the performance are the miss and false-alarm probabilities. We will investigate these. The procedure is conceptually straightforward and illustrated in the following example.

Example 2.5.3 (On-Off Modulation). Consider once again the binary communication system shown schematically in Figure 2.7. Suppose electron emissions from the photodetector form a homogeneous Poisson process with intensity $\bar{n}$ on hypothesis zero when the shutter is closed and with intensity $\bar{s}+\bar{n}$ on hypothesis one when the shutter is open. Here, $\bar{s}$ and $\bar{n}$ are the average number of signal and noise electrons, respectively, emanating per unit time from the detector. The total number N_T of electrons emitted during the interval $[0,T)$ is observed, and based on this observation, the receiver desires to guess whether the shutter is closed or open. By combining (2.111) and (2.102), the receiver's best strategy is defined by

$$N_T \underset{H_0}{\overset{H_1}{\gtrless}} \frac{\bar{s}T+\ln[P_0(C_{10}-C_{00})]-\ln[P_1(C_{01}-C_{11})]}{\ln(\bar{s}+\bar{n})-\ln(\bar{n})} \triangleq \gamma \tag{2.115}$$

The same test results for count-record data due to the homogeneity of the Poisson process on each hypothesis. The false-alarm and miss probabilities are given by

$$P_F=\Pr[N_T\geqslant\gamma|H_0]=1-\sum_{k=0}^{[\gamma]}(k!)^{-1}(\bar{n}T)^k\exp(-\bar{n}T) \tag{2.116}$$

and

$$P_M=\Pr[N_T<\gamma|H_1]=\sum_{k=0}^{[\gamma]}(k!)^{-1}(\bar{s}+\bar{n})^k\exp[-(\bar{s}+\bar{n})T] \tag{2.117}$$

where $[\gamma]$ denotes the greatest integer less than γ (for example, $[10.5]=10$ and $[10]=9$.) The minimum average cost per decision is determined by substituting these expressions for P_F and P_M into (2.113). The minimum probability of error is determined by setting $C_{00}=C_{11}=0$ and $C_{01}=C_{10}=1$ in the threshold constant γ and then substituting the expressions for P_F and P_M into (2.114). Little more can be said about the performance without giving specific values for the several parameters involved. Some graphs of the minimum probability of error as a function of $\bar{s}T$ and $\bar{n}T$ are given by T. F. Curran and M. Ross [8] for $P_1=0.1$ and $P_1=0.5$. The special case when $\bar{n}$ approaches zero is called the *quantum limit*. In this limit, the randomness in photoelectron emissions is the only deterrent to perfect detection. Suppose in addition to $\bar{n}$ being small that $C_{00}=C_{11}=0$, $C_{01}=C_{10}=1$, and $P_0=P_1=0.5$. The threshold constant γ will be small but nonzero when $\bar{n}$ is small but nonzero. According to (2.115), the strategy that minimizes the probability of error is to select H_1 if $N_T \geqslant 1$ and H_0 if $N_T=0$. The false-alarm and miss probabilities are zero and $\exp(-\bar{s}T)$, respectively, and the minimum error probability is given by

$$\Pr(\text{error})=0.5\exp(-\bar{s}T).$$

Consequently, in the absence of thermoelectrons, an average of twenty photoelectrons must be received in a T-second interval to guarantee that the error probability does not exceed 10^{-9}.

From (2.111) and (2.112), we see that the probability distribution for the test statistics

$$l(n_1,n_2,\ldots,n_k) \triangleq -\sum_{i=1}^{k}\left[m_i^{(1)}-m_i^{(0)}\right]+\sum_{i=1}^{k} n_i \ln\left[\frac{m_i^{(1)}}{m_i^{(0)}}\right], \tag{2.118}$$

where

$$m_i^{(j)} \triangleq E(n_i|H_j)=\Lambda_{t_i}^{(j)}-\Lambda_{t_{i-1}}^{(j)}$$

for $j=0$, 1, and

$$l(\{N_\sigma;t_0\leqslant\sigma<T\}) \triangleq -\int_{t_0}^{T}\left[\lambda_\sigma^{(1)}-\lambda_\sigma^{(0)}\right]d\sigma+\int_{t_0}^{T}\ln\left[\frac{\lambda_\sigma^{(1)}}{\lambda_\sigma^{(0)}}\right]dN_\sigma \tag{2.119}$$

will be needed on each hypothesis in order to evaluate P_F and P_M for histogram and count-record data, respectively. These distributions are known in principle for observations of a Poisson process. However, explicit expressions are difficult, if not impossible, to obtain. Even when available in simple situations like that of Example 2.5.3, the expressions for P_F and P_M are so complicated that the effect of parameter variations is obscured to the extent that the expressions are of limited use in design considerations. As a result of the analytic obstacles encountered in attempting to derive expressions for P_F and P_M, we shall seek instead to bound these conditional error probabilities. The bounds we give for both histogram and count-record observations of a Poisson process are specializations of a general bound introduced by H. Chernoff [6] and discussed by H. L. Van Trees [40, 116]. The bounds are in terms of the moment generating functions for the test statistics $l(n_1, n_2, \ldots, n_k)$ and $l(\{N_\sigma; t_0 \leqslant \sigma < T\})$, which can be easily evaluated.

Theorem 2.5.1 (Chernoff Bound). Let $\{N_t; t \geqslant t_0\}$ be a Poisson process with intensity $\lambda_t^{(1)}$ on hypothesis one and $\lambda_t^{(0)}$ on hypothesis zero. Then the false-alarm and miss probabilities associated with deciding between H_1 and H_0 based on histogram or count-record observations of N on $[t_0, T)$ satisfy the inequalities

$$P_F \leqslant \min_{s \geqslant 0} \exp[\mu(s) - s\gamma] \tag{2.120}$$

and

$$P_M \leqslant \min_{s \leqslant 1} \exp[\mu(s) - (s-1)\gamma], \tag{2.121}$$

where

$$\gamma = \ln[P_0(C_{10} - C_{00})] - \ln[P_1(C_{01} - C_{11})], \tag{2.122}$$

and where for histogram data

$$\mu(s) = \sum_{i=1}^{k} \left\{ [m_i^{(1)}]^s [m_i^{(0)}]^{1-s} - s m_i^{(1)} - (1-s) m_i^{(0)} \right\}, \tag{2.123}$$

and for count-record data

$$\mu(s) = \int_{t_0}^{T} \left\{ [\lambda_\sigma^{(1)}]^s [\lambda_\sigma^{(0)}]^{1-s} - s\lambda_\sigma^{(1)} - (1-s)\lambda_\sigma^{(0)} \right\} d\sigma. \tag{2.124}$$

Proof. Let $D_{t_0, T}$ denote histogram or count-record data derived from observations of N. From (2.111) and (2.112), the decision strategy is summarized by

$$l(D_{t_0, T}) \underset{H_0}{\overset{H_1}{\gtrless}} \gamma \tag{2.125}$$

where l is defined in (2.118) for histogram data and in (2.119) for count-record data. Let $M_{l|H_0}(s)$ and $M_{l|H_1}(s)$ be the moment generating functions for l on H_0 and H_1, respectively. By introducing the decision indicator function, we have for $s \geqslant 0$ that

$$\begin{aligned} M_{l|H_0}(s) &= E\left\{\exp\left[sl(D_{t_0,T})\right]|H_0\right\} \\ &\geqslant E\left\{\exp\left[sl(D_{t_0,T})\right]I_{H_1}(D_{t_0,T})|H_0\right\} \\ &\geqslant \exp(s\gamma)E\left[I_{H_1}(D_{t_0,T})|H_0\right] \\ &= \exp(s\gamma)P_F. \end{aligned} \tag{2.126}$$

We show below that $M_{l|H_0}(s) = \exp[\mu(s)]$, where $\mu(s)$ is given in (2.123) and (2.124) for histogram and count-record data, respectively. Hence, for arbitrary $s \geqslant 0$, $P_F \leqslant \exp[\mu(s) - s\gamma]$. The tightest bound on P_F results by selecting $s \geqslant 0$ to minimize the right side of this inequality. By the same procedure, we have for $s \leqslant 0$ that $P_M \leqslant \exp[\mu_1(s) - s\gamma]$, where $\mu_1(s) = \ln M_{l|H_1}(s)$. Consequently,

$$P_M \leqslant \min_{s \leqslant 0} \exp[\mu_1(s) - s\gamma] = \min_{s \leqslant 1} \exp[\mu_1(s-1) - (s-1)\gamma]. \tag{2.127}$$

We show below that $\mu_1(s-1) = \mu(s)$. This establishes (2.120) and (2.121). We now need to evaluate $\mu(s)$ and $\mu_1(s)$ for both histogram and count-record data. For histogram data, we have from (2.118) and the independent increments property of the Poisson process that

$$\begin{aligned} M_{l|H_0}(s) &= \prod_{i=1}^{k} \exp\left[-s\left(m_i^{(1)} - m_i^{(0)}\right)\right] E\left\{\left(\frac{m_i^{(1)}}{m_i^{(0)}}\right)^{sn_i} |H_0\right\} \\ &= \prod_{i=1}^{k} \exp\left[-s\left(m_i^{(1)} - m_i^{(0)}\right)\right] \\ &\quad \times \sum_{m=0}^{\infty} (m!)^{-1}\left(\left[m_i^{(1)}\right]^s\left[m_i^{(0)}\right]^{1-s}\right)^m \exp\left(-m_i^{(0)}\right) \\ &= \exp[\mu(s)], \end{aligned} \tag{2.128}$$

where $\mu(s)$ is given in (2.123). Similar calculation shows that

$$M_{l|H_1}(s) = \exp[\mu_1(s)] = \exp[\mu(s+1)].$$

For count-record data, we have from (2.119) that

$$M_{l|H_0}(s)=\exp\left[-s\int_{t_0}^{T}(\lambda_\sigma^{(1)}-\lambda_\sigma^{(0)})\,d\sigma\right]E\left\{\exp\left[s\int_{t_0}^{T}\ln\left[\frac{\lambda_\sigma^{(1)}}{\lambda_\sigma^{(0)}}\right]dN_\sigma\Big|H_0\right]\right\}$$

$$=\exp\left[-s\int_{t_0}^{T}(\lambda_\sigma^{(1)}-\lambda_\sigma^{(0)})\,d\sigma\right]E\left\{\prod_{i=1}^{N_T}\left[\lambda_{w_i}^{(1)}\right]^s\left[\lambda_{w_i}^{(0)}\right]^{-s}\Big|H_0\right\}. \quad (2.129)$$

We use the property of a Poisson process that the occurrence times have the same distribution as the order statistics of m independent identically distributed random variables given $N_T=m$. Using (2.53), we have

$$E\left\{\prod_{i=1}^{N_T}\left[\lambda_{w_i}^{(1)}\right]^s\left[\lambda_{w_i}^{(0)}\right]^{-s}\Big|H_0\right\}$$

$$=\left(\int_{t_0}^{T}\lambda_\sigma^{(0)}\,d\sigma\right)^{-1}E\left\{\left(\int_{t_0}^{T}\left[\lambda_\sigma^{(1)}\right]^s\left[\lambda_\sigma^{(0)}\right]^{1-s}d\sigma\right)^{N_T}\Big|H_0\right\}$$

$$=\sum_{m=0}^{\infty}(m!)^{-1}\left(\int_{t_0}^{T}\left[\lambda_\sigma^{(1)}\right]^s\left[\lambda_\sigma^{(0)}\right]^{1-s}d\sigma\right)^m\exp\left[-\int_{t_0}^{T}\lambda_\sigma^{(0)}\,d\sigma\right]$$

$$=\exp\left\{\int_{t_0}^{T}\left(\left[\lambda_\sigma^{(1)}\right]^s\left[\lambda_\sigma^{(0)}\right]^{1-s}-\lambda_\sigma^{(0)}\right)d\sigma\right\}.$$

Substitution of this into (2.129) shows that $M_{l|H_0}(s)=\exp[\mu(s)]$, where $\mu(s)$ is given in (2.124). Similar calculation shows $M_{l|H_1}(s)=\exp[\mu_1(s)]$ $=\exp[\mu(s+1)]$. This completes the proof of Theorem 2.5.1.

We now give an example illustrating the use of these Chernoff bounds.

Example 2.5.4 (Binary Pulse-Position Modulation). Suppose the binary communication system shown schematically in Figure 2.7 is modified so that on hypothesis zero the shutter is open during the first half and closed during the second half of the transmission interval allotted to sending one source digit. On hypothesis one, the shutter is closed during the first half and open during the second half of the interval. Assume electron emissions from the photodetector form a Poisson process during $[0,T)$ with intensity

$\lambda_t^{(0)}$ or $\lambda_t^{(1)}$ where

$$\lambda_t^{(0)} = \begin{cases} \bar{s}+\bar{n}, & 0 \leqslant t < \frac{1}{2}T \\ \bar{n}, & \frac{1}{2}T \leqslant t < T \end{cases}$$

and

$$\lambda_t^{(1)} = \begin{cases} \bar{n}, & 0 \leqslant t < \frac{1}{2}T \\ \bar{s}+\bar{n}, & \frac{1}{2}T \leqslant t < T. \end{cases}$$

Assume $P_0 = P_1 = 0.5$, $C_{00} = C_{11} = 0$, and $C_{01} = C_{10} = 1$. Count-record data are collected on the interval $[0, T)$ to decide H_0 or H_1. From (2.112), the strategy which minimizes the probability of error is summarized by

$$n_2 \underset{H_0}{\overset{H_1}{\gtrless}} n_1 \tag{2.130}$$

where n_1 and n_2 are the number of electrons observed on $[0, T/2)$ and $[T/2, T)$, respectively. It is easily seen from (2.124) that

$$\mu(s) = \tfrac{1}{2}T\left[(\bar{s}+\bar{n})^s(\bar{n})^{1-s} + \bar{n}^s(\bar{s}+\bar{n})^{1-s} - \bar{s} - 2\bar{n}\right].$$

It is evident that $\mu(s) = \mu(1-s)$, so $\mu(s)$ is symmetric about the point $s = 0.5$ where it attains its minimum value

$$\mu(0.5) = T\left\{\sqrt{\bar{n}(\bar{s}+\bar{n})} - 0.5\bar{s} - \bar{n}\right\}. \tag{2.131}$$

As $\gamma = 0$, we have from (2.120) and (2.121) that

$$P_F \leqslant \exp[\mu(0.5)]$$

and (2.132)

$$P_M \leqslant \exp[\mu(0.5)].$$

Furthermore, the minimum probability of error, which the test (2.130) achieves, is bounded according to

$$\Pr(\text{error}) = \tfrac{1}{2}P_F + \tfrac{1}{2}P_M \leqslant \exp[\mu(0.5)] \tag{2.133}$$

Plots of this bound on Pr(error) are shown in Figure 2.9 for various values of $T\bar{s}$ and $T\bar{n}$. We note, finally, that the probability of error for the test

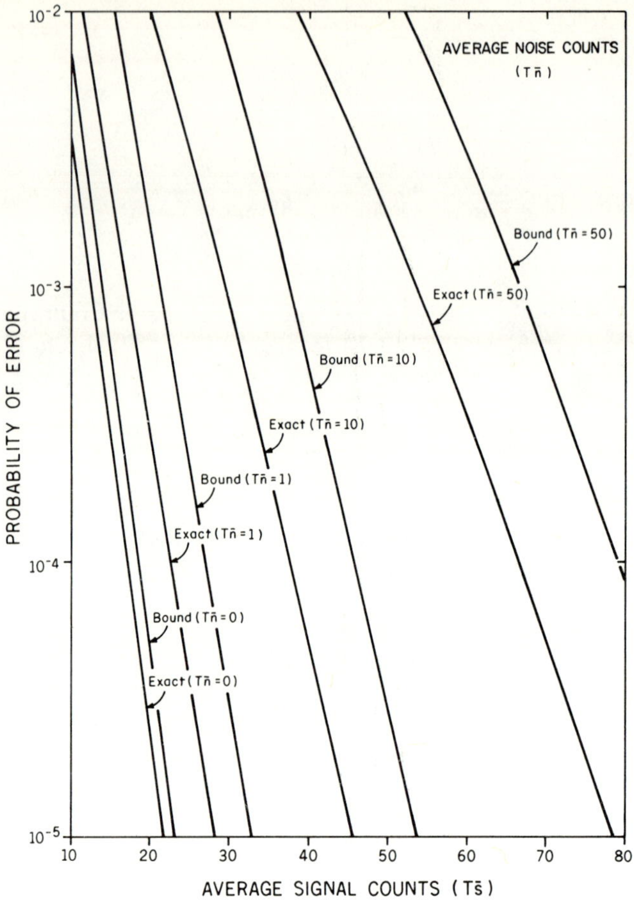

Figure 2.9. Probability of error for binary pulse-position modulation.

(2.130) is given by

$$\begin{aligned}\Pr(\text{error}) &= 0.5\Pr(n_2 \geqslant n_1|H_0) + 0.5\Pr(n_1 > n_2|H_1)\\ &= 0.5E[\Pr(n_2 \geqslant n_1|H_0, n_1)|H_0] + 0.5E[\Pr(n_1 > n_2|H_1, n_2)|H_1]\\ &= 0.5\sum_{i=0}^{\infty}\sum_{j=i}^{\infty}(i!j!)^{-1}[\tfrac{1}{2}T(\bar{s}+\bar{n})]^i(\tfrac{1}{2}T\bar{n})^j\exp[-\tfrac{1}{2}T(\bar{s}+2\bar{n})]\\ &\quad + 0.5\sum_{i=0}^{\infty}\sum_{j=i+1}^{\infty}(i!j!)^{-1}[\tfrac{1}{2}T(\bar{s}+\bar{n})]^i[\tfrac{1}{2}T\bar{n})^j\exp[-\tfrac{1}{2}T(\bar{s}+2\bar{n})].\end{aligned} \tag{2.134}$$

This probability of error is also plotted in Figure 2.9 for comparison with the bound (2.133).

M-ary Hypothesis Testing

The extension from deciding between two to $M \geqslant 2$ alternative mechanisms that govern an observed Poisson process is conceptually straightforward. There are now M^2 possible outcomes when performing the experiment of collecting data and subsequently making a decision about the underlying influence $H_1, H_2, \ldots, H_M$. These are to decide H_i when H_j is the influence, for $i,j = 1,2,\ldots,M$. With the Bayes' criterion, there is an $M \times M$ cost matrix that assigns a cost to each of the possible outcomes. The i,j-element, C_{ij}, of this matrix is the cost of deciding H_i when H_j is the influence. The conditional average cost of deciding in favor of H_i given an observation $D_{t_0,T}$ is given by

$$C(H_i|D_{t_0,T}) = \sum_{j=1}^{M} C_{ij} \Pr(H_j|D_{t_0,T}). \tag{2.135}$$

The receiver's strategy for deciding among the M alternative hypotheses given $D_{t_0,T}$ can be summarized by M decision indicator functions defined by

$$I_{H_i}(D_{t_0,T}) = \begin{cases} 1, & \text{if the decision for } D_{t_0,T} \text{ is } H_i \\ 0, & \text{otherwise} \end{cases} \tag{2.136}$$

for $i = 1,2,\ldots,M$. The conditional average cost per decision for a particular realization $D_{t_0,T}$ is

$$C(D_{t_0,T}) = \sum_{i=1}^{M} I_{H_i}(D_{t_0,T}) C(H_i|D_{t_0,T}), \tag{2.137}$$

and the average cost per decision is

$$\overline{C} = E\big[C(D_{t_0,T})\big] = \sum_{i=1}^{M} E\big[I_{H_i}(D_{t_0,T}) C(H_i|D_{t_0,T})\big]. \tag{2.138}$$

The receiver attempts to minimize $\overline{C}$. In parallel with our development for $M=2$, we see that the strategy that accomplishes this objective is to make the decision given $D_{t_0,T}$ which on the average costs the least; that is, the receiver should select H_i if $C(H_i|D_{t_0,T}) \leqslant C(H_j|D_{t_0,T})$ for $j = 1,2,\ldots,M$. In case more than one hypothesis has the same average cost given $D_{t_0,T}$, a selection among them can be made arbitrarily without affecting $\overline{C}$.

The special situation when $C_{ii}=0$ for $i=1,2,\ldots,M$ and $C_{ij}=1$ for $i\neq j$, $i,j=1,2,\ldots,M$ is of interest. Here, the relative costs for all types of errors are equal. The average cost per decision becomes

$$\overline{C}=\sum_{i=1}^{M}\sum_{\substack{j=1\\ j\neq i}}^{M} E\left[I_{H_i}(D_{t_0,T})\Pr(H_j|D_{t_0,T})\right]. \tag{2.139}$$

However,

$$\begin{aligned} E\left[I_{H_i}(D_{t_0,T})\Pr(H_j|D_{t_0,T})\right] &= E\left[I_{H_i}(D_{t_0,T})|H_j\right]\Pr(H_j)\\ &= \Pr[\text{decide } H_i|H_j]\Pr(H_j). \end{aligned}$$

Thus, the average cost per decision is the probability of error

$$\overline{C}=\Pr(\text{error}). \tag{2.140}$$

It is evident for this cost matrix that the procedure of making decisions which on the average cost the least is equivalent to selecting the most likely hypothesis given the data; that is, the receiver should select H_i if $\Pr(H_i|D_{t_0,T})\geqslant \Pr(H_j|D_{t_0,T})$ for $j=1,2,\ldots,M$. This procedure minimizes $\Pr(\text{error})$.

Example 2.5.5 (M-ary Pulse Position Modulation). Suppose the binary information source in Figure 2.7 is replaced by an M-ary source. For example, a source of letters of the English alphabet and a space would have $M=27$. Suppose the transmission interval allotted to sending a source letter is divided into M equal subintervals and that the shutter is open during the ith subinterval on the ith hypothesis H_i corresponding to the occurrence of the ith source letter. Assume electron emissions from the photodetector form a Poisson process during the transmission interval $[0,T)$ with intensity $\lambda_t^{(i)}$ on hypothesis H_i, where

$$\lambda_t^{(i)}=\begin{cases} \bar{n}, & 0\leqslant t<\dfrac{(i-1)T}{M}\\ \bar{s}+\bar{n}, & \dfrac{(i-1)T}{M}\leqslant t<\dfrac{iT}{M}\\ \bar{n}, & \dfrac{iT}{M}\leqslant t<T \end{cases}$$

Assume source letters occur with equal probability, $\Pr(H_i)=1/M$, and that the receiver desires to make decisions with a minimum probability of error. He collects count-record data $\{N_\sigma;0\leqslant\sigma<T\}$ to make these decisions. Thus, his strategy is to pick the most likely hypothesis given

$\{N_\sigma; 0 \leqslant t < T\}$. However,

$$\Pr(H_i | \{N_\sigma; 0 \leqslant \sigma < T\}) = \frac{p[\{N_\sigma; 0 \leqslant \sigma < T\} | H_i] \Pr(H_i)}{p[\{N_\sigma; 0 \leqslant \sigma < T\}]}.$$

The only factor on the right side that depends on H_i is the conditional sample-function density

$$p[\{N_\sigma; 0 \leqslant \sigma < T\} | H_i] = -\int_0^T \lambda_\sigma^{(i)}\, d\sigma + \int_0^T \ln \lambda_\sigma^{(i)}\, dN_\sigma$$

$$= -T\bar{n} - \frac{T\bar{s}}{M} + n_i\left(\frac{T}{M}\right)\ln(\bar{s}+\bar{n}) + \left(\frac{T}{M}\right)\sum_{\substack{j=1\\ j\neq i}}^{M} n_j \ln(\bar{n})$$

where n_i is the number of electrons observed in the ith interval $[(i-1)T/M, iT/M)$. By inspection of this, we see that the proedure that minimizes the probability of error is to select the hypothesis corresponding to the interval in which the greatest number of electron emissions occur. In case the same maximum number occurs in more than one interval, an arbitrary selection among the corresponding hypotheses can be made without affecting the probability of error. Because of the symmetry in the problem, we see that the minimum probability of error that results from this procedure is given by

$$\Pr(\text{error}) = \Pr(\text{error} | H_1)$$

$$= \Pr(n_1 < n_2 \text{ or } n_1 < n_3 \text{ or } \cdots \text{ or } n_1 < n_M | H_1)$$

$$\leqslant \sum_{i=2}^{M} \Pr(n_1 < n_i | H_1) = (M-1)\Pr(n_1 < n_2 | H_1).$$

The last equality results from the symmetry of the problem, and the inequality is called the *union bound*. The union bound provides an upper bound on Pr(error) that can be further extended using the Chernoff bound on $\Pr(n_1 < n_2 | H_1)$, which is the error probability for binary pulse position modulation on the interval $[0, 2T/M)$. Thus, from Example 2.5.4, we have that

$$\Pr(\text{error}) \leqslant (M-1)\exp[\mu(0.5)]$$

where

$$\mu(0.5) = \left(\frac{2T}{M}\right)\left[\sqrt{\bar{n}(\bar{s}+\bar{n})} - 0.5\bar{s} - \bar{n}\right].$$

Summary. In this section, we have developed the basic principles of deciding between alternative mechanisms that govern an observed point process. With the Bayes' criterion, we found that the best procedure is to select that mechanism which on the average costs the least for the given observations. For the minimum probability of error criterion, the best procedure is to select the most likely hypothesis given the observations. Both of these criteria can be implemented as a likelihood-ratio test. Additional criteria which can be similarly implemented are discussed by C. Helstrom [16, p. 84] and H. L. Van Trees [40, pp. 30–34].

These general results were specialized for both histogram and count-record observations of a Poisson process. The tests are given in (2.111) and (2.112), respectively. The sample-function density developed in Section 2.3 was shown to play an important role in defining the test for count-record data.

It was observed that the performance in binary hypothesis testing is determined by the miss and false-alarm probabilities. These conditional error probabilities are difficult to determine because of analytic complexities. Thus, we obtained Chernoff bounds on the miss and false-alarm probabilities. The Chernoff bounds for both histogram and count-record data given in Theorem 2.5.1 were easily obtained because the moment generating function for the test statistic can be evaluated analytically. H. L. Van Trees [40, p. 116] develops some approximations to the miss and false-alarm probabilities which, as the Chernoff bound derived here, are obtained from the moment generating function.

REFERENCES

1. M. Aoki, *Introduction to Optimization Techniques, Macmillan, New York, 1971.*
2. T. M. Apostol, *Mathematical Analysis*, Addison-Wesley, Reading, Mass., 1957.
3. L. Aroian, "The Type-B Gram-Charlier Series," *Ann. Math. Stat.*, **8**, 183–192 (1937).
4. E. W. Barankin, "Locally Best Unbiased Estimates," *Ann. Math. Stat.*, **20**, 477 (1949).
5. I. Bar David, "Communication Under the Poisson Regime," *IEEE Trans. Information Theory*, **IT-15**, No. 1, 31–37 (Jan. 1969).
6. H. Chernoff, "A Measure of Asymptotic Efficiency for Tests of Hypotheses Based on A Sum of Observations," *Ann. Math. Stat.*, **23**, 493–507 (1962).
7. D. R. Cox and P. A. W. Lewis, *The Statistical Analysis of Series of Events*, Methuen, London, 1966.
8. T. F. Curran and M. Ross, "Optimum Detection Thresholds in Optical Communications," *Proc. IEEE*, **53**, No. 11, 1770–1771 (Nov. 1965).
9. D. J. Daley, "Various Concepts of Orderliness for Point Processes," in *Stochastic Geometry and Analysis* (*Rollo Davidson Memorial Volume*), Wiley, New York, 1974.
10. D. J. Daley and D. Vere-Jones, "A Summary of the Theory of Point Processes," in

Stochastic Point Processes: Statistical Analysis, Theory, and Applications, P. A. W. Lewis, Ed., Wiley, New York, 1972.

11. R. D. Evans, *The Atomic Nucleus*, McGraw-Hill, New York, 1965.
12. M. Fisz, *Probability Theory and Mathematical Statistics*, Wiley, New York, 1963.
13. I. Gikhman and A. V. Skorokhod, *Introduction to the Theory of Random Processes*, Saunders, Philadelphia, 1965.
14. F. Haight, *Handbook of the Poisson Distribution*, Wiley, New York, 1967.
15. H. Hance, "Modulation/Demodulation Techniques for an Optical One-Gigabit/Sec. Intersatellite Transmission System," in *Proceedings of the International Telemetering Conference, Washington, D.C., 1971.*
16. C. Helstrom, "Estimation of Modulation Frequency of a Light Beam," Appendix E in *Optical Space Communication*, Proceedings of a Workshop held at Williams College, Ed. R. S. Kennedy and S. Karp, Williamstown, Mass., 1968.
17. G. F. Herrmann, "Optimum Versus Suboptimum Detection Under the Poisson Regime —Application to Optical Communication Using Biphase and Quadriphase Modulated Subcarriers," *IEEE Trans. Communications*, **COM-21**, No. 7, 800–809 (July 1973).
18. E. Hoversten and D. Snyder, "On the Performance of Pulse-Position Modulation in Direct-Detection Optical Communication Systems: Mean-Square Error and Threshold," *Proceedings of the 1972 International Information Theory Symposium, Pacific Grove, Calif., Jan. 1972.*
19. J. A. Jacques, "Tracer Kinetics," in *Principles of Nuclear Medicine*, Ed. H. N. Wagner, Jr., Saunders, Philadelphia, 1968.
20. B. Jansson, *Random Number Generators*, Victor Pettersons, Stockholm, 1966.
21. S. Karp, E. L. O'Neill, and R. M. Gagliardi, "Communication Theory for the Free Space Optical Channel," *Proc. IEEE*, **58**, No. 10, 1611–1626 (Oct. 1970).
22. A. Y. Khinchin, "On Poisson Sequences of Chance Events," **1**, No. 3, 291–297 (1956).
23. G. M. Lee, R. B. Fluchel, E. A. Paddon, and D. C. Torretta, "Bit Error Probability for Baseband and Subcarrier Optical Communication Systems," *IEEE Trans. on Communications*, **COM-22**, No. 1 (Jan. 1974).
24. W. M. Littlefield, "Investigation of the Linear Range of the Peripheral Auditory System," D. Sc. Thesis, Sever Institute of Washington University, St. Louis, Dec. 1973.
25. L. Mandel, "Fluctuations of Light Beams," in *Progress in Optics*, Vol. II, Ed. E. Wolf, Wiley, New York, 1963.
26. L. Mandel, "Fluctuations of Photon Beams and Their Correlations," *Proc. Phys. Soc. (London)*, **72-1**, 1037–1048 (1958).
27. J. Markham, D.L. Snyder, and J. R. Cox, Jr., "A Numerical Implementation of the Maximum-Likelihood Method of Parameter Estimation for Tracer-Kinetic Data," *J. Math. Biophysics*, (to appear).
28. J. Ortega and W. Rheinholdt, *Iterative Solution of Nonlinear Equations in Several Variables*, Academic, New York, 1970.
29. E. Parzen, *Stochastic Processes*, Holden-Day, San Francisco, 1962.
30. R. Peierls, "Statistical Errors in Counting Experiments," *Proc. Roy. Soc. (London)*, **A149**, 467–486 (1935).
31. A. Ralston and H. Wilf, *Mathematical Methods for Digital Computers*, Vol. 2, Wiley, New York, 1967, p. 249.
32. C. R. Rao, *Linear Statistical Inference and Its Applications*, Wiley, New York, 1965.

33. J. Riordan, "Moment Recurrence Relations for Binomial, Poisson, and Hypergeometric Frequency Distributions," *Ann. Math. Stat.*, **8**, 103–111 (1937).

34. E. Rutherford and H. Geiger, "The Probability Variations in the Distribution of α-Particles," *Phil. Mag. S6*, **20**, 698 (1910).

35. L. J. Savage, *The Foundations of Statistics*, Wiley, New York, 1954.

36. C. W. Sheppard, *Basic Principles of the Tracer Method*, Wiley, New York, 1962.

37. M. M. Ter-Pogossian, J. O. Eichling, D. O. Davis, and M. J. Welch, "The Measure in Vivo of Regional Cerebral Oxygen Utilization by means of Oxyhemoglobin Labelled with Radioactive Oxygen-15," *J. Clin. Invest.*, **49**, 381–391 (1970).

38. W. M. Siebert, "Stimulus Transformation in the Peripheral Auditory System," in *Recognizing Patterns*, Ed. P. A. Kolers and M. Eden, MIT Press, Cambridge, 1968.

39. W. M. Siebert, "Frequency Discrimination in the Auditory System: Place or Periodicity Mechanisms?," *Proc. IEEE*, **58**, No. 5, 723–730 (May 1970).

40. H. L. Van Trees, *Detection, Estimation, and Modulation Theory: Part I*, Wiley, New York, 1968.

41. J. B. West and C. T. Dollery, "Distribution of Blood Flow and Ventilation Perfusion Ratio in the Lung, Measured with Radioactive CO_2," *J. Appl. Physiol.*, **15**, 405–410 (1960).

42. J. B. West, C. T. Dollery, and P. Hugh-Jones, "The Use of Radioactive Carbon Dioxide to Measure Regional Blood Flow in the Lungs of Patients with Pulmonary Disease," *J. Clin. Invest.*, **40**, 1–12 (Jan. 1961).

43. S. S. Wilks, *Mathematical Statistics*, Wiley, New York, 1962.

PROBLEMS

2.2.1. The fact that points are distinct and that only a finite number occur in any finite interval is insufficient to guarantee that a point process is orderly. To verify this, consider a point process on the time inverval $[-1,1]$ constructed as follows. The process has two distinct points the first of which is selected uniformly at random on the interval $[-1,1]$. Call the location of the first point t_1. The second point is placed at $t=\frac{1}{2}t_1$ for $t_1 \neq 0$ and at $t=\frac{1}{2}$ for $t_1=0$.

a. Show the point process is orderly at $t=\frac{1}{2}$.
b. Show it is not orderly at $t=0$.

2.2.2. (Constructive Derivation of the Poisson Process). Suppose a process consisting of unit-area pulses of duration Δ is constructed in the following fashion. At time $t=0$, a decision is made as to whether or not a pulse is to occur in the interval $[0,\Delta)$. An affirmative decision is made if heads occurs on the toss of a biased coin for which the probability of getting heads is p and of getting tails is $q=1-p$. At time $t=\Delta$, a decision is made as to whether or not a pulse is to occur in $[\Delta,2\Delta)$. The decision is made by tossing the coin independently again. This procedure is repeated at the beginning of each interval. The decision for the interval $[k\Delta,(k+1)$

Δ) depends only on the result of the coin toss at time $t=k\Delta$ and not any previous tosses. Thus, pulse decisions are made without aftereffects.

a. Let $s=k\Delta$ and $t=(k+K)\Delta$, and denote by $N_{s,t}$ the number of pulses in $[s,t)$. Assume k and K are nonnegative integers so that $0\leqslant s\leqslant t$. Evaluate the mean and variance of $N_{s,t}$. Determine the probability that $N_{s,t}=n$ for $n=0,1,2,\ldots,K$.

b. Now examine what unfolds as Δ tends to zero. Let k and K increase in such a way that $s=k\Delta$ and $t=(k+K)\Delta$ remain fixed as Δ tends to zero. Also, let $p=\lambda\Delta$ so that the expected number of pulses in $[s,t)$ remains fixed. Show that $\Pr[N_{s,t}=n]$ tends to the Poisson distribution with parameter $\lambda(t-s)$ as Δ tends to zero. This is how in 1837 S. Poisson discovered this distribution now bearing his name. See F. Haight [14, p. 113] for an historical account of this discovery.

c. Define $N_t=N_0+N_{0,t}$, where $\Pr[N_0=0]=1$. Explain why $\{N_t;t\geqslant 0\}$ is a homogeneous Poisson process under the limiting conditions of part (b).

2.2.3. The moments $m_k\triangleq E(x^k)$, for $k=1,2,\ldots$, are one set of parameters that characterize a random variable x. There are other sets of parameters that arise in practice, one of which is the set of *cumulants* or *semiinvariants*, γ_k, defined by the series expansion

$$\Gamma_x(jv)\triangleq \ln E[\exp(jvx)]=\sum_{k=1}^{\infty}(k!)^{-1}(jv)^k\gamma_k.$$

The function $\Gamma_x(jv)$ is termed the semi-invariant generating function for x.

a. Derive the following relationships between the moments and cumulants:

$$m_1=\gamma_1,$$

$$m_2=\gamma_2+\gamma_1^2,$$

$$m_3=\gamma_3+3\gamma_1\gamma_2+\gamma_1^3,$$

$$m_4=\gamma_4+3\gamma_2^2+4\gamma_1\gamma_3+6\gamma_1^2\gamma_2+\gamma_1^4.$$

Conversely,

$$\gamma_1=m_1=\text{mean}(x),$$

$$\gamma_2=m_2-m_1^2=\text{variance}(x),$$

$$\gamma_3=m_3-3m_1m_2+2m_1^3,$$

$$\gamma_4=m_4-3m_2^2-4m_1m_3+12m_1^2m_2-6m_1^4.$$

b. Show that all the cumulants except γ_1 are invariant under translation; that is, show all the cumulants for $y = x + a$ are identical to those of x except for the first.

c. Determine all the cumulants when $x = N_{s,t}$, where $N_{s,t} = N_t - N_s$ is an increment of an inhomogeneous Poisson counting process on $[t_0, \infty)$ with intensity function $\{\lambda_t; t \geqslant t_0\}$.

d. Let $y = x_1 + x_2 + \cdots + x_n$ be the sum of n statistically independent random variables that are not necessarily identically distributed. Demonstrate that

$$\gamma_k(y) = \sum_{i=1}^{n} \gamma_k(x_i), \qquad k = 1, 2, \ldots .$$

That is, the kth cumulant is the sum of the kth cumulants.

e. Determine *all* the cumulants of y when the sequence $\{x_i\}_{i=1}^{n}$ in part (d) are identically distributed normal random variables having mean m and variance σ^2.

2.2.4. Let $\{N_t; t \geqslant t_0\}$ be an inhomogeneous Poisson process with intensity function λ_t for $t \geqslant t_0$.

a. Show that the first and second moments for the increment, $N_{s,t} = N_t - N_s$, are given by

$$E[N_{s,t}] = \int_s^t \lambda_\sigma \, d\sigma$$

and

$$E[N_{s,t}^2] = \int_s^t \lambda_\sigma \, d\sigma + \left(\int_s^t \lambda_\sigma \, d\sigma \right)^2,$$

respectively.

b. Show that higher-order moments of $N_{s,t}$ can be determined using the recursion formula

$$E(N_{s,t}^{m+1}) = \mu E(N_{s,t}^m) + \frac{\mu \, dE(N_{s,t}^m)}{d\mu},$$

where $\mu \triangleq \int_s^t \lambda_\sigma \, d\sigma$. This recursion formula is attributed by F. Haight [14, p. 6] to L. Aroian [3] and J. Riordan [33].

2.2.5 (I. Gikhman and A. Skorokhod [13]). Use (2.18) to show that the rth moment of $N_{s,t}$ is finite for all $r > 0$; that is, show $E(N_{s,t}^r) < \infty$. Hint: the series $\sum_{k=0}^{\infty} x^k (k+1)^r$ converges for all x such that $|x| < 1$.

2.2.6 (Birth Processes). Let $\{N_t; t \geqslant 0\}$ be a counting process described by the following assumptions:

i. $\{N_t; t \geqslant 0\}$ is conditionally orderly;

ii. for all $t \geqslant 0$, the limit of $\delta^{-1}\Pr[N_{t,t+\delta}=1|N_t=n]$ exists as δ tends to zero and the limit is a finite, integrable function of t and a function of n. Denote the limit by $\lambda_t(n)$; thus,

$$\lambda_t(n)=\lim_{\delta \downarrow 0}\delta^{-1}\Pr[N_{t,t+\delta}=1|N_t=n];$$

iii. $\Pr[N_0=1]=1$.

The difference between the process N and a Poisson counting process of Theorem 2.2.2 is contained in assumption (ii). Here we allow the intensity to depend on the number of events, so the process does not evolve without aftereffects. The process N is called a *pure birth process*. If N is the number of members of a population (of bacteria, particles in a chain reaction, etc.) then we allow the rate of population growth to vary with the size of the population. Note that initially there is one member in the population (we could just as easily have two) and there are no deaths (a birth-death process allows for this).

a. Let $P_n(t) \triangleq \Pr[N_t=n]$. Determine a differential equation with appropriate boundary conditions that $P_n(t)$ satisfies. Check your equation using the known equation for $\lambda_t(n) \equiv \lambda_t$ independent of n.

b. The simplest birth process is the homogeneous, linear birth process in which the intensity is independent of time and proportional to the population size; that is, $\lambda_t(n)=\lambda n$, where λ is a positive constant. Determine $P_1(t)$ for the homogeneous, linear birth process.

c. Demonstrate that $P_n(t)$ for the homogeneous linear birth process satisfies the recursion equation

$$P_n(t)=(n-1)\lambda\int_0^t P_{n-1}(\xi)e^{-\lambda n(t-\xi)}\,d\xi$$

for $n=2,3,\ldots$.

d. Let $\tilde{P}_n(s)=\int_0^\infty P_n(t)e^{-st}\,dt$ be the Laplace transform of $P_n(t)$ for the homogeneous linear birth process. Demonstrate that

$$\tilde{P}_n(s)=\sum_{k=0}^{n-1}(-1)^k\binom{n-1}{k}\frac{1}{s+(k+1)\lambda}$$

where $\binom{n-1}{k}$ is the binomial coefficient $(n-1)!/k!(n-1-k)!$.

e. By using either the result of (c) and induction or by inverting the result of (d), show

$$P_n(t) = e^{-\lambda t}[1-e^{-\lambda t}]^{n-1} \qquad \text{for } n = 1,2,\ldots.$$

This is called the Yule–Furry distribution.

f. The mean and variance of a homogeneous linear birth process N can be found by using $P_n(t)$ given in (e). However, in most general cases, $P_n(t)$ is not easily determined, in which cases it is convenient to use the following alternative procedure. Multiply the differential equation for $P_n(t)$ in (a) by n (or n^2) and sum on n to obtain a differential equation for $E[N_t]$ or (or $E[N_t^2]$). Solve the differential equation for the case of a homogeneous linear birth process in order to evaluate the mean and variance of N_t.

2.2.7 (Random Telegraph Wave). Let $\{N_t; t \geqslant t_0\}$ be an inhomogeneous Poisson counting process with intensity function λ_t. Let $\{x_t; t \geqslant t_0\}$ be the process defined by

$$x_t = x_0(-1)^{N_t}$$

where the initial state x_0 is either $+1$ or -1 with equal probability and independently of $\{N_t; t \geqslant t_0\}$. A typical sample function for x is shown in Figure P.2.2.7. This process is called a *random telegraph wave*. Let

$$M(ju,jv) = E[\exp(jux_s + jvx_t)]$$

be the joint characteristic function for x_s and x_t, where $t_0 \leqslant s \leqslant t$.

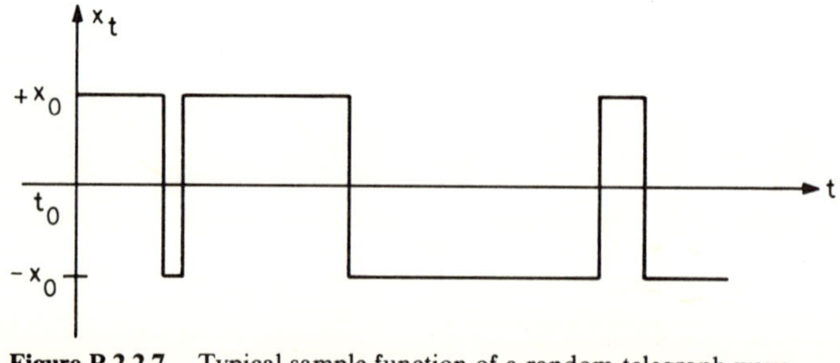

Figure P.2.2.7. Typical sample function of a random telegraph wave.

a. Show that the correlation function for x is given by

$$R(s,t) \triangleq E(x_s x_t) = \exp\left(-2\int_s^t \lambda_\sigma \, d\sigma\right)$$

where $t_0 \leqslant s \leqslant t$. Hint: establish and use the relation

$$\sum_{n \text{ even}} (n!)^{-1}\left(\int_s^t \lambda_\sigma \, d\sigma\right)^n = \frac{1}{2}\left[\exp\left(+\int_s^t \lambda_\sigma \, d\sigma\right) + \exp\left(-\int_s^t \lambda_\sigma \, d\sigma\right)\right].$$

b. Show that

$$M(ju, jv) = \cos(u)\cos(v) - R(s,t)\sin(u)\sin(v).$$

2.2.8. Suppose that the points of an inhomogeneous Poisson point process with intensity λ_t are deleted independently with probability p. Show that the resulting counting process is Poisson and specify its intensity. The counting process for the deleted point process is a possible model for a faulty counter in which points are not counted with probability p. Give an argument demonstrating that a Poisson point process subjected to systematic deletions (e.g., every other point is deleted) is no longer Poisson.

2.3.1. A homogeneous Poisson counting process in a three-dimensional Euclidean space is such that the number of points in a region A is Poisson distributed with parameter $\lambda v(A)$ where $v(A)$ denotes the volume of A and λ is a positive constant; that is,

$$\Pr[N(A) = n] = (n!)^{-1}(\lambda v(A))^n \exp(-\lambda v(A)).$$

Furthermore, the number of points in disjoint volumes is independent. The notion of *interarrival time* in a one-dimensional space is here replaced by the notion of *nearest-neighbor distance*. Suppose we model star centers as a three-dimensional, homogeneous Poisson point process with parameter $\lambda v(A)$. Determine the distribution function and the mean of the nearest-neighbor distance for stars.

2.3.2. The device shown in Figure P.2.3.2 is an interarrival time meter. For the situation shown, it is set to measure the second interarrival time t_2. Note, however, that at any given time t the output g_t may or may not be a good measure of the second interarrival time. Suppose the incident point

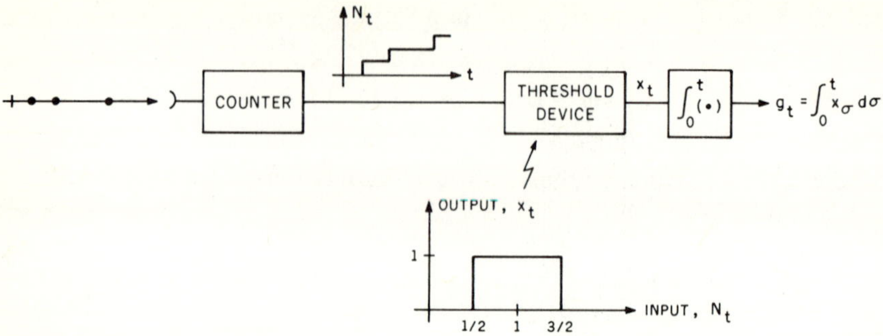

Figure P.2.3.2. Interarrival-time meter.

process is a homogeneous Poisson process with intensity λ.

a. Determine the probability that $x_t = 0$.

b. Evaluate the expected value of g_t.

c. What happens as $t \to \infty$?

2.3.3. Suppose that jobs submitted for processing at a computer facility are classified as long runs or short runs. Suppose further that the arrival of long and short runs are independent and that each can be modeled as a homogeneous Poisson process with arrival rates λ_L and λ_S, respectively.

a. Show that the probability of exactly m short runs arriving in a time interval between the arrival of two long runs is

$$\frac{\lambda_L}{\lambda_L+\lambda_S}\left(\frac{\lambda_S}{\lambda_L+\lambda_S}\right)^m$$

for $m = 0, 1, 2 \ldots$.

b. Suppose a total of n runs are known to arrive in some time interval. What is the probability that $k \leqslant n$ of them are short runs?

2.3.4 (Tests for a Poisson Process). Consider an interval $[0, T)$ in which n events are observed to occur at the sequence of instants $w_1, w_2, \ldots, w_n$ each measured from the beginning of the observation interval. This problem outlines two simple tests for deciding whether or not the events obey a homogeneous Poisson distribution with unknown mean rate λ.

a. *Waiting Time Test*: Using a central limit theorem argument, justify the assertion that *if* the events were governed by a Poisson law, then the probability is approximately 0.95 that $s_n = \Sigma_{j=i}^n w_j$ lies in the range

$\frac{1}{2}[n \pm 1.96(n/3)^{1/2}]$. Discuss briefly how this observation can be used as a possible test of whether or not the events obey a Poisson law.

b. *Interarrival Time Test*: Let $t_1 = w_1$, $t_2 = w_2 - w_1, \ldots, t_n = w_n - w_{n-1}$ be the experimentally determined interarrival times for the observed sequence of events. Let N be the number of interarrival times of duration exceeding U.

Demonstrate that *if* the events were governed by a Poisson law with parameter λ, then the mean and variance of the random variable N satisfy

$$E[N] = ne^{-\lambda U}$$

and

$$E[N] \pm \sigma_N = E[N]\left(1 \pm \sqrt{\frac{1-e^{-\lambda U}}{ne^{-\lambda U}}}\right)$$

Discuss briefly how a sketch of N versus U for the observed events and of the two equations derived from the Poisson assumption can be used to estimate λ and as a possible test of whether or not the events obey a Poisson law.

2.3.5. By a suitable time scaling, an inhomogeneous Poisson counting process can be converted into a homogeneous Poisson counting process, and vice versa. Let $\{N(t); t \geqslant 0\}$ be an inhomogeneous Poisson counting process with intensity $\{\lambda(t); t \geqslant 0\}$ such that the parameter function

$$\Lambda(t) = \int_0^t \lambda(\sigma)\, d\sigma$$

is nondecreasing. Let $\Lambda^{-1}(u)$ be the inverse function of Λ; that is, for $u > 0$, $\Lambda^{-1}(u)$ is the smallest value of t for which $\Lambda(t) \geqslant u$.

a. Show that the counting process defined by $M(u) = N(\Lambda^{-1}(u))$ is a homogeneous Poisson counting process for $u \geqslant 0$ with unit intensity.

b. Suppose $\lambda(t)$ is the periodic function shown in Figure P.2.3.5*a*. Sketch $\Lambda^{-1}(u)$. Also sketch the realization of $M(u)$ corresponding to the realizatiin of $N(t)$ shown in Figure P.2.3.5*b*.

c. Conversely, suppose $M(u)$ for $u \geqslant 0$ is a homogeneous Poisson process with unit intensity. Show that $N(t) = M(\Lambda(t))$ for $t \geqslant 0$ is an inhomogeneous Poisson counting process with intensity $\lambda(t)$.

2.3.6. Let $\{N_t; t \geqslant 0\}$ be a homogeneous Poisson process. Show that the conditional probability density for the kth occurrence time, w_k, given that

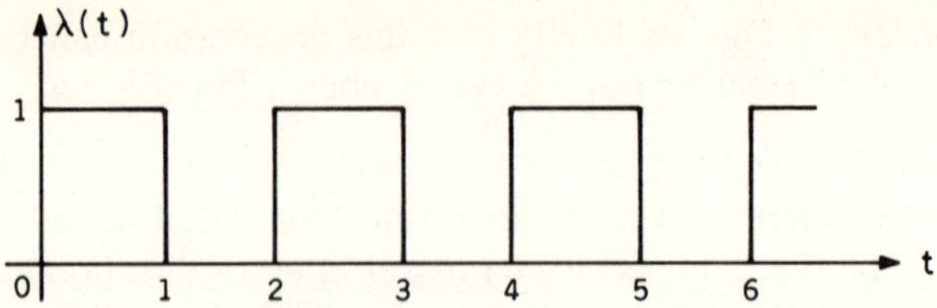

(a)

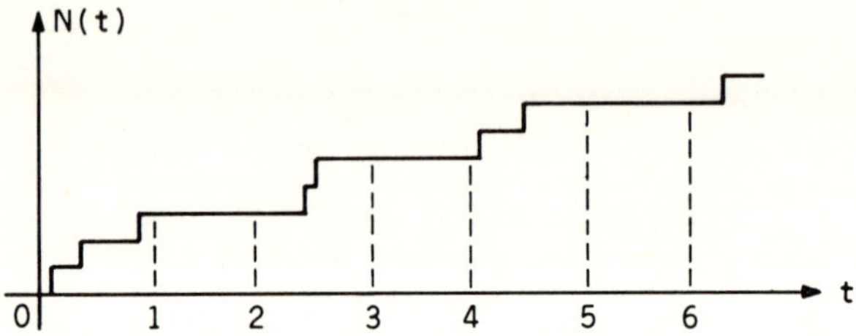

(b)

Figure P.2.3.5. Periodic intensity function for an inhomogeneous poisson process.

$N_T = n \geqslant k$ is given by

$$p_{w_k}(\tau | N_T = n) = n![(k-1)!(n-k)!]^{-1}\left(\frac{\tau}{T}\right)^{k-1}\left[1-\left(\frac{\tau}{T}\right)\right]^{n-k} T^{-1}$$

for $0 \leqslant \tau \leqslant T$. Thus w_k is a beta-distributed random variable given $N_T = n$.

2.3.7. Let $\{N_t; t \geqslant t_0\}$ be an inhomogeneous Poisson process with intensity function λ_t for $t \geqslant t_0$. Define the random variable y by

$$y \triangleq \int_{t_0}^{T} v_\sigma \, dN_\sigma = \begin{cases} 0, & N_T = 0 \\ \sum_{i=1}^{N_T} v_{w_i}, & N_T \geqslant 1. \end{cases}$$

Show the mean and variance of y are given by $\int_{t_0}^{T} v_\sigma \lambda_\sigma \, d\sigma$ and $\int_{t_0}^{T} v_\sigma^2 \lambda_\sigma \, d\sigma$, respectively. Hint: use the conditional properties of the occurrence times given that $N_T = n$.

2.3.8. The superposition of independent Poisson counting processes is a Poisson counting process. Let $\{N_i(t); t \geqslant t_0, i = 1, 2, \ldots, M\}$ be M independent Poisson counting processes with intensity functions $\{\lambda_i(t); t \geqslant t_0\}$. Show that $z(t) \triangleq \sum_{i=1}^{M} N_i(t)$ is a Poisson counting process and determine its intensity.

2.3.9. The difference of two Poisson processes obviously is not a Poisson process. Let $\{N_1(t); t \geq 0\}$ and $\{N_2(t); t \geq 0\}$ be independent, homogeneous Poisson counting processes with constant intensities λ_1 and λ_2, respectively. Let $\{z(t); t \geq 0\}$ be the difference

$$z(t) \triangleq N_1(t) - N_2(t)$$

a. Show z has stationary, independent increments.

b. Evaluate $\Pr[z_t - z_s = k]$ for $0 \leq s \leq t$ and for $k = 0, \pm 1, \pm 2, \ldots$.

c. Evaluate the limit $\lim_{t\to\infty} \Pr[|z(t)| < c]$ for $c > 0$.

2.3.10 (Khinchin Decomposition). Let $\{N_t; t \geq t_0\}$ be a Poisson counting process with parameter function Λ_t for $t \geq t_0$. Here, the function Λ may not possess a derivative so N may not have an intensity function.

a. Show that the characteristic functional for N on the interval $[t_0, T)$ is given by

$$\varphi_N(j\nu) \triangleq E\left[\exp\left(j\int_{t_0}^{T} \nu_\sigma dN_\sigma\right)\right]$$
$$= \exp\left[\int_{t_0}^{T} (e^{j\nu_\sigma} - 1)\, d\Lambda_\sigma\right].$$

b. The function Λ_t for $t \geq t_0$ can be expressed as the sum $\Lambda_t = \Lambda_t^{(c)} + \Lambda_t^{(d)}$, where $\Lambda_t^{(c)}$ is a continuous function of t and $\Lambda_t^{(d)}$ is a piecewise-constant function having only jump discontinuities. Use this decomposition and the characteristic functional of part (a) to conclude that N can be decomposed for $t \geq t_0$ into the sum of two independent Poisson processes as $N_t = N_t^{(c)} + N_t^{(d)}$, where the parameter functions for $N_t^{(c)}$ and $N_t^{(d)}$ are $\Lambda_t^{(c)}$ and $\Lambda_t^{(d)}$, respectively.

2.3.11 (Constructive Derivation of the Poisson Process). Suppose a point process on $[t_0, T)$ is constructed in the following manner. First, select a nonnegative integer N according to the Poisson distribution with parameter $\int_{t_0}^{T} \lambda_\sigma d\sigma$. Then select N numbers $u_1, u_2, \ldots, u_N$ independently over the interval $[t_0, T)$ each with probability density

$$p_{u_i}(U) = \lambda_U \left(\int_{t_0}^{T} \lambda_\sigma d\sigma\right)^{-1}.$$

Assign these numbers as coordinates for N points in $[t_0, T)$. Show that the resulting point process is an inhomogeneous Poisson process on $[t_0, T)$ with intensity λ_t.

2.3.12 (Constructive Derivation of the Poisson Process). The converse to Theorem 2.3.1 is established in this problem. Suppose $\{t_i\}$ is a sequence of independent, identically distributed random variables with the common distribution being exponential with parameter λ. Construct a point process by assigning these random variables successively as interarrival times starting from $t=0$. The conclusion to be drawn from the problem is that the point process so constructed is a homogeneous Poisson process with intensity λ.

a. Let $w_k = t_1 + t_2 + \cdots + t_k$ be the kth occurrence time of the constructed point process for $k=1,2,\ldots$. Show the joint probability density for w_n and w_{n+1} is given by

$$p_{w_n, w_{n+1}}(X, Y) = [(n-1)!]^{-1} \lambda^{n+1} X^{n-1} \exp(-\lambda Y)$$

for $0 \leqslant X \leqslant Y$ and $0 \leqslant Y$.

b. By using the result of part (a) and the identity of the events $\{N_t = n\}$ and $\{w_n < t, w_{n+1} \geqslant t\}$, show that N_t is Poisson distributed with parameter λt.

c. Show that

$$p_{w_1, w_2, \ldots, w_n}(W_1, W_2, \ldots, W_n | N_t = n) = n! \, t^{-n}$$

for $0 \leqslant W_1 \leqslant W_2 \leqslant \cdots \leqslant W_n \leqslant t$. Conclude, therefore, that the occurrence times for the constructed point process have the same distribution given $N_t = n$ as the order statistics of n independent random variables uniformly distributed on $[0, t]$.

d. Combine the results of Problem 2.3.11 and parts (b) and (c) to conclude that the constructed point process is a homogeneous Poisson process with intensity λ.

2.3.13. Let $\tilde{N}_t = N_t - \int_{t_0}^{t} \lambda_\sigma \, d\sigma$ where N is an inhomogeneous Poisson process with intensity function λ_t. Thus, $\tilde{N}$ is a centered (i.e., zero mean) version of N.

a. Show that the characteristic functional for $\tilde{N}$ is

$$\varphi_{\tilde{N}}(jv) = \exp\left[\int_{t_0}^{T} \lambda_\sigma [e^{jv_\sigma} - 1 - jv_\sigma] \, d\sigma \right]$$

b. Let $v_\sigma = \epsilon_1 v_{1\sigma} + \epsilon_2 v_{2\sigma}$ where ϵ_1 and ϵ_2 are constant parameters. Demonstrate for this choice of v that

$$\left.\frac{j^{-2}\partial^2 \varphi_{\tilde{N}}(jv)}{\partial\epsilon_1\,\partial\epsilon_2}\right|_{\substack{\epsilon_1=0\\ \epsilon_2=0}} = E\left\{\int_{t_0}^{T} v_{1\sigma}\,d\tilde{N}_\sigma \int_{t_0}^{T} v_{2\sigma}\,d\tilde{N}_\sigma\right\}$$

$$= \int_{t_0}^{T} v_{1\sigma} v_{2\sigma}\lambda_\sigma\,d\sigma$$

where the first equality can be deduced from the definition of $\varphi_{\tilde{N}}$ and the second can be deduced from the evaluation of $\varphi_{\tilde{N}}$ in part (a).

c. Now for $i=1$ and 2, let $v_{i\sigma}=1$ for $\sigma \in [t_0, t_i)$ and $v_{i\sigma}=0$ otherwise. Use the result of (b) to conclude that

$$E\left[\tilde{N}_{t_1}\tilde{N}_{t_2}\right] = \operatorname{cov}(N_{t_1}, N_{t_2}) = \min\left(\int_{t_0}^{t_1}\lambda_\sigma\,d\sigma, \int_{t_0}^{t_2}\lambda_\sigma\,d\sigma\right)$$

d. By taking $v_\sigma = \sum_{i=1}^{k}\epsilon_i v_{i\sigma}$, use this procedure to show

$$E\left[\tilde{N}_{t_1}\tilde{N}_{t_2}\tilde{N}_{t_3}\right] = \min\left(\int_{t_0}^{t_1}\lambda_\sigma\,d\sigma, \int_{t_0}^{t_2}\lambda_\sigma\,d\sigma \int_{t_0}^{t_3}\lambda_\sigma\,d\sigma\right)$$

and

$$E\left[\tilde{N}_{t_1}\tilde{N}_{t_2}\tilde{N}_{t_3}\tilde{N}_{t_4}\right] = \min\left(\int_{t_0}^{t_1}\lambda_\sigma\,d\sigma, \int_{t_0}^{t_2}\lambda_\sigma\,d\sigma, \int_{t_0}^{t_3}\lambda_\sigma\,d\sigma, \int_{t_0}^{t_4}\lambda_\sigma\,d\sigma\right)$$

$$+\operatorname{cov}(N_{t_1}, N_{t_2})\operatorname{cov}(N_{t_3}, N_{t_4})$$

$$+\operatorname{cov}(N_{t_1}, N_{t_3})\operatorname{cov}(N_{t_2}, N_{t_4})$$

$$+\operatorname{cov}(N_{t_1}, N_{t_4})\operatorname{cov}(N_{t_2}, N_{t_3})$$

Thus, conclude that the covariance function for a homogeneous Poisson counting process is identical to that of a Wiener process, but that this is not so for higher-order covariances.

2.4.1. Reproduced in the table below are data obtained in the historic experiment of Rutherford and Gieger [34] on the emission of α particles from a radioactive source.

k	0	1	2	3	4	5	6	7	8	9	10	11	Total
n_k	57	203	383	525	532	408	273	139	49	27	10	6	2612

In the table, k is the number of α-particles they observed in a unit time interval of duration $T=\frac{1}{8}$ minute, and n_k is the number of such intervals in which k α-particles were observed.

a. Assume α-particles are emitted as a homogeneous Poisson process with an intensity λ emissions per minute. Determine the maximum likelihood estimate $\hat{\lambda}_{\mathrm{ML}}$ of λ given the data observed by Rutherford and Gieger.

b. As a qualitative test of the assumption of Poisson distributed emissions, add two rows to the table. In one row, evaluate and enter $(k!)^{-1}(\hat{\lambda}_{\mathrm{ML}}T)^k\exp(-\hat{\lambda}_{\mathrm{ML}}T)$ for $k=0,1,\ldots$. In the second row enter $n_k(\sum_{k=0}^{11}n_k)^{-1}$, the relative frequency of k α-particles, for $k=0,1,\ldots$.

2.4.2. Consider the scaled inhomogeneous Poisson process of Example 2.4.5 but assume now that count-record data $\{N_\sigma; t_0\leqslant\sigma<T\}$ are observed on the interval $[t_0,T)$ rather than histogram data.

a. Determine the maximum likelihood estimate for X.

b. Demonstrate that the maximum likelihood estimate is efficient and evaluate its mean-square error.

2.4.3. Consider the model of Example 2.4.7, which is used in auditory electrophysiology.

a. Evaluate the mean and variance for the random variables A_c and A_s defined in (2.73). Hint: use the conditional properties of the occurrence times given N_T that are developed in Section 2.3.

b. Suppose $\mathbf{X}^*$ is an unbiased estimate of $\mathbf{X}$. Show that $E[(\mathbf{X}^*-\mathbf{X})(\mathbf{X}^*-\mathbf{X})'|\mathbf{X}]\geqslant\mathbf{F}^{-1}(\mathbf{X})$, where

$$\mathbf{F}(\mathbf{X})=T\begin{bmatrix} X_1^{-1}I_0(X_2) & I_1(X_2) & 0 \\ I_1(X_2) & \frac{1}{2}X_1[I_0(X_2)+I_2(X_2)] & 0 \\ 0 & 0 & \frac{1}{2}X_1X_2^2[I_0(X_2)-I_2(X_2)] \end{bmatrix},$$

where $I_n(X_2)$ is a modified Bessel function of the first kind of order n.

2.4.4. Suppose $\mathbf{X}^*$ is an unbiased estimate of $\mathbf{X}$ in terms of either histogram or count-record data. The mean square-error matrix then satisfies $\mathbf{\Sigma}(\mathbf{X})\geqslant\mathbf{F}^{-1}(\mathbf{X})$. Suppose now that the experiment of collecting data is repeated independently M times. Show that the mean square-error matrix for an unbiased estimate $\mathbf{X}_M^*$ of $\mathbf{X}$ in terms of the pooled data from the M experiments is lower bounded by $(1/M)\mathbf{F}^{-1}(\mathbf{X})$.

2.4.5. Let $\{N_t; t > -\infty\}$ be an inhomogeneous Poisson process with intensity $\lambda_t(X) = \mu(t - X)$, where $\mu(t)$ is the Gaussian pulse

$$\mu(t) = \frac{\alpha}{\sqrt{2\pi\tau}} \exp\left(-\frac{t^2}{2\tau}\right), \qquad -\infty < t < \infty.$$

Suppose the number N and occurrence times $(w_1, w_2, \ldots, w_N)$ for all points in $(-\infty, \infty)$ are observed.

a. Show that the maximum likelihood estimate of X is given by

$$\hat{X}_{\mathrm{ML}} = \frac{1}{N} \sum_{i=1}^{N} w_i.$$

Thus conclude that the maximum likelihood estimate of the pulse delay X is the arithmetic average of the observed occurrence times.

b. Determine the mean and variance of $\hat{X}_{\mathrm{ML}}$.

c. Is $\hat{X}_{\mathrm{ML}}$ unbiased? Is it efficient?

d. Evaluate $F(X)$, the Fisher information matrix.

2.4.6. At time $t = 0$, a quantity of radioactive material is placed in the field of view of an ideal detector and the number N and occurrence times $(w_1, w_2, \ldots, w_N)$ of detected photons are measured for all $t \geqslant 0$. Assume as a first approximation that detected photons form an inhomogeneous Poisson process on $[0, \infty)$ with a parametrized intensity $\lambda_t(\mathbf{X}) = X_1 \exp(-t/X_2)$, where X_1 depends on the quantity of material and the detector sensitivity and X_2 is a physical constant characteristic of the material.

a. Determine expressions for the maximum likelihood estimates of X_1 and X_2.

b. Determine lower bounds on the mean-square errors in estimating X_1 and X_2 using unbiased estimates.

R. Peierls [30] was evidently first to develop the expression of part (a). His interesting paper also contains material on the effect of a finite observation interval $[0, T)$, on the selection of T in repeated experiments, and on certain nonideal effects found in real detectors.

2.4.7 (Barankin Bound for Histogram Data). The Cramér–Rao inequality of Theorem 2.4.1 provides a lower bound on the mean square error. An alternative inequality developed by E. W. Barankin [4] provides the *greatest* lower bound on the mean square error. In this problem, we develop the

Barankin bound on the mean square error in estimating a single parameter using an unbiased estimate and histogram data. The method can be extended to count-record data and multiple parameters.

Let $\{N_t; t \geqslant t_0\}$ be an inhomogeneous Poisson process with intensity $\lambda_t(X)$. Observations taken to estimate X are in the form of counts N_{t_{i-1},t_i} occurring in k subintervals $[t_{i-1}, t_i)$ of an observation interval $[t_0, T)$, where $t_k = T$. Denote the counting probability in (2.65) by $P(n_1, n_2, \ldots, n_k | X)$. Let $X^* = X^*(n_1, n_2, \ldots, n_k)$ be an unbiased estimate of X in terms of the histogram data. Then

$$\sum_{n_1, n_2, \ldots, n_k} X^* P(n_1, n_2, \ldots, n_k | X) = X$$

for all $X \in \mathfrak{X}$, where $\mathfrak{X}$ denotes the set of possible parameter values. Now choose M values $X_1, X_2, \ldots, X_M$ for X, where for each m, $X_m \in \mathfrak{X}$. Also choose M numbers $\alpha_1, \alpha_2, \ldots, \alpha_M$. It follows that

$$\sum_{n_1, n_2, \ldots, n_k} X^* \left[\sum_{m=1}^{M} \alpha_m P(n_1, n_2, \ldots, n_k | X_m) \right] = \sum_{m=1}^{M} \alpha_m X_m$$

a. Let X_a denote the actual value of X. By subtracting $\sum_{m=1}^{M} \alpha_m X_a$ from both sides of the last expression, conclude that

$$\sum_{n_1, n_2, \ldots, n_k} (X^* - X_a) \left[\frac{\sum_{m=1}^{M} \alpha_m P(n_1, n_2, \ldots, n_k | X_m)}{P(n_1, n_2, \ldots, n_k | X_a)} \right] P(n_1, n_2, \ldots, n_k | X_a)$$

$$= \sum_{m=1}^{M} \alpha_m (X_m - X_a).$$

b. Using the result of part (a) and the Schwarz inequality, show that

$$E\left[(X^* - X_a)^2 | X_a \right] \geqslant \frac{\left[\sum_{m=1}^{M} \alpha_m (X_m - X_a) \right]^2}{\sum_{l=1}^{M} \sum_{m=1}^{M} \alpha_l G(X_l, X_m | X_a) \alpha_m}$$

where

$$G[X_l, X_m | X_a] \triangleq \sum_{n_1, n_2, \ldots, n_k} \frac{P(n_1, n_2, \ldots, n_k | X_l) P(n_1, n_2, \ldots, n_k | X_m)}{P(n_1, n_2, \ldots, n_k | X_a)}$$

This inequality holds for any choice of $X_1, X_2, \ldots, X_m$ and $\alpha_1, \alpha_2, \ldots, \alpha_M$. Consequently, the tightest bound is obtained by selecting these $2M$ parameters to maximize the right side. The result is the Barankin bound, which has been shown by Barankin to be the greatest lower bound.

c. Show that

$$\ln G[X_l, X_m | X_a] = -\int_{t_0}^{T} [\lambda_\sigma(X_l) + \lambda_\sigma(X_m) - \lambda_\sigma(X_a)]\, d\sigma$$
$$+ \sum_{i=1}^{k} \left[\int_{t_{i-1}}^{t_i} \lambda_\sigma(X_l)\, d\sigma \right] \left[\int_{t_{i-1}}^{t_i} \lambda_\sigma(X_m)\, d\sigma \right] \left[\int_{t_{i-1}}^{t_i} \lambda_\sigma(X_a)\, d\sigma \right]^{-1}$$

d. The Cramér–Rao bound for an unbiased estimate can be obtained as a special case of the Barankin bound by selecting $M=2$, $X_1 = X_a$, $X_2 = X_a + \delta$, $\alpha_1 = -\delta^{-1}$, and $\alpha_2 = \delta^{-1}$. Obtain the Cramér–Rao bound in this way by taking the limit as δ approaches zero.

2.5.1 (Binary Phase-Shift Keying). An optical communication system capable of accommodating a one megabit-per-second source of binary information is described by G. Lee, R. Fluchel, E. Paddon, and D. C. Torretta [23], H. Hance [15], and G. Herrmann [17]. In this system, the output of the information source modulates the phase of a high-frequency subcarrier. The modulated subcarrier then modulates a narrow beam of laser light that is pointed at the receiver. In order to decide the state, H_0 or H_1, of the information source, the receiver intercepts the light beam with a photodetector and observes the stream of electrons that emanate from it. Assume that electrons emanate according to a Poisson process with intensity $\lambda_t^{(i)}$ on hypothesis H_i, for $i=0$ and $i=1$, where

$$\lambda_t^{(i)} = A[1 + \alpha \sin(\omega t + \theta_i)],$$

in which $\theta_i = 0$ on H_0 and $\theta_i = \pi$ on H_1, and where A and α are constants such that $A > 0$ and $|\alpha| < 1$. Count-record data are collected during the interval $[0, T)$ to decide between H_0 and H_1. For simplicity, assume this interval contains an integral number of cycles of the subcarrier; that is, $\omega T = 2\pi k$ for some integer k. Source letters occur with equal probability.

a. Show the strategy that minimizes the probability of error given $\{N_\sigma; 0 \leqslant \sigma < T\}$ is defined by

$$\int_0^T \ln[1 + \alpha \sin(\omega\sigma)]\, dN_\sigma \underset{H_0}{\overset{H_1}{\gtrless}} \int_0^T \ln[1 - \alpha \sin(\omega\sigma)]\, dN_\sigma.$$

b. Let $P_{\min}(\text{error})$ denote the minimum probability of error that results by using the strategy of (a). Show

$$P_{\min}(\text{error}) \leqslant \exp\left\{-AT + \int_0^T A\sqrt{1-[\alpha\sin(\omega\sigma)]^2}\,d\sigma\right\}.$$

c. Conclude from (a) that if $\alpha \ll 1$, then a nearly equivalent strategy is given by

$$l \triangleq \int_0^T \sin(\omega\sigma)\,dN_\sigma \underset{H_0}{\overset{H_1}{\gtrless}} 0.$$

This suboptimal strategy is the one proposed in the one-megabit optical communication system. Let $\mu_l(s) = \ln E[\exp(sl)|H_0]$. Show that

$$\mu_l(s) = -AT + AT[I_0(s) + \alpha I_1(s)],$$

where $I_k(s)$ is a modified Bessel function of the first kind of order k.

d. Let $P(\text{error})$ denote the probability of error that results by using the suboptimal strategy of (c). Show

$$P(\text{error}) \leqslant \exp[\mu_l(s_m)],$$

where s_m is the value of s satisfying

$$I_0(s) + \left(\frac{2}{\alpha}\right)I_1(s) + I_2(s) = 0.$$

2.5.2. A number N of electrons is observed to emanate from a photodetector. On hypothesis H_0

$$\Pr(N=n|H_0) = (\alpha_0)^n(1+\alpha_0)^{-n-1}, \qquad n = 0,1,2,\ldots,$$

and on hypothesis H_1

$$\Pr(N=n|H_1) = (\alpha_1)^n(1+\alpha_1)^{-n-1}, \qquad n = 0,1,2,\ldots.$$

Thus, N has the Bose–Einstein distribution on each hypothesis. We will see in Chapter 6 that this distribution arises when incoherent light, such as from an incandescent lamp or astronomical object, is incident on the photodetector. Let P_0 and P_1 denote the prior probabilities of H_0 and H_1, respectively, and assume a cost matrix $\mathbf{C}$ is specified. The decision is to be made according to the Bayes' criterion.

a. Specify the test for deciding between H_0 and H_1 given an observation of N.

b. Show that the false-alarm and miss probabilities satisfy the Chernoff bounds in (2.120) and (2.121) and evaluate $\mu(s)$.

2.5.3. Let $\{N_t; t \geqslant t_0\}$ be a Poisson process with intensity $m\mu_t$ on hypothesis H_m for $m=1,2,\ldots,M$. Each hypothesis occurs with equal probability and count-record data are collected on the interval $[t_0, T)$ to decide between them.

a. Specify the test for deciding between $H_1, H_2, \ldots,$ or H_M with minimum probability of error given $\{N_\sigma; t_0 \leqslant \sigma < T\}$. Show that the number N_T of points in $[t_0, T)$ is a sufficient statistic for making the decision.

b. Indicate how to evaluate the minimum probability of error.

MARKED POINT PROCESSES: THE COMPOUND POISSON PROCESS

CHAPTER THREE

3.1 INTRODUCTION

Motivation for the model we develop in this chapter is provided by the low-frequency atmospheric-noise data shown in Figure 1.2. It is evident that a point process model can account for the pulse occurrence times. However, these times alone do not reflect all the significant features. In particular, the pulse amplitudes exhibit wide variation and have a strong influence on a radio receiver operating at low frequencies. Even a first-approximation model for low-frequency atmospheric noise should, therefore, include the amplitude as well as occurrence time of each pulse. It is this procedure of endowing each point with an ancillary variable, an amplitude in this instance, which characterizes the models of this chapter. More than one variable may be associated with each point. For example, both the velocity and acceleration of a vehicle passing an observer at a given instant might be important in a particular application.

A *marked point process* is a point process with an auxiliary variable, called a mark, associated with each point. These processes were introduced by D. König and K. Matthes [3] and K. Matthes [4,5]. As in the above example, the mark is used to identify random quantities associated with the point it accompanies.

The notation we use for a marked point process is indicated in Figure 3.1. Each occurrence time w_i in the basic point process has associated with it a mark $\mathbf{u}_i$ having random values in a mark space $\mathfrak{U}$. In our development, the mark space $\mathfrak{U}$ will be one of the following:

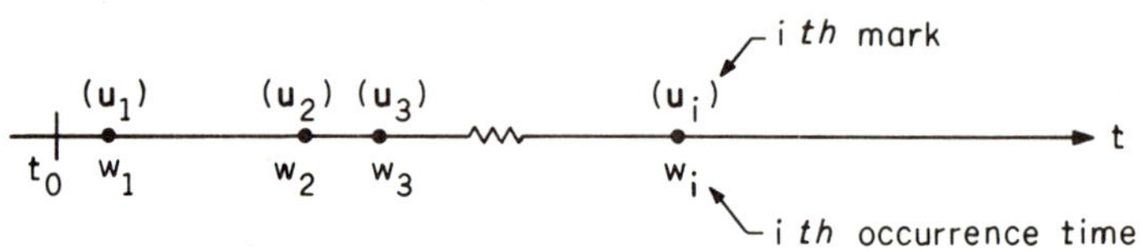

Figure 3.1. A marked point process.

a. A denumerable collection $\{\mathbf{U}_1, \mathbf{U}_2, \ldots, \mathbf{U}_n, \ldots\}$, where each $\mathbf{U}_n$ is a finite-dimensional vector of real numbers. A typical example is the collection $\{1, 2, \ldots, n, \ldots\}$ of positive integers. Here, the mark n on the ith point can indicate that n events occur simultaneously at the instant w_i. Certain point processes that fail to satisfy the conditions of orderliness can be included in this way as marked point processes with integer-valued marks.

b. A region of a finite-dimensional Euclidean space. A typical example is a semiinfinite real line representing the energies that may be released in a lightning discharge. Another example is a region in two dimensions representing possible velocities and accelerations of vehicles passing an observer.

More abstract mark spaces can be included, but such generality will not be needed here.

There is an analog of a counting process for marked point processes. We term this a *mark accumulator process*. It is defined as follows. Let $\{N_t; t \geqslant t_0\}$ be a counting process indicating the total number of points in the interval $[t_0, t)$ regardless of their mark. The mark accumulator process $\{\mathbf{x}_t; t \geqslant t_0\}$ is defined by

$$\mathbf{x}_t = \sum_{i=0}^{N_t} \mathbf{u}_i, \tag{3.1}$$

where we identify $\mathbf{u}_0 = \mathbf{0}$. Thus, $\mathbf{x}_t$ is the vector sum of the marks on all points occurring in $[t_0, t)$. The paths of $\mathbf{x}$ are piecewise constant with randomly occurring jumps. An example of a marked point process, its associated counting process N, and accumulator process x is shown in Figure 3.2; the mark space is the real line. It is evident that a marked point process and its accumulator process are equivalent whenever $\mathbf{0}$ is not an element of $\mathfrak{U}$.

In the general theory developed by K. Matthes and his colleagues, the marks $\{\mathbf{u}_i\}$ need not form an independent sequence of random variables. Nor is it required that the marks be independent of the counting process $\{N_t; t \geqslant t_0\}$ or the occurrence-time sequence $\{w_i\}$. Thus, general marked point processes encompass a very rich class of interesting and useful models. These are studied further in Chapter 7. Our objective in this chapter is more limited. We shall study a particular marked point process satisfying the following conditions:

i. $\{N_t; t \geqslant t_0\}$ is an inhomogeneous Poisson counting process with intensity function λ_t for $t \geqslant t_0$;
ii. $\{\mathbf{u}_i\}$ is a sequence of mutually independent, identically distributed random variables which are also independent of $\{N_t; t \geqslant t_0\}$.

A marked point process satisfying (i) and (ii) is termed a *compound Poisson process*. We use this same terminology for the associated mark accumulator process $\{\mathbf{x}_t; t \geqslant t_0\}$. The independence assumptions in (ii) are used repeatedly throughout the chapter. They imply that $\{\mathbf{x}_t; t \geqslant t_0\}$ has independent increments.

Our discussion of compound Poisson processes is divided into several topics. Statistics for a compound Poisson process are discussed in Section 3.2. The characteristic functional, which completely characterizes the pro-

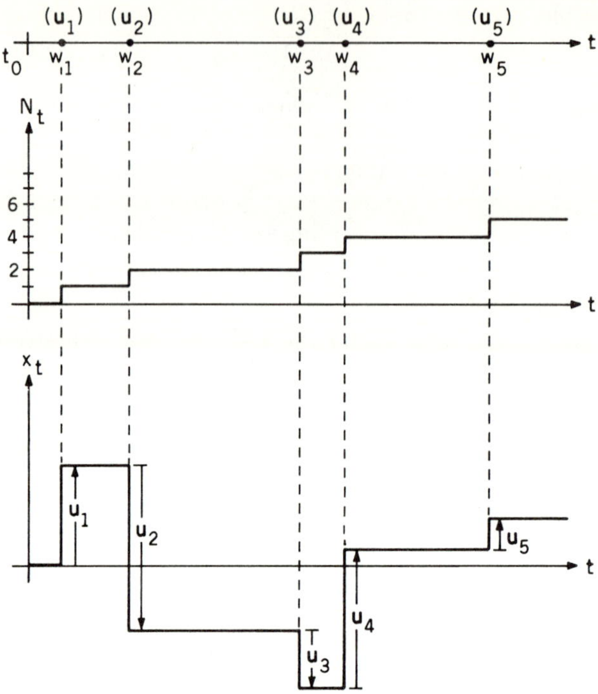

Figure 3.2. A marked point process and its mark accumulator process.

cess statistically, is derived, and some applications for it are given. The first- and second-order statistics are derived. In Section 3.3, we give some representations for a compound Poisson process in terms of sums of independent Poisson processes. These representations are useful not only for the insight they provide into the structure of the compound Poisson process but also in many applications where this process is used as a model. The sample-function density for a compound Poisson process is also derived. Its interpretation and use are entirely analogous to those of Chapter 2 for the sample-function density of a Poisson process. Problems of statistical inference for observed compound Poisson processes are examined as an application of the representations and the sample-function density.

3.2 STATISTICS OF A COMPOUND POISSON PROCESS

Let $\{\mathbf{x}_t; t \geqslant t_0\}$ be a compound Poisson process. Thus, $\mathbf{x}_t$ is given by (3.1), where $\{N_t; t \geqslant t_0\}$ and $\{\mathbf{u}_i\}$ satisfy the above properties (i) and (ii). In this section, we determine the statistics of $\{\mathbf{x}_t; t \geqslant t_0\}$. These include the characteristic functional associated with the random variables $\{\mathbf{x}_\sigma; t_0 \leqslant \sigma < T\}$ and the first- and second-order statistics of $\{\mathbf{x}_t; t \geqslant t_0\}$.

Characteristic Functional for a Compound Poisson Process

The characteristic functional for a Poisson process is defined in (2.55). We now generalize this to include a sequence of independent random variables assigned as marks to the basic point process. The characteristic functional associated with the random variables $\{\mathbf{x}_\sigma; t_0 \leqslant \sigma < T\}$ is defined by the expectation

$$\varphi_{\mathbf{x}}(j\mathbf{v}) = E\left[\exp\left(j\int_{t_0}^{T} \mathbf{v}'_\sigma \, d\mathbf{x}_\sigma\right)\right], \tag{3.2}$$

where the prime denotes the transpose operation, and where $\mathbf{v}_\sigma$ is an arbitrary vector-valued function of σ on $[t_0, T)$. The integral can be interpreted for now as a Riemann–Stieltjes integral having the evaluation

$$\int_{t_0}^{T} \mathbf{v}'_\sigma \, d\mathbf{x}_\sigma = \begin{cases} 0, & N_T = 0 \\ \sum_{i=1}^{N_T} \mathbf{v}'_{w_i} \mathbf{u}_i, & N_T \geqslant 1. \end{cases}$$

Integrals of this type are defined more carefully in Chapter 4. The expectation in (3.2) is easy to evaluate using: the properties of conditional expectation; the property of a Poisson process that the occurrence times have the same distribution given N_T as the order statistics of N_T independent, identically distributed random variables, the common distribution being that in (2.53); and the independence of the marks $\{\mathbf{u}_i\}$. Thus,

$$\varphi_{\mathbf{x}}(j\mathbf{v}) = \sum_{n=0}^{\infty} \Pr(N_T = n) E\left[\exp\left(j\int_{t_0}^{T} \mathbf{v}'_\sigma \, d\mathbf{x}_\sigma\right) \Big| N_T = n\right]$$

$$= \Pr(N_T = 0) + \sum_{n=1}^{\infty} \Pr(N_T = n) E\left[\exp\left(j\sum_{i=1}^{N_T} \mathbf{v}'_{w_i} \mathbf{u}_i\right) \Big| N_T = n\right]. \tag{3.3}$$

By conditioning on the occurrence times and using both the independence and identically distributed properties of the marks, we obtain

$$E\left[\exp\left(j\sum_{i=1}^{N_T}\mathbf{v}'_{w_i}\mathbf{u}_i\right)\Big|N_T=n\right]$$

$$=E\left\{E\left[\exp\left(j\sum_{i=1}^{N_T}\mathbf{v}'_{w_i}\mathbf{u}_i\right)\Big|N_T=n;w_1,w_2,\ldots,w_{N_T}\right]\Big|N_T=n\right\}$$

$$=E\left\{\prod_{i=1}^{N_T}M_{\mathbf{u}}(j\mathbf{v}_{w_i})\Big|N_T=n\right\}, \tag{3.4}$$

where $M_{\mathbf{u}}(j\mathbf{v}_{w_i})$ is the common characteristic function

$$M_{\mathbf{u}}(j\boldsymbol{\alpha})\triangleq E[\exp(j\boldsymbol{\alpha}'\mathbf{u}_i)], \tag{3.5}$$

for the mark random variables evaluated at $\boldsymbol{\alpha}=\mathbf{v}_{w_i}$. We now parallel the argument leading from (2.56) to (2.57) to obtain

$$E\left[\exp\left(j\sum_{i=1}^{N_T}\mathbf{v}'_{w_i}\mathbf{u}_i\right)\Big|N_T=n\right]=\left[\left(\int_{t_0}^{T}\lambda_\sigma\,d\sigma\right)^{-1}\int_{t_0}^{T}\lambda_\sigma M_{\mathbf{u}}(j\mathbf{v}_\sigma)\,d\sigma\right]^n. \tag{3.6}$$

Upon substituting this expression into (3.3) and using the Poisson distribution for N_T, we conclude that the characteristic functional for a compound Poisson process is given by

$$\varphi_{\mathbf{x}}(j\mathbf{v})=\exp\left[\int_{t_0}^{T}\lambda_\sigma(M_{\mathbf{u}}(j\mathbf{v}_\sigma)-1)\,d\sigma\right]. \tag{3.7}$$

We now illustrate the use of the characteristic functional of a compound Poisson process with several examples. These follow from particular selections of the function $\mathbf{v}$ and of the mark space $\mathfrak{U}$.

Example 3.2.1 (Characteristic Function). The characteristic function can be obtained by specialization of the characteristic functional. For illustration of this, take the function $\mathbf{v}$ in (3.2) to be

$$\mathbf{v}_\sigma=\begin{cases}\mathbf{0}, & t_0\leqslant\sigma<s\\ \boldsymbol{\alpha}, & s\leqslant\sigma<t\\ \mathbf{0}, & t\leqslant\sigma<T.\end{cases}$$

Then,

$$\int_{t_0}^{T} \mathbf{v}'_\sigma \, d\mathbf{x}_\sigma = \boldsymbol{\alpha}'(\mathbf{x}_t - \mathbf{x}_s),$$

and the characteristic functional (3.2) becomes the characteristic function for the increment $\mathbf{x}_t - \mathbf{x}_s$. Equation (3.7) becomes

$$\begin{aligned} \varphi_{\mathbf{x}}(j\mathbf{v}) &= M_{\mathbf{x}_t - \mathbf{x}_s}(j\boldsymbol{\alpha}) \\ &= \exp\left[\int_s^t \lambda_\sigma \, d\sigma \left(M_{\mathbf{u}}(j\boldsymbol{\alpha}) - 1 \right) \right]. \end{aligned} \tag{3.8}$$

The characteristic function for $\mathbf{x}_t$ is obtained upon setting $s = t_0$. Of course, the moments of $\mathbf{x}_t - \mathbf{x}_s$ can be determined by differentiating (3.8) and then setting $\boldsymbol{\alpha} = \mathbf{0}$. This is pursued below. It is generally difficult analytically to invert (3.8) to determine the probability distribution function for $\mathbf{x}_t - \mathbf{x}_s$ or $\mathbf{x}_t$. The Chernoff bound developed in Theorem 2.5.1 can sometimes be useful for bounding this probability when it cannot be determined analytically; see Problem 3.2.2.

Example 3.2.2 (Independent Increments). A compound Poisson process has independent increments. This is evident from its construction and can be rigorously established by specialization of the characteristic functional (3.7). For this purpose, let $[s_1, t_1)$, $[s_2, t_2), \ldots,$ and $[s_k, t_k)$ be a specified collection of k disjoint time intervals, and set

$$\mathbf{v}_\sigma = \begin{cases} \boldsymbol{\alpha}_k, & s_k \leqslant \sigma < t_k \\ 0, & \text{otherwise} \end{cases}$$

where $\boldsymbol{\alpha}_1$, $\boldsymbol{\alpha}_2, \ldots,$ and $\boldsymbol{\alpha}_k$ are given constants. This choice for $\mathbf{v}$ is a vector analog of that shown in Figure 2.5. Then (3.7) becomes

$$\varphi_{\mathbf{x}}(j\mathbf{v}) = \prod_{i=1}^{k} \exp\left[\int_{s_i}^{t_i} \lambda_\sigma \, d\sigma \left(M_{\mathbf{u}}(j\boldsymbol{\alpha}_i) - 1 \right) \right]. \tag{3.9}$$

which demonstrates that the compound Poisson process $\mathbf{x}$ has independent increments.

Example 3.2.3 (Poisson Process with Non-Unit Jumps). A Poisson counting process has unit jumps and can be viewed as a special case of a compound Poisson process when the mark space consists of a single point

$\mathfrak{U}=\{1\}$. In this case, $M_u(j\alpha)=\exp(j\alpha)$, and (3.7) reduces to the characteristic functional of a Poisson counting process (2.58). A Poisson process having all jumps of size $\mathbf{U}$ occurs when the mark space consists of the single point $\mathbf{U}$. For this process with non-unit jumps, we have

$$\varphi_{\mathbf{x}}(j\mathbf{v})=\exp\left[\int_{t_0}^{T}\lambda_\sigma(e^{j\mathbf{v}'_\sigma\mathbf{U}}-1)\,d\sigma\right]. \tag{3.10}$$

An alternative way to construct this process is to identify it with $\mathbf{U}N_t(\mathbf{U})$, where $\{N_t(\mathbf{U}); t\geq t_0\}$ is a Poisson counting process that counts jumps of size $\mathbf{U}$ in $\mathbf{x}$. This alternative construction leads to the representations in Section 3.3 for compound Poisson processes.

Example 3.2.4 (Denumerable Mark Space: Generalized Poisson Counting Process). Suppose the mark space is denumerable $\mathfrak{U}=\{\mathbf{U}_1,\mathbf{U}_2,\ldots,\mathbf{U}_k,\ldots\}$ and that the probability a given mark has value $\mathbf{U}_k$ is p_k for $k=1, 2,\ldots$. Then

$$M_{\mathbf{u}}(j\boldsymbol{\alpha})=\sum_{k=1}^{\infty}p_k\exp(j\boldsymbol{\alpha}'\mathbf{U}_k), \tag{3.11}$$

and (3.7) becomes

$$\varphi_{\mathbf{x}}(j\mathbf{v})=\exp\left[\int_{t_0}^{T}\lambda_\sigma\left(\sum_{k=1}^{\infty}p_k e^{j\mathbf{v}'_\sigma\mathbf{U}_k}-1\right)d\sigma\right]. \tag{3.12}$$

Let us examine two special cases of this. For the first, take $\mathfrak{U}=\{0,1\}$. The resulting compound Poisson process can model a defective counter in which points are missed and not counted with probability p. The mark 0 labels these deleted points. The mark accumulator process counts the number of points with unit mark. We see from (3.12) that

$$\varphi_x(jv)=\exp\left[\int_{t_0}^{T}(1-p)\lambda_\sigma(e^{jv_\sigma}-1)\,d\sigma\right]. \tag{3.13}$$

We conclude from this and (2.58) that the compound Poisson process resulting from a random deletion of points of a Poisson process with intensity λ_t is, in fact, a Poisson process with intensity $(1-p)\lambda_t$.

For the second special case, take $\mathfrak{U}=\{1,2,\ldots,k,\ldots\}$ so that marks are positive integers. The corresponding compound Poisson process is called a *generalized Poisson counting process* and is used as a model when points can occur simultaneously. Here, k points can occur at the same instant with probability p_k. In this way, the orderliness restriction of Chapter 2 is relaxed. We have

$$M_u(j\alpha)=\sum_{k=1}^{\infty} p_k \exp(j\alpha k), \tag{3.14}$$

and the characteristic functional for the mark accumulator process, which counts the total number of points including those that occur simultaneously, is given by

$$\varphi_x(jv)=\exp\left[\int_{t_0}^{T}\lambda_\sigma\left(\sum_{k=1}^{\infty} p_k e^{jv_\sigma k}-1\right)d\sigma\right]. \tag{3.15}$$

The characteristic function for x_t, the number of points in $[t_0,t)$, is

$$M_{x_t}(j\alpha)=\exp\left[\int_{t_0}^{t}\lambda_\sigma d\sigma\left(\sum_{k=1}^{\infty} p_k e^{j\alpha k}-1\right)\right]. \tag{3.16}$$

The characteristic function of (3.16) is usually difficult to invert to obtain an explicit expression for the counting distribution of x_t. Of course, this inversion is trivial for the special case of a Poisson counting process when $p_k=\delta_{1k}$, where δ_{1k} is zero unless $k=1$ and $\delta_{11}=1$. Another example where this inversion is possible, is obtained when $p_k=(-k\ln p)^{-1}(1-p)^k$ for $0<p<1$ and $k=1, 2,\ldots$. It is then straightforward to show that

$$M_{x_t}(j\alpha)=\left(\frac{p}{1-(1-p)\exp(j\alpha)}\right)^{\beta_t}, \tag{3.17}$$

where $\beta_t=-\int_{t_0}^{t}(-\ln p)^{-1}\lambda_\sigma\, d\sigma$; see Problem 3.2.4. This is the characteristic function for the negative binomial distribution. Consequently,

$$\Pr(x_t=n)=\frac{\Gamma(\beta_t+n)}{n!\,\Gamma(\beta_t)}p^{\beta_t}(1-p)^n \tag{3.18}$$

for $n=0,1,2,\ldots,$ where $\Gamma(\cdot)$ is the gamma function.

First- and Second-Order Statistics for a Compound Poisson Process

The first and second moments of a compound Poisson process $\{\mathbf{x}_t; t\geq t_0\}$ are easily determined using the moment generating properties of the characteristic function (3.8), namely,

$$E(\mathbf{x}_t)=\left.\frac{j^{-1}\partial M_{\mathbf{x}_t}(j\boldsymbol{\alpha})}{\partial\boldsymbol{\alpha}}\right|_{\boldsymbol{\alpha}=\mathbf{0}}$$

and

$$E(\mathbf{x}_t\mathbf{x}_t')=\left.\frac{j^{-2}\partial^2 M_{\mathbf{x}_t}(j\boldsymbol{\alpha})}{\partial\boldsymbol{\alpha}^2}\right|_{\boldsymbol{\alpha}=\mathbf{0}}$$

where $\partial M/\partial\boldsymbol{\alpha}$ denotes the gradient vector of M with respect to $\boldsymbol{\alpha}$ (the ith element is $\partial M/\partial\alpha_i$) $\partial^2 M/\partial\boldsymbol{\alpha}^2$ denotes the Hessian matrix of M with respect to $\boldsymbol{\alpha}$ (the i, jth element is $\partial^2 M/\partial\alpha_i\partial\alpha_j$), and the gradient and Hessian are evaluated at $\boldsymbol{\alpha}=\mathbf{0}$. By direct calculation using these relations and (3.8), we deduce that

$$E(\mathbf{x}_t)=E(\mathbf{u})\int_{t_0}^{t}\lambda_\sigma\,d\sigma \tag{3.19}$$

and

$$\begin{aligned}\boldsymbol{\Sigma}_t &\triangleq E(\mathbf{x}_t\mathbf{x}_t')-E(\mathbf{x}_t)E(\mathbf{x}_t')\\ &=E(\mathbf{u}\mathbf{u}')\int_{t_0}^{t}\lambda_\sigma\,d\sigma.\end{aligned} \tag{3.20}$$

We term $\boldsymbol{\Sigma}_t$ the *instantaneous covariance matrix* of $\mathbf{x}_t$. Thus, the expected value of $\mathbf{x}_t$ is the product of the mean value of the marks and the mean number of points in $[t_0,t)$. Similarly, the instantaneous covariance matrix of $\mathbf{x}_t$ is the product of the mean square value of the marks and the variance of the number of points in $[t_0,t)$.

The covariance matrix

$$\mathbf{K}_{\mathbf{x}}(s,t)\triangleq E(\mathbf{x}_s\mathbf{x}_t')-E(\mathbf{x}_s)E(\mathbf{x}_t') \tag{3.21}$$

is easily evaluated using the independent increments property of the compound Poisson process. Thus, for $t_0\leqslant s\leqslant t$, we have

$$\mathbf{K}_{\mathbf{x}}(s,t)=E[\mathbf{x}_s(\mathbf{x}_t-\mathbf{x}_s)']+E(\mathbf{x}_s\mathbf{x}_s')-E(\mathbf{x}_s)E(\mathbf{x}_t')=\boldsymbol{\Sigma}_s. \tag{3.22}$$

As $\mathbf{K}_{\mathbf{x}}(s,t)=\mathbf{K}_x'(t,s)$, we obtain for $t_0\leqslant t\leqslant s$ that

$$\mathbf{K}_{\mathbf{x}}(s,t)=\boldsymbol{\Sigma}_t.$$

Consequently, by putting these results together and using the fact that $\int_{t_0}^{t} \lambda_\sigma \, d\sigma$ is a nondecreasing function of t, we have

$$\mathbf{K}_{\mathbf{x}}(s,t) = \boldsymbol{\Sigma}_{\min(s,t)} = E(\mathbf{uu}') \min\left(\int_{t_0}^{s} \lambda_\sigma \, d\sigma, \int_{t_0}^{t} \lambda_\sigma \, d\sigma\right). \tag{3.23}$$

The first- and second-moment statistics for an inhomogeneous and a homogeneous Poisson process follow immediately from those of a compound Poisson process by specializing the mark space. These are summarized in Table 3.1.

Table 3.1. First- and Second-Order Statistics for Poisson Processes

	Type of Poisson Process		
	Compound	Inhomogeneous	Homogeneous
Mean, $E(\mathbf{x}_t)$	$E(\mathbf{u}) \int_{t_0}^{t} \lambda_\sigma \, d\sigma$	$\int_{t_0}^{t} \lambda_\sigma \, d\sigma$	$\lambda(t - t_0)$
Instantaneous Covariance, $\boldsymbol{\Sigma}_t$	$E(\mathbf{uu}') \int_{t_0}^{t} \lambda_\sigma \, d\sigma$	$\int_{t_0}^{t} \lambda_\sigma \, d\sigma$	$\lambda(t - t_0)$
Covariance, $\mathbf{K}_{\mathbf{x}}(s,t)$	$E(\mathbf{uu}') \min(\int_{t_0}^{s} \lambda_\sigma \, d\sigma, \int_{t_0}^{t} \lambda_\sigma \, d\sigma)$	$\min(\int_{t_0}^{s} \lambda_\sigma \, d\sigma, \int_{t_0}^{t} \lambda_\sigma \, d\sigma)$	$\lambda \min(s - t_0, t - t_0)$

Summary. The statistics of a compound Poisson process $\{\mathbf{x}_t; t \geqslant t_0\}$ have been developed in this section. The results are:

i. The characteristic functional completely characterizes a random process statistically. The evaluation of this important quantity is given in (3.2) for a compound Poisson process.

ii. The characteristic function for $\mathbf{x}_t$ is given in (3.8). The mean and instantaneous covariance matrix for $\mathbf{x}_t$ follow directly from this and are given in (3.19) and (3.20), respectively.

iii. The probability distribution function for $\mathbf{x}_t$ is generally difficult to determine. A Chernoff-type bound on it is developed in Problem 3.2.2, and a series expression is given in Problem 3.2.3.

iv. The characteristic functional for a generalized Poisson process is given in (3.15). This process relaxes the orderliness restriction of Chapter 2 in that points can occur simultaneously.

3.3 REPRESENTATION OF COMPOUND POISSON PROCESSES

A compound Poisson process can be represented as a superposition of independent Poisson counting processes. This representation not only provides insight into the structure of the process, but also it proves useful in applications. This is especially true in applications involving statistical inference for observed, compound Poisson processes. The representation will also be important in Chapter 4 where we study the influence of compound Poisson processes on dynamical systems.

We will first develop the representation when the mark space is denumerable. Thus, let $\{\mathbf{x}_t; t \geqslant t_0\}$ be a compound Poisson process with a denumerable mark space $\mathfrak{U} = \{\mathbf{U}_1, \mathbf{U}_2, \ldots, \mathbf{U}_k, \ldots\}$, where the kth mark $\mathbf{U}_k$ occurs with probability p_k. Suppose $t_0 = t_{0n} < t_{1n} < t_{2n} < \ldots < t_{nn} = t$ is a sequence of partitions of $[t_0, t)$ such that

$$\lim_{n \to \infty} \max_{1 \leqslant j \leqslant n} (t_{jn} - t_{j-1,n}) = 0.$$

Define the processes $\{\mathbf{x}_t(\mathbf{U}_k); t \geqslant t_0\}$, for $k = 1, 2, \ldots,$ by

$$\mathbf{x}_t(\mathbf{U}_k) = \lim_{n \to \infty} \sum_{j=1}^{n} (\mathbf{x}_{t_{jn}} - \mathbf{x}_{t_{j-1,n}}) I_{\{\mathbf{U}_k\}}[\mathbf{x}_{t_{jn}} - \mathbf{x}_{t_{j-1,n}}] \tag{3.24}$$

where $I_{\{\mathbf{U}\}}[\mathbf{x}]$ is a mark indicator function defined by

$$I_{\{\mathbf{U}\}}[\mathbf{x}] = \begin{cases} 1, & \mathbf{x} = \mathbf{U} \\ 0, & \mathbf{x} \neq \mathbf{U}. \end{cases}$$

Thus, $\mathbf{x}_t(\mathbf{U}_k)$ is the sum of the marks that occur during $[t_0, t)$ and have value $\mathbf{U}_k$. It is evident that for $t \geqslant t_0$ we almost surely have

$$\mathbf{x}_t = \sum_{k=1}^{\infty} \mathbf{x}_t(\mathbf{U}_k) = \sum_{k=1}^{\infty} \mathbf{U}_k N_t(\mathbf{U}_k), \tag{3.25}$$

where $N_t(\mathbf{U}_k)$ is the number of points that occur during $[t_0, t)$ and have mark $\mathbf{U}_k$. This leads to the decomposition of $\mathbf{x}_t$ shown in Figure 3.3. The important aspect of the decomposition is that $\{\mathbf{x}_t; t \geqslant t_0\}$ can be represented in terms of the counting processes $\{N_t(\mathbf{U}_1); t \geqslant t_0\}$, $\{N_t(\mathbf{U}_2); t \geqslant t_0\}$, and so forth. The question is: How are these counting processes characterized statistically? The answer is in the following theorem.

Theorem 3.3.1 (Representation for a Compound Poisson Process Having a Denumerable Mark Space). Let $\{\mathbf{x}_t; t \geqslant t_0\}$ be a compound Poisson process with a denumerable mark space $\mathfrak{U} = \{\mathbf{U}_1, \mathbf{U}_2, \ldots\}$. Points occur as a

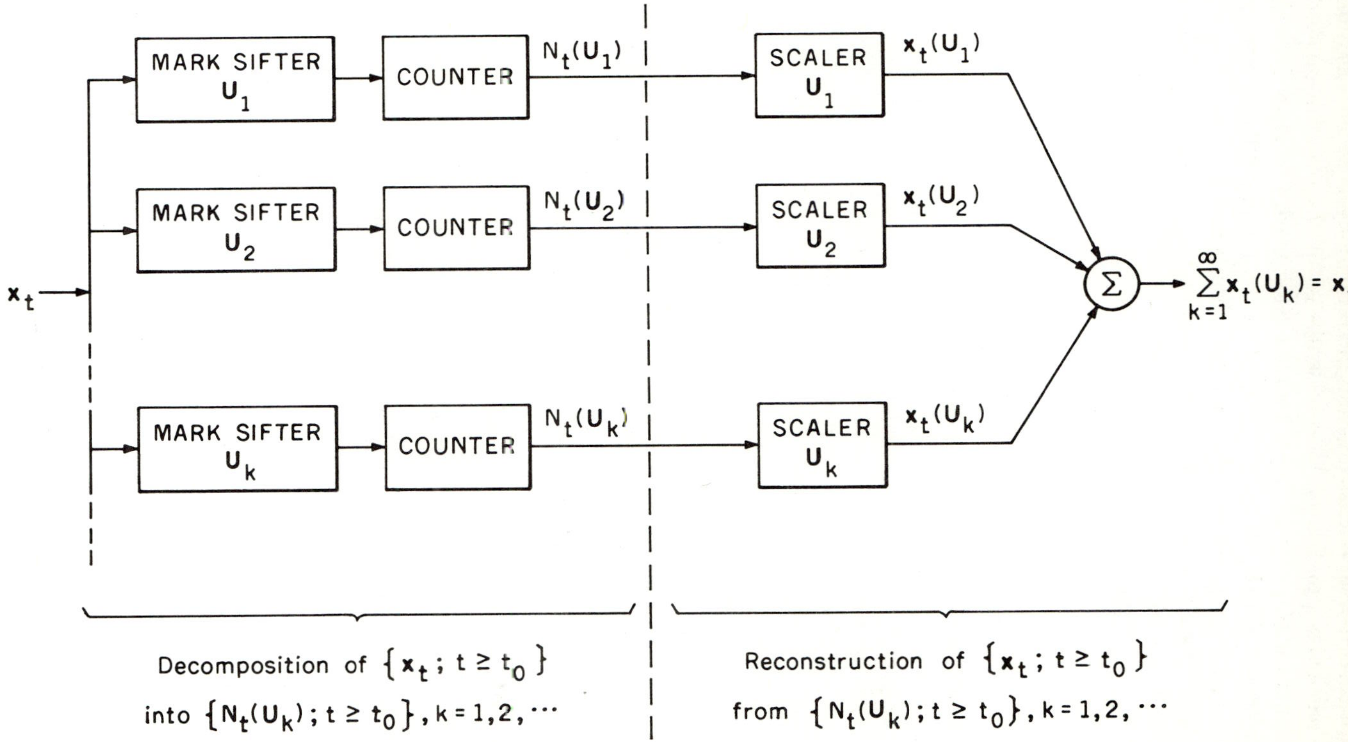

Figure 3.3. Decomposition of a compound Poisson process.

Poisson process with intensity λ_t, and the probability $\mathbf{U}_k$ occurs as a mark on a particular point is p_k independently of marks occurring on other points. Let $\{N_t(\mathbf{U}_k); t \geqslant t_0\}$ count points having mark $\mathbf{U}_k$. Then the counting processes $\{N_t(\mathbf{U}_k); t \geqslant t_0\}$, for $k = 1, 2, \ldots,$ are mutually independent Poisson counting processes for which the kth process has intensity $p_k\lambda_t$ for $t \geqslant t_0$. In terms of these, we have the representation of $\mathbf{x}_t$ as

$$\mathbf{x}_t = \sum_{k=1}^{\infty} \mathbf{U}_k N_t(\mathbf{U}_k) \tag{3.26}$$

for $t \geqslant t_0$.

Proof. We parallel a constructive argument due to L. Breiman [1, p. 311] to demonstrate that the processes defined in (3.24) are mutually independent Poisson processes of the type in Example 3.2.3 such that $\{\mathbf{x}_t(\mathbf{U}_k); t \geqslant t_0\}$ has jumps of size $\mathbf{U}_k$ and intensity $p_k\lambda_t$ for $t \geqslant t_0$. The representation in (3.26) is an immediate consequence of this. Thus, construct processes $\{\tilde{\mathbf{x}}_t(\mathbf{U}_k); t \geqslant t_0\}$, for $k = 1, 2, \ldots,$ which are mutually independent and such that $\{\tilde{\mathbf{x}}_t(\mathbf{U}_k); t \geqslant t_0\}$ has a characteristic functional on $[t_0, T)$ given by

$$\exp\left[\int_{t_0}^{T} p_k\lambda_\sigma\left(e^{j\mathbf{v}_\sigma'\mathbf{U}_k} - 1\right) d\sigma\right].$$

Hence, from (3.10), $\tilde{\mathbf{x}}_t(\mathbf{U}_k)$ is a Poisson process with jumps of size $\mathbf{U}_k$ and intensity $p_k\lambda_t$. It follows that $\{\mathbf{x}_t(\mathbf{U}_k); t \geqslant t_0\}$ equals $\{\tilde{\mathbf{x}}_t(\mathbf{U}_k); t \geqslant t_0\}$ with probability one for each k *separately*. Our objective is to show that this equality holds jointly for all k *simultaneously*. For this purpose, define

$$\tilde{\mathbf{x}}_t = \sum_{k=1}^{\infty} \tilde{\mathbf{x}}_t(\mathbf{U}_k).$$

From this construction and (3.12), it is evident that $\{\tilde{\mathbf{x}}_t; t \geqslant t_0\}$ has the same characteristic functional on $[t_0, T)$ as $\{\mathbf{x}_t; t \geqslant t_0\}$ for every $T \geqslant t_0$. Therefore, $\{\tilde{\mathbf{x}}_t; t \geqslant t_0\}$ and $\{\mathbf{x}_t; t \geqslant t_0\}$ have the same finite distributions. But by the construction,

$$\tilde{\mathbf{x}}_t(\mathbf{U}_k) = \lim_{n\to\infty} \sum_{j=1}^{n} \left(\tilde{\mathbf{x}}_{t_{jn}} - \tilde{\mathbf{x}}_{t_{n-1,n}}\right) I_{\{\mathbf{U}_k\}}\left(\tilde{\mathbf{x}}_{t_{jn}} - \tilde{\mathbf{x}}_{t_{j-1,n}}\right).$$

Thus, $\{\tilde{\mathbf{x}}_t(\mathbf{U}_k);\ t \geqslant t_0\}$ is the same function on $\{\tilde{\mathbf{x}}_t; t \geqslant t_0\}$ as $\{\mathbf{x}_t(\mathbf{U}_k); t \geqslant t_0\}$ is on $\{\mathbf{x}_t; t \geqslant t_0\}$. It follows that the processes $\{\mathbf{x}_t; t \geqslant t_0\}$, $\{\mathbf{x}_t(\mathbf{U}_k);\ t \geqslant t_0\}$, $k = 1, 2, \ldots,$ have the same joint distributions as the processes $\{\tilde{\mathbf{x}}_t; t \geqslant t_0\}$, $\{\tilde{\mathbf{x}}_t(\mathbf{U}_k); t \geqslant t_0\}$, $k = 1, 2, \ldots$. This establishes the theorem.

Example 3.3.1 (Multichannel Analysis). Suppose a quantity of radioactive material is observed using a scintillation detector. Light flashes in the scintillating crystal of such a detector have intensities that depend on the energies of the incident gamma photons emitted from the source. The spectrum of energies for these light pulses in a given detector is characteristic of the radioactive element. Examples of energy spectra for several elements are shown in Figure 3.4; these were obtained using an *Anger scintillation camera* by N. Mullani and H. Huang [6]. Denote the energy spectrum of an element by $f(E)$ and partition energy into disjoint energy bins $[E_k, E_k+\Delta E_k)$, $k=1, 2, \ldots$. We then model the sequence of light flashes in idealized form as a marked point process in which marks

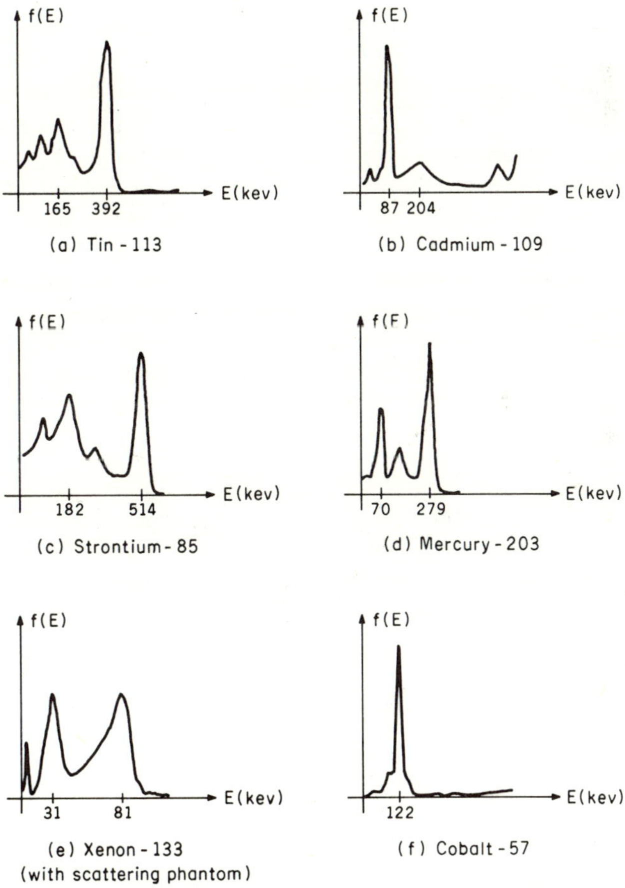

Figure 3.4. Energy spectra for several radioisotopes observed with an Anger scintillation camera.

are mutually independent and integer valued with integer k indicating a photon having energy in the kth energy bin. The probability a photon arrives having energy in the kth bin is

$$p_k \triangleq \left(\int_0^\infty f(E)\,dE\right)^{-1} \int_{E_k}^{E_k+\Delta E_k} f(E)\,dE.$$

Suppose light flashes occur at instants corresponding to a Poisson process with intensity λ_t for $t \geq 0$. Here, λ_t will be an exponentially decreasing function of time. The marked point process defined in this way is a compound Poisson process, and Theorem 3.3.1 implies that the counting processes $\{N_t([E_k, E_k+\Delta E_k)); t \geq 0\}$, $k=1,2,\ldots$, shown in Figure 3.5 are independent Poisson counting processes with corresponding intensities $p_k\lambda_t$, for $k=1, 2,\ldots$, and $t \geq 0$. In practice, the decomposition shown in Figure 3.5 is implemented using a device called a *multichannel analyzer* (see D. Ross and C. Harris [7, p. 160]). The device works by first generating narrow pulses having amplitudes proportional to energy. The energy filter is then simply an analog-to-digital converter, and for each k the counts $N_t([E_k, E_k+\Delta E_k))$ are accumulated in a register for display or they are assigned a memory location of a digital computer for numerical processing. In this way $2^{13}=8192$ parallel channels are easily achieved. The important consequence of Theorem 3.3.1 is that the collection of Poisson counting processes is equivalent to the marked point process at the input to the multichannel analyzer.

We now remove the restriction that the mark space be denumerable. Our objective is to obtain a continuous analog of (3.26) as a representation of a

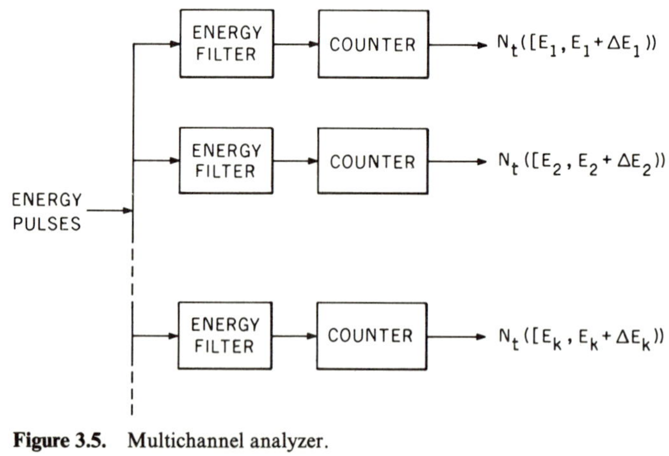

Figure 3.5. Multichannel analyzer.

general compound Poisson process. Thus, let $\{\mathbf{x}_t; t \geq t_0\}$ be a compound Poisson process with mark space $\mathfrak{U}$. For our purposes, the mark space may be taken as a possibly unbounded region of a finite-dimensional Euclidean space; thus, each mark is a vector of real numbers.

Jumps of $\{\mathbf{x}_t; t \geq t_0\}$ having amplitudes in disjoint regions of $\mathfrak{U}$ contribute independent, compound Poisson components to $\mathbf{x}_t$. This is seen as follows. Partition $\mathfrak{U}$ into disjoint regions $B_1, B_2, \ldots B_k, \ldots$ such that

$$\mathfrak{U} = \bigcup_{k=1}^{\infty} B_k.$$

Also, consider a sequence of partitions $t_0 = t_{0n} < t_{1n} < t_{2n} < \cdots < t_{nn} = t$ of $[t_0, t)$ such that

$$\lim_{n\to\infty} \max_{1 \leq j \leq n} (t_{jn} - t_{j-1,n}) = 0.$$

Now define the processes $\{\mathbf{x}_t(B_k); t \geq t_0\}$, for $k = 1, 2, \ldots$, according to

$$\mathbf{x}_t(B_k) = \lim_{n\to\infty} \sum_{j=1}^{n} (\mathbf{x}_{t_{jn}} - \mathbf{x}_{t_{j-1,n}}) I_{\{B_k\}}[\mathbf{x}_{t_{jn}} - \mathbf{x}_{t_{j-1,n}}], \tag{3.27}$$

where $I_{\{B_k\}}[\mathbf{x}]$ is a mark indicator function defined by

$$I_{\{B_k\}}[\mathbf{x}] = \begin{cases} 1, & \mathbf{x} \in B_k \\ 0, & \mathbf{x} \notin B_k. \end{cases}$$

These definitions generalize (3.24). The process $\{\mathbf{x}_t(B_k); t \geq t_0\}$ is a mark accumulator process for those points of the marked point process having a mark in B_k. It is evident that for $t \geq t_0$, we almost surely have

$$\mathbf{x}_t = \sum_{k=1}^{\infty} \mathbf{x}_t(B_k) \tag{3.28}$$

for $t \geq t_0$. By an argument entirely analogous to that used for the proof of Theorem 3.3.1, it follows that the processes $\{\mathbf{x}_t(B_k); t \geq t_0\}$, $k = 1, 2, \ldots$, are mutually independent, compound Poisson processes with corresponding intensities $p_k \lambda_t$, $k = 1, 2, \ldots$, where p_k is the probability that a mark takes on a value in B_k.

We now obtain a continuous analog of (3.26) for a compound Poisson process with a nondenumerable mark space. We will assume that the mark space is a finite-dimensional Euclidean space. I. Gikhman and A. Skorokhod [2, p. 270] develop a similar representation when the mark space is a more general, measurable space.

Let $\mathfrak{U}=\cup_{k=1}^{\infty}B_k$, where $B_1, B_2,\ldots,B_k,\ldots$ are disjoint regions with diameters not exceeding δ; the diameter of B_k is the maximum distance between any two points in B_k. Denote by $N_t(B_k)$ the number of points in $[t_0,t)$ having marks in B_k. By the above argument, the processes $\{N_t(B_k); t \geqslant t_0\}$, for $k=1, 2,\ldots$, are mutually independent Poisson counting processes with corresponding intensities $p_k\lambda_t$, for $k=1, 2,\ldots$. Let $\mathbf{U}_k$ denote an arbitrary point in B_k, and denote by $\|\mathbf{U}_k\|$ the largest distance from $\mathbf{U}_k$ to any other point in B_k. Then $\|\mathbf{U}_k\| \leqslant \delta$, and (3.28) implies that

$$|\mathbf{x}_t - \sum_{k=1}^{\infty} \mathbf{U}_k N_t(B_k)| \leqslant \sum_{k=1}^{\infty} \|\mathbf{U}_k\| N_t(B_k) \leqslant \delta N_t,$$

where N_t is the total number of points occurring in $[t_0,t)$ regardless of their mark. We note for the term on the right side that

$$E(\delta N_t)=\delta \int_{t_0}^{t} \lambda_\sigma \, d\sigma < \infty$$

and

$$\operatorname{var}(\delta N_t)=\delta^2 \int_{t_0}^{t} \lambda_\sigma \, d\sigma < \infty.$$

Consequently, for a partition $B_1, B_2,\ldots,B_k,\ldots$ of $\mathfrak{U}$ such that $\delta\to 0$, we have that

$$\lim_{\delta\to 0} E\left(|\mathbf{x}_t - \sum_{k=1}^{\infty} \mathbf{U}_k N_t(B_k)|^2\right)=0.$$

Thus, the sum $\Sigma_{k=1}^{\infty}\mathbf{U}_k N_t(B_k)$ converges to a limit in the mean-square sense as $\delta\to 0$. We denote this limit by $\int_{\mathfrak{U}}\mathbf{U}N_t(d\mathbf{U})$ and conclude that a compound Poisson process with a nondenumerable mark space can be represented as

$$\mathbf{x}_t = \int_{\mathfrak{U}} \mathbf{U} N_t(d\mathbf{U}), \tag{3.29}$$

which is the continuous analog of (3.26).

The interpretation of the term $N_t(d\mathbf{U})$ appearing in (3.29) is that $\{N_t(d\mathbf{U}); t\geqslant t_0\}$ is a Poisson counting process that counts the number of points for $t\geqslant t_0$ having marks in the elemental volume $[\mathbf{U},\mathbf{U}+d\mathbf{U})$ of $\mathfrak{U}$. The corresponding intensity function is $\lambda_t[P_{\mathbf{u}}(\mathbf{U}+d\mathbf{U})-P_{\mathbf{u}}(\mathbf{U})]$, where $P_{\mathbf{u}}(\mathbf{U})$ is the probability distribution function for the mark random variable. The processes $\{N_t(d\mathbf{U}_j);\ t\geqslant t_0\}$, $j=1, 2,\ldots$, are mutually independent for disjoint elemental volumes. Finally,

$$N_t(B_k)=\int_{B_k} N_t(d\mathbf{U}) \tag{3.30}$$

is a Poisson counting process associated with counting points having marks in B_k. The intensity for $\{N_t(B_k); t \geqslant t_0\}$ is $\lambda_t \int_{B_k} dP_{\mathbf{u}}(\mathbf{U}) \triangleq \lambda_t p_k$.

Additional interpretation for the representation in (3.29) can be given in terms of a time-space point process. Here, we view marked points as occurring in a multidimensional space $[t_0, \infty) \times \mathfrak{U}$; each point has a time coordinate in $[t_0, \infty)$ and a spatial coordinate in $\mathfrak{U}$. Let T be an interval of time in $[t_0, \infty)$ and B a region in $\mathfrak{U}$, and denote by $M(T, B)$ the number of points in $T \times B$; this is the number of points with time coordinates in T and marks in B. It follows from the above discussion that $\{M(T, B); T \subset [t_0, \infty), B \subset \mathfrak{U}\}$ is a time-space Poisson counting process on $[t_0, \infty) \times \mathfrak{U}$. That is, the number of points in disjoint regions of $[t_0, \infty) \times \mathfrak{U}$ are independent and the number in the region $T \times B$ is Poisson distributed with parameter $\int_T \int_B \lambda_t \, dt \, dP_{\mathbf{u}}(\mathbf{U})$. If the mark random variable is continuous so that a probability density $p_{\mathbf{u}}(\mathbf{U})$ for it exists, the intensity for this time-space Poisson process is $\lambda_t p_{\mathbf{u}}(\mathbf{U})$. Furthermore, $N_t(d\mathbf{U}) = \int_{t_0}^{t} M(d\sigma, d\mathbf{U})$ so that (3.29) can also be written as

$$\mathbf{x}_t = \int_{t_0}^{t} \int_{\mathfrak{U}} \mathbf{U} M(d\sigma, d\mathbf{U}).$$

The representation we have developed for a compound Poisson process having a nondenumerable mark space is summarized in the following theorem.

Theorem 3.3.2 (Representation for a Compound Poisson Process Having a Nondenumerable Mark Space). Let $\{\mathbf{x}_t; t \geqslant t_0\}$ be a compound Poisson process with a nondenumerable mark space $\mathfrak{U}$. Points occur as a Poisson process with intensity λ_t, and the probability distribution function for the marks is $P_{\mathbf{u}}(\mathbf{U})$. Then $\mathbf{x}_t$ can be represented as

$$\mathbf{x}_t = \int_{t_0}^{t} \int_{\mathfrak{U}} \mathbf{U} M(d\sigma, d\mathbf{U}), \tag{3.31}$$

where $\{M(T, B); T \subset [t_0, \infty), B \subset \mathfrak{U}\}$ is a time-space Poisson process such that

$$E[M(T, B)] = \int_T \int_B \lambda_t \, dt \, dP_{\mathbf{u}}(\mathbf{U}).$$

If the marks are continuous random variables with probability density $p_{\mathbf{u}}(\mathbf{U})$, the intensity function for M is $\lambda_t p_{\mathbf{u}}(\mathbf{U})$.

Applications to Statistical Inference

Problems of statistical inference for compound Poisson processes arise frequently in practice when these processes are used as models for observed data. We will now examine such problems briefly to indicate the usefulness of the representations of Theorems 3.3.1 and 3.3.2. Only maximum likelihood parameter estimation and testing of simple hypothesis are considered. The basic framework for solving these problems has been developed in Sections 2.4 and 2.5 of Chapter 2 where they were examined for observed Poisson processes. Recall that the sample-function density of the Poisson process played a central role in both parameter estimation and hypothesis testing as both the likelihood function and likelihood ratio are determined by it. The same is true here for the sample-function density of a compound Poisson process, so we begin by developing expressions for this important quantity.

Consider, first, a marked point process $\{\mathbf{x}_t; t \geqslant t_0\}$ with a denumerable mark space $\mathfrak{U} = \{\mathbf{U}_1, \mathbf{U}_2, \ldots\}$. Let $\{N_t; t \geqslant t_0\}$ count points of this process regardless of their mark. The quantity $p[\{\mathbf{x}_\sigma; t_0 \leqslant \sigma < t\}]$ defined below is the sample-function density for this process on the interval $[t_0, t)$:

$$p[\{\mathbf{x}_\sigma; t_0 \leqslant \sigma < t\}] = \begin{cases} \Pr(N_t = 0), & N_t = 0 \\ p_{\mathbf{w}}(\mathbf{W}, \mathbf{u}_1 = \xi_1, \ldots, \mathbf{u}_n = \xi_n, N_t = n), & N_t = n \geqslant 1 \end{cases} \tag{3.32}$$

where

$$p_{\mathbf{w}}(\mathbf{W}, \mathbf{u}_1 = \xi_1, \ldots, \mathbf{u}_n = \xi_n, N_t = n) \triangleq \Pr(N_t = n | \mathbf{w} = \mathbf{W}, \mathbf{u}_1 = \xi_1, \ldots, \mathbf{u}_n = \xi_n)$$

$$\times \Pr(\mathbf{u}_1 = \xi_1, \ldots, \mathbf{u}_n = \xi_n | \mathbf{w} = \mathbf{W}) p_{\mathbf{w}}^{(n)}(\mathbf{W}),$$

and where $\xi_i \in \mathfrak{U} = \{\mathbf{U}_1, \mathbf{U}_2, \ldots, \mathbf{U}_k, \ldots\}$ for each i, $1 \leqslant i \leqslant n$. The first factor in the product on the right is the conditional probability that $N_t = n$ given the n occurrence times and marks in $[t_0, t)$; the second factor is the conditional probability of the n marks given the n occurrence times; and the last factor is the joint occurrence density for the first n occurrence times. As with (2.49), the sample-function density (3.32) may be interpreted roughly as the probability of obtaining a particular realization of the marked point process on $[t_0, t)$ with $N_t = n$ points located at times $w_1 = W_1, \ldots, w_n = W_n$ and corresponding marks $\mathbf{u}_1 = \xi_1, \ldots, \mathbf{u}_n = \xi_n$.

The sample-function density for a compound Poisson process with a denumerable mark space is easily evaluated from the general definition in (3.32) because of the independence properties in the construction of this process. Let $P(\mathbf{U}_k)$ denote the probability mark $\mathbf{U}_k$ occurs on a point. Then these independence properties imply that

$$p_{\mathbf{w}}(\mathbf{W}, \mathbf{u}_1 = \xi_1, \ldots, \mathbf{u}_n = \xi_n, N_t = n) = \Pr(N_t = n | \mathbf{w} = \mathbf{W}) p_{\mathbf{w}}^{(n)}(\mathbf{W}) \prod_{i=1}^{n} P(\xi_i). \tag{3.33}$$

Furthermore, from (2.49), (2.50), and the definition of a compound Poisson process, we have that

$$p[\{\mathbf{x}_\sigma; t_0 \leqslant \sigma < t\}] = \begin{cases} \exp\left(-\int_{t_0}^{t} \lambda_\sigma \, d\sigma\right), & N_t = 0 \\ \left[\prod_{i=1}^{n} \lambda_{W_i} P(\xi_i)\right] \exp\left(-\int_{t_0}^{t} \lambda_\sigma \, d\sigma\right), & N_t = n \geqslant 1 \end{cases} \tag{3.34}$$

We find it convenient to incorporate into (3.34) a particular sample function of the marked point process on $[t_0, t)$; namely, the sample function for which $N_t = n$, and, for $n \geqslant 1$, $w_1 = W_1, \ldots, w_n = W_n$, $\mathbf{u}_1 = \xi_1, \ldots,$ and $\mathbf{u}_n = \xi_n$. Let $N_t(\mathbf{U}_i)$ be the number of marks among $\xi_1, \ldots, \xi_n$ equal to $\mathbf{U}_i$ for $i = 1, 2, \ldots,$ where $\sum_{i=1}^{\infty} N_t(\mathbf{U}_i) = n$. Then for this sample function (3.34) can be written alternatively as

$$p[\{\mathbf{x}_\sigma; t_0 \leqslant \sigma < t\}] = \exp\left\{-\int_{t_0}^{t} \lambda_\sigma \, d\sigma + \int_{t_0}^{t} \ln \lambda_\sigma \, dN_\sigma + \sum_{i=1}^{\infty} \ln[P(\mathbf{U}_i)] N_t(\mathbf{U}_i)\right\}, \tag{3.35}$$

where the second integral is the counting integral defined in (2.51). It is evident that the sample-function density for a compound Poisson process splits into two factors,

$$\exp\left(-\int_{t_0}^{t} \lambda_\sigma \, d\sigma + \int_{t_0}^{t} \ln \lambda_\sigma \, dN_\sigma\right)$$

and

$$\exp\left\{\sum_{i=1}^{\infty} \ln[P(\mathbf{U}_i)] N_t(\mathbf{U}_i)\right\}.$$

The first is contributed by the number of points and their occurrence times. This factor is just the sample-function density of (2.51) for a Poisson counting process. The second factor is contributed by the marks; it is the probability of the marks $\xi_1, \xi_2, \ldots, \xi_n$ occurring in the realization.

There is another way to write (3.35) that leads to an alternative and useful interpretation of the sample-function density for a compound Poisson process. By using

$$N_t = \sum_{i=1}^{\infty} N_t(\mathbf{U}_i) \quad \text{and} \quad \sum_{i=1}^{\infty} P(\mathbf{U}_i) = 1,$$

we rewrite (3.35) in the form

$$p[\{\mathbf{x}_\sigma; t_0 \leqslant \sigma < t\}] = \prod_{i=1}^{\infty} p[\{N_\sigma(\mathbf{U}_i); t_0 \leqslant \sigma < t\}], \tag{3.36}$$

where

$$p[\{N_\sigma(\mathbf{U}_i); t_0 \leqslant \sigma < t\}]$$
$$= \exp\left\{-\int_{t_0}^{t} P(\mathbf{U}_i)\lambda_\sigma \, d\sigma + \int_{t_0}^{t} \ln[P(\mathbf{U}_i)\lambda_\sigma] \, dN_\sigma(\mathbf{U}_i)\right\}.$$

Thus, the sample-function density for a compound Poisson process factors into the product of the sample-function densities for each of the independent components in the representation for $\{\mathbf{x}_t; t \geqslant t_0\}$ in Theorem 3.3.1.

Example 3.3.2 (Parameter Estimation for a Compound Poisson Process with a Denumerable Mark Space). Let $\{\mathbf{x}_t; t \geqslant t_0\}$ be a compound Poisson process with a denumerable mark space $\mathfrak{U} = \{\mathbf{U}_1, \mathbf{U}_2, \ldots, \mathbf{U}_k, \ldots\}$. Suppose points occur with intensity $\lambda_t(\boldsymbol{\theta})$ and that the probability mark $\mathbf{U}_k$ occurs is $P(\mathbf{U}_k|\boldsymbol{\theta})$, where $\boldsymbol{\theta}$ is a collection of unknown parameters. Observations of $\mathbf{x}$ are made on the interval $[t_0, T)$ in order to estimate these parameters. Suppose the data collected are the number, occurrence times, and marks for all points in $[t_0, T)$. From Section 2.4 of Chapter 2, the maximum likelihood estimate $\hat{\boldsymbol{\theta}}_{\text{ML}}$ of $\boldsymbol{\theta}$ in terms of observations $N_T = n$, $w_1 = W_1, \ldots, w_n = W_n$, $\mathbf{u}_1 = \xi_1, \ldots,$ and $\mathbf{u}_n = \xi_n$ is the value of $\boldsymbol{\theta}$ that maximizes the probability of having obtained these data. Just as in Section 2.4, we conclude that $\hat{\boldsymbol{\theta}}_{\text{ML}}$ is the value of $\boldsymbol{\theta}$ maximizing the sample-function density or, equivalently, its logarithm evaluated for the given data. Thus, from (3.35), $\hat{\boldsymbol{\theta}}_{\text{ML}}$ maximizes

$$l(\boldsymbol{\theta}) \triangleq \ln p[\{\mathbf{x}; t_0 \leqslant \sigma < T\}|\boldsymbol{\theta}]$$
$$= -\int_{t_0}^{T} \lambda_\sigma(\boldsymbol{\theta}) \, d\sigma + \int_{t_0}^{T} \ln \lambda_\sigma(\boldsymbol{\theta}) \, dN_\sigma + \sum_{i=1}^{\infty} \ln[P(\mathbf{U}_i|\boldsymbol{\theta})] N_t(\mathbf{U}_i) \tag{3.37}$$

as a function of $\boldsymbol{\theta}$, where the right side is evaluated for the observed realization of $\{\mathbf{x}_\sigma; t_0 \leqslant \sigma < T\}$. Except in especially simple situations, the maximization of $l(\boldsymbol{\theta})$ will need to be done numerically. We see from (3.37) that the marks can be disregarded and not measured if the mark distribution $P(\mathbf{U}_k|\boldsymbol{\theta})$ is, in fact, not a function of $\boldsymbol{\theta}$. Similarly, the occurrence times can be disregarded if $\lambda_t(\boldsymbol{\theta})$ is not a function of $\boldsymbol{\theta}$. An alternative expression for $l(\boldsymbol{\theta})$ that follows from (3.36) is

$$l(\boldsymbol{\theta}) = \sum_{i=1}^{\infty} l_i(\boldsymbol{\theta}), \tag{3.38}$$

where

$$l_i(\boldsymbol{\theta}) \triangleq -\int_{t_0}^{T} P(\mathbf{U}_i|\boldsymbol{\theta})\lambda_\sigma(\boldsymbol{\theta})\,d\sigma + \int_{t_0}^{T} \ln[P(\mathbf{U}_i|\boldsymbol{\theta})\lambda_\sigma(\boldsymbol{\theta})]\,dN_\sigma(\mathbf{U}_i).$$

Thus, the likelihood function $l(\boldsymbol{\theta})$ is the sum of the likelihood functions $l_i(\boldsymbol{\theta})$ for each independent component in the representation of Theorem 3.3.1. This decomposition leads to the implementation for generating $l(\boldsymbol{\theta})$ shown schematically in Figure 3.6. As discussed in Example 3.3.1, the initial sifting of points according to their mark is often accomplished in practice using a multichannel analyzer. An alternative way this sifting can occur is illustrated in the next example.

Example 3.3.3 (Optical Position Sensing). An optical communication system with a requirement for active tracking in order to ensure that the receiver looks directly at the transmitter needs a sensor to detect the position of the light beam arriving at the receiver. The electrical output of the sensor is used to generate feedback signals for controlling the orientation of telescopes and mirrors to maintain receiver-transmitter alignment even though the receiver and transmitter may be in relative motion. A simple device for sensing position consists of a photodetector having a photoemissive surface that is divided into four independent quadrants as illustrated in Figure 3.7. Photoelectron emissions are monitored in each quadrant during an observation interval $[0, T)$ and used to sense the position of an incident beam of light. We assume a static situation in which the intensity of light falling on the device has a circularly symmetric Gaussian distribution

$$I(a,b|\alpha,\beta) = I_0 \exp\left\{-\frac{1}{2\rho^2}\left[(a-\alpha)^2 + (b-\beta)^2\right]\right\}. \tag{3.39}$$

The parameters α and β are the beam position parameters, which are presumed unknown and desired. The parameters I_0 and ρ characterize the beam strength and width; these are assumed known. We assume photoemissions in a quadrant form a homogeneous Poisson process with an intensity proportional to the integral of $I(a,b|\alpha,\beta)$ over the active area of the detector in that quadrant; see (2.61). We incorporate the proportionality constant into I_0. The four quadrants of the detector are assumed to operate independently. Photoemissions for the entire sensor can therefore be viewed as a compound Poisson process with a denumerable mark space

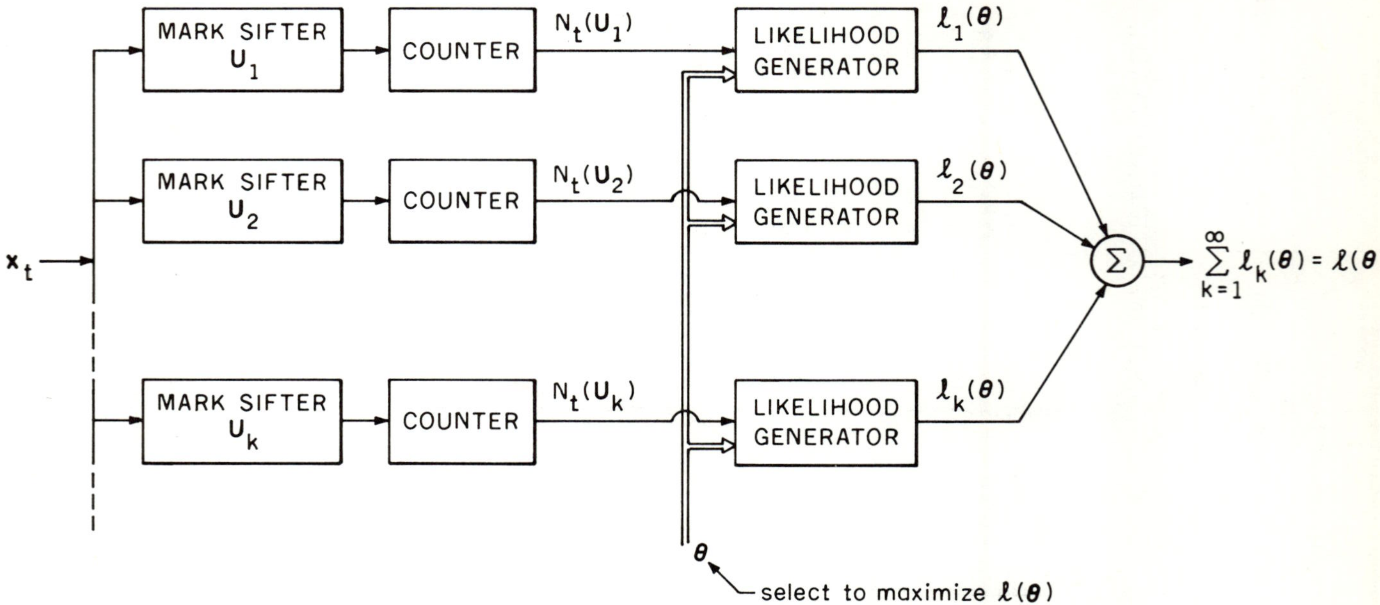

Figure 3.6. Likelihood function for a compound Poisson process with a denumerable mark space.

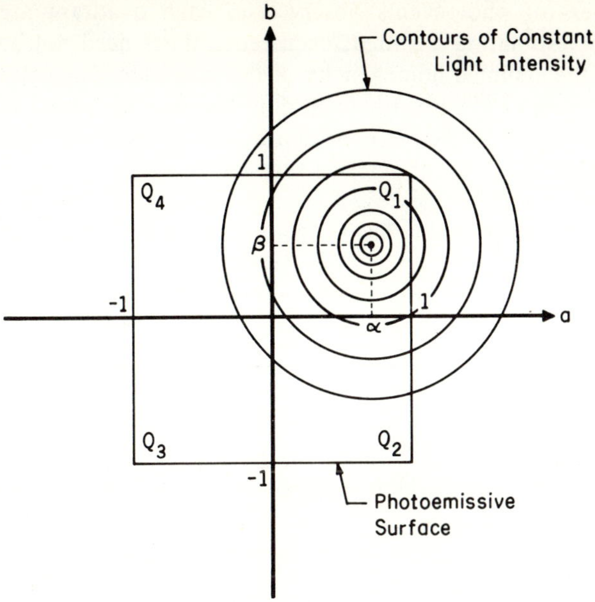

Figure 3.7. Four-quadrant position sensor.

$\mathfrak{U}=\{Q_1,Q_2,Q_3,Q_4\}$, where the mark Q_i identifies a point as originating in quadrant i. For use in (3.38), we have

$$P(Q_i|\alpha,\beta)\lambda_t(\alpha,\beta)=\int\int_{\mathcal{Q}_i} I(a,b|\alpha,\beta)\,da\,db, \tag{3.40}$$

where

$$\mathcal{Q}_1=\{a,b|0\leqslant a\leqslant 1, 0\leqslant b\leqslant 1\},$$

$$\mathcal{Q}_2=\{a,b|0\leqslant a\leqslant 1, -1\leqslant b\leqslant 0\},$$

$$\mathcal{Q}_3=\{a,b|-1\leqslant a\leqslant 0, -1\leqslant b\leqslant 0\},$$

$$\mathcal{Q}_4=\{a,b|-1\leqslant a\leqslant 0, 0\leqslant b\leqslant 1\}.$$

Substitution of these expressions into (3.38) results in

$$l(\alpha,\beta)=-T\int_{-1}^{1}\int_{-1}^{1} I(a,b|\alpha,\beta)\,da\,db$$
$$+\sum_{i=1}^{4} N_T(Q_i)\ln\left[\int\int_{\mathcal{Q}_i} I(a,b|\alpha,\beta)\,da\,db\right]. \tag{3.41}$$

The numbers of photoevents observed in each quadrant are sufficient statistics for estimating α and β; occurrence times need not be measured because of the assumed homogeneity of the processes. The values of α and β that maximize (3.41) give the maximum likelihood estimate of the beam position and might be used to control a mirror to center the beam over the origin of the four quadrant position sensor.

Example 3.3.4 (Hypothesis Testing for a Compound Poisson Process with a Denumerable Mark Space). Let $\{\mathbf{x}_t; t \geqslant t_0\}$ be a compound Poisson process with a denumerable mark space $\mathfrak{U} = \{\mathbf{U}_1, \mathbf{U}_2, \ldots, \mathbf{U}_k, \ldots\}$. Points and marks are governed by one of two underlying mechanisms, H_0 and H_1, with probabilities P_0 and P_1, respectively. Points occur with intensity $\lambda_t^{(i)}$, and mark $\mathbf{U}_k$ occurs with probability $P_i(\mathbf{U}_k)$ under mechanism H_i, $i=0,1$. Observations of $\mathbf{x}$ are made during the interval $[t_0, T)$ in order to decide which mechanism is in force. Suppose the data collected are the number, occurrence times, and marks for all points in $[t_0, T)$. From Section 2.5 of Chapter 2, and (2.102) in particular, we know that the strategy which minimizes the average cost is to evaluate the likelihood ratio

$$\Lambda[\{\mathbf{x}_\sigma; t_0 \leqslant \sigma < T\}] = \frac{\Pr(H_1|\mathbf{x}_\sigma; t_0 \leqslant \sigma < T)}{\Pr(H_0|\mathbf{x}_\sigma; t_0 \leqslant \sigma < T)}$$

$$= \frac{p[\{\mathbf{x}_\sigma; t_0 \leqslant \sigma < T\}|H_1]P_1}{p[\{\mathbf{x}_\sigma; t_0 \leqslant \sigma < T\}|H_0]P_0} \tag{3.42}$$

for the observed realization of $\{\mathbf{x}_\sigma; t_0 \leqslant \sigma < T\}$ and then compare the result to a precomputable constant, say η, depending only on the costs involved and not the observations. Thus, the sample-function density for the compound Poisson process plays a key role in hypothesis testing. By using (3.35) and (3.36), we obtain from (3.42) two alternative expressions for the test statistic,

$$\ln \Lambda[\{\mathbf{x}_\sigma; t_0 \leqslant \sigma < T\}] = -\int_{t_0}^{T} [\lambda_\sigma^{(1)} - \lambda_\sigma^{(0)}]\, d\sigma + \int_{t_0}^{T} \ln\left[\frac{\lambda_\sigma^{(1)}}{\lambda_\sigma^{(0)}}\right] dN_\sigma$$

$$+ \sum_{i=1}^{\infty} \ln\left[\frac{P_1(\mathbf{U}_i)}{P_0(\mathbf{U}_i)}\right] N_t(\mathbf{U}_i) \tag{3.43}$$

and

$$\ln \Lambda[\{\mathbf{x}_\sigma; t_0 \leqslant \sigma < T\}] = \sum_{i=1}^{\infty} \ln \Lambda[\{N_\sigma(\mathbf{U}_i); t_0 \leqslant \sigma < T\}], \tag{3.44}$$

where

$$\ln \Lambda[\{N_\sigma(\mathbf{U}_i); t_0 \leqslant \sigma < T\}] = -\int_{t_0}^{T} [P_1(\mathbf{U}_i)\lambda_\sigma^{(1)} - P_0(\mathbf{U}_i)\lambda_\sigma^{(0)}]\, d\sigma$$

$$+ \int_{t_0}^{T} \ln\left[\frac{P_1(\mathbf{U}_i)\lambda_\sigma^{(1)}}{P_0(\mathbf{U}_i)\lambda_\sigma^{(0)}}\right] dN_\sigma(\mathbf{U}_i).$$

The first two terms of (3.43) contain all the information about the underlying mechanism that is carried by the occurrence times, and the last term contains all the information carried by the marks. Thus, if $\lambda_t^{(1)} = \lambda_t^{(0)}$ for $t_0 \leqslant t < T$, the occurrence times do not contribute to the decision and can be disregarded without any penalty in performance. Similarly, if $P_1(\mathbf{U}_k) = P_0(\mathbf{U}_k)$ for some k, then the corresponding mark $\mathbf{U}_k$ can be disregarded. The alternative expression (3.44) indicates how the likelihood ratio is composed of the likelihood ratios of the independent components in the representation of Theorem 3.3.1. Namely, the logarithm of the likelihood ratio for $\{\mathbf{x}_\sigma; t_0 \leqslant \sigma < T\}$ is the sum of the logarithms of the likelihood ratios for the independent component processes $\{N_\sigma(\mathbf{U}_i); t_0 \leqslant \sigma < T\}$, $i = 1, 2, \ldots$.

Example 3.3.5 (Distinguishing Between Two Radioactive Elements). Consider again the model of Example 3.3.1 and suppose one of two radioactive elements is placed in the field of view of the scintillation detector. The problem is: Light pulses resulting from gamma photon absorptions are observed on the interval $[0, T)$; based on these, decide whether the radioactive element generating them has intensity $\lambda_t^0 \equiv \lambda_0$ and energy statistics $\{p_k^{(0)}\}$ or intensity $\lambda_t^1 \equiv \lambda_1$ and energy statistics $\{p_k^{(1)}\}$.

For this situation, we have

$$\ln \Lambda\left[\{N_\sigma([E_k, E_k + \Delta E_k)); t_0 \leqslant \sigma < T\}\right] = -T[p_k^{(1)}\lambda_1 - p_k^{(0)}\lambda_0]$$

$$+ N_T([E_k, E_k + \Delta E_k)) \ln\left[\frac{p_k^{(1)}\lambda_1}{p_k^{(0)}\lambda_0}\right],$$

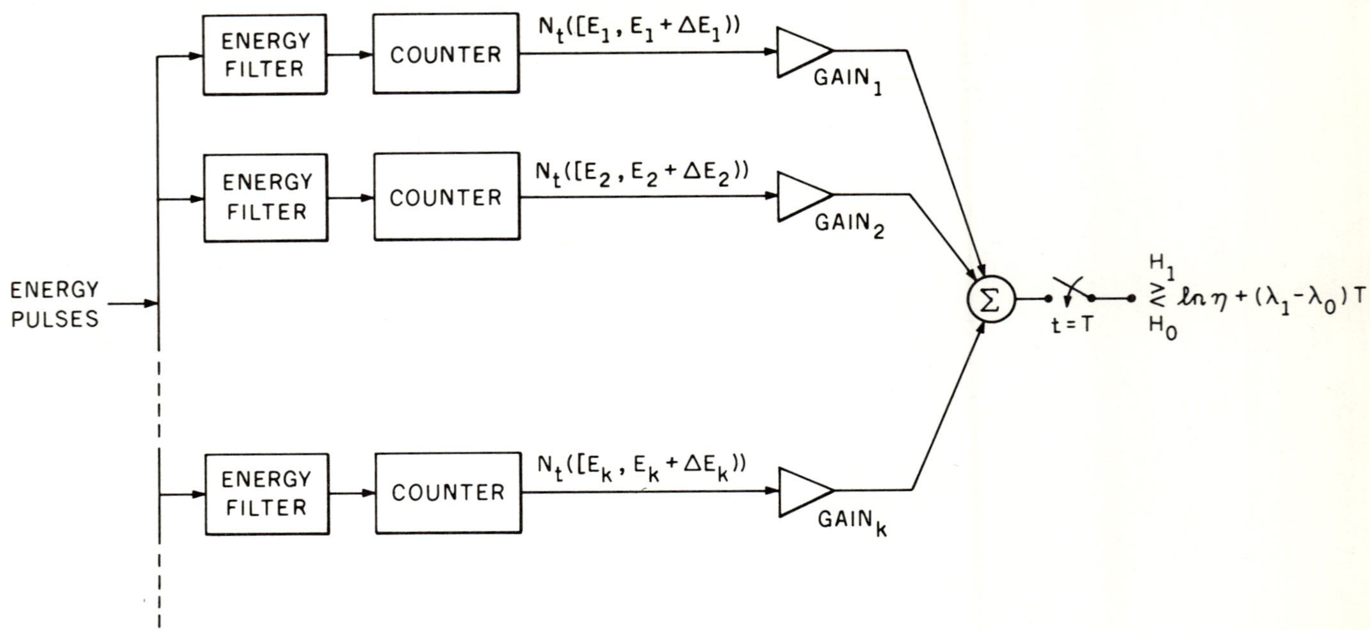

Figure 3.8. Detector for distinguishing between two radioactive elements.

and the detector is as shown in Figure 3.8. It is evident that the numbers of photons observed in each energy bin are sufficient statistics for making the decision.

Let us now remove the restriction that marks be denumerable. Thus, suppose $\{\mathbf{x}_t; t \geqslant t_0\}$ is a marked point process with mark space $\mathfrak{U}$, where $\mathfrak{U}$ is a region of a finite-dimensional Euclidean space. We assume that marks are continuous random variables with probability density $p_{\mathbf{u}}(\mathbf{U})$. Let $\{N_t; t \geqslant t_0\}$ count points of this process regardless of their mark. The sample function density for $\mathbf{x}$ on the interval $[t_0, t)$ is defined by

$$p[\{\mathbf{x}_\sigma; t_0 \leqslant \sigma < t\}] = \begin{cases} \Pr(N_t = 0), & N_t = 0 \\ p_{\mathbf{w}, \mathbf{u}_1, \ldots, \mathbf{u}_n}(\mathbf{W}, \mathbf{U}_1, \ldots, \mathbf{U}_n, N_t = n), & N_t = n \geqslant 1 \end{cases} \tag{3.45}$$

where

$$p_{\mathbf{w}, \mathbf{u}_1, \ldots, \mathbf{u}_n}(\mathbf{W}, \mathbf{U}_1, \ldots, \mathbf{U}_n, N_t = n) \triangleq \Pr(N_t = n | \mathbf{w} = \mathbf{W}, \mathbf{u}_1 = \mathbf{U}_1, \ldots, \mathbf{u}_n = \mathbf{U}_n)$$

$$\times p_{\mathbf{u}_1, \ldots, \mathbf{u}_n}(\mathbf{U}_1, \ldots, \mathbf{U}_n | \mathbf{w} = \mathbf{W}) p_{\mathbf{w}}^{(n)}(\mathbf{W}),$$

where the first factor in the product on the right is the conditional probability that $N_t = n$ given the n occurrence times and marks in $[t_0, t)$; the second factor is the conditional joint probability density of the first n marks given the n occurrence times; and the last factor is the joint occurrence density for the first n occurrence times. It is evident for a compound Poisson process that (3.45) becomes

$$p[\{\mathbf{x}_\sigma, t_0 \leqslant \sigma < t\}] = \begin{cases} \exp\left(-\int_{t_0}^{t} \lambda_\sigma \, d\sigma\right), & N_t = 0 \\ \left[\prod_{i=1}^{n} \lambda_{W_i} p_{\mathbf{u}}(\mathbf{U}_i)\right] \exp\left(-\int_{t_0}^{t} \lambda_\sigma \, d\sigma\right), & N_t = n \geqslant 1. \end{cases} \tag{3.46}$$

This is the analog of (3.34) for a nondenumerable mark space. In place of (3.35), we have

$$p[\{\mathbf{x}_\sigma; t_0 \leqslant \sigma < t\}] = \exp\left\{-\int_{t_0}^{t} \lambda_\sigma \, d\sigma + \int_{t_0}^{t} \ln \lambda_\sigma \, dN_\sigma + \int_{\mathfrak{U}} \ln[p_{\mathbf{u}}(\mathbf{U})] N_t(d\mathbf{U})\right\}, \tag{3.47}$$

where $N_t(d\mathbf{U})$ denotes the number of points in $[t_0,t)$ having marks in $[\mathbf{U},\mathbf{U}+d\mathbf{U})$; see (3.30). Finally, the analog of (3.36) for a nondenumerable mark space is

$$p[\{\mathbf{x}_\sigma;t_0\leqslant\sigma<t\}]=\exp\left\{-\int_{t_0}^{t}\lambda_\sigma\,d\sigma+\int_{t_0}^{t}\int_{\mathfrak{U}}\ln[p_{\mathbf{u}}(\mathbf{U})\lambda_\sigma]M(d\sigma,d\mathbf{U})\right\},\tag{3.48}$$

where M is the time-space Poisson counting process of the representation in Theorem 3.3.2. Equations (3.47) and (3.48) have interpretations analogous to those for (3.35) and (3.36), respectively. Furthermore, these equations are used in the same way as (3.35) and (3.36) for solving problems of statistical inference for observed compound Poisson processes having a nondenumerable mark space.

Example 3.3.6 (Optical Position Sensing, continued). Consider once again the model of Example 3.3.3, but suppose now that the photoemissive surface is not partitioned into four quadrants. Rather, the number, time of occurrence, and position of each photoevent are measured to determine the position of the light beam. The mark space corresponds to positions in the photoemissive surface: $\mathfrak{U}=\{a,b|-1\leqslant a\leqslant 1,-1\leqslant b\leqslant 1\}$. Photoevents form a compound Poisson process in which marks indicate the position of the photoevent in the detector. For use in (3.48), we have

$$\lambda_t(\alpha,\beta)=\int_{-1}^{1}\int_{-1}^{1}I(a,b|\alpha,\beta)\,da\,db$$

and

$$p_{\mathbf{u}}(a,b|\alpha,\beta)=I(a,b|\alpha,\beta)\left[\int_{-1}^{1}\int_{-1}^{1}I(U_1,U_2|\alpha,\beta)\,dU_1\,dU_2\right]^{-1},$$

where $I(a,b|\alpha,\beta)$ is defined in (3.39). From (3.48), the likelihood function is

$$\begin{aligned}l(\alpha,\beta)&=\ln p[\{\mathbf{x}_\sigma;t_0\leqslant\sigma<T\}|\alpha,\beta]\\&=-T\int_{-1}^{1}\int_{-1}^{1}I(a,b|\alpha,\beta)\,da\,db+\sum_{i=1}^{N_T}\ln I(a_i,b_i|\alpha,\beta),\end{aligned}\tag{3.49}$$

where N_T is the number of photoevents observed in $[0,T)$ and (a_i,b_i), for $i=1,2,\ldots,N_T$, their positions. The maximum likelihood estimate of the beam position is given by the values of α and β that maximize $l(\alpha,\beta)$. The search for these values will generally require numerical implementation, but there is one situation where an analytic solution is possible. This obtains when the beam size and range of possible positions are small in comparison to the size of the photoemissive surface. Then the first term in (3.49) will be independent of α and β to a good approximation and can be disregarded. It is then straightforward to verify that the maximum likelihood estimate of the beam position is given by

$$\left(\hat{\alpha}_{\mathrm{ML}},\hat{\beta}_{\mathrm{ML}}\right)=\left(\frac{1}{N_T}\sum_{i=1}^{N_T}a_i,\frac{1}{N_T}\sum_{i=1}^{N_T}b_i\right). \tag{3.50}$$

That is, the estimated position is at the arithmetic average of the observed photoevents.

Summary. In this section, we have developed a representation for compound Poisson processes in terms of a superposition of independent Poisson counting processes. Theorems 3.3.1 and 3.3.2 give the representations for denumerable and nondenumerable mark spaces, respectively. These representations provide insight into the structure of data processing needed to solve problems of statistical inference for observed compound Poisson processes. For example, they lead naturally to the use of multichannel analysis, which is commonly employed in practice. Expressions were derived in (3.35), (3.36), and (3.48) for the sample-function density of a compound Poisson process. This quantity plays a central role in solving problems of statistical inference.

REFERENCES

1. L. Breiman, *Probability*, Addison-Wesley, Reading, Mass., 1968.
2. I. I. Gikhman and A. V. Skorokhod, *Introduction to the Theory of Random Processes*, Saunders, Philadelphia, 1969.
3. D. König and K. Matthes, "Verallgemeinerungen den Erlangsche Formeln," *I. Math. Nachr.*, **26**, 45–56 (1963).
4. K. Matthes, "Stationare Zufällige Puntfolgen," *I. Jber. Deutsch. Math.-Verein*, **66**, 66–79 (1963).
5. K. Matthes, "Unbeschränkt Teilbare Verteilumsgestze Stationärer Zufälliger Puntfolgen," *Wiss. Z. Hochsch. Electro. Ilmenau*, **9**, 235–238 (1963).
6. N. Mullani and H. Huang, "Gamma Camera Calibration Procedures and Programs," Chapter 12 in Computer Processing of Dynamic Images from an Anger Scintillation Camera, Ed. K. B. Larson and J. R. Cox, Proceedings of a Workshop held at Washington University, St. Louis, Missouri, January 1971.
7. D. A. Ross and C. C. Harris, "Measurement of Radioactivity," Chapter V in *Principles of Nuclear Medicine*, Ed. H. N. Wagner, Jr., Saunders, Philadelphia, 1968.

PROBLEMS

3.2.1. Suppose a source of coherent light incident on a photomultiplier produces primary photoelectrons as a Poisson process with intensity λ_t for $t \geqslant t_0$. By a cascade process, each primary electron produces secondary electrons that are collected at the output of the photomultiplier. Suppose the number of secondary electrons produced independently by each primary electron is a Poisson distributed random variable with parameter μ. Determine the characteristic function, mean, and variance for the number of electrons appearing at the photomultiplier output during the interval $[s,t)$.

3.2.2. Let $\{x_t; t \geqslant t_0\}$ be a compound Poisson process. Assume points occur with intensity λ_t for $t \geqslant t_0$ and that the common characteristic function for the marks is $M_u(j\alpha)$. The characteristic function for x_t,

$$M_{x_t}(jv) = \exp\left[\int_{t_0}^{t} \lambda_\sigma \, d\sigma (M_u(j\alpha) - 1)\right],$$

is generally difficult to invert analytically in which case the probability distribution function $P_{x_t}(X) = \Pr(x_t \leqslant X)$ for x_t is not known. By paralleling the development of the Chernoff bounds in Theorem 2.5.1, establish that this probability satisfies

$$\max_{s \geqslant 0}\left[1 - M_{x_t}(s)e^{-sX}\right] \leqslant P_{x_t}(X) \leqslant \min_{s \leqslant 0} M_{x_t}(s)e^{-sX},$$

where

$$M_x(s) = \exp\left[\int_{t_0}^{t} \lambda_\sigma \, d\sigma (M_u(s) - 1)\right]$$

is the moment generating function for x_t.

3.2.3. Let $\{\mathbf{x}_t; t \geqslant t_0\}$ be a compound Poisson process. Assume points occur with intensity λ_t for $t \geqslant t_0$ and that the common probability distribution function for the marks is $P_{\mathbf{u}}(\mathbf{U})$. Let $P_{\mathbf{x}_t}(\mathbf{X}) = \Pr(\mathbf{x}_t \leqslant \mathbf{X})$ be the probability distribution function for $\mathbf{x}_t$; here, the vector inequality means that each component of $\mathbf{x}_t - \mathbf{X}$ is nonpositive. Show that

$$P_{\mathbf{x}_t}(\mathbf{X}) = \sum_{n=0}^{\infty} P_{\mathbf{u}}^{*(n)}(\mathbf{X})(n!)^{-1}\left(\int_{t_0}^{t} \lambda_\sigma \, d\sigma\right)^n \exp\left(-\int_{t_0}^{t} \lambda_\sigma \, d\sigma\right),$$

where $P_{\mathbf{u}}^{*(n)}(\mathrm{U})$ is the nth iterated convolution of $P_{\mathbf{u}}(\mathrm{U})$ with itself. Here

$$P_{\mathbf{u}}^{*(0)}(\mathrm{U}) = \begin{cases} 1, & \mathrm{U} \geqslant 0 \\ 0, & \mathrm{U} < 0, \end{cases}$$

$P_{\mathbf{u}}^{*(1)}(\mathrm{U}) = P_{\mathbf{u}}(\mathrm{U})$, and $P_{\mathbf{u}}^{*(n)}(\mathrm{U})$, for $n \geqslant 2$, can be determined from the recursive relation

$$P_{\mathbf{u}}^{*(n)}(\mathrm{U}) = \int_{\mathfrak{U}} P_{\mathbf{u}}^{*(n-1)}(\mathrm{U} - \xi)\, dP_{\mathbf{u}}(\xi).$$

3.2.4. Establish (3.17). Hint: The following series may be useful:

$$\ln(1-z) = -\sum_{n=1}^{\infty} \frac{z^n}{n} \quad \text{for } |z| \leqslant 1 \text{ and } z \neq 1.$$

3.2.5. The light source of Figure P.3.2.5 can operate at two levels of intensity. The levels occur with equal probability when the source is energized. Assume electrons are generated at the output of the photodetector as a homogeneous Poisson process at a rate $\lambda_1 = \lambda$ or $\lambda_2 = 2\lambda$ corresponding to the low- and high-level operation of the light source. Electrons are counted, but the counter is defective. Suppose an electron fails to be registered with probability p independently of all other electrons. Denote by x_t the total number of registered electrons in $[0, t)$.

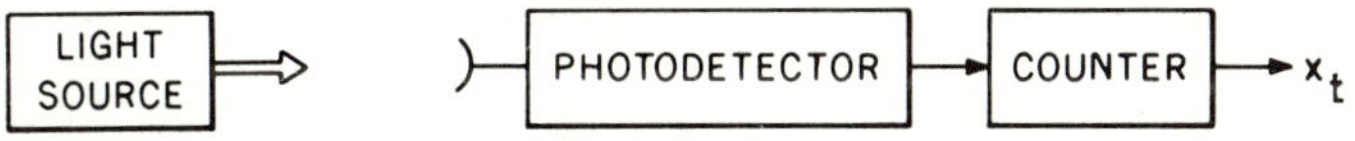

Figure P.3.2.5.

a. Determine the characteristic functional for x on $[0, T)$; that is evaluate

$$\phi_x(jv) \triangleq E\left[\exp\left(j\int_0^T v_\sigma\, dx_\sigma\right)\right]$$

b. Determine the characteristic function for x_t from the result of part (a).

c. Evaluate the mean and variance of x_t.

3.2.6. The following construction is used to generate a random staircase function in time. Let $\{u_n\}$ be a sequence of independent identically distributed random variables with the common distribution function

$$P_{u_n}(U)=\Pr[u_n \leqslant U]=\begin{cases} 1-e^{-\lambda U}, & U \geqslant 0 \\ 0, & U<0. \end{cases}$$

Let $\{x_t; t \geqslant 0\}$ be constructed as shown in Figure P.3.2.6. That is, odd variables in the sequence define the length of the treads of the staircase function x_t and even variables define the heights of the risers. Evaluate either the characteristic function or probability density for the height x_t of the staircase at time t. Determine the expected value and variance of this height.

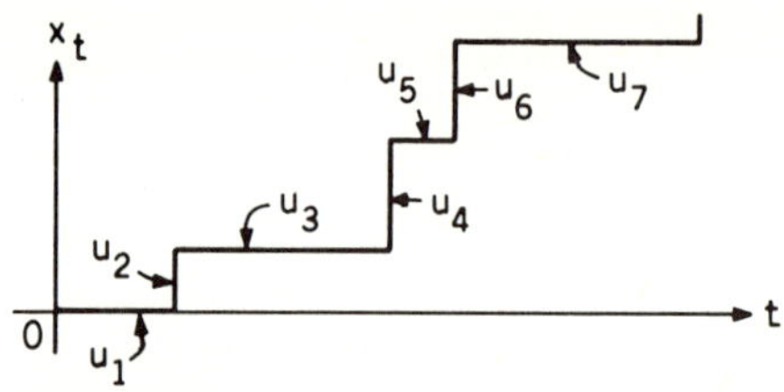

Figure P.3.2.6.

3.2.7. A random process $\{\mathbf{x}_t; t \geqslant t_0\}$ is said to be continuous in the mean-square sense if for all $t \geqslant t_0$

$$\lim_{\delta \to 0} E[|\mathbf{x}_{t+\delta} - \mathbf{x}_t|^2]=0.$$

Is a compound Poisson process continuous in this sense?

3.2.8. Consider the motion of a small particle as a function of time. Its movement is confined to one dimension and is influenced by random collisions with other particles. Let x_t denote the position of the particle at time t for $t \geqslant 0$. Assume:

i. the particle is initially at the origin, $x_0=0$;
ii. collisions occur at random instants corresponding to a homogeneous Poisson process with intensity λ;
iii. each collision independently, may be with either a type-A or type-B particle with probability $\frac{1}{3}$ and $\frac{2}{3}$, respectively;
iv. a collision with a type-A particle changes the position of the small particle by an amount either $+1$ or -1 unit with equal probability;

v. a collision with a type-B particle changes the position of the small particle by an amount $+2$ or -2 units with equal probability.

a. Determine the characteristic functional for x on $[0, T)$.

b. Determine the characteristic function for x_t from the result of part (a).

c. Determine the mean, variance, and covariance function for x.

d. Define

$$x_t^* = \frac{x_t - E(x_t)}{\sqrt{\operatorname{var}(x_t)}}.$$

Then x_t^* is the position of the particle normalized to have zero mean and unit variance. Determine the characteristic function for x_t^* and discuss what happens in the limit as $\lambda t \to \infty$.

3.3.1. Rederive equations (3.19), (3.20), and (3.23) for the mean, instantaneous covariance matrix, and covariance matrix of a compound Poisson process by starting with the representations (3.26) and (3.31).

3.3.2. Consider, again, the model of Problem 3.2.5.

a. Draw a block diagram of the likelihood ratio detector for deciding with minimum probability of error which light level is in force given observations of the counter output during $[0, T)$.

b. Derive an expression in the form of a summation for the minimum probability of error.

3.3.3. Suppose $\{x_t; t \geqslant 0\}$ is the staircase function constructed in Problem 3.2.6. Show that the likelihood function $l(\lambda)$ for estimating λ in terms of an observed path of x on $[0, T)$ is

$$l(\lambda) = -\lambda T + (2 \ln \lambda) N - \lambda x_T,$$

where N is the total number of treads and risers occurring on $[0, T)$. Derive an expression for the maximum likelihood estimate $\hat{\lambda}_{\text{ML}}$ of λ in terms of the observed path.

3.3.4 (Cramér–Rao Bound for Unbiased Estimates). Let $\{\mathbf{x}_t; t \geqslant t_0\}$ be a compound Poisson process with mark space $\mathfrak{U}$. Assume points occur with intensity $\lambda_t(\boldsymbol{\theta})$, where $\boldsymbol{\theta}$ is a collection of unknown parameters. If $\mathfrak{U}$ is denumerable, the mark occurrence probabilities $P(\mathbf{U}_k | \boldsymbol{\theta})$, for $k = 1, 2, \ldots,$

may also depend on $\boldsymbol{\theta}$. If $\mathfrak{U}$ is nondenumerable, assume a mark density $P_{\mathbf{u}}(\mathbf{U}|\boldsymbol{\theta})$ exists and may depend on $\boldsymbol{\theta}$. Let $\boldsymbol{\theta}^*$ denote any unbiased estimate of $\boldsymbol{\theta}$ in terms of $\{\mathbf{x}_\sigma; t_\sigma \leqslant \sigma < T\}$. Show that

$$E[(\boldsymbol{\theta}^* - \boldsymbol{\theta})(\boldsymbol{\theta}^* - \boldsymbol{\theta})'|\boldsymbol{\theta}] \geqslant \mathbf{F}^{-1}(\boldsymbol{\theta}),$$

where $\mathbf{F}(\boldsymbol{\theta})$ is a Fisher information matrix given for a denumerable mark space by

$$\mathbf{F}(\boldsymbol{\theta}) = \sum_{k=1}^{\infty} \int_{t_0}^{T} [\lambda_\sigma(\boldsymbol{\theta}) P(\mathbf{U}_k|\boldsymbol{\theta})]^{-1}$$

$$\times \left[\frac{\partial \lambda_\sigma(\boldsymbol{\theta}) P(\mathbf{U}_k|\boldsymbol{\theta})}{\partial \boldsymbol{\theta}} \right] \left[\frac{\partial \lambda_\sigma(\boldsymbol{\theta}) P(\mathbf{U}_k|\boldsymbol{\theta})}{\partial \boldsymbol{\theta}} \right]' d\sigma,$$

and for a nondenumerable mark space by

$$\mathbf{F}(\boldsymbol{\theta}) = \int_{\mathfrak{U}} \int_{t_0}^{T} [\lambda_\sigma(\boldsymbol{\theta}) p_{\mathbf{u}}(\mathbf{U}|\boldsymbol{\theta})]^{-1}$$

$$\times \left[\frac{\partial \lambda_\sigma(\boldsymbol{\theta}) p_{\mathbf{u}}(\mathbf{U}|\boldsymbol{\theta})}{\partial \boldsymbol{\theta}} \right] \left[\frac{\partial \lambda_\sigma(\boldsymbol{\theta}) p_{\mathbf{u}}(\mathbf{U}|\boldsymbol{\theta})}{\partial \boldsymbol{\theta}} \right]' d\sigma \prod_i dU_i,$$

where $\partial \lambda P / \partial \boldsymbol{\theta}$ and $\partial \lambda p / \partial \boldsymbol{\theta}$ denote gradients with respect to $\boldsymbol{\theta}$. Hint: Parallel the proof of Theorem 2.4.1.

3.3.5. Suppose $\{x_t; t \geqslant 0\}$ is the staircase function constructed in Problem 3.2.6. Use the result of Problem 3.3.4 to derive a lower bound on the mean-square error for any unbiased estimate of the parameter λ in terms of an observed path of x on $[0, T)$.

FILTERED POISSON PROCESSES

CHAPTER FOUR

There are numerous physical phenomena that can be reasonably modeled as a response to the points of a marked point process. The simplest models occur when the response can be expressed as a superposition of separate responses to each marked point. These models are developed in Section 4.1. Later, in Section 4.2, we remove the superposition requirement but impose the additional structure of Markov processes.

4.1 FILTERED POISSON PROCESSES GENERATED BY A SUPERPOSITION OF POINT RESPONSES

Let $\{\mathbf{x}_t; t \geqslant t_0\}$ be a marked point process as defined in Section 3.1 of Chapter 3. Denote the nth occurrence time and mark by τ_n and $\mathbf{u}_n$, respectively, and let $\{N_t; t \geqslant t_0\}$ be a counting process that counts points regardless of their mark. The processes of interest in this section have the form of the following superposition,

$$y_t = \begin{cases} 0, & N_t = 0 \\ \sum_{n=1}^{N_t} h(t, \tau_n : \mathbf{u}_n), & N_t \geqslant 1, \end{cases} \tag{4.1}$$

for $t \geqslant t_0$, where $h(t, \tau_n : \mathbf{u}_n)$ is termed the impulse-response or weighting function and is the response at time t to the nth marked point. A process in the form of (4.1) will be referred to as a *filtered point process*. In particular, we term $\{y_t; t \geqslant t_0\}$ a *filtered Poisson process* when $\{\mathbf{x}_t; t \geqslant t_0\}$ is a compound Poisson process. In this case $\{N_t; t \geqslant t_0\}$ is a Poisson counting process, and the marks $\{\mathbf{u}_n\}$ are mutually independent, independent of $\{N_t; t \geqslant t_0\}$, indentically distributed, vector-valued random variables.

We assume throughout the discussion that the weighting function h is causal; that is, $h(t, \tau : \mathbf{u}) = 0$ for $\tau > t$.

Some physical phenomena for which filtered point processes provide a reasonable model are given in the following examples.

Example 4.1.1 (Shot Noise). After being emitted from a heated cathode in a vacuum-tube diode, an electron travels to the anode. In doing so, it makes an incremental contribution to the anode current. Denote the anode current at time t by i_t. Then a reasonable first approximation model for $\{i_t; t \geqslant t_0\}$ is

$$i_t = \begin{cases} 0, & N_t = 0 \\ \sum_{n=1}^{N_t} h(t - \tau_n), & N_t \geqslant 1, \end{cases} \tag{4.2}$$

where t_0 is the time the diode is energized, N_t is the number of electrons emitted during $[t_0,t)$, $\{\tau_n\}$ are the emission times, and $h(t-\tau_n)$ is the contribution to i_t due to the nth electron emitted. The process $\{i_t; t \geqslant t_0\}$ is commonly called *shot noise.* The form of h is a function of the cathode–anode geometry, cathode temperature, and anode voltage; thus, the detailed form of h can be difficult to predict. It is of interest, therefore, that only more global aspects of h enter into certain calculations we make below. Specifically, the area of h is the electronic charge $e \cong 1.6 \times 10^{-19}$ coulombs, and the duration of h is approximately the cathode to anode transit time, which is typically on the order of 10^{-9} seconds. When the electron emission rate is temperature limited so that space-charge effects are negligible, $\{N_t; t \geqslant t_0\}$ can be reasonably modeled as a Poisson counting process. W. Davenport and W. Root [5, Chapter 7] give a detailed study of shot noise models as well as additional references.

Example 4.1.2 (Busy Telephone Channels). Denote the time at which a telephone call is initiated by τ and its duration by u. Then the number of telephone connections in use at time t in a telephone network energized at time t_0 and having an unlimited number of possible connections can be expressed in the form of (4.1) by setting

$$h(t,\tau_n:u_n)=\begin{cases} 1, & \tau_n \leqslant t < \tau_n + u_n \\ 0, & \text{otherwise} \end{cases} \tag{4.3}$$

where τ_n and u_n are the initiation time and duration for the nth call. In this case, y is a filtered point process involving a scalar mark. Also, we note that y is not linear in the mark variable. E. Parzen [21, p. 147] contains a detailed study of the situation described here when y is a homogeneous filtered Poisson process (see Problem 4.1.3).

Example 4.1.3 (Photoconductive Detector). Electron-hole pairs created in a bulk semiconductor by the absorption of light photons increase the conductivity of the material during the lifetime of the pairs. This effect, called photoconductivity, provides a basic mechanism for the detection of optical signals. Upon application of a bias voltage, the change in conductivity due to the absorption of a photon results in a measurable change in the current flowing through the device. To a reasonable approximation, the output current i_t can be modeled for $t \geqslant t_0$ as

$$i_t=\begin{cases} 0, & N_t=0 \\ \displaystyle\sum_{n=1}^{N_t} h(t-\tau_n:u_n), & N_t \geqslant 1 \end{cases}$$

where N_t is the number of photon absorptions that occur during $[t_0, t)$, $\{\tau_n\}$ are the times of these absorptions, and u_n is the duration of the incremental contribution $h(t-\tau_n : u_n)$ to i_t as shown in Figure 4.1. The duration u_n of h is a random variable depending on the lifetime of the electron–hole pair created by the photon absorption. Evidence suggests that this lifetime can be taken as exponentially distributed.

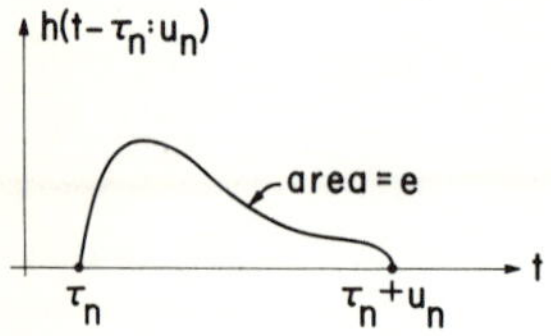

Figure 4.1. Incremental current contribution of a photon absorption.

Example 4.1.4 (ELF–VLF Atmospheric Noise). Atmospheric radio noise in the frequency bands below about 30 kHz (i.e., the extremely-low- and very-low-frequency bands) is mainly due to lightning discharges. The effect of such noise in a radio receiver can be modeled to a first approximation as a filtered point process of the form

$$y_t = \begin{cases} 0, & N_t = 0 \\ \sum_{n=1}^{N_t} u_n h(t-\tau_n), & N_t \geqslant 1 \end{cases} \tag{4.4}$$

where N_t denotes the number of discharges during the interval $[t_0, t)$ subsequent to the time t_0 at which the receiver is energized, $\{\tau_n\}$ denotes their occurrence times, and h is the response of the receiver to a discharge. Some sample functions of ELF noise recorded at the output of a narrow-band receiver and reported by J. Evans [7] are shown in Figure 4.2. Here, the amplitude marks $\{u_n\}$ are included to account for the obvious variability in the amplitude of the responses; this variability arises because all lightning strokes do not release the same amount of energy.

Example 4.1.5 (Radar Clutter, Scatter Communication Channels). Suppose a signal $s(t)$ is propagated on a radio channel consisting of small scatterers distributed in space. Examples are radar signals reflected from water droplets in rain and communication signals reflected from a cloud of small, needle-sized dipoles as dispersed, for example, in the Project West Ford experiment [23]. Let $s(t) = e(t)\cos(\omega_c t + \phi(t))$ for $t \geqslant t_0$. Here $e(t)$ and $\phi(t)$ are deterministic amplitude and phase modulation, respectively, of a carrier of frequency ω_c. A model described by R. Kennedy [16, Chapter 2]

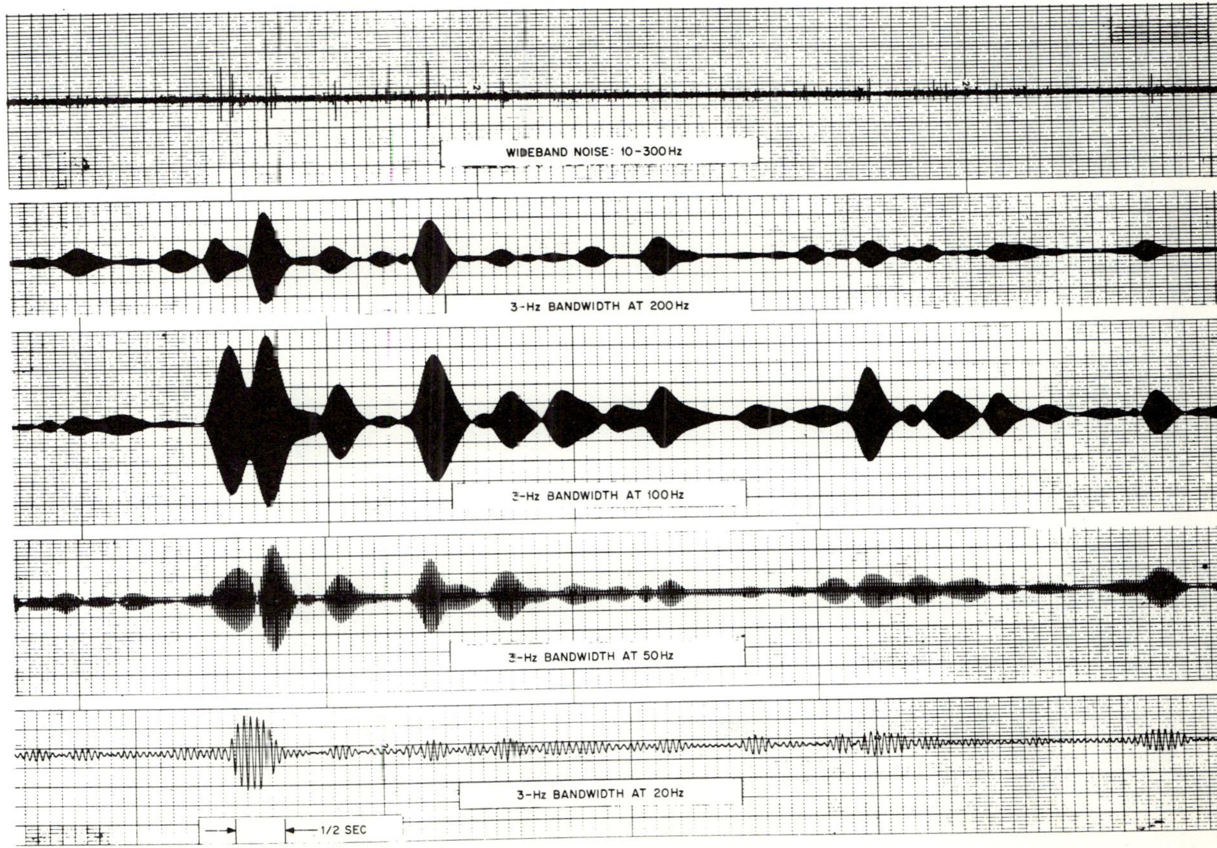

Figure 4.2. Sample function of ELF noise.

for the signal received $r(t)$ after scattering is

$$r(t)=\begin{cases} 0, & N_t=0 \\ \sum_{n=1}^{N_t} \rho_n e(t-\tau_n)\cos[\omega_c t+\phi(t-\tau_n)+\theta_n], & N_t \geqslant 1 \end{cases} \tag{4.5}$$

where N_t is the number of scattered signals received during $[t_0,t)$, $\{\tau_n\}$ are their arrival times at the receiver, and $\{\rho_n\}$ and $\{\theta_n\}$ are random attenuation and phase variables associated with each scatterer. Attenuation variables ρ_n are included because the amount of energy reflected from a scatterer depends on its shape and aspect. The phase variable θ_n can usually be taken as a uniformly distributed random variable because of the high frequencies, ω_c, normally used in radar and communication systems. Thus, $\{r(t);t\geqslant t_0\}$ is a filtered point process with a two-dimensional mark variable (ρ_n,θ_n).

It is evident from these examples that filtered point processes defined in (4.1) provide a reasonable model for a wide variety of phenomena. We now develop the statistics for a special class of these processes, namely, for filtered Poisson processes. The most important quantities we develop are the characteristic functional, mean, and covariance function for the filtered Poisson process. The characteristic functional is important because it provides a complete statistical characterization of the process in the same way we have seen for other processes in earlier chapters. The first moment and covariance function provide important information in applications, as we shall see.

STATISTICS FOR A FILTERED POISSON PROCESS

The characteristic functional for a filtered Poisson process is given in the following theorem. This functional is then used to evaluate various first- and second-order statistics of a filtered Poisson process.

Theorem 4.1 (Characteristic Functional for a Filtered Poisson Process). Let $\{y_t;t\geqslant t_0\}$ be a filtered Poisson process in which points occur with intensity λ_t for $t\geqslant t_0$. Then the characteristic functional

$$\phi_y(jv)=E\left[\exp\left(j\int_{t_0}^{T} y_\sigma\, dv_\sigma\right)\right] \tag{4.6}$$

associated with the random variables $\{y_\sigma; t_0 \leqslant \sigma < T\}$ has the evaluation

$$\phi_y(jv) = \exp\left\{ \int_{t_0}^{T} \lambda_\tau E\left[\exp\left(j \int_\tau^T h(\sigma, \tau : \mathbf{u})\, dv_\sigma \right) - 1 \right] d\tau \right\}, \tag{4.7}$$

where the expectation is with respect to the common distribution of the marks and v_σ is an arbitrary real-valued function of σ such that

$$\int_{t_0}^{T} \int_{t_0}^{T} f(\alpha, \beta)\, dv_\alpha\, dv_\beta < \infty \tag{4.8}$$

where f is defined by

$$f(\alpha, \beta) = \int_{t_0}^{\min(\alpha,\beta)} \lambda_\tau E[h(\alpha, \tau : \mathbf{u}) h(\beta, \tau : \mathbf{u})]\, d\tau$$

$$+ \int_{t_0}^{\alpha} \lambda_\tau E[h(\alpha, \tau : \mathbf{u})]\, d\tau \int_{t_0}^{\beta} \lambda_\tau E[h(\beta, \tau : \mathbf{u})]\, d\tau.$$

Proof. The proof of Theorem 4.1.1 is straightforward using conditional expectation and the properties of Poisson processes. We need to observe initially that because $N_T \geqslant N_\sigma$, for $\sigma \in [t_0, T)$, and because h is causal, we have

$$\phi_y(jv) = \Pr(N_T = 0) + \sum_{k=1}^{\infty} \Pr(N_T = k)$$

$$\times E\left\{ \exp\left[j \sum_{n=1}^{k} \int_{t_0}^{T} h(\sigma, \tau_n : \mathbf{u}_n)\, dv_\sigma \right] \Big| N_T = k \right\}. \tag{4.9}$$

The summation in the expectation is unchanged by a random reordering of the occurrence times. With this reordering, the occurrence times $\tau_1, \tau_2, \ldots, \tau_k$ given $N_T = k$ are independent and identically distributed, the common density being

$$p_{\tau_i}(\tau) = \lambda_\tau \left(\int_{t_0}^{T} \lambda_\sigma\, d\sigma \right)^{-1}$$

from (2.53). Since the mark variables are also independent and identically distributed, we obtain

$$E\left\{\exp\left[j\sum_{n=1}^{k}\int_{t_0}^{T}h(\sigma,\tau_n:\mathbf{u}_n)\,dv_\sigma\right]\Big|N_T=k\right\}$$

$$=\left\{\left(\int_{t_0}^{T}\lambda_\sigma\,d\sigma\right)^{-1}\int_{t_0}^{T}\lambda_\tau E\left[\exp\left(j\int_{t_0}^{T}h(\sigma,\tau:\mathbf{u})\,dv_\sigma\right]d\tau\right\}^{k},$$

where the expectation on the right is with respect to the distribution of the mark random variable $\mathbf{u}$. Substitution of this expression into (4.9) and a routine calculation with the Poisson distribution (2.1) results in (4.7). The restriction on v in (4.8) ensures that $E[(\int_{t_0}^{T}y_\sigma\,dv_\sigma)^2]<\infty$, as developed below.

We now illustrate the use of the characteristic functional (4.7) with several examples. These follow from particular selections of the function v.

Let us first examine the statistics for the instantaneous response y_t at time t. Take the function v in (4.6) to be

$$v_\sigma=\begin{cases}0, & t_0\leqslant\sigma<t\\ \alpha, & t\leqslant\sigma<T,\end{cases}$$

where α is a constant. Then

$$\phi_y(jv)=E[\exp(j\alpha y_t)]\triangleq M_{y_t}(j\alpha)$$

is the characteristic function for y at time t. We have

$$\int_{\tau}^{T}h(\sigma,\tau:\mathbf{u})\,dv_\sigma=\begin{cases}\alpha h(t,\tau:\mathbf{u}), & \tau\leqslant t\\ 0, & \tau>t.\end{cases}$$

Consequently, from (4.7) the evaluation of the characteristic function for y_t is

$$M_{y_t}(j\alpha)=\exp\left\{\int_{t_0}^{t}\lambda_\tau E[\exp(j\alpha h(t,\tau:\mathbf{u}))-1]\,d\tau\right\}. \tag{4.10}$$

This expression is generally difficult to invert to determine the probability distribution for y_t. It is useful, however, for evaluating the moments of y_t. For this purpose, let γ_n be the nth cumulant for the random variable y_t; the cumulants are defined in Problem 2.2.3. By using $j^n\gamma_n=(\partial^n\ln M_{y_t}/\partial\alpha^n)$

($\alpha=0$), we deduce from (4.10) that

$$\gamma_n=\int_{t_0}^{t}\lambda_\tau E[h^n(t,\tau:\mathbf{u})]\,d\tau. \tag{4.11}$$

The mean $E(y_t)$ and variance $\operatorname{var}(y_t)$ of the response at time t are easily obtained from (4.11) because $E(y_t)=\gamma_1$ and $\operatorname{var}(y_t)=\gamma_2$. Higher-order moments of y_t can be similarly obtained by using the results of Problem 2.2.3.

Example 4.1.6. As an example, suppose $h(t,\tau:\mathbf{u})=g(t-\tau)$; thus, the point process can be viewed as unmarked and exciting a linear, time-invariant filter with causal impulse response $g(t)$. Further, suppose $\lambda_t=\lambda$, a constant, so that the point process is homogeneous, and set $t_0=-\infty$. Then

$$\gamma_n=\lambda\int_{-\infty}^{t} g^n(t-\tau)\,d\tau=\lambda\int_0^\infty g^n(\alpha)\,d\alpha. \tag{4.12}$$

Consequently,

$$E[y_t]=\gamma_1(y_t)=\lambda\int_0^\infty g(\alpha)\,d\alpha=\lambda G(0), \tag{4.13}$$

where

$$G(f)=\int_0^\infty g(t)\exp(-j2\pi ft)\,dt$$

is the frequency response function of the filter. Thus, the expected value of the response y_t is time independent and equals the product of the average event rate, λ, and the zero-frequency response of the filter, $G(0)$. By Parseval's equation, we also have

$$\operatorname{var}(y_t)=\gamma_2=\lambda\int_0^\infty g^2(\alpha)\,d\alpha=\lambda\int_{-\infty}^{\infty}|G(f)|^2\,df. \tag{4.14}$$

It is evident that the variance of the response is the same as if the filter were excited by a process with a power spectral density of constant level λ; this is brought out more explicitly below in Example 4.1.7.

The joint statistics for the response at time t_1 and at time t_2 can be determined by taking the function v in (4.6) to be

$$v_\sigma=\begin{cases}0, & t_0\leqslant\sigma<t_1\\ \alpha_1, & t_1\leqslant\sigma<t_2\\ \alpha_1+\alpha_2, & t_2\leqslant\sigma<T\end{cases}$$

where α_1 and α_2 are constants. Then

$$\phi_y(jv) = E\left[\exp(j\alpha_1 y_{t_1} + j\alpha_2 y_{t_2})\right] = M_{t_1,t_2}(j\alpha_1, j\alpha_2)$$

is the joint characteristic function for y at times t_1 and t_2. Consequently, from (4.7) the evaluation this joint characteristic function for y_{t_1} and y_{t_2} is

$$M_{t_1,t_2}(j\alpha_1, j\alpha_2)$$
$$= \exp\left\{\int_{t_0}^{T} \lambda_\tau E(\exp[j\alpha_1 h(t_1,\tau:\mathbf{u}) + j\alpha_2 h(t_2,\tau:\mathbf{u})] - 1)\, d\tau\right\}. \quad (4.15)$$

We use (4.15) to evaluate the covariance function $K_y(t_1,t_2)$ for y, where

$$K_y(t_1,t_2) = E(y_{t_1} y_{t_2}) - E(y_{t_1})E(y_{t_2}).$$

Using the relation

$$j^{-2}\left(\frac{\partial^2 M_{t_1,t_2}}{\partial\alpha_1\,\partial\alpha_2}\right)(\alpha_1 = 0, \alpha_2 = 0) = E(y_{t_1} y_{t_2})$$

and (4.11) with $n=1$, we conclude that

$$K_y(t_1,t_2) = \int_{t_0}^{\min(t_1,t_2)} \lambda_\tau E[h(t_1,\tau:\mathbf{u}) h(t_2,\tau:\mathbf{u})]\, d\tau. \quad (4.16)$$

The function f in (4.8) is now seen to be the correlation function

$$f(t_1,t_2) = K_y(t_1,t_2) + E(y_{t_1})E(y_{t_2}).$$

Thus, condition (4.8) ensures that $\int_{t_0}^{T} y_\sigma\, dv_\sigma$ exists in the mean-square sense and that

$$E\left[\left(\int_{t_0}^{T} y_\sigma\, dv_\sigma\right)^2\right] = \int_{t_0}^{T}\int_{t_0}^{T} f(\sigma_1,\sigma_2)\, dv_{\sigma_1}\, dv_{\sigma_2} \quad (4.17)$$

holds and is finite. This completes the proof of Theorem 4.1.

Example 4.1.7 (Campbell's Theorem). For the situation of Example 4.1.6, equation (4.16) becomes

$$K_y(t_1,t_2) = \lambda \int_{-\infty}^{\min(t_1,t_2)} g(t_1-\tau) g(t_2-\tau)\, d\tau$$

$$= \lambda \int_{-\infty}^{\infty} g(t_1-\tau) g(t_2-\tau)\, d\tau.$$

Let $z=t_2-\tau$ and change variables of integration to get

$$K_y(t_1,t_2)=\lambda\int_{-\infty}^{\infty} g(t_1-t_2+z)g(z)\,dz.$$

Thus, the covariance function for y depends only on the difference of its arguments and not each independently; therefore, y is wide-sense stationary. Let $t=t_1-t_2$, and abbreviate $K_y(t,0)$ by $K_y(t)$. Then

$$K_y(t)=\lambda\int_{-\infty}^{\infty} g(t+z)g(z)\,dz \tag{4.18}$$

and also

$$S_y(f)=\int_{-\infty}^{\infty} K_y(t)\exp(-j2\pi ft)\,dt=\lambda|G(f)|^2, \tag{4.19}$$

where $S_y(f)$ is the power-density spectrum for $y-E(y)$. Thus, the power-density spectrum S_y is the same as if the filter were excited by a random process with constant power-density spectrum λ.

Equations (4.18) and (4.19) are commonly referred to as *Campbell's theorem*. In this terminology, (4.16) can be viewed as a generalized form of Campbell's theorem. A classic application for Campbell's theorem is the modeling of shot noise.

Example 4.1.8 (Shot Noise). Consider again the shot noise model of Example 4.1.1. Suppose a diode is connected to a steady bias voltage V and a load with frequency response function $Z(f)$ as shown in Figure 4.3. An equivalent circuit for the diode and biasing source can be obtained by use of Campbell's theorem. From Example 4.1.1, we have

$$i_t=\sum_{n=1}^{N_t} h(t-\tau_n).$$

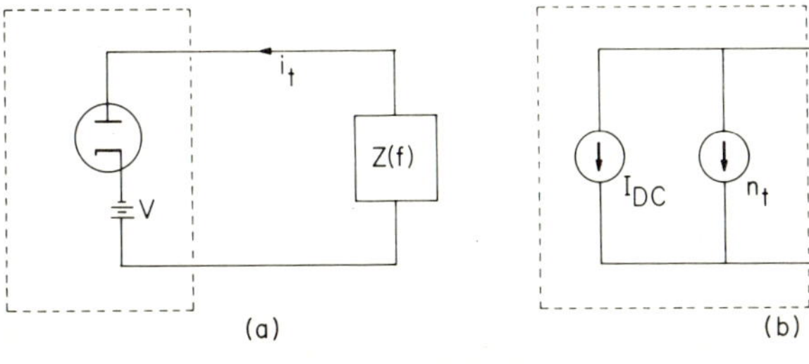

Figure 4.3. Diode circuit and its equivalent.

Assume $\{N_t; t > -\infty\}$ is a homogeneous Poisson counting process with constant intensity λ representing the rate of electron emissions from the cathode. The incremental contribution h of an electron to the total current i has area e, the electronic charge. Hence, $H(0) = e$, and from (4.13), we have

$$E(i_t) \triangleq I_{\mathrm{DC}} = \lambda e. \tag{4.20}$$

From elementary considerations, $I_{\mathrm{DC}} = V(R_d + Z(0))^{-1}$, where R_d is the diode resistance at the operating point. Thus, the emission rate λ can be determined from (4.20) when V, R_d, and $Z(0)$ are known. Furthermore, we have from Campbell's Theorem (4.19) that the power-density spectrum for $i_t - I_{\mathrm{DC}}$ is given by

$$S_i(f) = \lambda |H(f)|^2. \tag{4.21}$$

That i_t can be expressed as $I_{\mathrm{DC}} + n_t$, where $n_t = i_t - E(i_t)$, leads to the diode model in Figure 4.3*b*; here, n is a zero-mean, wide-sense stationary process with power-density spectrum $\lambda|H(f)|^2$. The power-density spectrum for the load voltage (minus its expectation) is $\lambda|H(f)|^2|Z(f)|^2$. Under usual conditions, H is a very slowly changing function of f by comparison to Z. For instance, the bandwidth of H is of the order of the inverse transit time, or about 10^9 Hz; the bandwidth of Z is ordinarily only a small fraction of this. Consequently,

$$\lambda|H(f)|^2|Z(f)| \cong \lambda|H(0)|^2|Z(f)|^2 = I_{\mathrm{DC}}e|Z(f)|^2.$$

In this case, n can be modeled as a process with constant power-density spectrum $I_{\mathrm{DC}}e$; this result is known as Schottky's formula and is widely used in models for devices exhibiting shot noise.

Example 4.1.9 (Signal-to-Noise Ratio in Photodetection). A model described by E. Hoversten [14] for a wide variety of photodetectors is shown in Figure 4.4. The output process $\{i_t; t \geqslant t_0\}$, here taken to be the current through the device, is comprised of two independent filtered Poisson processes, $\{i_t^s; t \geqslant t_0\}$ and $\{i_t^d; t \geqslant t_0\}$, and an independent Gaussian process $\{i_t^{\mathrm{th}}; t \geqslant t_0\}$ according to

$$i_t = i_t^s + i_t^d + i_t^{\mathrm{th}}.$$

The "signal" current i^s is given for $t \geqslant t_0$ by

$$i_t^s \triangleq \sum_{n=1}^{N_t^s} h_s(t, \tau_n : u_n)$$

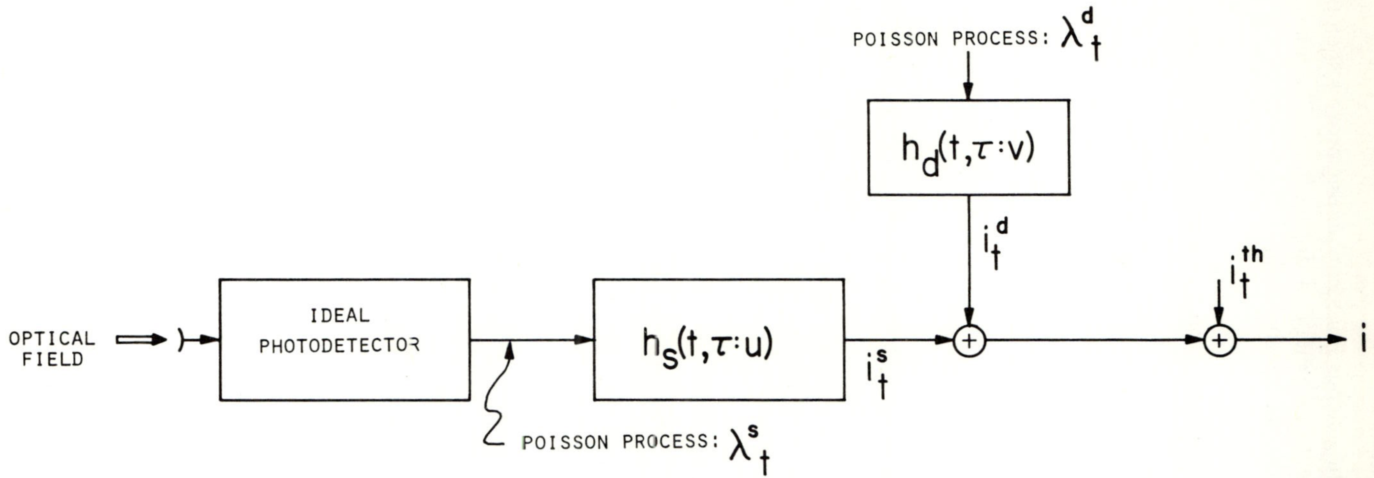

Figure 4.4. Photodetector model.

where $\{N_t^s; t \geqslant t_0\}$ is an inhomogeneous Poisson counting process such that N_t^s is the number of photoelectrons generated during $[t_0, t)$. The intensity λ_t^s is related to a narrow-band optical signal-field

$$s(t, \vec{r}) \triangleq \sqrt{2}\ \mathrm{Re}[S(t,\vec{r})\exp(j2\pi\nu t)]$$

incident on the active surface $\mathcal{A}$ of the detector according to

$$\lambda_t^s = \frac{\eta}{h\nu}\int_{\mathcal{A}} |S(t,\vec{r})|^2 d^2\vec{r}, \tag{4.22}$$

where η is a constant called the quantum efficiency of the detector, h is Planck's constant, ν is the unmodulated optical carrier frequency, and S is the complex envelope of the signal, which is assumed here to be deterministic. The filter response $h_s(t,\tau:u)$ has a form that depends on the particular type of photodetector used. Some examples are given in Table 4.1. In the presence of background radiation (for example, scattered sunlight), it is often assumed as an approximation that λ_t^s is simply replaced by $\lambda_t^s + \lambda^b$, where $\lambda^b = \eta P_b / h\nu$ and P_b is the background noise power incident on the active area $\mathcal{A}$ of the detector. Conditions for the validity of this assumption are discussed by E. Hoversten [14].

Table 4.1. Photodetector Models

$h_s(t,\tau:u)$	Remark	Type of Photodetector Modeled
$e\delta(t-\tau)$	e is the electronic charge and $\delta(t)$ is a Dirac delta function	Ideal photodetector (without gain)
$g(t-\tau)$	$g(t)$ is a deterministic impulse response	Photodiode
$ug(t-\tau)$	u is a random variable modeling random gain (e.g., number of secondary electrons generated for each primary photoelectron)	Photomultiplier, avalanche photodiode
$h(t-\tau:u)$	Response function shown in Figure 4.1	Photoconductor

The disturbance current $\{i_t^d; t \geqslant t_0\}$ is given by

$$i_t^d \triangleq \sum_{n=1}^{N_t^d} h_d(t, \tau_n : v_n)$$

where $\{N_t^d; t \geqslant t_0\}$ is an inhomogeneous Poisson counting process; i^d is commonly termed "dark current" and accounts for extraneous electrons generated during $[t_0, t)$. The intensity function λ_t^d is the instantaneous rate at which such spontaneous events occur. The filter responses h_s and h_d may or may not be identical depending on the type of photodetector used and degree of refinement desired in its mathematical model.

The disturbance current $\{i_t^{\text{th}}; t \geqslant t_0\}$ is a zero-mean Gaussian process modeling the combined effects of thermally generated noise in amplifiers or other circuits following the detector. In usual circumstances, i_t^{th} can be assumed stationary and having a constant power spectral density of level $2kT/R$, where k is Boltzmann's constant, T the temperature, and R the noise resistance. This thermal noise is especially significant for detectors that do not have internal gain.

As an application of this model and our results on the second-moment properties of filtered Poisson processes, we evaluate the signal-to-noise ratio at the output of an amplifier in response to the detector current i. For this purpose, we assume that the amplifier output $\{y_t; t \geqslant 0\}$ is given by

$$y_t = \sum_{n=1}^{N_t} h(t - \tau_n : u_n) + \int_0^t k(t-\tau) i_\tau^{\text{th}} \, d\tau,$$

where N_t is the total number of detection events during $[0, t)$ corresponding to signal, background, and dark current, (that is, $N_t = N_t^s + N_t^b + N_t^d$), and where $k(t-\tau)$ is the impulse response of the amplifier, and $h(t-\tau : u) = uek(t-\tau)$. We have assumed the detector response functions h_s and h_d both equal $ue\delta(t-\tau)$ corresponding to an ideal photodetector with a random gain u. Here, e is the electronic charge. By using (4.11), we deduce that the signal-to-noise ratio S/N defined by

$$\frac{S}{N} = \frac{E^2[y_t | i_t^b = i_t^d = i_t^{\text{th}} = 0]}{\text{var}(y_t)} \tag{4.23}$$

has the evaluation

$$\frac{S}{N} = \frac{e^2 E^2(u) \left[\int_0^\infty k(t-\tau) \lambda_\tau^s \, d\tau \right]^2}{e^2 E(u^2) \int_0^\infty k^2(t-\tau)(\lambda_\tau^s + \lambda^b + \lambda_\tau^d) \, d\tau + \left(\frac{2kT}{R} \right) \int_0^\infty k^2(t-\tau) \, d\tau}. \tag{4.24}$$

This expression for S/N can be simplified when the signal field corresponds to an unmodulated optical carrier for in this case we have $\lambda_t^s = \eta P_s / h\nu$, where P_s is the total carrier power incident on the active area $\mathscr{Q}$ of the detector. Assume further that the dark current rate λ_t^d is a constant λ^d. We then have

$$\frac{S}{N} = \frac{\left[\dfrac{\eta P_s}{h\nu}\right]^2}{2B\left[\dfrac{E(u^2)}{E^2(u)}\right]\left[\dfrac{\eta(P_s+P_b)}{h\nu}+\lambda^d\right]+2B\left[\dfrac{2kT}{e^2E^2(u)R}\right]}, \tag{4.25}$$

where

$$2B \triangleq \frac{\displaystyle\int_0^\infty k^2(t-\tau)\,d\tau}{\left(\displaystyle\int_0^\infty k(t-\tau)\,d\tau\right)^2}.$$

Here, B is the equivalent rectangular noise bandwidth of the filter k. We note that S/N can also be expressed with obvious definitions as

$$\frac{S}{N} = \frac{I_s^2}{2Be\left[\dfrac{E(u^2)}{E^2(u)}\right](I_s+I_b+I_d)+2B\left[\dfrac{2kT}{E^2(u)R}\right]} \tag{4.26}$$

in which I_s, I_b, and I_d are the signal, background, and dark currents, respectively. We note that for detectors without gain (that is, $u=1$), the last term in the denominator usually dominates S/N; the converse holds for photomultipliers and avalanche photodiodes, which have gain.

A Central Limit Theorem

A filtered Poisson process is a superposition of independent components. As such, it is not unexpected that under relatively weak conditions a filtered Poisson process can tend to a Gaussian process as certain parameters, such as the intensity of point occurrences, tend to a limit. This fact is the basis for the widespread use of Gaussian models for random phenomena that would at first appear to be better modeled as filtered point processes. Examples of phenomena where Gaussian statistics are frequently used are shot noise, photoelectron conversion noise, atmospheric noise above the VLF band, radar clutter, and communication in a scattering medium. In Theorem 4.1.2, we give conditions such that a normalized version of the response process y converges to a Gaussian random process on $[t_0, T)$.

Theorem 4.1.2 (Central Limit Theorem for Filtered Poisson Processes). Let

$$\left\{ y_t = \sum_{n=1}^{N_t} h(t, \tau_n : \mathbf{u}_n); t \geqslant t_0 \right\}$$

be a filtered Poisson process in which points occur with intensity $\{\lambda_t; t \geqslant t_0\}$. Suppose for $t \in [t_0, T)$ that

$$\gamma_1(t) = E(y_t) = \int_{t_0}^{t} \lambda_\tau E[h(t, \tau : \mathbf{u})] \, d\tau < \infty, \tag{4.27}$$

$$0 < \gamma_2(t) = \text{var}(y_t) = \int_{t_0}^{t} \lambda_\tau E[h^2(t, \tau : \mathbf{u})] \, d\tau < \infty$$

and that

$$[\gamma_2(t_1)\gamma_2(t_2)\gamma_2(t_3)]^{-1/2} \Gamma(t_1, t_2, t_3)$$

tends to zero uniformly on $[t_0, T) \times [t_0, T) \times [t_0, T)$ as certain parameters tend to prescribed limits, where

$$\Gamma(t_1, t_2, t_3) \triangleq \int_{t_0}^{\min(t_1, t_2, t_3)} \lambda_\tau E\{|h(t_1, \tau : \mathbf{u}) h(t_2, \tau : \mathbf{u}) h(t_3, \tau : \mathbf{u})|\} \, d\tau. \tag{4.28}$$

Then,

$$y_t^* \triangleq [y_t - \gamma_1(t)] \gamma_2^{-1/2}(t)$$

tends to a Gaussian process on $[t_0, T)$ with zero mean and covariance function $[\gamma_2(t_1)\gamma_2(t_2)]^{-1/2} K_y(t_1, t_2)$, where $K_y(t_1, t_2)$ is given in (4.16).

For the proof of Theorem 4.2, we need the following two lemmas.

Lemma 1. For some $\theta(\beta)$ satisfying $|\theta(\beta)| \leqslant 1$, there holds

$$\exp(j\beta) - 1 - j\beta + \tfrac{1}{2}\beta^2 = \tfrac{1}{6}\beta^3 \theta(\beta). \tag{4.29}$$

Proof. Integrate by parts to establish the identity

$$6(\exp(j\beta) - 1 - j\beta + \tfrac{1}{2}\beta^2)\beta^{-3} = 3j^3[\exp(j\beta)] \int_0^1 x^2 \exp(-j\beta x) \, dx.$$

The assertion of Lemma 1 follows by defining $\theta(\beta)$ as the right side of this equation.

Lemma 2 (Characteristic Functional for a Gaussian Process). Let $\{z_t; t \geqslant t_0\}$ be a Gaussian random process with zero mean and covariance function $\rho(t_1, t_2)$. Then the characteristic functional for z on $[t_0, T)$, defined by

$$\phi_z(jv) \triangleq E\left[\exp\left(j\int_{t_0}^{T} z_\sigma \, dv_\sigma\right)\right],$$

has the evaluation

$$\phi_z(jv) = \exp\left\{-\frac{1}{2}\int_{t_0}^{T}\int_{t_0}^{T}\rho(\sigma_1,\sigma_2)\, dv_{\sigma_1} dv_{\sigma_2}\right\}. \tag{4.30}$$

Proof. Let $x = \int_{t_0}^{T} z_\sigma \, dv_\sigma$; then x is a Gaussian random variable with zero mean and variance

$$\sigma_x^2 = \int_{t_0}^{T}\int_{t_0}^{T}\rho(\sigma_1,\sigma_2)\, dv_{\sigma_1} dv_{\sigma_2}.$$

Equation (4.30) then follows from the observation that $\phi_z(jv) = M_x(j1)$, where $M_x(j\alpha)$ is the characteristic function for x, which is given by $M_x(j\alpha) = \exp(-\frac{1}{2}\alpha^2\sigma_x^2)$.

Proof of Theorem 4.1.2. From the definition of y^* in terms of y, note that

$$\phi_{y^*}(jv) \triangleq E\left[\exp\left(j\int_{t_0}^{T} y_\sigma^* \, dv_\sigma\right)\right]$$

$$= E\left[\exp\left(j\int_{t_0}^{T} y_\sigma \gamma_2^{-1/2}(\sigma)\, dv_\sigma\right)\right]\exp\left(-j\int_{t_0}^{T}\gamma_1(\sigma)\gamma_2^{-1/2}(\sigma)\, dv_\sigma\right).$$

Thus, by using

$$y_t \gamma_2^{-1/2}(t) = \sum_{n=1}^{N_t} \gamma_2^{-1/2}(t) h(t, \tau_n : \mathbf{u}_n)$$

and (4.7), we have

$$\ln \phi_{y^*}(jv) = \int_{t_0}^{T} \lambda_\tau E\left(\exp\left[j \int_{t_0}^{T} \gamma_2^{-1/2}(\sigma) h(\sigma, \tau : \mathbf{u})\, dv_\sigma \right] - 1 \right) d\tau$$

$$-j \int_{t_0}^{T} \gamma_1(\sigma) \gamma_2^{-1/2}(\sigma)\, dv_\sigma. \tag{4.31}$$

Suppose now that z is a Gaussian process on $[t_0, T)$ with zero mean and covariance function

$$\rho(t_1, t_2) \triangleq [\gamma_2(t_1)\gamma_2(t_2)]^{-1/2} K_y(t_1, t_2)$$

where K_y is given in (4.16). Then subtracting the logarithm of ϕ_z in (4.30) from both sides of (4.31), using the expression for K_y in (4.16), and using the expression for γ_1 in (4.27) results in

$$\ln \phi_{y^*}(jv) + \frac{1}{2} \int_{t_0}^{T} \int_{t_0}^{T} [\gamma_2(\sigma_1)\gamma_2(\sigma_2)]^{-1/2} K_y(\sigma_1, \sigma_2)\, dv_{\sigma_1} dv_{\sigma_2}$$

$$= \int_{t_0}^{T} \lambda_\tau E\left\{ e^{j\beta(\tau : \mathbf{u})} - 1 - j\beta(\tau : \mathbf{u}) + \tfrac{1}{2}\beta^2(\tau : \mathbf{u}) \right\} d\tau$$

$$= \int_{t_0}^{T} \lambda_\tau E\left\{ \tfrac{1}{6}\beta^3(\tau, \mathbf{u}) \theta(\tau, \mathbf{u}) \right\} d\tau,$$

where

$$\beta(\tau : \mathbf{u}) \triangleq \int_{t_0}^{T} \gamma_2^{-1/2}(\sigma) h(\sigma, \tau : \mathbf{u})\, dv_\sigma$$

and the last equality follows from Lemma 1, where $|\theta(\tau, \mathbf{u})| \leqslant 1$. Upon expansion of β^3 and rearrangement of the integrals, the right side becomes

$$\frac{1}{6} \int_{t_0}^{T} \int_{t_0}^{T} \int_{t_0}^{T} [\gamma_2(\sigma_1)\gamma_2(\sigma_2)\gamma_2(\sigma_3)]^{-1/2}$$

$$\times E\left\{ \int_{t_0}^{T} \lambda_\tau h(\sigma_1, \tau : \mathbf{u}) h(\sigma_2, \tau : \mathbf{u}) h(\sigma_3, \tau : \mathbf{u}) \theta(\tau, \mathbf{u})\, d\tau \right\} dv_{\sigma_1} dv_{\sigma_2} dv_{\sigma_3}.$$

Taking the magnitude of both sides then yields

$$\left|\ln\phi_{y^*}(jv)+\frac{1}{2}\int_{t_0}^{T}\int_{t_0}^{T}[\gamma_2(\sigma_1)\gamma_2(\sigma_2)]^{-1/2}K_y(\sigma_1,\sigma_2)\,dv_{\sigma_1}\,dv_{\sigma_2}\right|$$

$$\leqslant\frac{1}{6}\int_{t_0}^{T}\int_{t_0}^{T}\int_{t_0}^{T}[\gamma_2(\sigma_1)\gamma_2(\sigma_2)\gamma_2(\sigma_3)]^{-1/2}\Gamma(\sigma_1,\sigma_2,\sigma_3)|dv_{\sigma_1}\,dv_{\sigma_2}\,dv_{\sigma_3}|,\tag{4.32}$$

where Γ is defined in (4.28). The integrand on the right tends uniformly to zero as certain parameters tend to prescribed limits. Consequently, the left side also tends to zero, and in the limit we conclude that y^* has the characteristic functional of z on $[t_0,T)$. Thus, in the limit, y^* has the same finite distributions as a Gaussian process with zero mean and covariance function $[\gamma_2(t_1)\gamma_2(t_2)]^{-1/2}K_y(t_1,t_2)$, where K_y is given in (4.16). This completes the proof.

Example 4.1.10 (Homogeneous Point Process). Suppose the point process underlying the generation of y is homogeneous with a constant intensity λ. Then (4.32) becomes

$$\left|\ln\phi_{y^*}(jv)+\frac{1}{2}\int_{t_0}^{T}\int_{t_0}^{T}\rho(\sigma_1,\sigma_2)\,dv_{\sigma_1}\,dv_{\sigma_2}\right|$$

$$\leqslant\tfrac{1}{6}\lambda^{-1/2}\int_{t_0}^{T}\int_{t_0}^{T}\int_{t_0}^{T}f(\sigma_1,\sigma_2,\sigma_3)\,dv_{\sigma_1}\,dv_{\sigma_2}\,dv_{\sigma_3},\tag{4.33}$$

where

$$\rho(\sigma_1,\sigma_2)\triangleq\frac{\int_{t_0}^{T}E[h(\sigma_1,\tau:\mathbf{u})h(\sigma_2,\tau:\mathbf{u})]\,d\tau}{\left\{\int_{t_0}^{T}E[h^2(\sigma_1,\tau:\mathbf{u})]\,d\tau\int_{t_0}^{T}E[h^2(\sigma_2,\tau:\mathbf{u})]\,d\tau\right\}^{1/2}}$$

and where

$$f(\sigma_1,\sigma_2,\sigma_3)$$

$$\triangleq\frac{\int_{t_0}^{T}E\{|h(\sigma_1,\tau:\mathbf{u})h(\sigma_2,\tau:\mathbf{u})h(\sigma_3,\tau:\mathbf{u})|\}\,d\tau}{\left\{\int_{t_0}^{T}E[h^2(\sigma_1,\tau:\mathbf{u})\,d\tau\int_{t_0}^{T}E[h^2(\sigma_2,\tau:\mathbf{u})\,d\tau\int_{t_0}^{T}E[h^2(\sigma_3,\tau:\mathbf{u})]\,d\tau\right\}^{1/2}}.$$

Consequently, for intervals where the triple integral in (4.33) is finite, we conclude that as λ tends to infinity, y^* tends to a Gaussian process with covariance function ρ.

Example 4.1.11 (Bounded Response Function). For the function v in (4.33), let us select

$$v_\sigma = \begin{cases} 0, & t_0 \leqslant \sigma < t \\ \alpha, & t \leqslant \sigma < T \end{cases}$$

where α is a constant. Then $\phi_{y^*}(jv)$ is the characteristic function $M_{y_t^*}(j\alpha)$ for y^* at time $t \in [t_0, T)$ and (4.33) becomes

$$|\ln M_{y_t^*}(j\alpha) + \tfrac{1}{2}\alpha^2| \leqslant \tfrac{1}{6}\lambda^{-1/2}\alpha^3 \frac{\int_{t_0}^{T} E\{|h^3(t,\tau:\mathbf{u})|\}\, d\tau}{\left(\int_{t_0}^{T} E\{h^2(t,\tau:\mathbf{u})\}\, d\tau\right)^{3/2}}. \tag{4.34}$$

Suppose, further, that the response function h has the form $h(t,\tau:\mathbf{u}) = ug(t,\tau)$ where u is a scalar mark variable and g is bounded as $|g| < C$. Then (4.34) implies

$$|\ln M_{y_t^*}(j\alpha) + \tfrac{1}{2}\alpha^2| \leqslant \tfrac{1}{6}\lambda^{-1/2}\alpha^3 E(u^3) E^{-3/2}(u^2) C \left(\int_{t_0}^{t} g^2(t,\tau)\, d\tau\right)^{-1/2} \tag{4.35}$$

We next use this expression to investigate the situation for a time-invariant, band-limited response function.

Example 4.1.12 (Time-Invariant and Band-limited Response Function). Suppose $h(t,\tau:\mathbf{u}) = ug(t-\tau)$ so that the response function is time invariant and linearly dependent on a mark variable u. Further, assume that $t_0 = -\infty$ and that

$$G(f) \triangleq \int_0^\infty g(t) \exp(-j2\pi f t)\, dt$$

is identically zero for $|f| > f_c/2$ so that the response is also band-limited. Then, by paralleling an argument due to A. Papoulis [20], we can show that g is bounded so that (4.35) can be used. From the Fourier inversion equation and the Schwarz inequality, we have that

$$|g(t)|^2 = \left| \int_{-f_c/2}^{f_c/2} G(f) \exp(j2\pi ft)\, df \right|^2$$

$$\leqslant f_c \int_{-f_c/2}^{f_c/2} |G(f)|^2 df \triangleq C^2.$$

By using Parseval's equation we also have

$$\int_{-\infty}^{t} g^2(t-\tau)\, d\tau = \int_0^\infty g^2(t)\, dt = \int_{-f_c/2}^{f_c/2} |G(f)|^2 df.$$

These results in (4.35) imply that

$$|\ln M_{y_t^*}(j\alpha) + \tfrac{1}{2}\alpha^2| \leqslant \tfrac{1}{6} E(u^3) E^{-3/2}(u^2) \left[\frac{f_c}{\lambda} \right]^{1/2} \alpha^3. \tag{4.36}$$

Thus, we conclude that under these conditions, y_t^* approaches a Gaussian random variable for $t \in (-\infty, \infty)$ as the bandwidth to rate ratio f_c/λ approaches zero.

It is difficult to assess from (4.36) how close the actual distribution for y_t^* is approximated by a Gaussian distribution for f_c/λ small but nonzero. A. Papoulis [20] shows for the case $u=1$ that

$$\left| \Pr(y_t^* \leqslant Y) - \frac{1}{\sqrt{2\pi}} \int_{-\infty}^{Y} \exp\left(\frac{-\xi^2}{2} \right) d\xi \right| \leqslant \tfrac{4}{3} \left(\frac{2\pi f_c}{\lambda} \right)^{1/2}, \tag{4.37}$$

which gives an upper bound on the error incurred.

Example 4.1.13 (Shot Noise). For shot noise, $f_c/2 \cong 10^9$ Hz, $\lambda^{-1} = eI_{\mathrm{DC}}^{-1}$, and for $I_{\mathrm{DC}} = 10^{-3}$ A, we have $(2\pi f_c/\lambda)^{1/2} \cong 1.4 \times 10^{-3}$. Consequently, the distribution function for y_t^* is uniformly within ± 0.002 of a zero-mean, unit-variance Gaussian distribution function. This justifies, in part, the common modeling of n in Figure 4.3 (b) as a Gaussian process.

Summary. In this section, we have studied a class of processes that arise as the response to the points of a marked point process. Filtered Poisson processes were emphasized. These are the superposition of the separate responses to each marked point of a compound Poisson process. Several applications where a filtered Poisson process provides a reasonable mathematical model were mentioned; additional applications are given by E. Parzen [21, Section 4.5]. Statistics for filtered Poisson processes were developed. The principle result is the characteristic functional given in Theorem 4.1.1, which provides a complete statistical characterization of the process. Second-moment properties of filtered Poisson processes are widely used in applications. Most particularly Campbell's theorem, which is developed in Example 4.1.7, is often cited. The central limit theorem given in Theorem 4.1.2 provides the mathematical justification for the frequent use of Gaussian models for processes that would appear more appropriately modeled by a filtered Poisson process.

We now turn to the development of a model in which the response need not be a superposition of individual responses to each marked point.

4.2 POISSON DRIVEN MARKOV PROCESSES

The response to a marked point process studied in Section 4.1 is comprised of the separate responses to each point. In this section, we develop a model in which the response is generally not a superposition of individual effects. The importance of the model is due to the wide variety of phenomena encountered in physics and engineering for which it provides an accurate mathematical representation.

For reasons that will become evident, we call the process developed in this section a *Poisson driven Markov process*. A Poisson driven Markov process satisfies an integral equation of the form

$$\mathbf{x}_t = \mathbf{x}_0 + \int_{t_0}^{t} \mathbf{a}_\sigma(\mathbf{x}_\sigma)\, d\sigma + \int_{t_0}^{t}\int_{\mathfrak{U}} \mathbf{b}_\sigma(\mathbf{x}_\sigma, \mathbf{U}) M(d\sigma, d\mathbf{U}), \tag{4.38}$$

where M is the time-space Poisson process introduced in Theorem 3.3.2 of Chapter 3 and $\mathfrak{U}$ is the mark space of the compound Poisson process associated with M. The functions $\mathbf{a}_t(\mathbf{x}_t)$ and $\mathbf{b}_t(\mathbf{x}_t, \mathbf{U})$ satisfy certain technical conditions to be stated later. As developed below, the last integral in (4.38) has an evaluation given by

$$\int_{t_0}^{t}\int_{\mathfrak{U}} \mathbf{b}_\sigma(\mathbf{x}_\sigma, \mathbf{U}) M(d\sigma, d\mathbf{U}) = \begin{cases} 0, & N_t = 0 \\ \displaystyle\sum_{n=1}^{N_t} \mathbf{b}_{\tau_n}(\mathbf{x}_{\tau_n}, \mathbf{u}_n), & N_t \geq 1, \end{cases}$$

where $N_t = \int_{t_0}^{t}\int_{\mathfrak{U}} M(d\sigma, d\mathbf{U})$ is the number of incident points during $[t_0, t)$ regardless of their mark and τ_n and $\mathbf{u}_n$ are the time of occurrence and mark of the nth point.

We also write the equation for $\{\mathbf{x}_t; t \geqslant t_0\}$ somewhat more concisely in differential form as

$$d\mathbf{x}_t = \mathbf{a}_t(\mathbf{x}_t)dt + \int_{\mathfrak{U}} \mathbf{b}_t(\mathbf{x}_t, \mathbf{U}) M(dt, d\mathbf{U}) \tag{4.39}$$

with the initial condition $\mathbf{x}_{t_0} = \mathbf{x}_0$. This stochastic differential equation can be viewed as defining a nonlinear transformation of the Poisson process M into the process $\mathbf{x}$. At $t = t_0$, $\mathbf{x}_{t_0} = \mathbf{x}_0$. The differential $d\mathbf{x}_t = \mathbf{x}_{t+dt} - \mathbf{x}_t$ denotes an infinitesimal increment in $\mathbf{x}$ which occurs during $[t, t+dt)$. If during this interval no points occur in the incident marked point process, $M(dt, d\mathbf{U}) = 0$ and the increment in $\mathbf{x}$ is $\mathbf{a}_t(\mathbf{x}_t)dt$. If a point having a mark $\mathbf{U}$ does occur during $[t, t+dt)$, the increment in $\mathbf{x}$ is $\mathbf{a}_t(\mathbf{x}_t)dt + \mathbf{b}_t(\mathbf{x}_t, \mathbf{U})$, which is dominated by $\mathbf{b}_t(\mathbf{x}_t, \mathbf{U})$. It is evident that $\mathbf{x}$ will have discontinuities at the occurrence times of the incident points and that the size of the discontinuity for a point occurring at time t with mark $\mathbf{U}$ is $\mathbf{b}_t(\mathbf{x}_t, \mathbf{U})$.

We will refer to $\mathbf{x}_t$ as the *state* of the Poisson driven Markov process at time t. The state takes values in the *state space* $\mathfrak{X}$ of the process; this is an n-dimensional Euclidean space.

The following examples motivate our study of Poisson driven Markov processes and provide additional interpretation of (4.39).

Example 4.2.1 (Brownian Motion). Consider the motion in one dimension of a small particle of mass m immersed in a fluid. Thermally agitated molecules of the fluid strike the particle causing it to move chaotically. The motion is damped to some extent depending upon the viscosity of the fluid. Let v_t denote the velocity of the particle at time $t \geqslant t_0$. Then, in the absence of any external force, such as a gravitational force, the equation of motion is

$$m\left(\frac{dv_t}{dt}\right) = -\alpha v_t + f_t, \tag{4.40}$$

where α is a friction coefficient that depends on the viscosity of the fluid and f_t represents fluctuation forces due to molecular collisions. This equation is called *Langevin's equation* in honor of P. Langevin [17], who first proposed it as a model for Brownian motion. If it is supposed that the time of contact during a collision is very short, the resulting force may be assumed impulsive. In this case, the velocity will not be differentiable and (4.40) must be replaced by

$$m\,dv_t = -\alpha v_t\,dt + du_t, \tag{4.41}$$

where $\{u_t; t \geqslant t_0\}$ is the accumulator process of a marked point process.

The occurrence times and marks of this point process correspond to the instants collisions take place and the force induced on the particle, respectively. If we suppose further that $\{u_t; t \geqslant t_0\}$ is a compound Poisson process and use the representation for these processes in Theorem 3.3.2, Langevin's equation becomes

$$dv_t = -m^{-1}\alpha v_t\,dt + \int_{\mathrm{U}} m^{-1} U M(dt, dU). \tag{4.42}$$

This is in the form of (4.39) with $a_t(v_t) = -m^{-1}\alpha v_t$ and $b_t(v_t, U) = m^{-1}U$. It is straightforward to verify that the solution to (4.42) is

$$v_t = v_0 \exp[-m^{-1}\alpha(t-t_0)] + \int_{t_0}^{t}\int_{\mathrm{U}} m^{-1} U \exp[-m^{-1}\alpha(t-\sigma)] M(d\sigma, dU), \tag{4.43}$$

where v_0 is the velocity of the particle at time t_0. A solution can be obtained in this example due to the linearity of (4.42). Explicit solution of (4.39) is not generally possible when the equation is nonlinear. The last integral in (4.43) is zero if no impacts occur during $[t_0, t)$. Otherwise, its evaluation is

$$\sum_{n=1}^{N_t} h(t, \tau_n : u_n)$$

where τ_n and u_n are the time of occurrence and force induced by the nth impact and

$$h(t, \tau : u) = \begin{cases} m^{-1} u \exp[-m^{-1}\alpha(t-\tau)], & t \geqslant \tau \\ 0, & t < \tau. \end{cases}$$

Thus, if the initial velocity v_0 is zero, $\{v_t; t \geqslant t_0\}$ is a filtered Poisson process of the type studied in Section 4.1.

Example 4.2.2 (Linear Dynamical Systems). Linear dynamical systems are employed as models in all branches of engineering. These are systems which can be described by ordinary, linear differential equations. Langevin's equation (4.40) is an especially simple example. More generally, the equations take the form

$$\frac{d\mathbf{x}_t}{dt} = \mathbf{A}_t \mathbf{x}_t + \mathbf{B}_t \mathbf{f}_t, \qquad \mathbf{x}_{t_0} = \mathbf{x}_0 \tag{4.44}$$

$$\mathbf{y}_t = \mathbf{C}_t \mathbf{x}_t, \tag{4.45}$$

where $\mathbf{f}$ is the input or excitation function, $\mathbf{x}$ is termed the *state* of the system, and $\mathbf{y}$ is the output or response function. If $\mathbf{f}$, $\mathbf{x}$, and $\mathbf{y}$ are vectors of dimension p, n, and r, respectively, then $\mathbf{A}$, $\mathbf{B}$, and $\mathbf{C}$ are matrices of dimension $n\times n$, $n\times p$, and $r\times n$, respectively. Equations (4.44) and (4.45) are called the *state* and *output* equations, respectively. There are a number of ways for reducing an nth order linear differential equation to the form of (4.44) and (4.45). These are described by L. Zadeh and C. DeSoer [27] and R. Brockett [4]. If the excitation $\mathbf{f}$ in (4.44) is impulsive with the $\mathbf{n}$th impulse occurring at time τ_n and carrying weight $\mathbf{u}_n$, then $\mathbf{x}$ will not be differentiable and the state equation must be rewritten as

$$d\mathbf{x}_t=\mathbf{A}_t\mathbf{x}_t\,dt+\mathbf{B}_t\,d\mathbf{u}_t, \qquad \mathbf{x}_{t_0}=\mathbf{x}_0 \tag{4.46}$$

where $\mathbf{u}$ is a piecewise constant function of t. Jumps in $\mathbf{u}$ occur at the impulse times of $\mathbf{f}$ and have a size equal to the corresponding impulse weight. Suppose $\{\mathbf{u}_t; t\geqslant t_0\}$ is a compound Poisson process. Then, by using the representation in Theorem 3.3.2 we can write (4.46) as

$$d\mathbf{x}_t=\mathbf{A}_t\mathbf{x}_t\,dt+\int_{\mathrm{U}}\mathbf{B}_t\mathrm{U}M(d\sigma,d\mathrm{U}), \qquad \mathbf{x}_{t_0}=\mathbf{x}_0. \tag{4.47}$$

Equation (4.47) is in the form of (4.39) with $\mathbf{a}_t(\mathbf{x}_t)=\mathbf{A}_t\mathbf{x}_t$ and $\mathbf{b}_t(\mathbf{x}_t,\mathrm{U})=\mathbf{B}_t\mathrm{U}$. This equation can be solved explicitly for $\mathbf{x}_t$ due to its linearity. Thus, it is easily verified that

$$\mathbf{x}_t=\mathbf{\Phi}(t,t_0)\mathbf{x}_0+\int_{t_0}^{t}\int_{\mathrm{U}}\mathbf{\Phi}(t,\sigma)\mathbf{B}_\sigma\mathrm{U}M(d\sigma,d\mathrm{U}), \tag{4.48}$$

where $\mathbf{\Phi}(t,t_0)$ is an $n\times n$ matrix, called the *state-transition matrix*, satisfying

$$\frac{\partial\mathbf{\Phi}(t,t_0)}{\partial t}=\mathbf{A}_t\mathbf{\Phi}(t,t_0), \qquad \mathbf{\Phi}(t_0,t_0)=\mathbf{I}. \tag{4.49}$$

According to (4.48), the state is a superposition of responses to each marked point. If $\mathbf{x}_{t_0}=\mathbf{0}$, then $\{\mathbf{x}_t; t\geqslant t_0\}$ is a vector-valued, filtered Poisson process of the type studied in Section 4.1.

Example 4.2.3 (Phase-Tracking Loop Excited by Shot Noise). Suppose an optical communication system is used to establish a phase or timing reference signal at a remote receiver. For this purpose, a subcarrier, $\cos[\omega_r t+\theta_r(t)]$, of reference frequency ω_r and phase θ_r is used to modulate the intensity of a coherent light source, such as a laser, which is pointed at the receiver. As shown by W. Pratt [22, p. 178], the term $|S(t,\vec{\mathrm{r}})|^2$ appearing in (4.22) is of the form

$$|S(t,\vec{r})|^2 = P\{1+\alpha\cos(\omega_r t+\theta_r(t))\},$$

where P is a nonnegative constant and α is the subcarrier modulation index with $|\alpha|\leqslant 1$. As shown in Figure 4.5, the modulated optical field is incident on a photodetector with gain, such as a photomultiplier or avalanche photodiode. We take the detector response to be an impulse $u\delta(t-\tau)$, where u is a nonnegative random variable modeling the detector gain. The photodetector output $\{u_t; t\geqslant t_0\}$ is a compound Poisson process given by

$$u_t = \begin{cases} 0, & N_t = 0 \\ \sum_{n=1}^{N_t} u_n, & N_t \geqslant 1, \end{cases}$$

where the intensity of the Poisson counting process $\{N_t; t\geqslant t_0\}$ is

$$\lambda_t = \lambda_t^s + \lambda_0 = A\{1+m\cos[\omega_r t+\theta_r(t)]\}$$

where $A \triangleq (\eta\mathcal{Q}P/h\nu)+\lambda_0$, $Am \triangleq \eta\mathcal{Q}P\alpha/h\nu$, and λ_0 models extraneous counts due to background radiation and dark current (that is, $\lambda_0 = \lambda^b + \lambda^d$). The parameters η, $\mathcal{Q}$, h, and ν are as defined in Example 4.1.9.

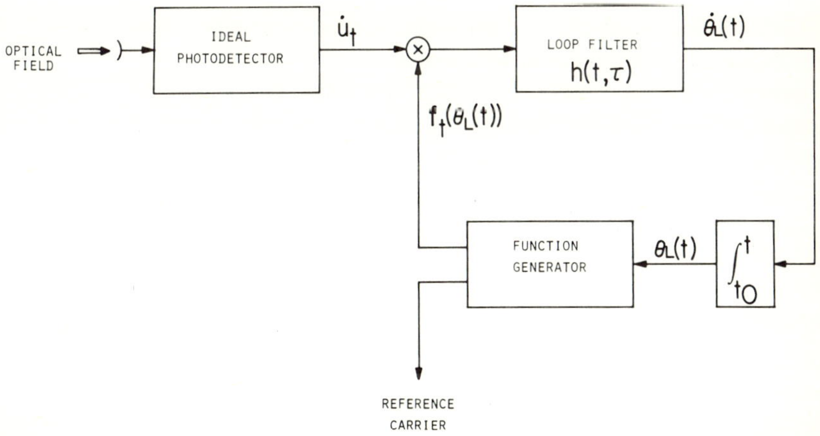

Figure 4.5. Phase-tracking loop.

The output of the photodetector is the input to a tracking loop that generates a receiver reference carrier, $\cos[\omega_L t + \theta_L(t)]$, of loop frequency ω_L and phase θ_L, where θ_L is the solution to

$$\theta_L(t) = \begin{cases} 0, & N_t = 0 \\ \sum_{n=1}^{N_t} u_n f_{\tau_n}[\theta_L(\tau_n)] h(t, \tau_n), & N_t \geq 1. \end{cases}$$

Alternatively, $\theta_L(t)$ can be written as

$$\theta_L(t) = \int_{t_0}^{t} \int_{\mathfrak{U}} U f_\tau[\theta_L(\tau)] h(t, \tau) M(d\tau, dU). \tag{4.50}$$

Various forms for the function f are of interest in practice. The most common form is

$$f_t[\theta_L(t)] = \sin[\omega_L t + \theta_L(t)] \tag{4.51}$$

for which the tracking loop is termed a *phase-locked loop*; in this case, the feedback function f and the reference carrier in Figure 4.5 are easily realized using a voltage-controlled oscillator. Phase-locked loops and their application in communication systems are discussed by A. Viterbi [25, Chapters 2–4] and W. Lindsey and M. Simon [18, Chapter 2]. Another useful form for f is

$$f_t[\theta_L(t)] = \frac{\sin[\omega_L t + \theta_L(t)]}{1 + m\cos[\omega_L t + \theta_L(t)]}$$

for which the tracking loop is termed a *tanlock loop*; motivation for the use of this loop will be given in Chapter 6.

Equation (4.50) cannot be solved explicitly for $\theta_L(t)$ and, therefore, cannot be placed in the form of the superposition (4.1). It follows that an analysis of the behavior of the loop cannot be accomplished using the procedures of Section 4.1. However, by describing θ_L in the form of (4.39), R. Gagliardi and M. Haney [10] and R. Forrester and D. Snyder [9] have analyzed the loop behavior to some extent; see Example 4.2.12 below. For this purpose, we assume that the loop filter h has a finite-dimensional state realization as

$$\begin{aligned} d\mathbf{z}_t &= \mathbf{A}_t \mathbf{z}_t\,dt + \int_{\mathfrak{U}} U \mathbf{b}_t f_t[\theta_L(t)] M(dt, dU), \\ d\theta_L(t) &= \mathbf{c}_t' \mathbf{z}_t\,dt + \int_{\mathfrak{U}} U d_t f_t[\theta_L(t)] M(dt, dU). \end{aligned} \tag{4.52}$$

That is, h is described by the matrix $\mathbf{A}$, the vectors $\mathbf{b}$ and $\mathbf{c}$, the scalar d, and the state vector $\mathbf{z}$. To place (4.52) in the form of (4.39), define $\mathbf{x}$ by

$$\mathbf{x}_t \triangleq \begin{bmatrix} \mathbf{z}_t \\ \theta_L(t) \end{bmatrix}$$

Then, in (4.39), we have

$$\mathbf{a}_t(\mathbf{x}_t) = \begin{bmatrix} \mathbf{A}_t & \mathbf{0} \\ \mathbf{c}'_t & \mathbf{0} \end{bmatrix}$$

and

$$\mathbf{b}_t(\mathbf{x}_t, \mathbf{U}) = U \begin{bmatrix} \mathbf{b}_t \\ d_t \end{bmatrix} f_t[\mathbf{h}'\mathbf{x}_t]$$

where $\mathbf{h}' = [0 \quad 0 \quad 0 \quad \cdots \quad 0 \quad 1]$.

In the remainder of this section, we give an introduction to the theory of Poisson driven Markov processes; that is, we will study solutions of (4.38) and (4.39). The calculus for Poisson driven Markov processes is developed in Section 4.2.1. Our first consideration is with a more careful treatment of the last integral in (4.38), which we term a *counting integral.* A counting integral is a special form of integral introduced by K. Itô [15] in his fundamental memoir on stochastic differential equations. The scalar version of the integral is discussed by J. Doob [6, p. 436] and the vector version by A. Skorokhod [24, Chapter 2] and I. Gihman and A. Skorokhod [12, Part II, Chapter 2]. Properties of the counting integral will be used to establish conditions for the existence and uniqueness of solutions to (4.38). The rule for differentiating a function of a Poisson driven Markov process is also given in Section 4.2.1. This rule is important here as well as in later chapters, particularly in Chapter 6 where it is used repeatedly. The differential rule we give for Poisson driven Markov processes is analogous to the Itô differential rule for differentiating a function of a Markov diffusion process; see A. Skorokhod [24, p. 24] or E. Wong [26, p. 148] for a statement of the Itô differential rule. In Section 4.2.2, we investigate the statistics of solutions to (4.38). For the definition of the counting integral adopted in Section 4.2.1, $\{\mathbf{x}_t; t \geqslant t_0\}$ is a Markov process, which motivates our terminology. Markov processes will be defined and reviewed briefly; they are developed further in Chapter 6.

4.2.1 Calculus for Poisson Driven Markov Processes

Integration. We have so far let the last integral in (4.38) pass without detailed comment in order to motivate its natural occurrence in applications where nonlinear systems are excited by marked point processes. We now give a definition for the integral and indicate its evaluation.

The integral of interest is of the general form

$$\mathbf{I}_t(A,\omega)=\int_{t_0}^{t}\int_{A}\mathbf{b}_\sigma(\mathbf{U},\omega)M(d\sigma,d\mathbf{U},\omega), \tag{4.53}$$

where $A\subseteq\mathfrak{U}$ and $\mathbf{b}_\sigma(\mathbf{U},\omega)$ is a random function of σ and $\mathbf{U}$. To emphasize the randomness of $\mathbf{b}$, we have included the dependency in (4.53) on the sample-space variable ω of an underlying probability space. In (4.38), this randomness in $\mathbf{b}$ enters through the influence of $\mathbf{x}$, which, in turn, is random because of the influence of M. We will henceforth suppress the explicit dependence on ω and write (4.53) simply as

$$\mathbf{I}_t(A)=\int_{t_0}^{t}\int_{A}\mathbf{b}_\sigma(\mathbf{U})M(d\sigma,d\mathbf{U}). \tag{4.54}$$

It is the possible dependence of $\mathbf{b}$ on M that makes the definition and evaluation of $\mathbf{I}$ somewhat delicate. As an example illustrating the potential difficulties, suppose $\mathbf{b}$ does not depend on the marks $\mathbf{U}$, so $\mathbf{I}_t(\mathfrak{U})=\int_{t_0}^{t}\mathbf{b}_\sigma\,dN_\sigma$, where $N_t=\int_{t_0}^{t}\int_{\mathfrak{U}}M(d\sigma,d\mathbf{U})$ is the number of points on $[t_0,t)$ regardless of their mark. Suppose, further, that $\mathbf{b}$ is real valued and $b_\sigma=N_\sigma$. Then $I_t=\int_{t_0}^{t}N_\sigma\,dN_\sigma$. The evaluation of I_t according to the usual rules of integration would suggest for this example that $I_t=N_t^2/2$. However, closer examination indicates, first, that the integral $\int_{t_0}^{t}N_\sigma\,dN_\sigma$ cannot exist in the usual Riemann–Stieltjes sense because of the simultaneous jumps in the integrand $b_\sigma=N_\sigma$ and the integrator, N_σ (see T. Apostol [1, Theorem 9-28].) Secondly, an effort to evaluate $\int_{t_0}^{t}N_\sigma\,dN_\sigma$ from first principles can lead to more than one answer. To see this, consider a sequence of partitions $t_0=t_{0n}\leqslant t_{1n}\leqslant t_{2n}\leqslant\cdots\leqslant t_{nn}=t$ of $[t_0,t)$ such that $\max_{1\leqslant i\leqslant n}(t_{in}-t_{i-1,n})\to 0$ as $n\to\infty$, and examine the sum

$$S_n(\alpha)=\sum_{i=1}^{n}N_{t'_{in}}\left(N_{t_{in}}-N_{t_{i-1,n}}\right)$$

as $n\to\infty$, where $t'_{in}=\alpha t_{i-1,n}+(1-\alpha)t_{in}$, for $0\leqslant\alpha\leqslant 1$, is a point in the subinterval $[t_{i-1},t_i]$. The choice of α affects the evaluation of this limit, which is not the case for the usual Riemann–Stieltjes integral. First set

$\alpha = 1$ so that t'_{in} is the left end $t_{i-1,n}$ of the subinterval. We have

$$\lim_{n\to\infty} S_n(1) = \lim_{n\to\infty} \sum_{i=1}^{n} N_{t_{i-1,n}}\left(N_{t_{in}} - N_{t_{i-1,n}}\right)$$

$$= \frac{1}{2}\lim_{n\to\infty} \sum_{i=1}^{n} \left(N_{t_{in}}^2 - N_{t_{i-1,n}}^2\right) - \frac{1}{2}\lim_{n\to\infty} \sum_{i=1}^{n} \left(N_{t_{in}} - N_{t_{i-1,n}}\right)^2$$

$$= \tfrac{1}{2}N_t^2 - \tfrac{1}{2}N_t.$$

Similarly, for $\alpha = 0$, so that $t_{in} = t'_{in}$ is the right end of the subinterval, we have

$$\lim_{n\to\infty} S_n(0) = \lim_{n\to\infty} \sum_{i=1}^{n} N_{t_{in}}\left(N_{t_{in}} - N_{t_{i-1,n}}\right)$$

$$= \frac{1}{2}\lim_{n\to\infty} \sum_{i=1}^{n} \left(N_{t_{in}}^2 - N_{t_{i-1,n}}^2\right) + \frac{1}{2}\lim_{n\to\infty} \sum_{i=1}^{n} \left(N_{t_{in}} - N_{t_{i-1,n}}\right)^2$$

$$= \tfrac{1}{2}N_t^2 + \tfrac{1}{2}N_t.$$

And for $0 < \alpha < 1$, we have

$$\lim_{n\to\infty} S_n(\alpha) = \lim_{n\to\infty} \sum_{i=1}^{n} N_{\alpha t_{i-1,n} + (1-\alpha)t_{in}}\left(N_{t_{in}} - N_{t_{i-1,n}}\right)$$

$$= \lim_{n\to\infty} \sum_{i=1}^{n} N_{\alpha t_{i-1,n} + (1-\alpha)t_{in}}\left(N_{t_{in}} - N_{\alpha t_{i-1,n} + (1-\alpha)t_{in}}\right)$$

$$+ \lim_{n\to\infty} \sum_{i=1}^{n} N_{\alpha t_{i-1,n} + (1-\alpha)t_{in}}\left(N_{\alpha t_{i-1,n} + (1-\alpha)t_{in}} - N_{t_{i-1,n}}\right)$$

$$= \tfrac{1}{2}N_t^2. \tag{4.55}$$

Thus, we conclude

$$\int_{t_0}^{t} N_\sigma \, dN_\sigma = \begin{cases} \tfrac{1}{2}N_t^2 + \tfrac{1}{2}N_t, & \alpha = 0 \\ \tfrac{1}{2}N_t^2, & 0 < \alpha < 1 \\ \tfrac{1}{2}N_t^2 - \tfrac{1}{2}N_t, & \alpha = 1. \end{cases} \tag{4.56}$$

Only for α in the range $0<\alpha<1$ does the integral $\int_{t_0}^{t} N_t\, dN_t$ have the usual evaluation $N_t^2/2$ of an ordinary integral, and the selection of α in this range for the definition of the integral in (4.56) would appear natural. However, there is strong motivation to forgo this familiar evaluation and use $\alpha=1$ so that t_{in}' is at the left end of the interval. The reason is that for this choice alone, the term $N_{t_{in}'}$ does not anticipate the increment $N_{t_{in}} - N_{t_{i-1,n}}$ because $\{N_t; t \geqslant t_0\}$ has independent increments. Furthermore, with the choice $\alpha=1$, the highly developed theory of *Martingales* can be used to study the statistical properties of the integral; see Problem 4.2.1. While we will not use the Martingale theory explicitly, we will follow Itô's precedent and select $\alpha=1$ in the definition of $\mathbf{I}_t(A)$ in (4.54). A consequence of this selection is that the counting integral that results is incompatible with the integration rules of ordinary calculus, as is evident in the above simple example. Thus, some care is needed in the evaluation of counting integrals. We will give an explicit evaluation consistent with that of previous sections where counting integrals have been used formally.

The following conditions will be assumed to hold throughout the discussion.

a. M is a Poisson process on $[t_0, \infty) \times \mathfrak{U}$. This is the process of Theorem 3.3.2 for representing a compound Poisson process on $[t_0, \infty)$ with mark space $\mathfrak{U}$. Here, $\int_{t_1}^{t_2}\int_A M(dt, d\mathbf{U})$ is the random number of points occurring in the interval $[t_1, t_2)$ with marks in $A \subseteq \mathfrak{U}$. We take $\mathfrak{U}$ to be an m-dimensional Euclidean space.

b. For all finite intervals $[t_1, t_2)$ and $A \subseteq \mathfrak{U}$, there holds

$$E\left[\int_{t_1}^{t_2}\int_A M(d\sigma, d\mathbf{U})\right] = \Pr(\mathbf{u} \in A)\int_{t_1}^{t_2} \lambda_\sigma\, d\sigma < \infty.$$

This assumption implies that regardless of their marks, the number of points in $[t_1, t_2)$ is finite with probability one.

c. The process $\{\mathbf{b}_t(\mathbf{U}); t \geqslant t_0, \mathbf{U} \in \mathfrak{U}\}$ does not anticipate the underlying point process. We mean by this that $\mathbf{b}_t(\mathbf{U}')$ is statistically independent of $\int_{t_1}^{t_2}\int_A M(d\sigma, d\mathbf{U})$ for any choice of $\mathbf{U}' \in \mathfrak{U}$ and $A \subseteq \mathbf{U}$ and for all triples $t \leqslant t_1 \leqslant t_2$.

d. Almost every realization of the process $\mathbf{b}_t(\mathbf{U})$ is continuous in $\mathbf{U}$, left continuous in t, and bounded for $(t, \mathbf{U}) \in [t_1, t_2) \times A$, where $[t_1, t_2)$ is any bounded interval in $[t_0, \infty)$ and A is any bounded set in $\mathfrak{U}$.

These several conditions will ensure the existence of the counting integral we introduce next.

Definition 4.1.1 (Counting Integral). Let A be a set in $\mathfrak{U}$ and consider a sequence of partitions of $[t_0,t)\times A$ obtained as follows. Let $A_{1p}, A_{2p},\ldots,$ for $p=1,2,\ldots,$ be a sequence of pairwise disjoint partitions of A such that $A_{kp}\cap A_{jp}=\phi$ for $k\neq j$, $A=\cup_k A_{kp}$, and $\max_k\|A_{kp}\|\to 0$ as $p\to\infty$, where $\|A_{kp}\|$ denotes the diameter of the set A_{kp}.† Similarly, let $t_0=t_{0q}\leqslant t_{1q}\leqslant\cdots\leqslant t_{qq}=t$, for $q=1, 2,\ldots,$ be a sequence of partitions of $[t_0,t)$ such that $\max_{1\leqslant j\leqslant q}|t_{jq}-t_{j-1,q}|\to 0$ as $q\to\infty$. Define

$$\mathbf{I}_t^{(p,q)}(A)=\sum_{j=1}^{q}\sum_{k}\mathbf{b}_{t_{j-1,q}}(\mathbf{U}'_{kp})\int_{t_{j-1,q}}^{t_{jq}}\int_{A_{kp}}M(d\sigma,d\mathbf{U}), \tag{4.57}$$

where $\mathbf{U}'_{kp}$ is an arbitrary point in A_{kp}. If independently of the partition used, $I_t^{(p,q)}(A)$ converges with probability one to a unique finite limit as p and q tend to infinity, then we shall say that the counting integral of $\mathbf{b}$ with respect to M exists on $[t_0,t)\times A$ and denote the limit by

$$\mathbf{I}_t(A)=\int_{t_0}^{t}\int_A\mathbf{b}_\sigma(\mathbf{U})M(d\sigma,d\mathbf{U}). \tag{4.58}$$

The following theorem shows that $\mathbf{I}_t(A)$ exists under the conditions (a)–(d) given above. The evaluation of the integral and some of its moment properties are also given.

Theorem 4.2.1 (Existence and Some Properties of Counting Integrals). Under the conditions (a)–(d) given above, the counting integral $\mathbf{I}_t(A)$ exists and has the evaluation

$$\mathbf{I}_t(A)=\int_{t_0}^{t}\int_A\mathbf{b}_\sigma(\mathbf{U})M(d\sigma,d\mathbf{U})$$

$$=\begin{cases}0, & N_t-0\\ \sum_{n=1}^{N_t}b_{\tau_n}(\mathbf{u}_n), & N_t\geqslant 1\end{cases} \tag{4.59}$$

where $N_t=\int_{t_0}^{t}\int_A M(d\sigma,d\mathbf{U})$ is the total number of points in $[t_0,t)$ with mark in A and τ_n and $\mathbf{u}_n$ are the occurrence time and mark for the nth point. Furthermore, if

$$\tilde{\mathbf{I}}_t(A)=\mathbf{I}_t(A)-\int_{t_0}^{t}\int_A\mathbf{b}_\sigma(\mathbf{U})\lambda_\sigma\,d\sigma\,dP_{\mathbf{u}}(\mathbf{U})$$

†Recall from Section 3.3 that the diameter of a set is the maximum distance between any two points in the set.

then

$$E\left[\tilde{\mathbf{I}}_t(A)\right]=E[\mathbf{I}_t(A)]-\int_{t_0}^{t}\int_A E[\mathbf{b}_\sigma(\mathbf{U})]\lambda_\sigma\,d\sigma\,dP_{\mathbf{u}}(\mathbf{U})=\mathbf{0} \tag{4.60}$$

and

$$E\left[|\tilde{\mathbf{I}}_t(A)|^2\right]=\int_{t_0}^{t}\int_A E[|\mathbf{b}_\sigma(\mathbf{U})|^2]\lambda_\sigma\,d\sigma\,dP_{\mathbf{u}}(\mathbf{U}), \tag{4.61}$$

where $P_{\mathbf{u}}(\mathbf{U})$ is the probability distribution function for the mark random variable and $|\mathbf{v}|$ denotes the length of the vector $\mathbf{v}$.

Proof. $\mathbf{I}_t(A)$ exists and (4.59) holds trivially if $N_t=0$. Therefore, suppose $N_t\geqslant 1$, and choose p and q sufficiently large in (4.57) that each cell $[t_{j-1,q},t_{jq})\times A_{kp}$ in the partition of $[t_0,t)\times A$ contains no more than one point. This can be done for almost every realization of the marked point process because it is uniformly orderly and the total number of points in $[t_0,t)\times A$ is finite with probability one. Then,

$$\mathbf{b}_{t_{j-1,q}}(\mathbf{U}'_{kp})\int_{t_{j-1,q}}^{t_{jq}}\int_{A_{kp}}M(d\sigma,d\mathbf{U})=\mathbf{b}_{t_{j-1,q}}(\mathbf{U}'_{kp}) \tag{4.62}$$

if $(\tau_n,\mathbf{u}_n)\in[t_{j-1,q},t_{jq})\times A_{kp}$ for some n, where $1\leqslant n\leqslant N_t$. Otherwise,

$$\mathbf{b}_{t_{j-1,q}}(\mathbf{U}'_{kp})\int_{t_{j-1,q}}^{t_{jq}}\int_{A_{kp}}M(d\sigma,d\mathbf{U})=\mathbf{0}.$$

Hence, the sum on k in (4.57) is well defined because only finitely many terms are nonzero with probability one. Denote the N_t cells in the partition of $[t_0,t)\times A$ for which (4.62) holds by $[t_{j_n-1,q},t_{j_nq})\times A_{k_np}$ for $n=1,2,\ldots,N_t$, where these are ordered such that $(\tau_n,\mathbf{u}_n)\in[t_{j_n-1,q},t_{j_nq})\times A_{k_np}$. These N_t cells are bounded for p sufficiently large and any q. Hence, from condition (d), (4.57) becomes

$$\mathbf{I}_t^{(p,q)}(A)=\sum_{n=1}^{N_t}\mathbf{b}_{t_{j_n-1,q}}(\mathbf{U}'_{k_np}).$$

Consequently,

$$\left|\mathbf{I}_t^{(p,q)}(A)-\sum_{n=1}^{N_t}\mathbf{b}_{\tau_n}(\mathbf{u}_n)\right|\leqslant\sum_{n=1}^{N_t}\left|\mathbf{b}_{t_{j_n-1,q}}(\mathbf{U}'_{k_np})-\mathbf{b}_{\tau_n}(\mathbf{u}_n)\right|.$$

By selecting p and q sufficiently large, the right side can be made arbitrarily small with probability one due to conditions (b) and (d). These conditions also imply that $\Sigma_{n=1}^{N_t} \mathbf{b}_{\tau_n}(\mathbf{u}_n)$ is finite with probability one. Thus, we conclude that $\mathbf{I}_t^{(p,q)}(A)$ converges with probability one to a finite limit as p and q tend to infinity and that the limit $\mathbf{I}_t(A)$ has the evaluation in (4.59).

Let us next establish the moment relation in (4.60). Due to conditions (a) and (c), we have

$$E\left[\mathbf{I}_t^{(p,q)}(A)\right] = \sum_{j=1}^{q} \sum_{k=1}^{p} E\left[\mathbf{b}_{t_{j-1,q}}(\mathbf{U}'_{kp})\right] \int_{t_{j-1,q}}^{t_{jq}} \int_{A_{kp}} \lambda_\sigma \, d\sigma \, dP_{\mathbf{u}}(\mathbf{U}).$$

Thus,

$$\begin{aligned} E\left[\mathbf{I}_t(A)\right] &= E\left[\lim_{p,q\to\infty} \mathbf{I}_t^{(p,q)}(A)\right] \\ &= \lim_{p,q\to\infty} E\left[\mathbf{I}_t^{(p,q)}(A)\right] \\ &= \int_{t_0}^{t} \int_A E\left[\mathbf{b}_\sigma(\mathbf{U})\right] \lambda_\sigma \, d\sigma \, dP_{\mathbf{u}}(\mathbf{U}), \end{aligned}$$

which establishes (4.60). The interchange of the limit and expectation may be justified by the finiteness of $I_t^{(p,q)}$, for all p and q, the finiteness of I_t, and the bounded convergence theorem, Theorem 1.4.1. The second moment in (4.6.1) is evaluated in a similar fashion; see Problem 4.2.2. The proof of Theorem 4.2.1 is now complete.

Examination of the proof of Theorem 4.2.1 shows that the existence and evaluation of the counting integral do not depend upon condition (a). The underlying marked point process does not need to be a compound Poisson process for the validity of (4.59). What is required is that the point process be orderly and, with probability one, have a finite number of points in a finite interval $[t_1, t_2)$. This observation will be important in subsequent chapters when we study point processes that do not have Poisson statistics. Of course, the moment relations (4.60) and (4.61) do require condition (a) for their validity.

The continuity and boundedness restrictions of condition (d) can be relaxed for alternative definitions of the counting integral. Under the condition

$$\int_{t_0}^{t} \int_A E\left[|\mathbf{b}(\mathbf{U})|^2\right] \lambda_\sigma \, d\sigma \, dP_{\mathbf{u}}(\mathbf{U}) < \infty \tag{4.63}$$

the sequence $\mathbf{I}_t^{(p,q)}(A)$ of (4.57) is shown by I. Gihman and A. Skorokhod [12, p. 254] to converge in the mean square sense to a limit that is then taken as the definition for $\mathbf{I}_t(A)$. Properties (4.59), (4.60), and (4.61) will hold. A Skorokhod [24, p. 34] discusses counting integrals under still weaker conditions.

A special case of Theorem 4.2.1 will appear repeatedly in later chapters. This occurs when the mark space reduces to a single point, $\mathfrak{U}=\{1\}$, so that all marks are unity. Equation (4.59) becomes

$$\mathbf{I}_t = \int_{t_0}^{t} \mathbf{b}_\sigma \, dN_\sigma = \begin{cases} 0, & N_t = 1 \\ \sum_{n=1}^{N_t} \mathbf{b}_{\tau_n}, & N_t \geq 1, \end{cases} \tag{4.64}$$

where N_t is the number of points which occur during $[t_0, t)$.

Example 4.2.4 Consider again the integral $I_t = \int_{t_0}^{t} N_\sigma \, dN_\sigma$ appearing in (4.56). According to (4.64), the evaluation of this integral is

$$I_t = \begin{cases} 0, & N_t = 0 \\ \sum_{n=1}^{N_t} N_{\tau_n}, & N_t \geq 1. \end{cases}$$

Thus,

$$I_t = \begin{cases} 0, & N_t = 0 \\ \sum_{n=1}^{N_t} (n-1), & N_t \geq 1, \end{cases} \tag{4.65}$$

which follows because N_t is a left-continuous function of t; that is, N_{τ_n} is the number of points during $[t_0, \tau_n)$ and does not include the point at τ_n. Consequently, by evaluating the sum in (4.65), we conclude that

$$I_t = \int_{t_0}^{t} N_\sigma \, dN_\sigma = \tfrac{1}{2} N_t^2 - \tfrac{1}{2} N_t \tag{4.66}$$

in correspondence with the choice $\alpha = 1$ in (4.56).

Differentiation. Differentiation of random processes defined by counting integrals does not obey the rules of ordinary calculus. As an example, consider the random process $\{\phi_t; t \geq t_0\}$ defined by

$$\phi_t = \tfrac{1}{2} N_t + \int_{t_0}^{t} N_\sigma \, dN_\sigma.$$

The differential of ϕ_t is $d\phi_t = (\frac{1}{2} + N_t) dN_t$. However, from (4.66), we see that $\phi_t = \frac{1}{2} N_t^2$. Consequently, $d\phi_t$ does not equal $N_t dN_t$, which is contrary to the rules of ordinary calculus. The explanation for this departure from ordinary calculus may be seen roughly as follows. As $\phi_t = \frac{1}{2}N_t^2$, we have

$$
\begin{aligned}
d\phi_t = \phi_{t+dt} - \phi_t &= \tfrac{1}{2}\left[N_{t+dt}^2 - N_t^2\right] \\
&= \tfrac{1}{2}\left[(dN_t + N_t)^2 - N_t^2\right] \\
&= \tfrac{1}{2}(dN_t)^2 + N_t dN_t.
\end{aligned}
$$

In ordinary calculus, the term $\frac{1}{2}(dN_t)^2$ is of second order compared to $N_t dN_t$ and, therefore, it is neglected. Here, however, $(dN_t)^2 = dN_t$ because dN_t is, with high probability, one or zero corresponding to whether or not a point occurs in $[t, t+dt)$. Replacing $(dN_t)^2$ by dN_t results in $d\phi_t = (\frac{1}{2} + N_t)dN_t$. It follows from this simple example that some care must be exercised when differentiating random processes defined by counting integrals. A general differential rule is given in the following theorem.

Theorem 4.2.2 (Differential Rule). Let $\{\zeta_t; t \geqslant t_0\}$ be a random process defined in terms of a counting integral by

$$
\zeta_t = \zeta_0 + \int_{t_0}^{t} \mathbf{a}_\sigma \, d\sigma + \int_{t_0}^{t} \int_{\mathfrak{U}} \mathbf{b}_\sigma(\mathbf{U}) M(d\sigma, d\mathbf{U}), \tag{4.67}
$$

where the initial condition ζ_0 is a random variable and where the integral $\int_{t_0}^{t} \mathbf{a}_\sigma \, d\sigma$ exists as an ordinary integral for almost every realization of the random process $\{\mathbf{a}_t; t \geqslant t_0\}$. The vectors ζ_t, ζ_0, $\mathbf{a}_t$ and $\mathbf{b}_t$ are all of dimension n. Assume conditions (a)–(d) hold, so the counting integral exists according to Theorem 4.2.1. Let the function $\phi_t(\mathbf{X})$ be bounded for t and $\mathbf{X}$ finite and have continuous first derivatives with respect to t, $X_1, X_2, \ldots,$ and X_n. Then with probability one the random process $\{\phi_t(\zeta_t), t \geqslant t_0\}$ satisfies

$$
\begin{aligned}
\phi_t(\zeta_t) = \phi_0 &+ \int_{t_0}^{t} \left[\frac{\partial \phi_\sigma(\zeta_\sigma)}{\partial \sigma} + \left\langle \mathbf{a}_\sigma, \frac{\partial \phi_\sigma(\zeta_\sigma)}{\partial \zeta_\sigma} \right\rangle \right] d\sigma \\
&+ \int_{t_0}^{t} \int_{\mathfrak{U}} \left[\phi_\sigma(\zeta_\sigma + \mathbf{b}_\sigma(\mathbf{U})) - \phi_\sigma(\zeta_\sigma)\right] M(d\sigma, d\mathbf{U}),
\end{aligned} \tag{4.68}
$$

where $\phi_0 = \phi_{t_0}(\zeta_0)$, $\partial\phi_\sigma(\zeta_\sigma)/\partial\zeta_\sigma$ denotes the gradient of $\phi_\sigma(\zeta_\sigma)$ with respect to ζ_σ (the ith element of the gradient is the derivative of ϕ with respect to the ith element of ζ), and $\langle \mathbf{a}_\sigma, \partial\phi_\sigma(\zeta_\sigma)/\partial\zeta_\sigma \rangle$ denotes the usual dot product of the vectors $\mathbf{a}_\sigma$ and $\partial\phi_\sigma(\zeta_\sigma)/\partial\zeta_\sigma$; that is

$$\left\langle \mathbf{a}_\sigma, \frac{\partial\phi_\sigma(\zeta_\sigma)}{\partial\zeta_\sigma} \right\rangle = \sum_{i=1}^{n} \frac{a_{i\sigma}\partial\phi_\sigma(\zeta_\sigma)}{\partial\zeta_{i\sigma}}.$$

In differential form, (4.67) and (4.68) imply that if

$$d\zeta_t = \mathbf{a}_t dt + \int_{\mathfrak{U}} \mathbf{b}_t(\mathbf{U}) M(dt, d\mathbf{U}) \tag{4.69}$$

then

$$d\phi_t(\zeta_t) = \left[\frac{\partial\phi_t(\zeta_t)}{\partial t} + \left\langle \mathbf{a}_t, \frac{\partial\phi_t(\zeta_t)}{\partial\zeta_t} \right\rangle \right] dt$$

$$+ \int_{\mathfrak{U}} [\phi_t(\zeta_t + \mathbf{b}_t(\mathbf{U})) - \phi_t(\zeta_t)] M(dt, d\mathbf{U}). \tag{4.70}$$

Proof. (I. Gihman and A. Dorogovcev [11], I. Gihman and A. Skorokhod [12, p. 263], P. Fishman [8]). We first note that as ζ_t defined by (4.67) is a left-continuous function of t, so is $\phi_t(\zeta_t)$ defined by (4.68). Similarly, $\phi_t(\zeta_t + \mathbf{b}_t(\mathbf{U}))$ is a left-continuous function of t and a continuous function of $\mathbf{U}$. Therefore, the counting integral in (4.68) exists according to Theorem 4.2.1. Now consider a sequence of partitions $t_0 = t_{0q} < t_{1q} < \cdots < t_{qq} = t$ of $[t_0, t)$ such that $\max_{1 \leqslant j \leqslant q} |t_{jq} - t_{j-1,q}|$ tends to zero as q tends to infinity. We have

$$\phi_t(\zeta_t) - \phi_0 = \sum_{j=1}^{q} \left[\phi_{t_{jq}}(\Delta\zeta_{t_{jq}} + \zeta_{t_{j-1,q}}) - \phi_{t_{j-1,q}}(\zeta_{t_{j-1,q}}) \right], \tag{4.71}$$

where from (4.67)

$$\Delta\zeta_{t_{jq}} = \zeta_{t_{jq}} - \zeta_{t_{j-1,q}} = \Delta\boldsymbol{\alpha}_{jq} + \Delta\boldsymbol{\beta}_{jq},$$

where we define

$$\Delta\boldsymbol{\alpha}_{jq} = \int_{t_{j-1,q}}^{t_{jq}} \mathbf{a}_\sigma \, d\sigma$$

and

$$\Delta\boldsymbol{\beta}_{jq} = \int_{t_{j-1,q}}^{t_{jq}} \int_{\mathfrak{U}} \mathbf{b}_\sigma(\mathbf{U}) M(d\sigma, d\mathbf{U}).$$

The sum (4.71) can be rewritten as

$$\phi_t(\zeta_t)-\phi_0=\Sigma_{1q}+\Sigma_{2q}+\Sigma_{3q},$$

where

$$\Sigma_{1q}=\sum_{j=1}^{q}\left[\phi_{t_{jq}}(\Delta\zeta_{t_{jq}}+\zeta_{t_{j-1,q}})-\phi_{t_{j-1,q}}(\Delta\zeta_{t_{jq}}+\zeta_{t_{j-1,q}})\right]$$

$$\Sigma_{2q}=\sum_{j=1}^{q}\left[\phi_{t_{j-1,q}}(\Delta\zeta_{t_{jq}}+\zeta_{t_{j-1,q}})-\phi_{t_{j-1,q}}(\zeta_{t_{j-1,q}}+\Delta\boldsymbol{\beta}_{jq})\right]$$

and

$$\Sigma_{3q}=\sum_{j=1}^{q}\left[\phi_{t_{j-1,q}}(\zeta_{t_{j-1,q}}+\Delta\boldsymbol{\beta}_{jq})-\phi_{t_{j-1,q}}(\zeta_{t_{j-1,q}})\right]$$

Now, by the mean value theorem, we can write

$$\Sigma_{1q}=\sum_{j=1}^{q}\dot{\phi}_{t'_{jq}}(\zeta_{t_{jq}})[t_{jq}-t_{j-1,q}],$$

where $\dot{\phi}_t(\zeta_t)=\partial\phi_t(\zeta_t)/\partial t$ and $t_{j-1,q}\leqslant t'_{jq}\leqslant t_{jq}$ for $j=1,2,\ldots,q$. Thus, in the limit $q\to\infty$, Σ_{1q} converges to the Riemann integral $\int_{t_0}^{t}[\partial\phi_\sigma(\zeta_\sigma)/\partial\sigma]d\sigma$. Similarly, by the mean value theorem for several variables, Σ_{2q} converges in the limit $q\to\infty$ to $\int_{t_0}^{t}\langle\mathbf{a}_\sigma,\partial\phi_\sigma(\zeta_\sigma)/\partial\zeta_\sigma\rangle d\sigma$. By an argument identical to that in the proof of Theorem 4.2.1, the quantities $\Delta\boldsymbol{\beta}_{jq}$ in Σ_{3q} are either nonzero or zero for q sufficiently large in correspondence with whether or not a point occurs in $[t_{j-1,q},t_{jq})$. Denote the N_t intervals where $\Delta\boldsymbol{\beta}_{jq}$ is nonzero by $[t_{j_n-1,q},t_{j_nq})$ for $n=1,2,\ldots,N_t$, where these intervals are ordered such that $(\tau_n,\mathbf{u}_n)\in[t_{j_n-1,q},t_{j_nq})\times\mathfrak{U}$. Here, $(\tau_n,\mathbf{u}_n)$ are the occurrence time and mark for the nth point in the underlying point process. Then for q sufficiently large

$$\Sigma_{3q}=\begin{cases}0, & N_t=0\\ \displaystyle\sum_{n=1}^{N_t}\left[\phi_{t_{j_n-1,q}}(\zeta_{t_{j_n-1,q}}+\Delta\boldsymbol{\beta}_{j_nq})-\phi_{t_{j_n-1,q}}(\zeta_{t_{j_n-1,q}})\right], & N_t\geqslant 1.\end{cases}$$

In the limit $q\to\infty$, this sum then becomes

$$\Sigma_{3\infty}=\begin{cases}0, & N_t=0\\ \displaystyle\sum_{n=1}^{N_t}\left[\phi_{\tau_n}(\zeta_{\tau_n}+\mathbf{b}_{\tau_n}(\mathbf{u}_n))-\phi_{\tau_n}(\zeta_{\tau_n})\right], & N_t\geqslant 1,\end{cases}$$

where we have used the continuity of ϕ with respect to t and ζ and the left continuity of ζ. According to Theorem 4.2.1, we therefore have

$$\Sigma_{3\infty}=\int_{t_0}^{t}\int_{\mathfrak{U}}[\phi_\sigma(\zeta_\sigma+\mathbf{b}_\sigma(\mathbf{U}))-\phi_\sigma(\zeta_\sigma)]M(d\sigma,d\mathbf{U}).$$

Combining these expressions for $\Sigma_{1\infty}$, $\Sigma_{2\infty}$, and $\Sigma_{3\infty}$ yields (4.68), and the proof of Theorem 4.2.2 is complete.

Examination of the proof of Theorem 4.2.2 shows that it does not depend upon condition (a). The underlying point process does not need to be a compound Poisson process for the validity of (4.68) and (4.70). What is required is that the point process be uniformly orderly and have a finite number of points in a finite interval. This observation will be important in later chapters when we study point processes that do not have Poisson statistics.

A special case of Theorem 4.2.2 will appear repeatedly in later chapters. This occurs when the mark space reduces to a single point, $\mathfrak{U}=\{1\}$, so that all the marks are unity. In this case, (4.69) and (4.70) become

$$d\zeta_t=\mathbf{a}_t dt+\mathbf{b}_t dN_t \tag{4.72}$$

and

$$d\phi_t(\zeta_t)=\left[\frac{\partial\phi_t(\zeta_t)}{\partial t}+\left\langle \mathbf{a}_t,\frac{\partial\phi_t(\zeta_t)}{\partial\zeta_t}\right\rangle\right]dt+[\phi_t(\zeta_t+\mathbf{b}_t)-\phi_t(\zeta_t)]dN_t, \tag{4.73}$$

respectively, where N_t is the number of points which occur during $[t_0,t)$.

The following are some examples illustrating the use of the differential rule. These will be useful in later calculations.

Example 4.2.5 (Power Law). Let $\{\zeta_t; t\geq t_0\}$ be a real-valued process satisfying

$$d\zeta_t=a_t dt+\int_{\mathfrak{U}}b_t(\mathbf{U})M(dt,d\mathbf{U}),\qquad \zeta_0=0.$$

Suppose $\phi(\zeta_t)=\zeta_t^\nu$. Then

$$d\zeta_t^\nu=\nu\zeta_t^{\nu-1}a_t dt+\int_{\mathfrak{U}}[(\zeta_t+b_t(\mathbf{U}))^\nu-\zeta_t^\nu]M(dt,d\mathbf{U}). \tag{4.74}$$

Alternatively,

$$\zeta_t^\nu = \int_{t_0}^t \nu\zeta_\sigma^{\nu-1} a_\sigma \, d\sigma + \int_{t_0}^t \int_{\mathfrak{U}} \left[(\zeta_\sigma + b_\sigma(\mathbf{U}))^\nu - \zeta_\sigma^\nu \right] M(d\sigma, d\mathbf{U}). \tag{4.75}$$

In particular, for $\nu=2$, $a_t=0$, and $b_t(\mathbf{U})=1$, we have $\zeta_t=N_t$ and

$$dN_t^2 = \left[(N_t+1)^2 - N_t^2 \right] dN_t.$$

From this, we conclude that

$$\int_{t_0}^t N_\sigma \, dN_\sigma = \tfrac{1}{2} N_t (N_t - 1).$$

This emphasizes again the care with which counting integrals must be handled. Their evaluation differs from ordinary integrals.

Example 4.2.6 (Component Product). Let $\{\boldsymbol{\zeta}_t; t \geqslant t_0\}$ be an n-dimensional process satisfying

$$d\boldsymbol{\zeta}_t = \mathbf{a}_t \, dt + \int_{\mathfrak{U}} \mathbf{b}_t(\mathbf{U}) M(dt, d\mathbf{U}). \tag{4.76}$$

Denote by $\mathbf{e}_i$ an n-dimensional unit vector with ith element equal to one and all other elements equal to zero. Define $\phi_{ij}(\boldsymbol{\zeta}_t) = \mathbf{e}_i' \boldsymbol{\zeta}_t \boldsymbol{\zeta}_t' \mathbf{e}_j$. Then $\phi_{ij}(\boldsymbol{\zeta}_t)$ is the product of the ith and jth components of $\boldsymbol{\zeta}_t$. From (4.70), we have

$$d\phi_{ij}(\boldsymbol{\zeta}_t) = \langle \mathbf{a}_t, (\mathbf{e}_i\mathbf{e}_j' + \mathbf{e}_j\mathbf{e}_i')\boldsymbol{\zeta}_t \rangle dt$$
$$+ \mathbf{e}_i' \int_{\mathfrak{U}} \left[\mathbf{b}_t(\mathbf{U})\boldsymbol{\zeta}_t' + \boldsymbol{\zeta}_t \mathbf{b}_t'(\mathbf{U}) + \mathbf{b}_t(\mathbf{U})\mathbf{b}_t'(\mathbf{U}) \right] M(dt, d\mathbf{U}) \mathbf{e}_j. \tag{4.77}$$

This expression will be useful below when we examine the second moment statistics for a Poisson driven Markov process.

Example 4.2.7 (Exponential). Suppose $\boldsymbol{\zeta}_t$ satisfies (4.76) and let $\phi(\boldsymbol{\zeta}_t) = \exp[j\langle \mathbf{v}, \boldsymbol{\zeta}_t \rangle]$, where $\mathbf{v}$ is a constant n-dimensional vector. Then $\partial\phi(\boldsymbol{\zeta}_t)/\partial\boldsymbol{\zeta}_t = j\mathbf{v}\exp[j\langle \mathbf{v}, \boldsymbol{\zeta}_t \rangle]$, and we have from (4.70) that

$$d\exp[j\langle \mathbf{v}, \boldsymbol{\zeta}_t \rangle] = \exp[j\langle \mathbf{v}, \boldsymbol{\zeta}_t \rangle] \Big\{ j\langle \mathbf{v}, \mathbf{a}_t \rangle dt$$
$$+ \int_{\mathfrak{U}} \left(\exp[j\langle \mathbf{v}, \mathbf{b}_t(\mathbf{U}) \rangle] - 1 \right) M(dt, d\mathbf{U}) \Big\}. \tag{4.78}$$

This expression will be useful below when we examine the characteristic function for a Poisson driven Markov process.

Example 4.2.8 (Logarithm). Suppose ζ_t satisfies (4.76) and let $\phi(\zeta_t) = \ln[\langle \mathbf{v}, \zeta_t \rangle]$, where $\mathbf{v}$ is a constant n-dimensional vector. Then, we have from (4.70) that

$$d\ln[\langle \mathbf{v}, \zeta_t \rangle] = \frac{\langle \mathbf{v}, \mathbf{a}_t \rangle}{\langle \mathbf{v}, \zeta_t \rangle} dt + \ln\left[1 + \frac{\langle \mathbf{v}, \mathbf{b}_t \rangle}{\langle v, \zeta_t \rangle}\right] dN_t. \tag{4.79}$$

Existence and Uniqueness of Solutions to (4.38). Equation (4.38) is a nonlinear integral equation defining the Poisson driven Markov process $\{\mathbf{x}_t; t \geqslant t_0\}$. With Theorem 4.2.1, we have developed an interpretation for the last integral on the right in (4.38), a counting integral. And with Theorem 4.2.2, we have developed a rule for differentiating functions of $\mathbf{x}_t$. These results will be useful below when we study the statistics of $\{\mathbf{x}_t; t \geqslant t_0\}$. However, we have not yet given conditions on $\mathbf{a}_t(\mathbf{x}_t)$ and $\mathbf{b}_t(\mathbf{x}_t, \mathbf{U})$ so that a solution to (4.38) exists and is unique. These technical considerations are covered in the following theorem.

Theorem 4.2.3 (Conditions for the Existence and Uniqueness of a Poisson Driven Markov Process). Let

$$\{M([t_1, t_2), A); [t_1, t_2) \subseteq [t_0, \infty), A \subseteq \mathfrak{U}\}$$

be the time-space Poisson process defined in Theorem 3.3.2 for representing a compound Poisson process on $[t_0, \infty)$ with mark space $\mathfrak{U}$. Let $[t_0, T)$ be a finite interval, and assume that

$$\sup_{t_0 \leqslant t < T} \lambda_t < \infty. \tag{4.80}$$

Denote n-dimensional Euclidean space by R^n. Suppose $\mathbf{x}_0$ and the functions $\mathbf{a}_t(\mathbf{X})$ and $\mathbf{b}_t(\mathbf{X}, \mathbf{U})$, for $t \in [t_0, T)$, $\mathbf{X} \in R^n$, and $\mathbf{U} \in \mathfrak{U}$, satisfy the following conditions:

i. $(T - t_0)E(|\mathbf{x}_0|^2) < \infty$;
ii. $\mathbf{b}_t(\mathbf{X}, \mathbf{U})$ is a continuous function of $\mathbf{X}$ and $\mathbf{U}$ and a left-continuous function of t;
iii. for all $\mathbf{X} \in R^n$ and $t \in [t_0, T)$, there exist finite nonnegative constants K_1 and K_2 such that

$$(T - t_0)|\mathbf{a}_t(\mathbf{X}) + \lambda_t E[\mathbf{b}_t(\mathbf{X}, \mathbf{u})]|^2 \leqslant K_1(1 + |\mathbf{X}|^2)$$

and

$$E[|\mathbf{b}_t(\mathbf{X},\mathbf{u})|^2] \leqslant K_2(1+|\mathbf{X}|^2);$$

iv. for all $\mathbf{X} \in R^n$, $\mathbf{Y} \in R^n$, and $t \in [t_0, T)$, there exist finite nonnegative constants L_1 and L_2 such that

$$(T-t_0)|\mathbf{a}_t(\mathbf{X})-\mathbf{a}_t(\mathbf{Y})+\lambda_t E[\mathbf{b}_t(\mathbf{X},\mathbf{u})-\mathbf{b}_t(\mathbf{Y},\mathbf{u})]|^2 \leqslant L_1|\mathbf{X}-\mathbf{Y}|^2$$

and

$$E[|\mathbf{b}_t(\mathbf{X},\mathbf{u})-\mathbf{b}_t(\mathbf{Y},\mathbf{u})|^2] \leqslant L_2|\mathbf{X}-\mathbf{Y}|^2.$$

Under these conditions, there exists with probability one a solution $\{\mathbf{x}_t; t_0 \leqslant t < T\}$ to the stochastic integral equation

$$\mathbf{x}_t = \mathbf{x}_0 + \int_{t_0}^{t} \mathbf{a}_\sigma(\mathbf{x}_\sigma)\,d\sigma + \int_{t_0}^{t}\int_{\mathfrak{U}} \mathbf{b}_\sigma(\mathbf{x}_\sigma,\mathbf{U})M(d\sigma,d\mathbf{U}). \tag{4.81}$$

Furthermore, the solution satisfies $\int_{t_0}^{T} E(|\mathbf{x}_t|^2)\,dt < \infty$ and is unique in the sense that if $\mathbf{x}$ and $\mathbf{y}$ are solutions on $[t_0, T)$, then $\Pr[\sup_{t_0 \leqslant t < T}|\mathbf{x}_t - \mathbf{y}_t| > 0] = 0$.

The expectation in condition (i) is with respect to the distribution of $\mathbf{x}_0$, and those in conditions (iii) and (iv) are with respect to the mark distribution $P_{\mathbf{u}}(\mathbf{U})$; for example,

$$E[|\mathbf{b}_t(\mathbf{X},\mathbf{u})|^2] = \int_{\mathfrak{U}} |\mathbf{b}_t(\mathbf{X},\mathbf{U})|^2\,dP_{\mathbf{u}}(\mathbf{U}).$$

Condition (iii) is a growth condition; we will show in the proof of Theorem 4.2.3 that this implies a boundedness condition on $\mathbf{b}$. Condition (iv) is termed a *uniform Lipschitz condition.*

Proof. (A. Skorokhod [24, p. 47], A. Bharucha-Reid [3, p. 230].) Let Ξ be the complete, normed, linear space of left-continuous, R^n-valued random functions $\{\xi_t; t_0 \leqslant t < T\}$ such that

$$\sup_{t_0 \leqslant t < T} E(|\xi_t|^2) < \infty. \tag{4.82}$$

We take the norm on the space Ξ to be

$$\|\xi\| = \left[\sup_{t_0 \leqslant t < T} E(|\xi_t|^2)\right]^{1/2}.$$

We demonstrate first that the counting integral

$$\int_{t_0}^{t}\int_{\mathfrak{U}} \mathbf{b}_\sigma(\xi_\sigma, \mathbf{U}) M(d\sigma, d\mathbf{U}) \tag{4.83}$$

exists for all $t \in [t_0, T)$, where $\xi \in \Xi$. It follows from condition (ii) and the left continuity of functions in Ξ that $\mathbf{b}_t(\xi_t, \mathbf{U})$ is left continuous in t and continuous in $\mathbf{U}$ for $\xi \in \Xi$. Assumption (4.80) implies that

$$E\big[M([t_0, T), \mathfrak{U})\big] = \int_{t_0}^{T} \lambda_\sigma \, d\sigma < \infty.$$

To use Theorem 4.2.1, we must also show that almost every realization of $\mathbf{b}_t(\zeta_t, \mathbf{u})$ is bounded. Now, from condition (iii), we have

$$E\big(|\mathbf{b}_t(\xi_t, \mathbf{u})|^2\big) \leqslant K_2\big[1 + E\big(|\xi_t|^2\big)\big]. \tag{4.84}$$

Thus, $\sup_{t_0 \leqslant t \leqslant T} E(|\mathbf{b}_t(\xi_t, \mathbf{u})|^2) < \infty$, which implies that $\Pr(\sup_{t_0 \leqslant t < T} |\mathbf{b}_t(\xi_t, \mathbf{u})| < \infty) = 1$. It follows from Theorem 4.2.1 that the counting integral (4.83) exists.

As the counting integral (4.83) exists, we can define a mapping $\mathbf{S}$ of R^n-valued functions from Ξ into R^n-valued functions by

$$(\mathbf{S}\xi)_t = \mathbf{x}_0 + \int_{t_0}^{t} \mathbf{a}_\sigma(\xi_\sigma) \, d\sigma + \int_{t_0}^{t}\int_{\mathfrak{U}} \mathbf{b}_\sigma(\xi_\sigma, \mathbf{U}) M(d\sigma, d\mathbf{U}). \tag{4.85}$$

We rewrite $\mathbf{S}$ in terms of the zero-mean process defined by

$$\tilde{M}([t_1, t_2), A) = M([t_1, t_2), A) - \int_{t_1}^{t_2}\int_{A} \lambda_\sigma \, d\sigma \, dP_{\mathbf{u}}(\mathbf{U})$$

as

$$(\mathbf{S}\xi)_t = \mathbf{x}_0 + \int_{t_0}^{t} \left\{ \mathbf{a}_\sigma(\xi_\sigma) + \lambda_\sigma \int_{\mathfrak{U}} \mathbf{b}_\sigma(\xi_\sigma, \mathbf{U}) \, dP_{\mathbf{u}}(\mathbf{U}) \right\} d\sigma$$

$$+ \int_{t_0}^{t}\int_{\mathfrak{U}} \mathbf{b}_\sigma(\xi_\sigma, \mathbf{U}) \tilde{M}(d\sigma, d\mathbf{U}). \tag{4.86}$$

The map $\mathbf{S}$ has the property that it maps functions of Ξ into functions of Ξ. This can be seen by noting first that

$$E\big[|(\mathbf{S}\xi)_t|^2\big] \leqslant 3E(|\mathbf{x}_0|^2) + 3(t - t_0) \int_{t_0}^{t} E\Big(\Big|\mathbf{a}_\sigma(\xi_\sigma) + \lambda_\sigma \int_{\mathfrak{U}} \mathbf{b}_\sigma(\xi_\sigma, \mathbf{U}) \, dP_{\mathbf{u}}(\mathbf{U})\Big|^2 \Big) d\sigma$$

$$+ 3\int_{t_0}^{t}\int_{\mathfrak{U}} E(|\mathbf{b}_\sigma(\xi_\sigma, \mathbf{U})|^2) \lambda_\sigma \, d\sigma \, dP_{\mathbf{u}}(\mathbf{U}), \tag{4.87}$$

where we have used (4.61) and the Cauchy–Schwarz inequalities

$$\left|\sum_{i=1}^{k} \boldsymbol{\alpha}_i\right|^2 \leqslant k \sum_{i=1}^{k} |\boldsymbol{\alpha}_i|^2$$

and

$$\left|\int_{t_0}^{t} \boldsymbol{\alpha}_\sigma \, d\sigma\right|^2 \leqslant (t-t_0)\int_{t_0}^{t} |\boldsymbol{\alpha}_\sigma|^2 \, d\sigma.$$

Now, from (4.87) and condition (iii), we obtain

$$E[|(\mathbf{S}\xi)_t|^2] \leqslant 3E(|\mathbf{x}_0|^2) + 3K_1\int_{t_0}^{t} E(1+|\xi_\sigma|^2)\,d\sigma + 3K_2\int_{t_0}^{t} E(1+|\xi_\sigma|^2)\lambda_\sigma\,d\sigma.$$

Hence,

$$\|\mathbf{S}\xi\| \leqslant 3E(|\mathbf{x}_0|^2) + 3K(1+\|\xi\|)\int_{t_0}^{T}(1+\lambda_\sigma)\,d\sigma < \infty,$$

where $K=\max(K_1,K_2)$. Moreover, $(\mathbf{S}\xi)_t$ is a left-continuous function of t because of its definition (4.85) and the properties of counting integrals. Thus $\mathbf{S}\xi\in\Xi$ if $\xi\in\Xi$. From this property of $\mathbf{S}$, it is easily seen that for the iterated map $\mathbf{S}^k\xi$, defined by $\mathbf{S}^k\xi=\mathbf{S}(\mathbf{S}^{k-1}\xi)$ for $k\geqslant 1$ with $\mathbf{S}^0\xi=\xi$, we also have $\mathbf{S}^k\xi\in\Xi$ if $\xi\in\Xi$.

We show next that the map $\mathbf{S}^k$ is a contraction for sufficiently large k. For this purpose, let $\xi\in\Xi$ and $\zeta\in\Xi$, and note that

$$\begin{aligned} E[|(\mathbf{S}\xi)_t-(\mathbf{S}\zeta)_t|^2] \leqslant 2(t-t_0)\int_{t_0}^{t} E\Bigg(\Bigg|&\mathbf{a}_\sigma(\xi_\sigma)-\mathbf{a}_\sigma(\zeta_\sigma) \\ &+\lambda_\sigma\int_{\mathfrak{U}}[\mathbf{b}_\sigma(\xi_\sigma,\mathbf{U})-\mathbf{b}_\sigma(\zeta_\sigma,\mathbf{U})]\,dP_{\mathbf{u}}(\mathbf{U})\Bigg|^2\Bigg)\,d\sigma \\ &+2\int_{t_0}^{t}\int_{\mathfrak{U}} E(|\mathbf{b}_\sigma(\xi_\sigma,\mathbf{U})-\mathbf{b}_\sigma(\zeta_\sigma,\mathbf{U})|^2)\lambda_\sigma\,d\sigma\,dP_{\mathbf{u}}(\mathbf{U}). \end{aligned}$$

Then, using condition (iv), we obtain the inequality

$$E[|(\mathbf{S}\xi)_t-(\mathbf{S}\zeta)_t|^2] \leqslant L\int_{t_0}^{t} E(|\xi_\sigma-\zeta_\sigma|^2)\,d\sigma,$$

where we define

$$L=2\left[L_1+L_2\sup_{t_0\leqslant t<T}\lambda_t\right].$$

Iteration of this inequality yields

$$E[|(\mathbf{S}^k\xi)_t-(\mathbf{S}^k\zeta)_t|^2]$$

$$\leqslant L^k\int_{t_0}^{t}\int_{t_0}^{\sigma_1}\cdots\int_{t_0}^{\sigma_{k-2}}\int_{t_0}^{\sigma_{k-1}}E\left(|\xi_{\sigma_k}-\zeta_{\sigma_k}|^2\right)d\sigma_k\,d\sigma_{k-1}\cdots d\sigma_2\,d\sigma_1$$

$$\leqslant\frac{L^k(t-t_0)^{k-1}}{(k-1)!}\int_{t_0}^{T}E\left(|\xi_{\sigma_k}-\zeta_{\sigma_k}|^2\right)d\sigma_k$$

$$\leqslant\frac{L^k(T-t_0)^k}{(k-1)!}\|\xi-\zeta\|.$$

Therefore, for every $\xi\in\Xi$ and $\zeta\in\Xi$, there holds

$$\|\mathbf{S}^k\xi-\mathbf{S}^k\zeta\|\leqslant\frac{L^k(T-t_0)^k}{(k-1)!}\|\xi-\zeta\|. \tag{4.88}$$

Thus, $\mathbf{S}^k$ is a contraction mapping in the space Ξ for k sufficiently large that the constant coefficient $L^k(T-t_0)^k[(k-1)!]^{-1}$ is less than unity.

As $\mathbf{S}^k$ maps functions of Ξ into functions of Ξ and is a contraction for k sufficiently large, there exists a unique function, say $\mathbf{x}$, in Ξ such that $\mathbf{S}^k\mathbf{x}=\mathbf{x}$ for k sufficiently large because of the contraction mapping theorem (see L. Liusternik and V. Sobolev [19, p. 27]). We next show that for this $\mathbf{x}$, there holds $\mathbf{S}\mathbf{x}=\mathbf{x}$, which establishes the existence of a solution to (4.81). For $k_0\geqslant 1$, and k sufficiently large that $\mathbf{S}^k$ is a contraction mapping, we have $\mathbf{S}^{kk_0}\mathbf{x}=\mathbf{x}$, and from the inequality (4.88), we then have

$$\|\mathbf{x}-\mathbf{S}\mathbf{x}\|=\|\mathbf{S}^{kk_0}\mathbf{x}-\mathbf{S}^{kk_0+1}\mathbf{x}\|\leqslant\frac{L^{kk_0}(T-t_0)^{kk_0}}{(kk_0-1)!}\|\mathbf{x}-\mathbf{S}\mathbf{x}\|. \tag{4.89}$$

However, for $k_0\geqslant 1$, we have $L^{kk_0}(T-t_0)^{kk_0}[(kk_0-1)!]^{-1}<1$, which with (4.89) implies $\mathbf{S}\mathbf{x}=\mathbf{x}$ in the sense that $\|\mathbf{x}-\mathbf{S}\mathbf{x}\|=0$. Consequently,

$$\Pr\left[\sup_{t_0\leqslant t<T}|\mathbf{x}_t-(\mathbf{S}\mathbf{x})_t|>0\right]=0.$$

Finally, $\mathbf{x}$ is a unique solution to $\mathbf{S}\mathbf{x}=\mathbf{x}$ because any such solution is also a solution to $\mathbf{S}^k\mathbf{x}=\mathbf{x}$, which has a unique solution because $\mathbf{S}^k$ is a contraction. This uniqueness is in the sense that if $\mathbf{S}\mathbf{x}=\mathbf{x}$ and $\mathbf{S}\mathbf{y}=\mathbf{y}$, then

$\|\mathbf{x}-\mathbf{y}\|=0$, which implies $\Pr[\sup_{t_0\leqslant t<T}|\mathbf{x}_t-\mathbf{y}_t|>0]=0$. This completes the proof of Theorem 4.2.3.

We refer to A. Skorokhod [24, Chapter 3] and I. Gihman and A. Skorokhod [12, Part II, Chapter 2] for further results on the existence and uniqueness of solutions to the stochastic integral equation (4.81). In particular, these authors relax the continuity condition (ii) and establish the existence and uniqueness of solutions to (4.81) in a mean-square sense. Furthermore, they consider stochastic integral equations that have an additional term on the right side of (4.81) that is derived from a Wiener process. Recall that a Wiener process is a zero-mean random process with independent, normally distributed increments. The form of the more general stochastic integral equation considered by A. Skorokhod [24] and I. Gihman and A. Skorokhod [12] is

$$\mathbf{x}_t=\mathbf{x}_0+\int_{t_0}^{t}\mathbf{a}_\sigma(\mathbf{x}_\sigma)\,d\sigma+\int_{t_0}^{t}\mathbf{B}_\sigma(\mathbf{x}_\sigma)\,d\mathbf{w}_\sigma+\int_{t_0}^{t}\int_{\mathrm{U}}\mathbf{b}_\sigma(\mathbf{x}_\sigma,\mathbf{U})\,M(d\sigma,d\mathbf{U}), \tag{4.90}$$

where $\{\mathbf{w}_t;t\geqslant t_0\}$ is a vector of independent Wiener processes normalized so that $E[\mathbf{w}_s\mathbf{w}'_t]=\mathbf{I}\min(s,t)$, where $\mathbf{I}$ is the identity matrix. Also, $\{\mathbf{w}_t;$ $t\geqslant t_0\}$ is independent of the underlying point process. Several authors develop the theory for the second integral on the right side of (4.90); see, for example, E. Wong [26, Chapter 4], A. Bharucha-Reid [3, Chapter 7], J. Doob [6, Chapter 6], and I. Gikhman and A. Skorokhod [13, Chapter 8]. We mention, finally, that the differential rule of Theorem 4.2.2 can be generalized for processes satisfying (4.90). In particular, I. Gihman and A. Dorogovcev [11] and I. Gihman and A. Skorokhod [12] show that if $\{\mathbf{x}_t;t\geqslant t_0\}$ satisfies (4.90), then $\{\phi_t(\mathbf{x}_t);t\geqslant t_0\}$ satisfies

$$\begin{aligned}\phi_t(\mathbf{x}_t)=\phi_0+\int_{t_0}^{t}\Bigg\{&\frac{\partial\phi_\sigma(\mathbf{x}_\sigma)}{\partial\sigma}+\left\langle\mathbf{a}_\sigma(\mathbf{x}_\sigma),\frac{\partial\phi_\sigma(\mathbf{x}_\sigma)}{\partial\mathbf{x}_\sigma}\right\rangle\\&+\tfrac{1}{2}\,\mathrm{tr}\left[\mathbf{B}_\sigma(\mathbf{x}_\sigma)\mathbf{B}'_\sigma(\mathbf{x}_\sigma)\frac{\partial^2\phi_\sigma(\mathbf{x}_\sigma)}{\partial\mathbf{x}_\sigma^2}\right]\Bigg\}\,d\sigma+\int_{t_0}^{t}\left\langle\frac{\partial\phi_\sigma(\mathbf{x}_\sigma)}{\partial\mathbf{x}_\sigma},\mathbf{B}_\sigma(\mathbf{x}_\sigma)\,d\mathbf{w}_\sigma\right\rangle\\&+\int_{t_0}^{t}\int_{\mathrm{U}}[\phi_\sigma(\mathbf{x}_\sigma+\mathbf{b}_\sigma(\mathbf{x}_\sigma,\mathbf{U}))-\phi_\sigma(\mathbf{x}_\sigma)]\,M(d\sigma,d\mathbf{U}),\end{aligned} \tag{4.91}$$

where $\phi_0=\phi_{t_0}(\mathbf{x}_0)$, $\mathrm{tr}(\cdot)$ denotes the trace operator, and $\partial^2\phi(\mathbf{x})/\partial\mathbf{x}^2$ is a matrix of second derivatives of ϕ (the ij-element is $\partial^2\phi(\mathbf{x})/\partial x_i\,\partial x_j$).

We now examine the statistics of processes defined by (4.81).

4.2.2 Statistics for Poisson Driven Markov Processes

We assume throughout this subsection that the process $\{\mathbf{x}_t; t \geq t_0\}$ is defined by

$$\mathbf{x}_t = \mathbf{x}_0 + \int_{t_0}^{t} \mathbf{a}_\sigma(\mathbf{x}_\sigma)\,d\sigma + \int_{t_0}^{t}\int_{\mathfrak{U}} \mathbf{b}_\sigma(\mathbf{x}_\sigma, \mathbf{U}) M(d\sigma, d\mathbf{U}). \tag{4.92}$$

Under the conditions of Theorem 4.2.3, there exists a unique solution to this nonlinear integral equation. These conditions are assumed to hold for (4.92). In addition, we assume the initial condition $\mathbf{x}_0$ does not anticipate the underlying point process on $[t_0, \infty)$, by which we mean that $\mathbf{x}_0$ is statistically independent of $\int_{t_1}^{t_2}\int_{\mathfrak{U}} M(d\sigma, d\mathbf{U})$ for all t_1 and t_2 such that $t_0 \leq t_1 \leq t_2$.

The purpose of this subsection is to investigate the statistics of $\{\mathbf{x}_t; t \geq t_0\}$. The principal conclusion to be emphasized is that $\{\mathbf{x}_t; t \geq t_0\}$ is a Markov process, which motivates our terminology for processes satisfying (4.92) as Poisson driven Markov processes. The importance of this fact is that $\{\mathbf{x}_t; t \geq t_0\}$ can be completely characterized statistically by two quantities, the distribution for $\mathbf{x}_0$ and the conditional distribution for $\mathbf{x}_t$ given $\mathbf{x}_s$ for $t_0 \leq s \leq t$. The distribution for $\mathbf{x}_0$ is assumed known, and we shall investigate the conditional distribution of $\mathbf{x}_t$ given $\mathbf{x}_s$.

Let us recall, first, that a random process $\{\xi_t; t \geq t_0\}$ is by definition a *Markov process* if for any time s such that $t_0 \leq s \leq t$, the conditional distribution of ξ_t given the random variables $\{\xi_\sigma; t_0 \leq \sigma \leq s\}$ is the same as the conditional distribution of ξ_t given ξ_s alone. In other words, the process is a Markov process if ξ_t is conditionally independent of $\{\xi_\sigma; t_0 \leq \sigma < s\}$ given ξ_s.

We have assumed that the initial condition $\mathbf{x}_0$ in (4.92) is statistically independent of $\int_{t_1}^{t_2}\int_A M(d\sigma, d\mathbf{U})$ for any choice of t_1 and t_2 such that $t_0 \leq t_1 \leq t_2$ and any $A \subseteq \mathfrak{U}$. This nonanticipatory property of the initial condition reproduces itself in time with the important implication that the process $\{\mathbf{x}_t; t \geq t_0\}$ defined by (4.92) is a Markov process. To verify these assertions, let s split the interval $[t_0, t)$ into two subintervals $[t_0, s)$ and $[s, t)$. It is evident from (4.92) that $\mathbf{x}_t$ can be written in terms of $\mathbf{x}_s$ as

$$\mathbf{x}_t = \mathbf{x}_s + \int_{s}^{t} \mathbf{a}_\sigma(\mathbf{x}_\sigma)\,d\sigma + \int_{s}^{t}\int_{\mathfrak{U}} \mathbf{b}_\sigma(\mathbf{x}_\sigma, \mathbf{U}) M(d\sigma, d\mathbf{U}), \tag{4.93}$$

where

$$\mathbf{x}_s = \mathbf{x}_0 + \int_{t_0}^{s} \mathbf{a}_\sigma(\mathbf{x}_\sigma)\,d\sigma + \int_{t_0}^{s}\int_{\mathfrak{U}} \mathbf{b}_\sigma(\mathbf{x}_\sigma, \mathbf{U}) M(d\sigma, d\mathbf{U}). \tag{4.94}$$

Equation (4.93) can be regarded as a stochastic integral equation for

$\{\mathbf{x}_\sigma; s \leqslant \sigma < t\}$ with $\mathbf{x}_s$ the initial condition. But from (4.94), this initial condition $\mathbf{x}_s$ is a function only of $\mathbf{x}_0$ and the underlying compound Poisson process on $[t_0, s)$. As the compound Poisson process has independent increments, $\mathbf{x}_s$ is, therefore, statistically independent of $\int_{t_1}^{t_2}\int_A M(d\sigma, d\mathbf{U})$ for all t_1 and t_2 such that $s \leqslant t_1 \leqslant t_2$ and any $A \subseteq \mathfrak{U}$. In other words, $\mathbf{x}_s$ does not anticipate the point process on $[s, \infty)$. Furthermore, according to (4.93), $\mathbf{x}_t$ is a function only of $\mathbf{x}_s$ and the underlying compound Poisson process on $[s, t)$. We conclude that because $\mathbf{x}_s$ is determined by $\{\mathbf{x}_\sigma; t_0 \leqslant \sigma < s\}$ and because the point process on $[s, t)$ is independent of $\{\mathbf{x}_\sigma; t_0 \leqslant \sigma < s\}$ by virtue of its independent increments, the conditional distribution for $\mathbf{x}_t$ given $\{\mathbf{x}_\sigma; t_0 \leqslant \sigma \leqslant s\}$ is the same as the conditional distribution for $\mathbf{x}_t$ given $\mathbf{x}_s$ alone. Hence, $\{\mathbf{x}_t; t \geqslant t_0\}$ is a Markov process.

An immediate consequence of $\{\mathbf{x}_t; t \geqslant t_0\}$ being a Markov process is that it can be completely characterized statistically by only two quantities, the distribution of $\mathbf{x}_0$ and the conditional distribution of $\mathbf{x}_t$ given $\mathbf{x}_s$ for any pair of times t and s such that $t_0 \leqslant s \leqslant t$. Recall that $\{\mathbf{x}_t; t \geqslant t_0\}$ is completely characterized statistically when the joint probability

$$\Pr(\mathbf{x}_{t_1} \in B_1, \mathbf{x}_{t_2} \in B_2, \ldots, \mathbf{x}_{t_K} \in B_K) \tag{4.95}$$

is known for any countable collection of ordered times $t_1, t_2, \ldots, t_K$ in $[t_0, \infty)$ and any choice of regions $B_1, B_2, \ldots, B_K$ of R^n for which the probability is defined. But by virtue of $\{\mathbf{x}_t; t \geqslant t_0\}$ being a Markov process, (4.95) can be written as

$$\Pr(\mathbf{x}_{t_1} \in B_1, \mathbf{x}_{t_2} \in B_2, \ldots, \mathbf{x}_{t_K} \in B_K) = \Pr(\mathbf{x}_{t_1} \in B_1) \prod_{k=2}^{K} \Pr(\mathbf{x}_{t_k} \in B_k | \mathbf{x}_{t_{k-1}} \in B_{k-1}) \tag{4.96}$$

Thus, the probability (4.95) can be determined knowing only the distribution of $\mathbf{x}_t$ for any $t \geqslant t_0$ and the conditional distribution of $\mathbf{x}_t$ given $\mathbf{x}_s$ for any pair of times s and t, $t_0 \leqslant s \leqslant t$. We term this conditional distribution the *transition distribution* of the process. The distribution of $\mathbf{x}_t$ can, in turn, be determined from the distribution of $\mathbf{x}_0$ and the transition distribution by use of the relation

$$\Pr(\mathbf{x}_{t_1} \in B_1) = \int_{R^n} \Pr(\mathbf{x}_{t_1} \in B_1 | \mathbf{x}_0 = \mathbf{X}) \, dP_{\mathbf{x}_0}(\mathbf{X}). \tag{4.97}$$

Thus, $\Pr(\mathbf{x}_{t_1} \in B_1, \mathbf{x}_{t_2} \in B_2, \ldots, \mathbf{x}_{t_k} \in B_k)$ can be determined from the initial distribution, which we have assumed is known, and the transition distribution, which we characterize in the following theorem.

Theorem 4.2.4 (Transition Statistics for a Poisson Driven Markov Process). Let $\{\mathbf{x}_t; t \geqslant t_0\}$ be a Poisson driven Markov process satisfying (4.92) and the conditions of Theorem 4.2.3. For $t_0 \leqslant s \leqslant t$, define

$$M_{\mathbf{x}_t|\mathbf{x}_s}(j\mathbf{v}|\mathbf{W}) = E(\exp[j\langle \mathbf{v}, \mathbf{x}_t\rangle]|\mathbf{x}_s = \mathbf{W}). \tag{4.98}$$

This quantity, which we term the *transition characteristic function* of $\{\mathbf{x}_t; t \geqslant t_0\}$, satisfies

$$dM_{\mathbf{x}_t|\mathbf{x}_s}(j\mathbf{v}|\mathbf{W}) = E\{\exp[j\langle \mathbf{v}, \mathbf{x}_t\rangle]\Psi_t(\mathbf{v}|\mathbf{x}_t)|\mathbf{x}_s = \mathbf{W}\}\, dt \tag{4.99}$$

$$M_{\mathbf{x}_s|\mathbf{x}_s}(j\mathbf{v}|\mathbf{W}) = \exp[j\langle \mathbf{v}, \mathbf{W}\rangle],$$

where Ψ_t is given by

$$\Psi_t(\mathbf{v}|\mathbf{x}_t) = j\langle \mathbf{v}, \mathbf{a}_t(\mathbf{x}_t)\rangle + \lambda_t \int_{\mathcal{U}} \{\exp[j\langle \mathbf{v}, \mathbf{b}_t(\mathbf{x}_t, \mathbf{U})\rangle] - 1\}\, dP_{\mathbf{u}}(\mathbf{U}). \tag{4.100}$$

Proof. The proof of Theorem 4.2.4 is straightforward using (4.78), which for (4.92) becomes

$$d\exp[j\langle \mathbf{v}, \mathbf{x}_t\rangle] = \exp[j\langle \mathbf{v}, \mathbf{x}_t\rangle]\bigg\{ j\langle \mathbf{v}, \mathbf{a}_t(\mathbf{x}_t)\rangle\, dt + \int_{\mathcal{U}} (\exp[j\langle \mathbf{v}, \mathbf{b}_t(\mathbf{x}_t, \mathbf{U})\rangle] - 1) M(dt, d\mathbf{U})\bigg\}.$$

Now take the conditional expectation of this expression given $\mathbf{x}_s = \mathbf{W}$ to get

$$dM_{\mathbf{x}_t|\mathbf{x}_s}(j\mathbf{v}|\mathbf{W}) = E\bigg\{ \exp[j\langle \mathbf{v}, \mathbf{x}_t\rangle] E\bigg\{ j\langle \mathbf{v}, \mathbf{a}_t(\mathbf{x}_t)\rangle\, dt + \int_{\mathcal{U}} (\exp[j\langle \mathbf{v}, \mathbf{b}_t(\mathbf{x}_t, \mathbf{U})\rangle] - 1) M(dt, d\mathbf{U})|\mathbf{x}_t, \mathbf{x}_s = \mathbf{W}\bigg\} \bigg| \mathbf{x}_s = \mathbf{W}\bigg\}$$

The interchange of differentiation and expectation can be justified using the bounded convergence theorem, Theorem 1.4.1. This proves Theorem 4.2.4.

The function Ψ appearing in Theorem 4.2.4 can be related qualitatively to the infinitesimal transition statistics of the process $\{\mathbf{x}_t; t \geqslant t_0\}$ of (4.92). For this purpose, note that the infinitesimal increment $\Delta\mathbf{x}_t = \mathbf{x}_{t+\Delta t} - \mathbf{x}_t$ has either the value

$$\Delta\mathbf{x}_t = \mathbf{a}_t(\mathbf{x}_t)\Delta t + o(\Delta t)$$

with probability $1-\lambda_t\Delta t+o(\Delta t)$, where $o(\Delta t)/\Delta t\to 0$ as $\Delta t\to 0$, or the value

$$\Delta \mathbf{x}_t=\mathbf{a}_t(\mathbf{x}_t)\Delta t+\mathbf{b}_t(\mathbf{x}_t,\mathbf{U})+o(\Delta t)$$

with probability $\lambda_t\Delta t\Delta P_{\mathbf{u}}(\mathbf{U})+o(\Delta t)$. Thus

$$\begin{aligned}E\{\exp[j\langle\mathbf{v},\Delta\mathbf{x}_t\rangle]-1|\mathbf{x}_t\}&=(\exp[j\langle\mathbf{v},\mathbf{a}_t(\mathbf{x}_t)\rangle\Delta t]-1)(1-\lambda_t\Delta t)\\&\quad+\lambda_t\Delta t\int_{\mathcal{U}}(\exp[j\langle\mathbf{v},\mathbf{a}_t(\mathbf{x}_t)\Delta t+\mathbf{b}_t(\mathbf{x}_t,\mathbf{U})\rangle]-1)\,dP_{\mathbf{u}}(\mathbf{U})+o(\Delta t)\end{aligned}$$

The limit as $\Delta t\to 0$ of the right side of this equation multiplied by $(\Delta t)^{-1}$ exists and equals $\Psi_t(\mathbf{v}|\mathbf{x}_t)$. Consequently, the limit exists on the left side, and we obtain

$$\Psi_t(\mathbf{v}|\mathbf{x}_t)=\lim_{\Delta t\to 0}(\Delta t)^{-1}E\{\exp[j\langle\mathbf{v},\Delta\mathbf{x}_t\rangle]-1|\mathbf{x}_t\}. \tag{4.101}$$

Because of this relation, Ψ_t is termed the *characteristic form for the differential generator* of the Markov process **x**.

The following examples illustrate the use of (4.99).

Example 4.2.9 (Poisson Counting Process). As the simplest example illustrating the use of (4.99), suppose $\{\mathbf{x}_t; t\geqslant t_0\}$ is a real-valued process satisfying (4.92) with $a=0$ and $b=1$. Then $x_t=N_t$, where N_t is the number of points on $[t_0,t)$ regardless of their mark. We easily have $\Psi_t(v|N_t)=\lambda_t[\exp(jv)-1]$, and (4.99) becomes

$$\begin{aligned}dM_{N_t|N_s}(jv|k)&=E\{\exp(jvN_t)\lambda_t[\exp(jv)-1]|N_s=k\}\,dt\\&=M_{N_t|N_s}(jv|k)\lambda_t[\exp(jv-1)]\,dt\end{aligned}$$

$$M_{N_s|N_s}(jv|k)=\exp(jvk).$$

The solution is

$$M_{x_t|x_s}(jv|k)=\exp(jvk)\exp\left\{\int_s^t\lambda_\sigma\,d\sigma[\exp(jv)-1]\right\}. \tag{4.102}$$

The reader should compare this expression to the characteristic function for the increment N_t-N_s as given in equation (2.35) of Chapter 2.

Example 4.2.10 (Linear Dynamical System). Linear dynamical systems were introduced in Example 4.2.2. These are Poisson driven Markov processes satisfying the linear differential equation (4.47). We write this equation as

$$d\mathbf{x}_t = \mathbf{A}_t\mathbf{x}_t\,dt + \int_{\mathfrak{U}} \mathbf{b}_t(\mathbf{U})M(dt,d\mathbf{U}), \qquad \mathbf{x}_{t_0} = \mathbf{x}_0 \tag{4.103}$$

where we set $\mathbf{b}_t(\mathbf{U}) = B_t\mathbf{U}$ for (4.47). For our present considerations, there is no need to assume $\mathbf{b}_t(\mathbf{U})$ is a linear function of $\mathbf{U}$. From (4.45), the output, or response function, is

$$\mathbf{y}_t = \mathbf{C}_t\mathbf{x}_t. \tag{4.104}$$

We will need the following two properties of the state transition matrix $\boldsymbol{\Phi}(t,t_0)$ defined in (4.49): (1) the inverse $\boldsymbol{\Phi}^{-1}(t,t_0)$ exists and satisfies

$$\boldsymbol{\Phi}^{-1}(t,t_0) = \boldsymbol{\Phi}(t_0,t);$$

and (2)

$$\boldsymbol{\Phi}(t,t_0)\boldsymbol{\Phi}(t_0,\tau) = \boldsymbol{\Phi}(t,\tau).$$

These properties are established by R. Brockett [4, p. 27].

We want to determine the transition characteristic function for $\mathbf{x}_t$ given that $\mathbf{x}_s = \mathbf{W}$. As an intermediate step, it is convenient to evaluate the transition characteristic function for the process $\mathbf{z}_t = \boldsymbol{\Phi}(t_0,t)\mathbf{x}_t$ given that $\mathbf{z}_s = \boldsymbol{\Phi}(t_0,s)\mathbf{W}$; the desired characteristic functions will follow easily from this because $\mathbf{x}_t = \boldsymbol{\Phi}(t,t_0)\mathbf{z}_t$.

By using (4.48) and the definition of $\mathbf{z}_t$, we have that $\{\mathbf{z}_t; t \geqslant t_0\}$ satisfies

$$d\mathbf{z}_t = \int_{\mathfrak{U}} \boldsymbol{\Phi}(t_0,t)\mathbf{b}_t(\mathbf{U})M(dt,d\mathbf{U}), \qquad \mathbf{z}_{t_0} = \mathbf{x}_0. \tag{4.105}$$

The characteristic form for the differential generator of this process is, therefore,

$$\Psi_t(\boldsymbol{\alpha}|z_t) = \lambda_t \int_{\mathfrak{U}} \{\exp[j\langle\boldsymbol{\alpha}, \boldsymbol{\Phi}(t_0,t)\mathbf{b}_t(\mathbf{U})\rangle] - 1\}\,dP_{\mathbf{u}}(\mathbf{U}),$$

and by Theorem 4.2.4 we have after an elementary calculation that

$$M_{\mathbf{z}_t|\mathbf{z}_s}(j\boldsymbol{\alpha}|\boldsymbol{\Phi}(t_0,s)\mathbf{W}) = \exp[j\langle\boldsymbol{\alpha}, \boldsymbol{\Phi}(t_0,s)\mathbf{W}\rangle]$$

$$\times \exp\left\{\int_s^t \int_{\mathfrak{U}} \lambda_\sigma \{\exp[j\langle\boldsymbol{\alpha}, \boldsymbol{\Phi}(t_0,\sigma)\mathbf{b}_\sigma(\mathbf{U})\rangle] - 1\}\,d\sigma\,dP_{\mathbf{u}}(\mathbf{U})\right\}. \tag{4.106}$$

Observing that $\langle\mathbf{v},\mathbf{x}_t\rangle = \langle\boldsymbol{\Phi}'(t,t_0)\mathbf{v},\mathbf{z}_t\rangle$, we then deduce the desired result that the transition characteristic function for $\mathbf{x}_t$ given that $\mathbf{x}_s = \mathbf{W}$ is given

by

$$M_{\mathbf{x}_t|\mathbf{x}_s}(j\mathbf{v}|\mathbf{W}) = \exp[j\langle \mathbf{v}, \boldsymbol{\Phi}(t,s)\mathbf{W}\rangle]$$

$$\times \exp\left\{\int_s^t \int_{\mathrm{U}} \lambda_\sigma \{\exp[j\langle \mathbf{v}, \boldsymbol{\Phi}(t,\sigma)\mathbf{b}_\sigma(\mathbf{U})\rangle] - 1\}\, d\sigma\, dP_{\mathbf{u}}(\mathbf{U})\right\}. \tag{4.107}$$

The conditional characteristic function for $\mathbf{y}_t$, the response (4.104), given that $\mathbf{y}_s = \mathbf{C}_s\mathbf{W}$ can be similarly obtained from (4.106) by noting that

$$\langle \mathbf{v}, \mathbf{y}_t\rangle = \langle \boldsymbol{\Phi}'(t,t_0)\mathbf{C}_t'\mathbf{v}, \mathbf{z}_t\rangle.$$

Thus,

$$M_{\mathbf{y}_t|\mathbf{y}_s}(j\mathbf{v}|\mathbf{C}_s\mathbf{W}) = \exp[j\langle \mathbf{v}, \mathbf{C}_t\boldsymbol{\Phi}(t,s)\mathbf{W}\rangle]$$

$$\times \exp\left\{\int_s^t \int_{\mathrm{U}} \lambda_\sigma \{\exp[j\langle \mathbf{v}, \mathbf{h}(t,\sigma:\mathbf{U})\rangle] - 1\}\, d\sigma\, dP_{\mathbf{u}}(\mathbf{U})\right\}, \tag{4.108}$$

where we define $\mathbf{h}(t,\sigma:\mathbf{U}) = \mathbf{C}_t\boldsymbol{\Phi}(t,\sigma)\mathbf{b}_\sigma(\mathbf{U})$, which is the response at time t to a point at time σ having a mark $\mathbf{U}$.

The reader should compare (4.108) with (4.10).

Example 4.2.11 (Covariance for a Linear Dynamical System). The covariance of the state of a linear dynamical system can be determined by using (4.107) and the moment generating properties of characteristic functions. An alternative approach is to use the result of Example 4.2.6. Thus, suppose $\{\mathbf{x}_t; t \geqslant t_0\}$ is a Poisson driven Markov process satisfying (4.103) and that the initial state $\mathbf{x}_0$ has mean $E(\mathbf{x}_0)$ and covariance

$$\Lambda_0 = E[(\mathbf{x}_0 - E(\mathbf{x}_0))(\mathbf{x} - E(\mathbf{x}_0))'].$$

It is convenient to form a zero-mean version of $\{\mathbf{x}_t; t \geqslant t_0\}$ by defining $\tilde{\mathbf{x}}_t = \mathbf{x}_t - E(\mathbf{x}_t)$ for $t \geqslant t_0$. It is easily seen from (4.103) that

$$d\tilde{\mathbf{x}}_t = (\mathbf{A}_t\tilde{\mathbf{x}}_t - \lambda_t\bar{\mathbf{b}}_t)\,dt + \int_{\mathrm{U}} \mathbf{b}_t(\mathbf{U})M(dt,d\mathbf{U}), \tag{4.109}$$

$$\tilde{\mathbf{x}}_{t_0} = \mathbf{x}_0 - E(\mathbf{x}_0),$$

where $\bar{\mathbf{b}}_t = E[\mathbf{b}_t(\mathbf{u})]$. Furthermore, in parallel with (4.48), there holds

$$\tilde{\mathbf{x}}_t = \boldsymbol{\Phi}(t,\tau)\tilde{\mathbf{x}}_\tau - \int_\tau^t \boldsymbol{\Phi}(t,\sigma)\lambda_\sigma\bar{\mathbf{b}}_\sigma\, d\sigma + \int_\tau^t \int_{\mathrm{U}} \boldsymbol{\Phi}(t,\sigma)\mathbf{b}_\sigma(\mathbf{U})M(d\sigma,d\mathbf{U}), \tag{4.110}$$

where $\mathbf{\Phi}(t,\tau)$ is the state-transition matrix for (4.109).

First, we want to derive an equation for the instantaneous covariance matrix for $\mathbf{x}_t$, which is defined by

$$\operatorname{cov}(\mathbf{x}_t,\mathbf{x}_t)=\mathbf{\Lambda}_t=E(\tilde{\mathbf{x}}_t\tilde{\mathbf{x}}_t').$$

As in Example 4.2.6, let $\mathbf{e}_i$ denote an n-dimensional unit vector with ith element equal to one and all other elements equal to zero. Also, let $\phi_{ij}(\mathbf{X})=\mathbf{e}_i'\mathbf{X}\mathbf{X}'\mathbf{e}_j$. Then from (4.109) and (4.77), we have

$$\begin{aligned} d\phi_{ij}(\tilde{\mathbf{x}}_t) &= \langle(\mathbf{e}_i\mathbf{e}_j'+\mathbf{e}_j\mathbf{e}_i')\tilde{\mathbf{x}}_t, \mathbf{A}_t\tilde{\mathbf{x}}_t-\lambda_t\bar{\mathbf{b}}_t\rangle\,dt \\ &\quad +\mathbf{e}_i'\int_{\mathfrak{U}}[\mathbf{b}_t(\mathbf{U})\tilde{\mathbf{x}}_t+\tilde{\mathbf{x}}_t\mathbf{b}_t'(\mathbf{U})+\mathbf{b}_t(\mathbf{U})\mathbf{b}_t'(\mathbf{U})]M(dt,d\mathbf{U})\mathbf{e}_j. \end{aligned} \tag{4.111}$$

Noting that the (i,j)-element, Λ_{ij}, of $\mathbf{\Lambda}_t$ equals $E[\phi_{ij}(\tilde{\mathbf{x}}_t)]$ and taking the expectation of both sides of (4.111), we have that

$$d\Lambda_{ij}=\sum_{k=1}^{n}(A_{ik}\Lambda_{kj}+\Lambda_{ik}A_{jk})\,dt+\lambda_t\mathbf{e}_i'E[\mathbf{b}_t(\mathbf{u})\mathbf{b}_t'(\mathbf{u})]\mathbf{e}_j\,dt.$$

Consequently, the instantaneous covariance matrix for $\mathbf{x}_t$ satisfies the linear, matrix differential equation

$$\frac{d\mathbf{\Lambda}_t}{dt}=\mathbf{A}_t\mathbf{\Lambda}_t+\mathbf{\Lambda}_t\mathbf{A}_t'+\lambda_tE[\mathbf{b}_t(\mathbf{u})\mathbf{b}_t'(\mathbf{u})], \tag{4.112}$$

$$\mathbf{\Lambda}_{t_0}=\mathbf{\Lambda}_0.$$

Let us next determine the covariance matrix $\mathbf{K}_{\mathbf{x}}(t,\tau)$ for $\{\mathbf{x}_t; t\geq t_0\}$, where $\mathbf{K}_{\mathbf{x}}(t,\tau)=E(\tilde{\mathbf{x}}_t\tilde{\mathbf{x}}_\tau')$. For $t\geq\tau$, we have from (4.110) that $\mathbf{K}_{\mathbf{x}}(t,\tau)=\mathbf{\Phi}(t,\tau)\mathbf{\Lambda}_\tau$ because the expectation

$$\begin{aligned} &E\left\{\left[-\int_\tau^t\Phi(t,\sigma)\lambda_\sigma\bar{\mathbf{b}}_\sigma\,d\sigma+\int_\tau^t\int_{\mathfrak{U}}\mathbf{\Phi}(t,\sigma)\mathbf{b}_\sigma(\mathbf{U})M(d\sigma,d\mathbf{U})\right]\mathbf{x}_\tau'\right\} \\ &\quad =E\left\{E\left[-\int_\tau^t\mathbf{\Phi}(t,\sigma)\lambda_\sigma\bar{\mathbf{b}}_\sigma\,d\sigma+\int_\tau^t\int_{\mathfrak{U}}\mathbf{\Phi}(t,\sigma)\mathbf{b}_\sigma(\mathbf{U})M(d\sigma,d\mathbf{U})|\mathbf{x}_\tau\right]\mathbf{x}_\tau'\right\} \end{aligned}$$

is zero. A similar procedure for $t\leq\tau$ yields

$$\mathbf{K}_{\mathbf{x}}(t,\tau)=\begin{cases}\mathbf{\Phi}(t,\tau)\mathbf{\Lambda}_\tau, & t\geq\tau\\ \mathbf{\Lambda}_t\mathbf{\Phi}'(\tau,t), & t\leq\tau.\end{cases} \tag{4.113}$$

Finally, let us note that if $\mathbf{y}_t = \mathbf{C}_t \mathbf{x}_t$, then the covariance matrix for $\{\mathbf{y}_t; t \geqslant t_0\}$ is

$$\mathbf{K}_{\mathbf{y}}(t,\tau) = \mathbf{C}_t \mathbf{K}_{\mathbf{x}}(t,\tau)\mathbf{C}'_\tau. \tag{4.114}$$

By using the fact that the solution to (4.112) can be expressed in terms of $\mathbf{\Phi}(t,\tau)$ as

$$\Lambda_t = \mathbf{\Phi}(t,t_0)\Lambda_0\mathbf{\Phi}'(t,t_0) + \int_{t_0}^{t} \mathbf{\Phi}(t,\tau)\lambda_\tau E[\mathbf{b}_\tau(\mathbf{u})\mathbf{b}'_\tau(\mathbf{u})]\mathbf{\Phi}'(t,\tau)\,d\tau, \tag{4.115}$$

the reader should verify the equivalence of the expressions for the covariance of $\{\mathbf{y}_t; t \geqslant t_0\}$ given in (4.114) and (4.16) when $\Lambda_0 = \mathbf{0}$ and the dimension of $\mathbf{y}$ is unity.

The stochastic integral equations associated with the above Examples 4.2.9 and 4.2.10 are *linear*. It is for this reason that an explicit expression for the transition characteristic function of the state $\{\mathbf{x}_t; t \geqslant t_0\}$ can be found by solving (4.99). More generally, when the integral equation is not linear, the solution of (4.99) poses an insurmountable problem analytically. Even so, (4.99) still provides a useful point of departure for the *approximate* statistical analysis of the influence of a compound Poisson process on a nonlinear system when an exact analysis in infeasible. The appropriate approximations will, of course, depend strongly on the functions involved in the particular application, so we cannot give any general rules to follow. As one instance illustrating that (4.99) can be useful in engineering analyses even when it cannot be solved explicitly, we continue the investigation begun in Example 4.2.3 of a phase-tracking loop excited by shot noise.

Example 4.2.12 (Phase-Tracking Loop Excited by Shot Noise, Continued). Consider, again, the model of Example 4.2.3, and make the following simplifying assumptions: (1) $\omega_L = \omega_R = \omega$; (2) $\theta_R(t) = 0$; (3) $h(t,\tau) = g\delta(t-\tau)$; and (4) $U = e$, where e is the electronic charge. Assumption (1) is that the receiver loop frequency ω_L and transmitted reference frequency ω_R are equal, the common frequency being ω. The transmitter and receiver are, therefore, assumed to be frequency synchronized. Assumption (2) is that the transmitted reference phase is a constant, namely zero, independent of time. In this case, the loop phase $\theta_L(t)$ is ideally also zero, and any departure from zero is the phase tracking error. Our interest is in the statistics of the phase error process $\{\phi_t; t \geqslant t_0\}$, where $\phi_t = \theta_L(t) - \theta_R(t) = \theta_L(t)$. Assumption (3) is that the loop filter shown in Figure 4.5 is simply an amplifier with constant gain g. The phase-tracking loop for this choice of filter is termed *first order*. Assumption (4) is that the photodetector has no gain (see Table 4.1). Under assumptions (1)–(4), equation (4.50) describing the operation of the loop becomes

$$d\phi_t = f(\psi_t)dN_t, \qquad \phi_{t_0} = \phi_0 \tag{4.116}$$

where we define $\psi_t = \omega t + \phi_t$, and where the function $f(\psi_t)$ is defined for a phase-lock loop by

$$f(\psi_t) = ge\sin(\psi_t) \tag{4.117}$$

and for a tanlock loop by

$$f(\psi_t) = ge[1 + m\cos(\psi_t)]^{-1}\sin(\psi_t). \tag{4.118}$$

Under assumption (2), the Poisson counting process $\{N_t; t \geqslant t_0\}$ in (4.116) has intensity

$$\lambda_t = A[1 + m\cos(\omega t)]$$

for $t \geqslant t_0$.

Let $M_{\phi_t}(jv) = E[\exp(jv\phi_t)]$ be the characteristic function for the phase-tracking error ϕ_t at time $t \geqslant t_0$. As there holds

$$M_{\phi_t}(jv) = E\{E[\exp(jv\phi_t)|\phi_0]\},$$

we have from (4.99) that

$$\frac{dM_{\phi_t}(jv)}{dt} = A[1 + m\cos(\omega t)]E\{\exp(jv\phi_t)[\exp(jvf(\psi_t)) - 1]\}. \tag{4.119}$$

It does not appear feasible to solve (4.119) for an explicit expression for $M_{\phi_t}(jv)$. Consequently, we resort to an approximate analysis based on the additional assumption that the phase-error process $\{\phi_t; t \geqslant t_0\}$ approaches a steady state as t becomes large in the sense that the distribution for ϕ_t becomes independent of t. This is a reasonable assumption from a practical viewpoint. Then, for t large, (4.119) becomes

$$0 \doteq A[1 + m\cos(\omega t)]E\{\exp(jv\phi_t)[\exp(jvf(\psi_t)) - 1]\}, \tag{4.120}$$

where the symbol $\doteq$ is used to indicate that equality holds in the usual sense for t large when all terms on the right side that depend explicitly on t are isolated and removed; these terms presumably average to zero as t approaches infinity if a steady-state distribution for ϕ_t exists.

Now, it appears to be a reasonable approximation to replace $\exp[jvf(\psi_t)]$ by the first few terms in its Taylor-series expansion if $|f(\psi_t)|$ is sufficiently small, say less than some prescribed number ϵ, which for three terms might be on the order of 0.1. It is easy to verify from (4.117) and (4.118) that for a phase-lock loop

$$\max_{\psi_t} |f(\psi_t)| = |ge|,$$

and for a tanlock loop

$$\max_{\psi_t} |f(\psi_t)| = \frac{|ge|}{1-m^2}.$$

Thus, if for a phase-lock loop, the condition $|ge| < \epsilon$ is satisfied or if for a tanlock loop, the condition $|ge|/(1-m^2) < \epsilon$ is satisfied, we will use the approximation

$$\exp[jvf(\psi_t)] - 1 \approx jvf(\psi_t) + \tfrac{1}{2}(jv)^2 f^2(\psi_t) \tag{4.121}$$

The right side of (4.121) is a periodic function of ψ_t; the first and second terms can be expanded in a Fourier sine and cosine series, respectively. Thus,

$$\exp[jvf(\psi_t)] - 1 \approx a_0 + a_1 \cos\psi_t + b_1 \sin\psi_t + a_2 \cos 2\psi_t + \cdots \tag{4.122}$$

in which a_0, $a_1, \ldots$ are the Fourier coefficients. Upon multiplication of (4.122) by λ_t, we deduce that

$$\lambda_t\{\exp[jvf(\psi_t)]\} \approx Aa_0 + \tfrac{1}{2}Ama_1 \cos\phi_t + \tfrac{1}{2}Amb_1 \sin\phi_t$$
$$+ (\text{terms depending explicitly on } t) \tag{4.123}$$

Hence, for use in (4.120), we see that only three coefficients a_0, a_1, and b_1 are required. These are found for a phase-lock loop to be

$$\begin{aligned} a_0 &= \tfrac{1}{4}(jv)^2(ge)^2, \\ a_1 &= 0, \\ b_1 &= jv(ge), \end{aligned} \tag{4.124}$$

and for a tanlock loop,

$$\begin{aligned} a_0 &= \tfrac{1}{2}(jv)^2\left[(1-m^2)^{-1/2} - 1\right]m^{-2}(ge)^2 \\ a_1 &= (jv)^2\left[1-(1-m^2)^{1/2}\right]\left[1-(1-m^2)^{-1/2}\right]m^{-3}(ge)^2 \\ b_1 &= (jv)2\left[1-(1-m^2)^{1/2}\right]m^{-2}(ge). \end{aligned} \tag{4.125}$$

By introducing (4.123) and (4.124) into (4.120), we see for a phase-lock loop that

$$0=E\left\{\left[(jv)^2+(jv)\alpha\sin(\phi_\infty)\right]\exp(jv\phi_\infty)\right\}, \tag{4.126}$$

where $\phi_\infty=\lim_{t\to\infty}\phi_t$ and where we define

$$\alpha=\frac{2m}{ge}. \tag{4.127}$$

Inverse Fourier transformation of (4.126) yields

$$0=\frac{d^2p_{\phi_\infty}(\Phi)}{d\Phi^2}+\frac{d\alpha\sin(\Phi)p_{\phi_\infty}(\Phi)}{d\Phi}. \tag{4.128}$$

This second-order differential equation also arises in the study of phase-lock loops operating in a wideband Gaussian noise; see A. Viterbi [25, p. 89]. It must be solved on the interval $[-\pi,\pi]$ and subject to the boundary condition

$$p_{\phi_\infty}(\pi)=p_{\phi_\infty}(-\pi)$$

and the normalization

$$\int_{-\pi}^{\pi}p_{\phi_\infty}(\Phi)\,d\Phi=1.$$

The solution obtained is

$$p_{\phi_\infty}(\Phi)=\frac{\exp[\alpha\cos(\Phi)]}{2\pi I_0(\alpha)}, \qquad -\pi\leqslant\Phi\leqslant\pi \tag{4.129}$$

where $I_0(\alpha)$ is a zero-order Bessel function of the first kind. Equation (4.129) was first obtained by R. Gagliardi and M. Haney [10] for phase-lock loops operating in shot noise. The single parameter α of (4.127) is seen to determine the steady-state phase tracking performance under the condition $|ge|<\epsilon\approx0.1$. This parameter can be interpreted qualitatively as follows. The constant $B_L=Amge/4$ is shown by A. Viterbi [25, p. 36] to be the "linearized-loop noise bandwidth" of a phase-lock loop with forward gain g and input signal $Ame\cos(\omega t)$. The signal $Ame\cos(\omega t)$ is the instantaneous average current associated with the component $Am\cos(\omega t)$ of λ_t conveying synchronization information. As it is usually desired to operate the loop with a specified bandwidth, it is necessary to adjust the gain g so that $g=4B_L/Ame$, in which case $\alpha=Am^2/2B_L$. Hence, multiplication of the numerator and denominator by Ae^2 results in

$$\alpha = \frac{(Ame)^2/2}{Ae^2 B_L}. \tag{4.130}$$

The constant $(Ame)^2/2$ is the average power in the synchronization signal $Ame\cos(\omega t)$. And from Example 4.1.8, the constant Ae^2 is the power spectral density of the input noise signal Ae; the signal Ae is the instantaneous average current associated with the component A of λ_t conveying no synchronization information. The constant Ae^2B_L is, therefore, the noise power in the noise bandwidth of the loop. We conclude that α may be interpreted qualitatively as the signal-to-shot noise power ratio in the noise bandwidth of the loop.

By following the same procedure used to derive (4.128), we find from (4.123), (4.125), and (4.120) that for a tanlock loop

$$0 = \frac{d^2[1+\Gamma\cos(\Phi)]p_{\phi_\infty}(\Phi)}{d\Phi^2} + \frac{d\Lambda\sin(\Phi)p_{\phi_\infty}(\Phi)}{d\Phi}, \tag{4.131}$$

where

$$\Gamma = (1-m^2)^{1/2} - 1$$

and

$$\Lambda = \frac{2m(1-m^2)^{1/2}}{ge}.$$

Hence,

$$p_{\phi_\infty}(\Phi) = C[1+\Gamma\cos(\Phi)]^{(\Lambda/\Gamma)-1}, \tag{4.132}$$

where C is the normalization constant

$$C = \left\{\int_{-\pi}^{\pi}[1+\Gamma\cos(\Phi)]^{(\Lambda/\Gamma)-1}\,d\Phi\right\}^{-1}.$$

Equation (4.132) was first obtained by R. Forrester and D. Snyder [9] for tanlock loops operating in shot noise.

Summary. Several important ideas have been developed in this section. Our first concern was with the calculus for treating integral and differential equations excited by a marked point process. The most important ideas in connection with this calculus are:

1. *Counting integrals.* Care is needed in the definition of integrals

with respect to counting processes as they do not exist as ordinary Riemann–Stieltjes integrals. Counting integrals were defined as a limit of the partial sum (4.57) involving a forward increment of the counting process. The choice of the forward increment results in integrals that do not obey the integration rules of ordinary calculus so that care must be exercised in their evaluation. The evaluation of counting integrals is given in Theorem 4.2.1.

2. *Differentiation.* Differentiation of functions of processes defined by counting integrals does not obey the rules of ordinary calculus. For this reason, a special differential rule is developed in Theorem 4.2.2.

Poisson driven Markov processes are defined by counting integrals in which the counting process is derived from a compound Poisson process. Poisson driven Markov processes may be viewed as the response to a compound Poisson process of a system that can be described by an ordinary differential equation. The differential equation must be of finite order but may be both time dependent and nonlinear. Statistics for Poisson driven Markov processes were developed. The important ideas about these statistics are:

1. Under certain technical conditions given in Theorem 4.2.3, the solution to a nonlinear integral or differential equation excited by a compound Poisson process is a Markov process. This means that the solution can be completely characterized statistically by two quantities, the distribution of the initial condition and the transition distribution, which relates the statistics of the solution at one time to an earlier time.
2. The transition characteristic function satisfies a differential equation given in Theorem 4.2.4. While this differential equation can be solved only in certain special circumstances, it still provides a useful point of departure for the approximate statistical analysis of a nonlinear system excited by a compound Poisson process.

REFERENCES

1. T. M. Apostol, *Mathematical Analysis*, Addison-Wesley, Reading, Mass., 1957.
2. I. Bar-David, "Communication Under the Poisson Regime," *IEEE Trans. Information Theory*, **IT-15**, 31–37 (Jan. 1969).
3. A. T. Bharucha-Reid, *Random Integral Equations*, Academic, New York, 1972.
4. R. W. Brockett, *Finite Dimensional Linear Systems*, Wiley, New York, 1970.
5. W. B. Davenport, Jr. and W. L. Root, *An Introduction to the Theory of Random Signals and Noise*, McGraw-Hill, New York, 1958.

6. J. L. Doob, *Stochastic Processes*, Wiley, New York, 1953.
7. J. Evans, "Preliminary Analysis of ELF Noise," Tech. Note 1969-18, M.I.T. Lincoln Laboratory, Lexington, Mass., March 1969.
8. P. M. Fishman, "Statistical Inference for Space-Time Point Processes," D.Sc. Thesis, Sever Institute, Washington University, St. Louis, Mo., June 1974.
9. R. H. Forrester and D. L. Snyder, "Phase-Tracking Performance of Direct-Detection Optical Receivers," *IEEE Trans. Communications*, **COM-21**, No. 9, 1037–1039 (Sept. 1973).
10. R. Gagliardi and M. Haney, "Optical Synchronization—Phase Locking with Shot Noise Processes," Tech. Rpt. USCEE—396, Dept. of Electrical Engineering, Univ. of Southern California, Los Angeles, Calif., August 1970.
11. I. I. Gihman and A. Ja. Dorogovcev, "On Stability of Solutions of Stochastic Differential Equations," *Ukrain. Mat. Z.*, **17**, No. 6, 229–250 (1965).
12. I. I. Gihman and A. V. Skorokhod, *Stochastic Differential Equations*, Springer-Verlag, New York, 1972.
13. I. Gikhman and A. Skorokhod, *Introduction to the Theory of Random Processes*, Saunders, Philadelphia, 1969.
14. E. V. Hoversten, "Optical Communication Theory," in *Laser Handbook*, Eds. F. T. Arecchi and E. O. Schultz, North Holland, Amsterdam, 1972, pp. 1805–1862.
15. K. Itô, "On Stochastic Differential Equations," *Mem. Am. Math. Soc.*, No. 4, 1–51 (1951).
16. R. S. Kennedy, *Fading Dispersive Communication Channels*, Wiley, New York, 1969.
17. P. Langevin, "Sur La Théorie Du Mouvement Brownien," *C. R. Acad. Sci. Paris*, **146**, 530–533 (1908).
18. W. C. Lindsey and M. K. Simon, *Telecommunication Systems Engineering*, Prentice-Hall, Englewood Cliffs, N. J., 1973.
19. L. Liusternik and V. Sobolev, *Elements of Functional Analysis*, Ungar, New York, 1961.
20. A. Papoulis, "High Density Shot Noise and Gaussianity," *J. Appl. Prob.*, **8**, No. 1, 118–127 (March 1971).
21. E. Parzen, *Stochastic Processes*, Holden-Day, San Francisco, 1962.
22. W. K. Pratt, *Laser Communications Systems*, Wiley, New York, 1969.
23. "Project West Ford," *Proc. IEEE* (Special Issue), **52**, No. 5, (May 1964).
24. A. V. Skorokhod, *Studies in the Theory of Random Processes*, Addison-Wesley, Reading, Mass., 1964.
25. A. J. Viterbi, *Principles of Coherent Communication*, McGraw-Hill, New York, 1966.
26. E. Wong, *Stochastic Processes in Information and Dynamical Systems*, McGraw-Hill, New York, 1961.
27. L. A. Zadeh and C. A. DeSoer, *Linear System Theory*, McGraw-Hill, New York, 1963.

PROBLEMS

4.1.1. Consider a vacuum-tube diode biased with a constant voltage. Neglecting the effects of any space charge, it can be assumed that electrons are emitted from the cathode with zero velocity and as a homogeneous

Poisson process with intensity λ electrons per second. The incremental contribution of an electron emitted at time τ to the anode current is

$$h(t,\tau)=\begin{cases}\dfrac{2e(t-\tau)}{T^2}, & \tau\leqslant t\leqslant T+\tau\\ 0, & \text{otherwise}\end{cases}$$

where e is the electronic charge and T is the time for the electron to traverse from the cathode to anode.

a. Evaluate the nth cumulant of the current i_t at time t, where

$$i_t=\begin{cases}0, & N_t=0\\ \sum_{n=1}^{N_t} h(t,\tau_n), & N_t\geqslant 1.\end{cases}$$

b. Use the result of part (a) to determine the mean and variance of i_t.

c. Determine the covariance function for the current process.

4.1.2. Suppose a sinusoidal bias voltage is applied to the diode of Problem 4.1.1. Assume for this situation that electrons are emitted with zero velocity and as an inhomogeneous Poisson process with intensity $\lambda_t=\lambda[1+m\sin(\omega t)]$. Repeat parts (a)–(c) of Problem 4.1.1.

4.1.3. Let

$$y_t=\begin{cases}0, & N_t=0\\ \sum_{n=1}^{N_t} h(t,\tau_n:u_n), & N_t\geqslant 1\end{cases}$$

be a filtered homogeneous Poisson process of rate λ for $t\geqslant 0$. Assume the weighting function h has the form

$$h(t,\tau_n:u_n)=\begin{cases}1, & \tau_n\leqslant t\leqslant \tau_n+u_n\\ 0, & \text{otherwise.}\end{cases}$$

Here, y_t can be interpreted as the number of busy telephone lines when τ_n is time the nth call of duration u_n is placed, where $u_n\geqslant 0$.

a. Verify that the characteristic function for y_t is

$$M_{y_t}(j\alpha)\triangleq E[\exp(j\alpha y_t)]=\exp\left\{\lambda(e^{j\alpha}-1)\int_0^t(1-P_u(U))\,dU\right\},$$

where $P_u(U)$ is the probability distribution function for u. Thus, conclude that y_t is Poisson distributed with parameter $\lambda\int_0^t(1-P_u(U))dU$. Is $\{y_t;\ t \geq 0\}$ a Poisson process?

b. Verify that $E(u)=\int_0^\infty(1-P_u(U))dU$. Hence, conclude for large t that y_t is Poisson distributed with parameter $\lambda E(u)$.

4.1.4. For the system shown in Figure P.4.1.4, u_t for $t \geq t_0$ is a deterministic control signal applied to a plant with impulse response $g(t,\tau)$ and $\{y_t; t \geq t_0\}$ is a filtered Poisson process defined by

$$y_t = \begin{cases} 0, & N_t = 0 \\ \sum_{n=1}^{N_t} u_n h(t,\tau_n), & N_t \geq 1. \end{cases}$$

You can view y as a model for lightning induced VLF atmospheric noise that additively disturbs the control signal. Then N_t is the number of lightning strokes at time t, $\{\tau_n\}$ are the stroke occurrence times, and $\{u_n\}$ are energy marks. Assume strokes occur with rate λ_t for $t \geq t_0$. Determine the characteristic functional of the output x on the interval $[0,T)$. What is the mean value function $m(t) \triangleq E(x_t)$ and covariance function $K(t,\tau) \triangleq E\{(x_t - m(t))(x_\tau - m(\tau))\}$ for the output process?

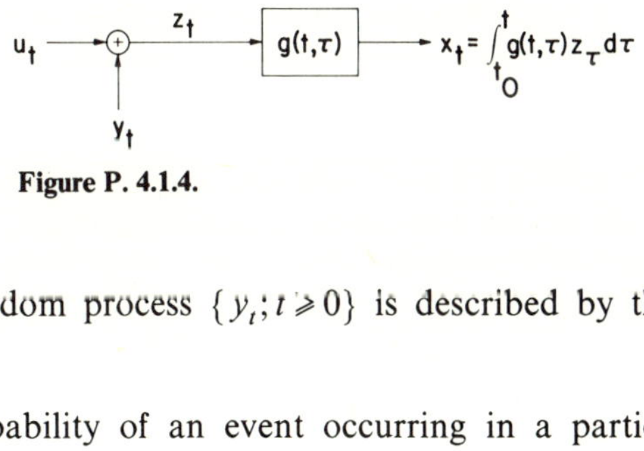

Figure P. 4.1.4.

4.1.5. The random process $\{y_t; t \geq 0\}$ is described by the following assumptions:

a. The probability of an event occurring in a particular interval is independent of the probability of an event occurring in any disjoint interval.

b.
$$\lim_{\Delta t \downarrow 0} \left\{ \frac{\Pr[1 \text{ event in } [t, t+\Delta t)]}{\Delta t} \right\} = \lambda.$$

$$\lim_{\Delta t \downarrow 0} \left\{ \frac{\Pr[2 \text{ or more events in } [t, t+\Delta t)]}{\Pr[1 \text{ event in } [t, t+\Delta t)]} \right\} = 0.$$

c. If an event occurs at t, the amplitude of the process increases by 1 unit immediately and then increases by 1 more unit Δ seconds later. A typical sample function shown in Figure P.4.1.5 illustrates this behavior.

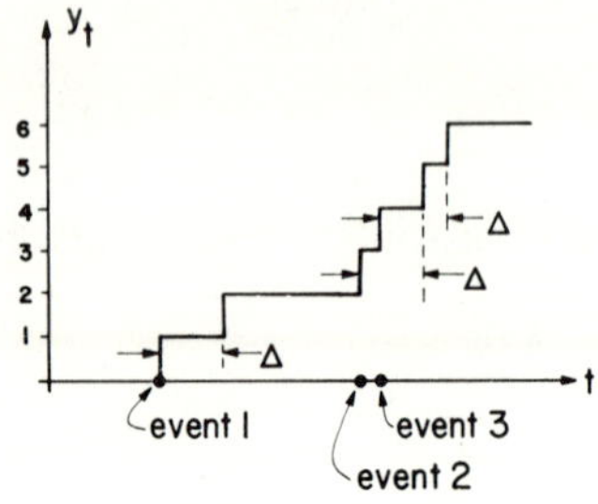

Figure P. 4.1.5.

d. The process starts at 0 at $t=0$.

i. Find either $P_{y_t}(Y;t)$, the probability distribution function for y_t at time t, for all $t \geqslant 0$, or $M_{y_t}(jv)$ the corresponding characteristic function.

ii. Is y_t an independent increment process? Justify your answer *very* briefly.

4.1.6. Let $\{y_t; t > -\infty\}$ be a filtered Poisson process defined by

$$y_t = \sum_{n=-\infty}^{\infty} h(t-\tau_n)$$

where $\{\tau_n\}$ are the occurrence times of a homogeneous Poisson process with constant intensity λ. Assume

$$h(t) = \begin{cases} |t|^{-1}, & t \geqslant 0 \\ -|t|^{-1}, & t < 0 \end{cases}$$

Show that y_t has a Cauchy distribution.

4.1.7 (An Important Generalization: Random Response Function). For $t \geqslant t_0$, let

$$y_t = \begin{cases} 0, & N_t = 0 \\ \sum_{n=1}^{N_t} h_n(t, \tau_n), & N_t \geqslant 1, \end{cases}$$

where

i. $\{N_t; t \geqslant t_0\}$ is a Poisson counting process with intensity λ_t and occurrence times $\{\tau_n\}$.

ii. $\{h_n(t,\tau); t \geqslant t_0, \tau \geqslant t_0\}$, for $n = 1,2,\ldots$, are mutually independent, identically distributed random processes all having the same distribution as the process $\{h(t,\tau); t \geqslant t_0, \tau \geqslant t_0\}$. Also, the processes $\{h_n(t,\tau); t \geqslant t_0, t \geqslant \tau\}$ are independent of $\{N_t; t \geqslant t_0\}$.

iii. The characteristic functional for $h(t,\tau)$ for $t \in [t_0, T)$ and τ fixed is defined by

$$\phi_h(jv;\tau) \triangleq E\left[\exp\left\{j\int_{t_0}^{T} h(\sigma,\tau)\,dv_\sigma\right\}\right].$$

Assume this to be known.

iv. Assume $h_n(t,\tau) = 0$ for $t < \tau$.

a. Show that the characteristic functional for y on the interval $[t_0, T)$ is given by

$$\phi_y(jv) \triangleq E\left[\exp\left\{\int_{t_0}^{T} y_\sigma\,dv_\sigma\right\}\right]$$

$$= \exp\left\{\int_{t_0}^{T} \lambda_\tau[\phi_h(jv:\tau) - 1]\,d\tau\right\}.$$

b. Determine the mean and variance for y_t, where $t \in [t_0, T)$.

c. Determine the covariance function for y; that is, evaluate $K_y(t_1,t_2) = E[y_{t_1} y_{t_2}] - E[y_{t_1}]E[y_{t_2}]$.

4.1.8. Let $\{y_t; t \geqslant t_0\}$ be a process of the type defined in Problem 4.1.7. Assume $\lambda_t \equiv \lambda$, a constant. Suppose now that $h(t,\tau)$ is an independent increment process in t for each fixed τ, that $h(t,\tau) = 0$ for $t < \tau$, that $h(\tau,\tau) = 1$, and

$$\Pr[h(t,\tau) = k] = \frac{[\gamma(t-\tau)]^{k-1}}{(k-1)!}\exp[-\gamma(t-\tau)], \qquad \gamma > 0$$

for $t \geqslant \tau$ and $k = 1,2,\ldots$. Thus h is a homogeneous Poisson counting process that starts at time τ with unit count. The process y is a reasonable model for a photomultiplier. In this case, N_t and $\{\tau_n\}$ represent the number and occurrence times of primary photoelectrons and $h_n(t,\tau_n)$ models the number of secondary electrons out of the photomultiplier due to the nth photoelectron; typically $\gamma \cong 10^6$. Determine the characteristic function for y_t:

$$M_{y_t}(j\alpha) = E[\exp(j\alpha y_t)].$$

4.1.9. Consider a process in which events happen in clusters rather than singly. An example is the production of showers of particles by cosmic rays. Each cosmic ray may cause the simultaneous release of a number of particles from the active surface of a gamma-ray detector. If the rate of cosmic ray arrivals is high and the energy distribution of the rays is such that showers of k particles may occur for $k=1,2,\ldots$, we might model the situation as follows.

For $k=1,2,\ldots$, suppose the event that k particles are released occurs in accordance with a simple Poisson counting process $N_t(k)$ with intensity λ_k. At each event of this process k particles are released, so the total number of particles released in $[0,t)$ due to this process is $kN_t(k)$.

Assume the processes $\{N_t(k)\}$ are mutually independent. Let N_t be the total number of particles that have been produced in $[0,t)$; thus,

$$N_t = N_t(1) + 2N_t(2) + 3N_t(3) + \cdots + kN_t(k) + \cdots.$$

a. Show that

$$E[N_t] = \mu_1 t$$

and

$$\mathrm{Var}[N_t] = \mu_2 t$$

where

$$\mu_1 = \sum_{k=1}^{\infty} k\lambda_k, \qquad \mu_2 = \sum_{k=1}^{\infty} k^2\lambda_k.$$

b. Show that $\{N_t, t \geqslant 0\}$ may be represented as a compound Poisson process.

c. Suppose that the showers consist of particles which communicate equal charges to an electrometer whose deflection at time t is

$$y_t = \begin{cases} 0, & N_t = 0 \\ \sum_{n=1}^{N_t} h(t-\tau_n), & N_t \geqslant 1, \end{cases}$$

where $\{\tau_n\}$ are the times at which the changes are observed and $h(t)$ is the impulse response of the electrometer. Show that

$$E[y_t]=\mu_1\int_0^t h(\sigma)\,d\sigma,$$

$$\mathrm{Var}[y_t]=\mu_2\int_0^t h^2(\sigma)\,d\sigma.$$

4.1.10. Suppose $\{y_t; t\geqslant 0\}$ has the form

$$y_t=\begin{cases}0, & N_t=0\\ \displaystyle\sum_{n=1}^{N_t} u_n \exp\left[\frac{-(t-\tau_n)}{RC}\right], & N_t\geqslant 1\end{cases}$$

in which $\{\tau_n\}$ are the times of occurrence of points of a homogeneous Poisson process with constant intensity λ, $\{u_n\}$ are independent and identically distributed random variables with finite second moments, and R and C are positive constants.

a. Evaluate the mean-value function $m(t)=E(y_t)$ for $t\geqslant 0$. What happens as $t\to\infty$?

b. Evaluate the covariance function $K(t,v)=E[y(t)y(v)]-m(t)m(v)$ of y_t for $t,v\geqslant 0$.

c. Determine the characteristic function $M(j\alpha)=E[\exp(j\alpha y_t)]$ for y_t for $t\geqslant 0$.

d. Show that

$$y_t^*=\frac{y_t-m(t)}{\sqrt{\mathrm{var}(y_t)}}$$

converges in distribution to a zero-mean, unit variance Gaussian random variable as $RC\to\infty$.

4.1.11. Let $\{y_t; t\geqslant t_0\}$ be the process defined in Problem 4.1.7. Define

$$\gamma_1(t)=\int_{t_0}^t \lambda_\tau E[h(t,\tau)]\,d\tau,$$

$$\gamma_2(t)=\int_{t_0}^t \lambda_\tau E[h^2(t,\tau)]\,d\tau,$$

and

$$\Gamma(t_1,t_2,t_3,)=\int_{t_0}^{\min(t_1,t_2,t_3,)}\lambda_\tau E[|h(t_1,\tau)h(t_2,\tau)h(t_3,\tau)|]\,d\tau.$$

Show that if $[\gamma_2(t_1)\gamma_2(t_2)\gamma_3(t_3)]^{-1/2}\Gamma(t_1,t_2,t_3)$ tends uniformly to zero on $[t_0,T)\times[t_0,T)\times[t_0,T)$ as certain parameters tend to prescribed limits, then the normalized process

$$y_t^* = \frac{y_t-\gamma_1(t)}{\gamma_2^{1/2}(t)}$$

tends to a Gaussian random process on $[t_0,T)$. Determine the mean and covariance function for $\{y_t^*; t\geqslant t_0\}$.

4.1.12. A typical sample function for a *nonlinearly* filtered Poisson process is shown in Figure P.4.1.12. Starting at $t=0$, y_t decreases exponentially from 1 until τ_1, the first arrival time. At $t=\tau_1$, y returns to 1 and then decreases exponentially with the same decay constant until τ_2, the second arrival time. At $t=\tau_2$, y returns to 1, and so on. Assume points arrive as a homogeneous Poisson process with constant intensity λ. Evaluate the mean and variance of y at any time $t\geqslant 0$.

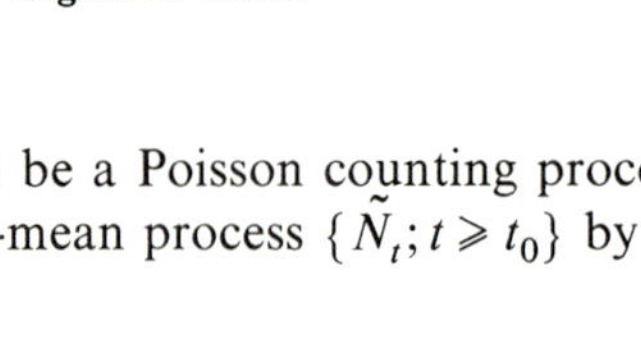

Figure P. 4.1.12.

4.2.1. Let $\{N_t; t\geqslant t_0\}$ be a Poisson counting process with intensity λ_t for $t\geqslant t_0$. Define the zero-mean process $\{\tilde{N}_t; t\geqslant t_0\}$ by $\tilde{N}_t=N_t-\int_{t_0}^t\lambda_\sigma\,d\sigma$. Consider the integral

$$\tilde{I}_t(\alpha)=\int_{t_0}^t N_\sigma\,d\tilde{N}_\sigma=\lim_{n\to\infty}\sum_{i=1}^n N_{t'_{in}}\left(\tilde{N}_{t_{in}}-\tilde{N}_{t_{i-1,n}}\right),$$

where $t_0=t_{0n}<t_{1n}<\cdots<t_{nn}=t$ is a sequence of partitions of $[t_0,t)$ such that $\max_{1\leqslant i\leqslant n}|t_{in}-t_{i-1,n}|$ tends to zero as n tends to infinity and where $t'_{in}=\alpha t_{i-1,n}+(1-\alpha)t_{in}$ for $0\leqslant\alpha\leqslant 1$.

a. Evaluate $\tilde{I}_t(\alpha)$ as a function of α.

b. Evaluate the mean and variance of $\tilde{I}_t(\alpha)$ as a function of α.

c. Evaluate $E[\tilde{I}_t(\alpha)|N_\sigma; t_0 \leqslant \sigma \leqslant s]$ as a function of α, where $s \leqslant t$.

d. A random process $\{m_t; t \geqslant t_0\}$ is termed a *martingale, submartingale,* or *supermartingale* adapted to $\{N_t; t \geqslant t_0\}$ if $E[m_t|N_\sigma; t_0 \leqslant \sigma \leqslant s]$ is equal, greater, or less than m_s, respectively. As a function of α, classify $\{\tilde{I}_t(\alpha); t \geqslant t_0\}$ as a martingale, submartingale, or supermartingale adapted to $\{N_t; t \geqslant t_0\}$.

4.2.2 [Derivation of (4.61)]. Let $\{\mathbf{I}_t^{(p,q)}(A); t \geqslant t_0\}$ be defined in (4.57), and suppose

$$\tilde{\mathbf{I}}_t^{(p,q)}(A) = \mathbf{I}_t^{(p,q)}(A) - \sum_{j=1}^{q} \sum_{k=1}^{p} \mathbf{b}_{t_{j-1,q}}(\mathbf{U}'_{kp}) \int_{t_{j-1,q}}^{t_{jq}} \int_{A_{kp}} \lambda_\sigma \, d\sigma \, dP_{\mathbf{u}}(\mathbf{U}).$$

a. Evaluate $E[|\tilde{\mathbf{I}}_t^{(p,q)}(A)|^2]$.

b. Deduce (4.61) from the result of (a) as p and q tend to infinity.

4.2.3. Suppose $\{\zeta_t; t \geqslant t_0\}$ satisfies

$$\zeta_t = 1 + \int_{t_0}^{t} \zeta_\sigma f_\sigma \, dN_\sigma,$$

where f_t is a deterministic function of t for $t \geqslant t_0$ and $\{N_t; t \geqslant t_0\}$ is a Poisson process with intensity λ_t for $t \geqslant t_0$.

a. Show that $\zeta_t = \exp[\int_{t_0}^{t} \ln(1+f_\sigma)\, dN_\sigma]$. Hint: let $\phi(\zeta_t) = \ln \zeta_t$ and use the result of Example 4.2.8.

b. Show $E(\zeta_t) = \exp[\int_{t_0}^{t} f_\sigma \lambda_\sigma \, d\sigma]$.

c. Show $\zeta_t^2 = 1 + \int_{t_0}^{t} \zeta_\sigma^2 f_\sigma (2 + f_\sigma)\, dN_\sigma$.

d. Evaluate the variance of ζ_t.

4.2.4 (Formula for Integration by Parts). The formula of ordinary calculus for integration by parts is

$$\int_{t_0}^{t} \Phi_1(\sigma)\, d\Phi_2(\sigma) = \int_{t_0}^{t} d[\Phi_1(\sigma)\Phi_2(\sigma)] - \int_{t_0}^{t} \Phi_2(\sigma)\, d\Phi_1(\sigma),$$

where $\Phi_1(t)$ and $\Phi_2(t)$ are deterministic functions of t. The analog of this formula when $\Phi_1(t)$ and $\Phi_2(t)$ are defined by counting integrals is the following. Suppose

$$\Phi_i(t)=\int_{t_0}^{t}\int_{\mathfrak{U}}\phi_i(\sigma,\mathbf{U})M(d\sigma,d\mathbf{U}), \qquad i=1,2$$

where $\{\phi_i(t,\mathbf{U}); t\geqslant t_0, \mathbf{U}\in\mathfrak{U}\}$ satisfy the conditions of Theorems 4.2.1 and 4.2.3. Show that

$$\int_{t_0}^{t}\Phi_1(\sigma)\,d\Phi_2(\sigma)=\int_{t_0}^{t}d[\Phi_1(\sigma)\Phi_2(\sigma)]-\int_{t_0}^{t}\Phi_2(\sigma)\,d\Phi_1(\sigma)$$

$$-\int_{t_0}^{t}\int_{\mathfrak{U}}\phi_1(\sigma,\mathbf{U})\phi_2(\sigma,\mathbf{U})M(d\sigma,d\mathbf{U}).$$

4.2.5 [Derivation of (2.52)]. Let $\{\zeta_t; t\geqslant t_0\}$ be defined by

$$\zeta_t=-\int_{t_0}^{t}\lambda_\sigma\,d\sigma+\int_{t_0}^{t}\ln\lambda_\sigma\,dN_\sigma,$$

where $\{N_t; t\geqslant t_0\}$ is a Poisson process with intensity λ_t for $t\geqslant t_0$. Recall from (2.51) that $\phi(\zeta_t)=\exp(\zeta_t)$ is the sample-function density for N on the interval $[t_0,t)$.

a. Show

$$\phi^k(\zeta_t)=1-k\int_{t_0}^{t}\phi^k(\zeta_\sigma)\lambda_\sigma\,d\sigma+\int_{t_0}^{t}\phi^k(\zeta_\sigma)[\lambda_\sigma^k-1]\,dN_\sigma.$$

b. Use the result of part (a) to evaluate $E[\phi^k(\zeta_t)]$.

4.2.6. Let $\{\zeta_t; t\geqslant t_0\}$ be defined by

$$\zeta_t=-\int_{t_0}^{t}\lambda_\sigma\,d\sigma+\int_{t_0}^{t}\int_{\mathfrak{U}}\ln[p_{\mathbf{u}}(\mathbf{U})\lambda_\sigma]M(d\sigma,d\mathbf{U}),$$

where M is the time-space Poisson process of the representation of a compound Poisson process in Theorem 3.3.2, and $p_{\mathbf{u}}(\mathbf{U})$ is the probability density function of the mark random variable $\mathbf{u}$. Recall from (3.48) that $\phi(\zeta_t)=\exp(\zeta_t)$ is the sample function density for the compound Poisson process on the interval $[t_0,t)$.

a. Derive an integral equation for $\phi^k(\zeta_t)$.

b. Evaluate $E[\phi^k(\zeta_t)]$.

4.2.7 (I. Bar-David's Expectation Formula).

a. Let $\{N_t; t \geq t_0\}$ be a Poisson counting process with intensity λ_t for $t \geq t_0$. Denote occurrence times by $\tau_1, \tau_2, \ldots$, and let f_t be a deterministic function of t for $t \geq t_0$. Define the process $\{\zeta_t; t \geq t_0\}$ by $\zeta_t = \int_{t_0}^t \ln f_\sigma \, dN_\sigma$, and let

$$\phi(\zeta_t) = \exp(\zeta_t) = \begin{cases} 1, & N_t = 0 \\ \prod_{i=1}^{N_t} f_{\tau_i}, & N_t \geq 1. \end{cases}$$

Show that

$$E[\phi(\zeta_t)] = \exp\left[\int_{t_0}^t (f_\sigma - 1)\lambda_\sigma \, d\sigma\right].$$

This expectation formula is due to I. Bar-David [2]. Hint: Establish and use the fact that

$$\phi(\zeta_t) = 1 + \int_{t_0}^t \phi(\zeta_\sigma)[f_\sigma - 1]\, dN_\sigma.$$

b. As an extension, consider a compound Poisson process on $[t_0, t)$ with occurrence times $\tau_1, \tau_2, \ldots$, and marks $\mathbf{u}_1, \mathbf{u}_2, \ldots$. Let $f_t(\mathbf{U})$ be a deterministic function of t and $\mathbf{U}$ for $t \geq t_0$ and $\mathbf{U} \in \mathfrak{U}$. Define

$$\zeta_t = \int_{t_0}^t \int_{\mathfrak{U}} \ln f_\sigma(\mathbf{U})\, M(d\sigma, d\mathbf{U}).$$

Show that

$$E[\phi(\zeta_t)] = \exp\left\{\int_{t_0}^t \int_{\mathfrak{U}} [f_\sigma(\mathbf{U}) - 1]\lambda_\sigma \, d\sigma \, dP_{\mathbf{u}}(\mathbf{U})\right\},$$

where

$$\phi(\zeta_t) = \exp(\zeta_t) = \begin{cases} 1, & N_t = 0 \\ \prod_{i=1}^{N_t} f_{\tau_i}(\mathbf{u}_i), & N_t \geq 1. \end{cases}$$

SELF-EXCITING POINT PROCESSES

CHAPTER FIVE

5.1 INTRODUCTION

A temporal point process must be both orderly and without aftereffects to be a Poisson process. The orderliness restriction that points be isolated from one another is relaxed in Chapter 3 where a generalized Poisson process is defined and studied. The restriction that the process be without aftereffects, that is, having the past without influence on the future, is removed in this chapter dealing with *self-exciting point processes*.

We first relax completely the restriction that the point process evolve without aftereffects. Thus, in the most general self-exciting point process, the entire past—that is, the number N_t and occurrence times $w_1, w_2, \ldots, w_{N_t}$ for all points realized before some time t—can influence the number and occurrence times for all points subsequent to t. This influence of the past on the future is characterized by the influence on only the microscopic future: the foremost quantity required in the development is defined in terms of the conditional probability of a single point occurring in the infinitesimal interval $[t, t+\Delta t)$ given the number and time of occurrence for all points realized up to time t. A general self-exciting point process can be conceived of as a modified inhomogeneous Poisson process in which the intensity is not only a function of time but also the entire past of the point process.

An example of a phenomenon that may be reasonably modeled as a self-exciting point process occurs in the operation of a electronic vacuum tube under space-charge limited conditions. Electron emissions from the cathode are suppressed by electrons emitted previously and still in transit to the anode. In fact, the cloud of transiting electrons can have a total charge of sufficient magnitude that further electron emissions are nearly extinguished. Thus, emission times form a self-exciting point process. This phenomenon is investigated in detail by W. Davenport and W. Root [5, Chapter 7] and S. Srinivasan and R. Vasudevan [24, Chapter 5].

General self-exciting point processes are examined in Section 5.2. Expressions are derived for the counting and occurrence-time statistics. Perhaps the most important quantity developed for applications is the sample-function density for self-exciting point processes. This quantity plays the same role in estimation and hypothesis testing as in previous chapters. Moreover, the sample-function density will prove useful in Chapter 6 where we study the doubly-stochastic Poisson process, which is a form of self-exciting point process.

For many applications, the point process does not depend on its entire past but, rather, only on a small fraction of it such as the total number of accumulated counts N_t or the most recent occurrence time w_{N_t}. The most notable of these processes are *Markov birth processes* and *renewal processes*, both of which have been extensively studied. These self-exciting point processes with limited memory are investigated in Section 5.3.

It is perhaps worth alerting the reader in advance that many of the results obtained in the chapter have the nature of characterizations rather than explicit formulas. Thus, for example, we derive but do not solve a differential equation for the probability $\Pr(N_t = n)$ of the number of points in the interval $[t_0, t)$ for a general self-exciting point process. Indeed, it is only in very special circumstances that explicit expressions for $\Pr(N_t = n)$ can be obtained, as for instance when $\{N_t; t \geqslant t_0\}$ is a Poisson process, which is the most degenerate of self-exciting point processes. The renewal process of Section 5.3 is in many ways the simplest non-Poisson self-exciting point process we will encounter. Even so, a neat, closed-form expression for $\Pr(N_t = n)$, as well as other quantities of interest, is generally unavailable. The characterizations we develop do provide considerable insight into the nature of point processes that evolve without aftereffects.

The approach we have adopted in the chapter is largely due to I. Rubin [21, 22]. We have preferred this approach over others because it is based on a natural and intuitively appealing modification of the Poisson process. Moreover, it turns out that important expressions, such as for the sample-function density, bear a remarkable similarity to those of a Poisson process. As a consequence, qualitative interpretations we have already developed for the Poisson process often carry over unchanged to a self-exciting point process.

We refer to J. Moyal [16], H. Wold [25], J. McFadden [15], O. Macchi [13, 14], A. Ramakrishnan [18, 19], and D. Daley and D. Vere-Jones [4] for details and additional references on alternative approaches for general self-exciting point processes. One characterization used first by S. Rice [20] in his 1944 studies of the zero crossings of a continuous random process and thoroughly developed by A. Ramakrishnan [18, 19] beginning in 1950 and P. Kuznetsov and R. Stratonovich [12] in 1956 is outlined in Problems 5.2.6 and 5.2.7. This characterization appears especially well suited for the study of filtered, self-exciting point processes. There are numerous treatments of Markov birth processes and renewal processes, both of which are special forms of self-exciting point processes with limited memory. A. Bharucha-Reid [1] and W. Smith [23] may be consulted on these specialized, but important models.

5.2 GENERAL SELF-EXCITING POINT PROCESSES

We first introduce some assumptions, notation, and terminology that will be used throughout the chapter. Thus, let $\{N_t; t \geqslant t_0\}$ be a counting process and assume:

a. $\{N_t; t \geqslant t_0\}$ is conditionally orderly;
b. the limit of the function $a(\Delta t, N_t)$ defined by

$$a(\Delta t, N_t) = \begin{cases} (\Delta t)^{-1} \Pr[N_{t,t+\Delta t} = 1 | N_t], & \text{for } N_t = 0 \\ (\Delta t)^{-1} \Pr[N_{t,t+\Delta t} = 1 | N_t; w_1, w_2, \ldots, w_{N_t}], & \text{for } N_t \geqslant 1 \end{cases} \tag{5.1}$$

exists as Δt tends to zero for almost every realization of $\{N_t; t \geqslant t_0\}$;
c. $\Pr[N_{t_0} = 0] = 1$.

The property of *conditional orderliness* in assumption (a) is defined in Chapter 2. Qualitatively, it means that the conditional probability of two or more points in $[t, t+\Delta t)$ can be made an arbitrarily small fraction of there being only one point by choosing Δt sufficiently small. We denote the limits in assumption (b) by $\mu_t(0)$ and $\mu_t(N_t; w_1, w_2, \ldots, w_{N_t})$ for $N_t = 0$ and $N_t \geqslant 1$, respectively; thus,

$$\mu_t(0) = \lim_{\Delta t \downarrow 0} (\Delta t)^{-1} \Pr[N_{t,t+\Delta t} = 1 | N_t], \qquad \text{for } N_t = 0 \tag{5.2}$$

and

$$\mu_t(N_t; w_1, w_2, \ldots, w_{N_t})$$

$$= \lim_{\Delta t \downarrow 0} (\Delta t)^{-1} \Pr[N_{t,t+\Delta t} = 1 | N_t; w_1, w_2, \ldots, w_{N_t}], \qquad \text{for } N_t \geqslant 1.$$

These limiting functions characterize the infinitesimal or microscopic behavior of the point process conditioned on its previous global behavior. It is convenient for us to introduce a random process $\{\lambda_t; t \geqslant t_0\}$ defined by

$$\lambda_t \triangleq \begin{cases} \mu_t(0), & t_0 \leqslant t \leqslant w_1 \\ \mu_t(N_t; w_1, w_2, \ldots, w_{N_t}), & w_{N_t} < t \leqslant w_{N_t+1} \end{cases} \tag{5.3}$$

A typical sample function for this process is shown in Figure 5.1. It is continuous from the left, and we assume for $t \geqslant t_0$ that

$$E(\lambda_t) < \infty. \tag{5.4}$$

We term N a *self-exciting counting process* if λ depends on more than t. In correspondence with the terminology in Chapter 2 for Poisson processes, λ will be called the intensity or more appropriately here, the

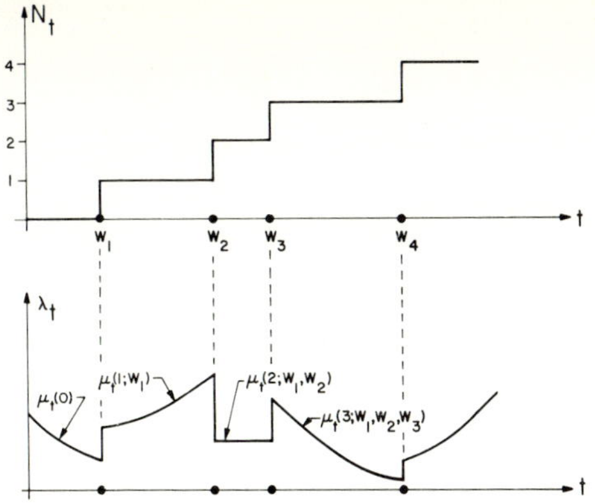

Figure 5.1. Typical sample function for the intensity process $\{\lambda_t; t \geqslant t_0\}$.

intensity process, the latter because λ is itself a random process whose sample path depends on the realization of the counting process N.

It is evident by examination of (2.27) in Theorem 2.2.2 that if λ does not depend in any way on the realization of N, the point process is not self-exciting, and $\{N_t; t \geqslant t_0\}$ is an inhomogeneous Poisson counting process. If λ depends on the number of points on $[t_0, t)$ but not on their times of occurrence, N is the pure birth process described in Problem 2.2.6. More generally, of course, λ can depend on one or more of the occurrence times as well as the number of points.

The most immediate interpretation of λ is that the probability of the occurrence of one point in the infinitesimal interval $[t, t+\Delta t)$ given that the number of points in $[t_0, t)$ is n and that their times of occurrence are $W_1, W_2, \ldots, W_n$ is approximated to order Δt by $\lambda_t \Delta t$, where λ_t is the value at time t of that sample function of the intensity process corresponding to the realized count n and occurrence times $W_1, W_2, \ldots, W_n$ on $[t_0, t)$. Because N is conditionally orderly, the conditional probability of there being no point in $[t, t+\Delta t)$ is approximated to order Δt by $1-\lambda_t \Delta t$. Further interpretation of $\mu_t(0)$ and $\mu_t(N_t; W_1, W_2, \ldots, W_{N_t})$, and therefore of λ_t, is provided by the following observations.

Suppose $N_s = 0$, and consider the conditional probability of there being no points in the interval $[s, t+\Delta t)$, where $t_0 \leqslant s \leqslant t < t+\Delta t$. It follows from (5.2) that

$$\begin{aligned}\Pr(N_{s,t+\Delta t}=0 \mid N_s=0) &= \Pr(N_{s,t}=0 \mid N_s=0)\Pr(N_{t,t+\Delta t}=0 \mid N_t=0)\\ &= \Pr(N_{s,t}=0 \mid N_s=0)[1-\mu_t(0)\Delta t + o(\Delta t)].\end{aligned}$$

We deduce from this that

$$\frac{d\Pr(N_{s,t}=0|N_s=0)}{dt} = -\mu_t(0)\Pr(N_{s,t}=0|N_s=0),$$

with the initial condition $\Pr(N_{s,s}=0|N_s=0)=1$. Hence,

$$\Pr(N_{s,t}=0|N_s=0)=\exp\left[-\int_s^t \mu_\sigma(0)\,d\sigma\right]. \tag{5.5}$$

Similarly, for $N_s \geqslant 1$, we obtain

$$\Pr(N_{s,t}=0|N_s; w_1, w_2, \ldots, w_{N_s})=\exp\left[-\int_s^t \mu_\sigma(N_s; w_1, w_2, \ldots, w_{N_s})\,d\sigma\right]. \tag{5.6}$$

Thus, the functions $\mu_t(0)$ and $\mu_t(N_t; w_1, w_2, \ldots, w_{N_t})$, defined by conditional occurrence probabilities for infinitesimal intervals in (5.2), specify and are specified by conditional probabilities of there being no points in the global interval $[s,t)$ in (5.5) and (5.6).

Counting Statistics for Self-Exciting Point Processes

In this subsection, we characterize the counting statistics for a self-exciting point process. In particular, we are interested in the distribution of the number N_t of points occurring in the interval $[t_0,t)$. A central role in specifying this distribution is played by the conditional expectation $E(\lambda_t|N_t)$ of λ_t given the total number of points in $[t_0,t)$ but not their locations. This conditional expectation is by definition a function of the random variable N_t which we denote by

$$\hat{\lambda}_t(N_t)=E(\lambda_t|N_t). \tag{5.7}$$

We term $\hat{\lambda}_t(N_t)$ the *count-conditional intensity*. The principle conclusion we draw is that N_t has the same distribution as a pure birth process (see Problem 2.2.6), having a birthrate equal to the count-conditional intensity $\hat{\lambda}_t(N_t)$. This assertion is established in Theorem 5.2.1 by deriving a differential-difference equation for the probability of the number of points in $[t_0,t)$.

The following lemma motivates the interpretation of $\hat{\lambda}_t(N_t)$ as an intensity and provides a result needed for the proof of Theorem 5.2.1.

Lemma 5.2.1. Let $\{N_t; t \geqslant t_0\}$ be a self-exciting point process with intensity process $\{\lambda_t; t \geqslant t_0\}$ defined in (5.3). Assume that $\{N_t; t \geqslant t_0\}$ is conditionally orderly and that $E(\lambda_t) < \infty$ for $t \geqslant t_0$. Then the function $\hat{\lambda}_t(N_t) = E(\lambda_t | N_t)$ satisfies

$$\hat{\lambda}_t(N_t) = \lim_{\Delta t \downarrow 0} (\Delta t)^{-1} \Pr(N_{t,t+\Delta t} \geqslant 1 | N_t) \tag{5.8}$$

and

$$\hat{\lambda}_t(N_t) = \lim_{\Delta t \downarrow 0} (\Delta t)^{-1} \Pr(N_{t,t+\Delta t} = 1 | N_t). \tag{5.9}$$

Proof.† Suppose first that $N_t = 0$. From (5.5), we see that

$$\lim_{\Delta t \downarrow 0} (\Delta t)^{-1} \Pr(N_{t,t+\Delta t} \geqslant 1 | N_t = 0) = \mu_t(0).$$

Thus, (5.8) holds for $N_t = 0$ because $\mu_t(0) = E(\lambda_t | N_t = 0)$ from (5.3). Equation (5.9) is obtained from (5.8) using the conditional orderliness of $\{N_t; t \geqslant t_0\}$, which implies that for any $\epsilon > 0$, there is a sufficiently small Δt that

$$\Pr[N_{t,t+\Delta t} > 1 | N_t = 0] \leqslant \epsilon \Pr[N_{t,t+\Delta t} = 1 | N_t = 0]$$

For this choice of ϵ and Δt, there holds

$$\begin{aligned} &(1+\epsilon)^{-1} (\Delta t)^{-1} \Pr(N_{t,t+\Delta t} \geqslant 1 | N_t = 0) \\ &\leqslant (\Delta t)^{-1} \Pr(N_{t,t+\Delta t} = 1 | N_t = 0) \leqslant (\Delta t)^{-1} \Pr(N_{t,t+\Delta t} \geqslant 1 | N_t = 0). \end{aligned}$$

Hence, for any $\epsilon > 0$,

$$(1+\epsilon)^{-1} \mu_t(0) \leqslant \lim_{\Delta t \downarrow 0} (\Delta t)^{-1} \Pr(N_{t,t+\Delta t} = 1 | N_t = 0) \leqslant \mu_t(0).$$

Equation (5.9) follows from this for $N_t = 0$ by virtue of the arbitrariness of ϵ.

Suppose next that $N_t \geqslant 1$. Let $\{\rho_n\}$ be a sequence of positive numbers converging to zero, and define

$$f_n = \rho_n^{-1} \Pr[N_{t,t+\rho_n} \geqslant 1 | N_t; w_1, w_2, \ldots, w_{N_t}]$$

†This proof is largely due to P. Fishman [8].

and

$$g_n = \rho_n^{-1} \int_t^{t+\rho_n} \mu_\sigma(N_t, w_1, w_2, \ldots, w_{N_t})\, d\sigma.$$

The sequence $\{g_n\}$ converges with probability one to $\mu_t(N_t; w_1, w_2, \ldots, w_{N_t})$ and $E(g_n) < \infty$ for each n because $E[\mu_t(N_t; w_1, w_2, \ldots, w_{N_t})] < \infty$ by assumption. From (5.6), it follows that the sequence $\{f_n\}$ converges with probability one to $\mu_t(N_t; w_1, w_2, \ldots, w_{N_t})$ and also that

$$f_n = \rho_n^{-1}\left\{1 - \exp\left[-\int_t^{t+\rho_n} \mu_\sigma(N_t; w_1, w_2, \ldots, w_{N_t})\, d\sigma\right]\right\}$$

$$\leqslant \rho_n^{-1} \int_t^{t+\rho_n} \mu_\sigma(N_t; w_1, w_2, \ldots, w_{N_t})\, d\sigma = g_n.$$

Since

$$E\left[\mu_t(N_t; w_1, w_2, \ldots, w_{N_t})\right] = \lim_{n\to\infty} \rho_n^{-1} \int_t^{t+\rho_n} E\left[\mu_\sigma(N_t; w_1, w_2, \ldots, w_{N_t})\right] d\sigma$$

$$= \lim_{n\to\infty} E(g_n),$$

it follows from the generalized bounded convergence theorem, Theorem 1.4.1, that

$$E\left[\mu_t(N_t; w_1, w_2, \ldots, w_{N_t}) | N_t\right] = \lim_{n\to\infty} E(f_n | N_t)$$

$$= \lim_{n\to\infty} \rho_n^{-1} \Pr[N_{t,t+\rho_n} \geqslant 1 | N_t],$$

which proves (5.8) for $N_t \geqslant 1$. The proof of (5.9) for $N_t \geqslant 1$ parallels the proof above when $N_t = 0$ and will not be repeated. This establishes Lemma 5.2.1.

It is evident from Lemma 5.2.1 that to order Δt, the conditional probabilities of zero, one, and more than one point in $[t, t+\Delta t)$ given the number of points in $[t_0, t)$, and not their times of occurrence, are $1 - \hat{\lambda}_t(N_t)\Delta t$, $\hat{\lambda}_t(N_t)\Delta t$, and zero, respectively. As a consequence, we have the following theorem characterizing the counting statistics for a self-exciting point process.

Theorem 5.2.1 (Counting Statistics for a Self-Exciting Point Process). Let $\{N_t; t \geqslant t_0\}$ be a self-exciting point process with intensity process $\{\lambda_t; t \geqslant t_0\}$ defined in (5.3). Assume $\{N_t; t \geqslant t_0\}$ is conditionally orderly, $E(\lambda_t) < \infty$, and $\Pr(N_{t_0} = 0) = 1$. Define the count-conditional intensity process $\{\hat{\lambda}_t(N_t); t \geqslant t_0\}$ as in (5.7). Then the probability $\Pr[N_t = n]$ that the number of points in $[t_0, t)$ is n satisfies

$$\frac{\partial \Pr[N_t=0]}{\partial t} = -\hat{\lambda}_t(0)\Pr[N_t=0] \tag{5.10}$$

for $n=0$ and

$$\frac{\partial \Pr[N_t=n]}{\partial t} = -\hat{\lambda}_t(n)\Pr[N_t=n]+\hat{\lambda}_t(n-1)\Pr[N_t=n-1] \tag{5.11}$$

for $n \geqslant 1$, with the initial condition $\Pr[N_{t_0}=0]=1$.

Proof. Equation (5.10) is a consequence of (5.5) because $\hat{\lambda}_t(0)=\mu_t(0)$. We leave the derivation of (5.11) as an exercise for the reader as it closely parallels that for Theorem 2.2.2; see Problem 5.2.1.

Comparison of (5.10) and (5.11) with the result of Problem 2.2.6(a) shows that the counting statistics for a self-exciting counting process are identical with those of a pure birth process with a population dependent birth rate $\hat{\lambda}_t(n)$, where $\hat{\lambda}_t(n)$ is the count-conditional expectation of the intensity of the self-exciting counting process. Equation (5.11) can in principle be solved sequentially for the counting probabilities $\Pr[N_t=1]$, $\Pr[N_t=2],\ldots$. For this purpose, we note from (5.10) or (5.5) that

$$\Pr[N_t=0]=\exp\left[-\int_{t_0}^{t}\mu_\sigma(0)\,d\sigma\right] \tag{5.12}$$

and from (5.11) that

$$\Pr[N_t=n]=\int_{t_0}^{t}\hat{\lambda}_\tau(n-1)\Pr[N_\tau=n-1]\exp\left[-\int_{\tau}^{t}\hat{\lambda}_\sigma(n)\,d\sigma\right]d\tau \tag{5.13}$$

for $n=1,2,\ldots$.

Example 5.2.1. Let $\{N_t; t\geqslant 0\}$ be a self-exciting counting process with conditional infinitesimal occurrence probabilities defined according to

$$\mu_t(0)=\lambda_0$$

and

$$\mu_t(N_t; w_1, w_2, \ldots, w_{N_t})=\begin{cases}\lambda_1, & N_t=1,3,5,\ldots \quad \text{(odd)}\\ \lambda_0, & N_t=2,4,6,\ldots \quad \text{(even)}.\end{cases}$$

where $\lambda_1>\lambda_0$. Thus, $\{\lambda_t; t>0\}$ is a two-state process as indicated in Figure

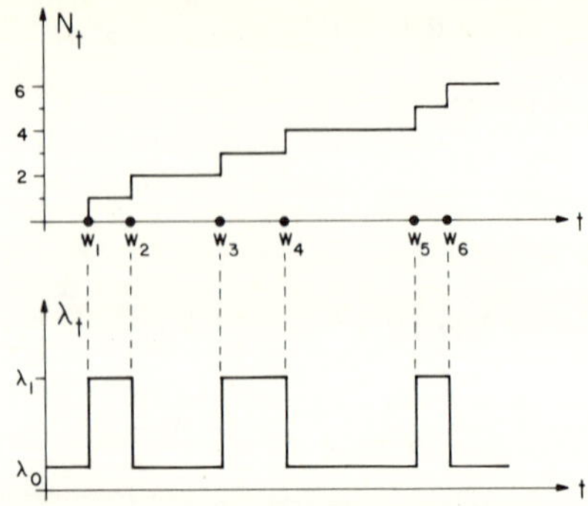

Figure 5.2. Typical sample function for $\{N_t; t \geqslant 0\}$ and $\{\lambda_t; t \geqslant 0\}$.

5.2. From (5.7), the count-conditional intensity is given by

$$\hat{\lambda}_t(n) = \begin{cases} \lambda_1, & n = 1, 3, 5, \ldots \quad \text{(odd)} \\ \lambda_0, & n = 0, 2, 4, \ldots \quad \text{(even)}. \end{cases}$$

The probability that the number of points at time t, for $t \geqslant 0$, is zero is from (5.12)

$$\Pr[N_t = 0] = \exp(-\lambda_0 t).$$

By substituting this expression into (5.13) and integrating, we obtain

$$\Pr[N_t = 1] = \left(\frac{\lambda_0}{\lambda_1 - \lambda_0}\right)[\exp(-\lambda_0 t) - \exp(-\lambda_1 t)].$$

Similarly, by substituting this expression into (5.13) and integrating, we obtain

$$\Pr[N_t = 2] = \left(\frac{\lambda_1 \lambda_0}{\lambda_1 - \lambda_0}\right) t \exp(-\lambda_0 t) - \left(\frac{\lambda_1 \lambda_0}{(\lambda_1 - \lambda_0)^2}\right)[\exp(-\lambda_0 t) - \exp(-\lambda_1 t)].$$

This procedure can in principle be continued to determine $Pr[N_t = n]$ for any $n \geqslant 0$. These counting probabilities are identical to those of a pure birth-process in which the birth rate depends on whether n is even or odd in correspondence with $\hat{\lambda}_t(n)$.

Time Statistics for Self-Exciting Point Processes

In this subsection, we derive various expressions associated with the statistics of the occurrence times of a self-exciting point process. These expressions are used in this and the next chapter to answer questions of decision and estimation involving observed, self-exciting point processes. The quantities of interest are the *survival probability*, the *joint occurrence density*, and the *sample-function density*. Of these, the sample-function density will be the most important. Its use in decision and estimation problems for self-exciting point processes will be identical to that in Chapters 2 and 3 for Poisson and compound Poisson processes.

Let $w_1 w_2, \ldots$ denote the occurrence times of a self-exciting point process with intensity process $\{\lambda_t; t \geqslant t_0\}$. Denote by

$$\mathcal{P}_{w_{n+1}|w_1, w_2, \ldots, w_n}(t | W_1, W_2, \ldots, W_n)$$

the conditional probability that $w_{n+1} \geqslant t$ given that $w_i = W_i$ for $i = 1, 2, \ldots, n$. If the occurrence times correspond to some anomalous behavior in a machine (or process, message, etc.), then $\mathcal{P}$ is the conditional probability that the machine survives beyond time t before the $(n+1)$st anomaly occurs, given that previous anomalies occurred at times $W_1, W_2, \ldots, W_n$. Motivated by this interpretation, we refer to $\mathcal{P}$ as the *conditional survival probability* for the $(n+1)$st point.

Because of the identity of the events $\{N_t = 0\}$ and $\{w_1 \geqslant t\}$, the survival probability for the first occurrence time is easily determined from (5.12) to be

$$\mathcal{P}_{w_1}(t) = Pr[w_1 \geqslant t] = \exp\left(- \int_{t_0}^{t} \mu_\sigma(0)\, d\sigma \right). \tag{5.14}$$

More generally, the expression for the conditional survival probability for the $(n+1)$st point is given for $t \geqslant W_n$ and $n \geqslant 1$ by

$$\mathcal{P}_{w_{n+1}|w_1, w_2, \ldots, w_n}(t | W_1, W_2, \ldots, W_n) = \exp\left(- \int_{W_n}^{t} \mu_\sigma(n; W_1, W_2, \ldots, W_n)\, d\sigma \right). \tag{5.15}$$

Equation (5.15) can be established by straightforward manipulation as follows. We write the conditional probability of the $(n+1)$st event being in the interval $[t, t+\Delta t)$ in two ways. First, we have

$$\begin{aligned} &\Pr[t \leqslant w_{n+1} < t + \Delta t | N_t = n;\ w_1 = W_1,\ w_2 = W_2, \ldots, w_n = W_n] \\ &\quad = \Pr[N_{t, t+\Delta t} = 1 | N_t = n;\ w_1 = W_1,\ w_2 = W_2, \ldots, w_n = W_n] \\ &\quad = \mu_t(n;\ W_1, W_2, \ldots, W_n)\Delta t + o(\Delta t). \end{aligned}$$

Secondly, for $t \geqslant W_n$, we also have

$$\Pr[t \leqslant w_{n+1} < t+\Delta t | N_t = n;\ w_1 = W_1,\ w_2 = W_2, \ldots, w_n = W_n]$$

$$= \frac{\Pr[t \leqslant w_{n+1} < t+\Delta t | w_1 = W_1,\ w_2 = W_2, \ldots, w_n = W_n]}{\Pr[w_{n+1} \geqslant t | w_1 = W_1,\ w_2 = W_2, \ldots, w_n = W_n]},$$

where we have used the identity between the events $\{N_t = n\}$ and $\{w_{n+1} \geqslant t\}$ for $t \geqslant W_n$. Equating these expressions, dividing by Δt, and taking the limit as Δt tends to zero results in

$$\mu_t(n;\ W_1, W_2, \ldots, W_n) = \frac{p_{w_{n+1}|w_1, w_2, \ldots, w_n}(t | W_1, W_2, \ldots, W_n)}{\mathcal{P}_{w_{n+1}|w_1, w_2, \ldots, w_n}(t | W_1, W_2, \ldots, W_n)}, \quad (5.16)$$

where the numerator on the right side is the conditional probability density for the $(n+1)$st occurrence time. Thus, we deduce the following relation between the survival probability and the intensity

$$\mu_t(n;\ W_1, W_2, \ldots, W_n) = \frac{-\partial \ln \mathcal{P}_{w_{n+1}|w_1, w_2, \ldots, w_n}(t | W_1, W_2, \ldots, W_n)}{\partial t}, \quad (5.17)$$

from which (5.15) follows.

Example 5.2.2 (Dead Time in Gamma-Photon Detectors). Practical devices used for detecting gamma photons from a radioactive source have the undesirable property that they become inoperative for a brief period of time, which may be deterministic or random, following each detected photon. During this time, termed the *dead time*, any incident photons go undetected because they cannot be observed with the device. If incident photons arriving during a dead time further extend the inoperative period, the detector is said to be *paralyzable*; otherwise it is *nonparalyzable*.

We consider a nonparalyzable detector in which the successive dead times $\tau_1, \tau_2, \ldots$ are independent and identically distributed as a nonnegative random variable τ having a probability distribution function $P_\tau(T) = Pr[\tau \leqslant T]$. With reference to Figure 5.3, let the incident point process be an inhomogeneous Poisson process with intensity ν_t for $t \geqslant t_0$. Let N_t be the number of detected points on $[t_0, t)$ and let $w_1, w_2, \ldots, w_{N_t}$ be their occurrence times.

Clearly, $\{N_t;\ t \geqslant t_0\}$ is a self-exciting counting process. Its intensity process $\{\lambda_t;\ t \geqslant t_0\}$ can be determined from the conditional probability of there being a detected point in the infinitesimal interval $[t, t+\Delta t)$ given the point process on $[t_0, t)$. If the detector is initially not in a dead time and,

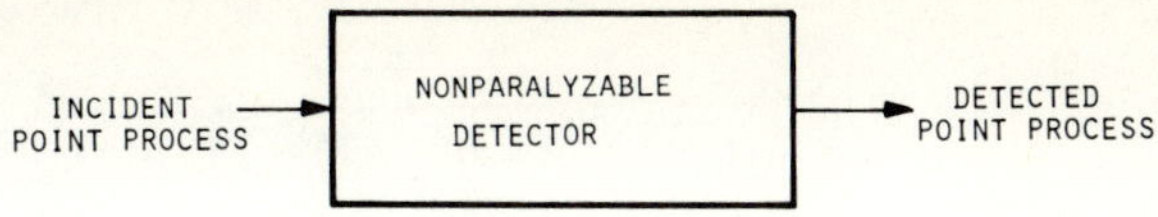

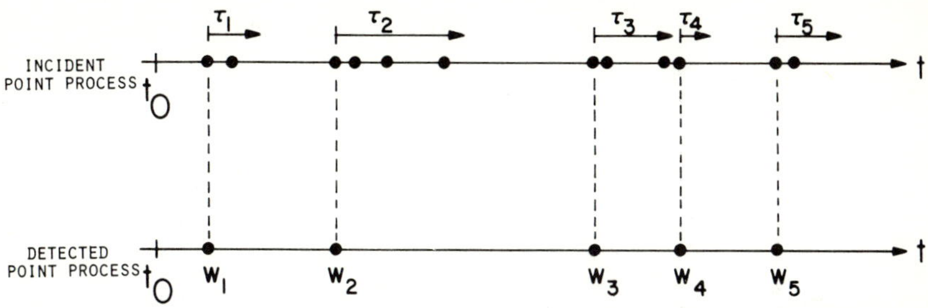

Figure 5.3. Typical realization for an incident and detected point process in a nonparalyzable detector.

therefore, free to detect an incident point, we obtain $\mu_t(0)=\nu_t$. Similarly, for $n \geqslant 1$, we obtain

$$\mu_t(n;\, W_1, W_2, \ldots, W_n) = \nu_t P_\tau(t - W_n). \tag{5.18}$$

The intensity process is then determined using (5.3), (5.17), and (5.18). In particular, for a detector with a fixed dead time t_d,

$$P_\tau(T) = \begin{cases} 0, & T < t_d \\ 1, & T \geqslant t_d \end{cases}$$

and

$$\mu_t(n;\, W_1, W_2, \ldots, W_n) = \begin{cases} 0, & W_n \leqslant t < W_n + t_d \\ \nu_t, & W_n + t_d \leqslant t. \end{cases} \tag{5.19}$$

The corresponding intensity process is shown in Figure 5.4 for the incident intensity $\nu_t = \exp(-t)$.

From (5.14) and (5.15), the survival probability for the nth point is given by

$$\mathcal{P}_{w_1}(t) = \exp\left(-\int_{t_0}^{t} \nu_\sigma \, d\sigma \right) \tag{5.20}$$

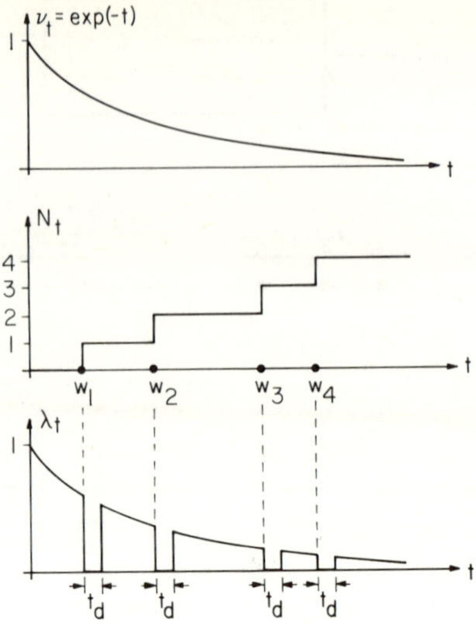

Figure 5.4. Example of an intensity process for a detector having a fixed dead time.

for $n=1$ and

$$\mathcal{P}_{w_{n+1}|w_1,w_2,\ldots,w_n}(t|W_1,W_2,\ldots,W_n)=\exp\left[-\int_{W_n}^{t}\nu_\sigma P_\tau(\sigma-W_n)\,d\sigma\right] \quad (5.21)$$

for $n\geqslant 2$ and $t\geqslant W_n$.

Denote by $p_{\mathbf{w}}^{(n)}(\mathbf{W})$ the joint probability density function for the first n occurrence times. This *joint occurrence density* for a self-exciting point process is given by

$$p_{w_1}{}^{(1)}(W_1)=\mu_{W_1}(0)\exp\left[-\int_{t_0}^{W_1}\mu_\sigma(0)\,d\sigma\right] \quad (5.22)$$

for $n=1$ and

$$p_{\mathbf{w}}^{(n)}(\mathbf{W})=\mu_{W_1}(0)\left\{\prod_{i=2}^{n}\mu_{W_i}(i-1;\,W_1,W_2,\ldots,W_{i-1})\right\}$$
$$\times\exp\left[-\int_{t_0}^{W_1}\mu_\sigma(0)\,d\sigma-\sum_{i=2}^{n}\int_{W_{i-1}}^{W_i}\mu_\sigma(i-1;\,W_1,W_2,\ldots,W_{i-1})\,d\sigma\right] \quad (5.23)$$

for $n \geqslant 2$ and $t_0 \leqslant W_1 < W_2 < \cdots < W_n$. For the proof of (5.22), simply differentiate (5.14) with respect to t. For the proof of (5.23), use the relation

$$p_{\mathbf{w}}^{(n)}(\mathbf{W}) = p_{w_1}(W_1) \prod_{i=2}^{n} p_{w_i|w_1, w_2, \ldots, w_{i-1}}(W_i | W_1, W_2, \ldots, W_{i-1})$$

in which p_{w_1} is the probability density for the first occurrence time w_1, and $p_{w_i|w_1, w_2, \ldots, w_{i-1}}$ is the conditional density for the ith occurrence time. These densities can be determined by differentiation of (5.14) and (5.15) with respect to t, and then (5.23) follows.

The *sample-function density* for a self-exciting point process plays a central role in decision and estimation problems in which the entire path of the associated counting process is accessible for observation. We now derive and give applications for this important quantity.

The sample-function density for a self-exciting point process is the joint probability of the number of points in the interval $[t_0, T)$ and the density of their occurrence times. As in Section 2.3, this quantity is denoted by $p[\{N_\sigma;\ t_0 \leqslant \sigma < T\}]$, where

$$p[\{N_\sigma;\ t_0 \leqslant \sigma < T\}] = \begin{cases} \Pr[N_T = 0], & N_T = 0 \\ p_{\mathbf{w}}(\mathbf{W}, N_T = n), & N_T = n \geqslant 1 \end{cases} \tag{5.24}$$

where

$$p_{\mathbf{w}}(\mathbf{W}, N_T = n) = \Pr[N_T = n | w_1 = W_1,\ w_2 = W_2, \ldots, w_n = W_n] p_{\mathbf{w}}^{(n)}(\mathbf{W}).$$

The counting probability on the right side can be expressed in terms of the conditional survival probability of the $(n+1)$st point as

$$\begin{aligned} &\Pr[N_T = n | w_1 = W_1,\ w_2 = W_2, \ldots, w_n = W_n] \\ &\quad = \mathcal{S}_{w_{n+1}|w_1, w_2, \ldots, w_n}(T | W_1, W_2, \ldots, W_n) \end{aligned}$$

for $n \geqslant 1$. Thus, on combining (5.15), (5.22), and (5.23) in (5.24), we see that the sample-function density is given by

$$p[\{N_\sigma;\ t_0 \leqslant \sigma < T\}] = \begin{cases} \exp\left[-\int_{t_0}^{T} \mu_\sigma(0), d\sigma \right], & N_T = 0 \\ \mu_{W_1}(0) \exp\left[-\int_{t_0}^{W_1} \mu_\sigma(0)\, d\sigma - \int_{W_1}^{T} \mu_\sigma(1;\ W_1)\, d\sigma \right], & N_T = 1 \end{cases} \tag{5.25}$$

for $N_T=0$ and $N_T=1$ and by

$$
\begin{aligned}
p[\{N_\sigma;\ t_0\leqslant\sigma<T\}]
&=\mu_{W_1}(0)\left\{\prod_{i=2}^{n}\mu_{W_i}(i-1;\ W_1,W_2,\dots,W_{i-1})\right\}\\
&\quad\times\exp\left[-\int_{t_0}^{W_1}\mu_\sigma(0)\,d\sigma-\sum_{i=2}^{n}\int_{W_{i-1}}^{W_i}\mu_\sigma(i-1;\ W_1,W_2,\dots,W_{i-1})\,d\sigma\right.\\
&\qquad\left.-\int_{W_n}^{T}\mu_\sigma(n;\ W_1,W_2,\dots,W_n)\,d\sigma\right]
\end{aligned}
\tag{5.26}
$$

for $n\geqslant 2$ and $t_0\leqslant W_1<W_2<\cdots<W_n<T$.

This result can be placed in a more compact and suggestive form by incorporating into (5.25) and (5.26) the intensity process $\{\lambda_t\colon t\geqslant t_0\}$ defined in (5.3) and the counting integrals of Chapter 4. We note first that

$$
\int_{t_0}^{T}\lambda_\sigma\,d\sigma=
\begin{cases}
\displaystyle\int_{t_0}^{T}\mu_\sigma(0)\,d\sigma, & N_T=0\\
\displaystyle\int_{t_0}^{w_1}\mu_\sigma(0)\,d\sigma+\int_{w_1}^{T}\mu_\sigma(1;\ w_1)\,d\sigma, & N_T=1\\
\displaystyle\int_{t_0}^{w_1}\mu_\sigma(0)\,d\sigma+\sum_{i=2}^{N_T}\int_{w_{i-1}}^{w_i}\mu_\sigma(i-1;\ w_1,w_2,\dots,w_{i-1})\,d\sigma & \\
\displaystyle\qquad+\int_{w_{N_T}}^{T}\mu_\sigma(N_T;\ w_1,w_2,\dots,w_{N_T})\,d\sigma, & N_T\geqslant 2.
\end{cases}
\tag{5.27}
$$

Similarly, for the counting integrals defined in Section 4.2.1, we have that

$$
\int_{t_0}^{T}\ln\lambda_\sigma\,dN_\sigma=
\begin{cases}
0, & N_T=0\\
\ln\mu_{w_1}(0), & N_T=1\\
\displaystyle\ln\mu_{w_1}(0)+\sum_{i=2}^{N_T}\ln\mu_{w_i}(i-1;\ w_1,w_2,\dots,w_{i-1}), & N_T\geqslant 2.
\end{cases}
\tag{5.28}
$$

By combining (5.25) to (5.28), we conclude that the sample-function density can be expressed in terms of the counting process $\{N_\sigma;\ t_0\leqslant\sigma<T\}$ and its associated intensity process $\{\lambda_\sigma;\ t_0\leqslant\sigma<T\}$ as in the following theorem.

Theorem 5.2.2 (Sample-Function Density for a Self-Exciting Point Process). Let $\{N_t;\ t\geqslant t_0\}$ be a self-exciting counting process with an intensity process $\{\lambda_t;\ t \geqslant t_0\}$. The sample-function density associated with the random variables $\{N_\sigma;\ t_0 \leqslant \sigma < T\}$ is given by

$$p[\{N_\sigma;\ t_0 \leqslant \sigma < T\}] = \exp\left(-\int_{t_0}^{T} \lambda_\sigma\, d\sigma + \int_{t_0}^{T} \ln \lambda_\sigma\, dN_\sigma\right). \tag{5.29}$$

When λ is a deterministic function of time only, $\{N_t;\ t \geqslant t_0\}$ is a Poisson counting process. In this case (5.29) and (2.51) are identical, as expected. More generally, it is interesting to note that when λ depends on the counting path so that N is self-exciting, the form of the sample-function density in (5.29) remains unchanged. The use of (5.29) in estimation and decision problems involving observed self-exciting point processes is precisely the same as the use of (2.51) for observed Poisson processes. This is outlined in the following examples.

Example 5.2.3 (Parameter Estimation for Self-Exciting Point Processes). The problem of estimating parameters arises frequently in real situations where self-exciting processes are used as the mathematical model for the observed process. This problem is discussed in Section 2.4 of Chapter 2 for Poisson processes, and the situation is entirely analogous here. We consider as observed data the entire counting path $\{N_t;\ t_0 \leqslant t < T\}$ of the self-exciting point process. The intensity process $\{\lambda_t(\mathbf{X});\ t_0 \leqslant t < T\}$ is assumed now to be a function of a set of nonrandom parameters $\mathbf{X}$ taking values in a parameter set $\mathfrak{X}$. As in (2.70), the likelihood function $l(\mathbf{X})$ associated with an observed counting record is given from (5.29) by

$$l(\mathbf{X}) = -\int_{t_0}^{T} \lambda_\sigma(\mathbf{X})\, d\sigma + \int_{t_0}^{T} \ln \lambda_\sigma(\mathbf{X})\, dN_\sigma \tag{5.30}$$

where $\{\lambda_t(\mathbf{X});\ t_0 \leqslant t < T\}$ is that sample function of the intensity process corresponding to the observed counting path on $[t_0, T)$ and assuming the parameters to have the value $\mathbf{X}$. The maximum likelihood estimate $\hat{\mathbf{X}}_{\mathrm{ML}}$ of $\mathbf{X}$ in terms of the observed counting path $\{N_t;\ t_0 \leqslant t < T\}$ maximizes $l(\mathbf{X})$ with respect to $\mathbf{X}$ subject to the constraint that $\hat{\mathbf{X}}_{\mathrm{ML}} \in \mathfrak{X}$. A particular application of this general principle is given in the next example.

Example 5.2.4 (Correction for Counting Losses Due to Fixed Dead Time). Suppose a homogeneous Poisson process having a constant intensity ν is incident on a detector having a fixed dead time t_d. Here, ν is an unknown, nonnegative parameter to be estimated in terms of the point process observed at the output of the detector on the interval $[0, T)$. Let $\{N_t;\ t \geqslant 0\}$ be the counting process associated with the output point process. This is a self-exciting point process for which $\mu_t(0) = \nu$ and, from

(5.19),

$$\mu_t(n; W_1, W_2, \dots, W_n) = \begin{cases} 0, & W_n \leqslant t < W_n + t_d \\ \nu, & W_n + t_d \leqslant t. \end{cases}$$

The intensity process $\{\lambda_t(\nu);\ t \geqslant 0\}$ associated with N is a two-state process having an instantaneous jump from ν to zero following each detected point and then returning instantaneously to ν after t_d seconds. From (5.30) the likelihood function $l(\nu)$ for estimating ν in terms of an observed counting path with $N_T = n$, $w_1 = W_1$, $w_2 = W_2, \dots,$ and $w_n = W_n$ is given by

$$l(\nu) = \begin{cases} -\nu[W_n - (n-1)t_d] + n \ln \nu, & \text{for } W_n \leqslant T < W_n + t_d \\ -\nu[T - nt_d] + n \ln \nu, & \text{for } W_n + t_d \leqslant T. \end{cases} \tag{5.31}$$

By setting the derivative of $l(\nu)$ with respect to ν equal to zero, we deduce that

$$\hat{\nu}_{\mathrm{ML}} = \begin{cases} \bar{n}\left[\dfrac{W_n + t_d}{T} - \bar{n}t_d\right]^{-1}, & \text{for } W_n \leqslant T < W_n + t_d \\ \bar{n}[1 - \bar{n}t_d]^{-1}, & \text{for } W_n + t_d \leqslant T, \end{cases} \tag{5.32}$$

where $\bar{n} = n/T$ is the time-average rate at which points are observed. If $\bar{n}T \gg 1$, the end effect in (5.32) can be neglected so that

$$\hat{\nu}_{\mathrm{ML}} = \bar{n}[1 - \bar{n}t_d]^{-1}. \tag{5.33}$$

The factor $[1 - \bar{n}t_d]^{-1}$ in (5.33) can be regarded as the correction to be applied to the average rate at which points are observed in order to obtain the true maximum likelihood estimate of the average rate of incident points. We note finally that the entire counting path is not needed to form the estimate of ν in (5.31); only the total number of detected points n is required. If end effects cannot be neglected, then the last occurrence time W_n is also needed so that (5.32) can be used.

Example 5.2.5 (Hypothesis Testing for Self-Exciting Point Processes). We consider a simple binary hypothesis testing problem in which a self-exciting counting process $\{N_t;\ t \geqslant t_0\}$ is observed. The intensity process for N is either $\{\lambda_t^{(0)};\ t \geqslant t_0\}$ or $\{\lambda_t^{(1)};\ t \geqslant t_0\}$, on hypotheses H_0 and H_1, respectively, and the problem is to decide which on the basis of an observed path of N on $[t_0, T)$. This problem is discussed in Section 2.5 of Chapter 2 for Poisson processes, and the situation here is entirely analogous. Paralleling the arguments leading to (2.112) and using (5.29), we see that the likelihood ratio test for deciding which intensity underlies the observed path becomes

$$-\int_{t_0}^{T}[\lambda_\sigma^{(1)}-\lambda_\sigma^{(0)}]\,d\sigma+\int_{t_0}^{T}\ln\left(\frac{\lambda_\sigma^{(1)}}{\lambda_\sigma^{(0)}}\right)dN_\sigma \underset{H_0}{\overset{H_1}{\gtrless}} \ln\left[\frac{P_0(C_{10}-C_{00})}{P_1(C_{01}-C_{11})}\right], \tag{5.34}$$

where P_0 and P_1 are the prior probabilities of H_0 and H_1, respectively, and $C_{01}-C_{11}$ and $C_{10}-C_{00}$ are the relative costs of a miss and false alarm, respectively. A block diagram realization for the detector corresponding to (5.34) is shown in Figure 5.5. Notice that the only modification of the realization in Figure 2.8 for a Poisson process is that the output of the function generators for the intensities on each hypothesis now depends on the observed point process. A particular application of this general principle is given in the next example.

Example 5.2.6. Suppose $\{N_t;\ t\geqslant 0\}$ is a self-exciting counting process having one of two intensity processes $\{\lambda_t^{(1)};\ t\geqslant 0\}$ or $\{\lambda_t^{(0)};\ t\geqslant 0\}$ of the two-state form in Example 5.2.1. Assume that $\lambda_t^{(i)}$, for $i=0,1$, switches between levels a_i and b_i, $b_i\geqslant a_i$, according to

$$\lambda_t^{(i)}=\begin{cases} a_i, & N_t=0,2,4,\dots \quad \text{(even)}\\ b_i, & N_t=1,3,5,\dots \quad \text{(odd)}.\end{cases}$$

For this situation, the left side of (5.34) becomes

$$(a_0-a_1)T_e+(b_0-b_1)T_0+N_{e\to o}\ln\left(\frac{a_1}{a_0}\right)+N_{o\to e}\ln\left(\frac{b_1}{b_0}\right),$$

where T_e is the amount of time in $[0,T)$ that N_t is even, $T_o=T-T_e$ is the amount of time N_t is odd, $N_{o\to e}$ is the number of odd to even transitions in the path of N on $[0,T)$, and $N_{e\to o}=N_T-N_{o\to e}$ is the number of even to odd transitions.

A Limit Theorem for Sums of Self-Exciting Point Processes

In this subsection, we briefly examine the situation indicated in Figure 5.6 in which a point process is obtained by pooling the points of many independent point processes. The independent components may, in general, be self-exciting and inhomogeneous. A theorem is given which indicates that if the component processes are sufficiently sparse that almost

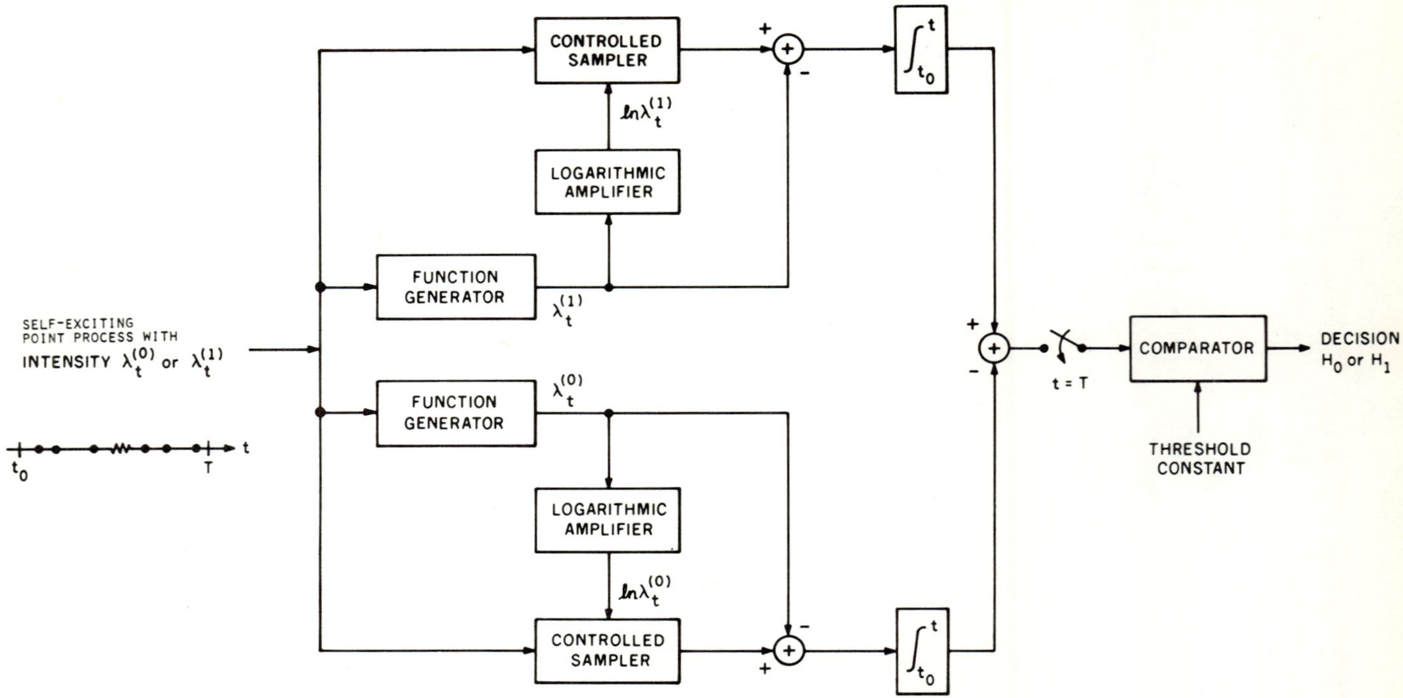

Figure 5.5. Optimum test for count-record observations of self-exciting point process.

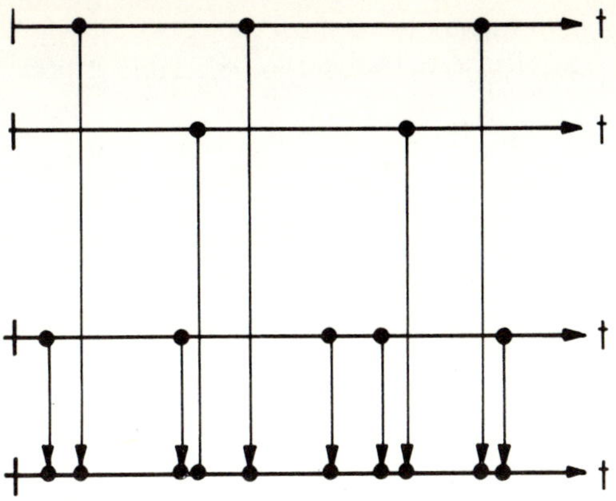

Figure 5.6. A pooled point process.

every one contributes no more than a single point to the pooled process, then the pooled process is approximately a Poisson process. This result explains, at least in part, the relative frequency with which the Poisson process is found to be a reasonable model in applications. The theorem may be viewed as the analog for point processes of the central limit theorem for sums of random variables.

Let $\{N_{ik}(t);\ t \geqslant t_0\}$ be the counting process associated with the ith component point process for $i = 1, 2, \ldots, k$. Denote by $\{M_k(t);\ t \geqslant t_0\}$ the counting process for the pooled point process obtained from the k component processes. Thus, for $t \geqslant t_0$,

$$M_k(t) = \sum_{i=1}^{k} N_{ik}(t).$$

We will say that the component processes are *uniformly sparse* if for t finite there holds

$$\lim_{k\to\infty} \sup_{1 \leqslant i \leqslant k} \Pr[N_{ik}(t) \geqslant 1] = 0.$$

The interpretation of this expression is that the probability of any component contributing a point to the pooled process tends to zero on any finite interval as the number of components tends to infinity.

Under relatively weak additional conditions, $\{M_k(t);\ t \geqslant t_0\}$ tends to a Poisson process as k tends to infinity. The principle result is contained in the following theorem, due to P. Franken [9] and B. Grigelionis [11], which we state without proof.

Theorem 5.2.3 (Limit Theorem for Pooled Point Processes). Suppose the component counting processes $\{N_{ik}(t);\ t \geqslant t_0\}$ for $i = 1, 2, \ldots, k$ are mutually independent and uniformly sparse for t finite. Then the pooled counting process $\{M_k(t);\ t \geqslant t_0\}$ converges in distribution to a Poisson process with intensity $\{\lambda_t;\ t \geqslant t_0\}$ if and only if both

$$\lim_{k \to \infty} \sum_{i=1}^{k} \Pr(N_{ik}(t) > 1) = 0$$

and

$$\lim_{k \to \infty} \sum_{i=1}^{k} \Pr(N_{ik}(t) = 1) = \int_{t_0}^{t} \lambda_\sigma \, d\sigma$$

for $t_0 \leqslant t < \infty$.

Theorem 5.4.1 is proved and discussed in a comprehensive review of limit theorems for pooled point processes by E. Cinlar [2].

Summary. Statistical characterizations for general self-exciting point processes were developed in this section. The following important observations were made:

1. The counting statistics for a self-exciting point process are identical to those of a pure birth process having a birth rate equal to the count-conditional intensity of the self-exciting point process.
2. The sample function density for a self-exciting point process has the same form as that for an inhomogeneous Poisson process. Its use in solving decision and estimation problems for observed self-exciting point processes is also the same as that developed in Chapter 2 for the Poisson process.
3. A point process obtained by pooling the points of independent and sparse self-exciting processes converges under weak conditions to a Poisson process.

5.3. SELF-EXCITING POINT PROCESSES WITH LIMITED MEMORY

The future evolution of the self-exciting point processes introduced in Section 5.2 can, in general, depend on the occurrence times of *all* past points as well as their total number. This is overly complicated for many applications, since it often happens that relatively only a few of the most recent occurrence times, and possibly the total count, influence the future evolution. In this section, we introduce a class of self-exciting point processes having limited memory in the sense that their future evolution depends at most on a fixed number of recent occurrence times. These *m-memory self-exciting point processes* are defined as follows.

m-Memory Self-Exciting Point Processes. A self-exciting point process is by definition *0-memory* if for all $N_t \geqslant 1$, the function $\mu_t(N_t;\ w_1, w_2, \ldots, w_{N_t})$ of (5.2) is independent of any of the occurrence times $w_1, w_2, \ldots, w_{N_t}$. It is *m-memory*, for $m = 1, 2, \ldots$, if for all $N_t \geqslant m$, $\mu_t(N_t;\ w_1, w_2, \ldots, w_{N_t})$ depends only on t, N_t, and the m most recent occurrence times $w_{N_t - m + 1}, \ldots w_{N_t - 1}, w_{N_t}$ and is independent of the $N_t - m$ earlier times.

In the following subsections, we characterize m-memory self-exciting point processes. The characterizations we state in Theorems 5.3.1 and 5.3.3 are due to I. Rubin [21].

Characterization of m-Memory Point Processes for $m = 0$

Zero-memory self-exciting point processes are characterized in the following theorem.

Theorem 5.3.1 (Characterization of 0-memory Self-Exciting Point Processes). Assume $\{N_t;\ t \geqslant t_0\}$ is a self-exciting counting process for which the hypotheses of Theorem 5.2.1 hold. Then the underlying point process is 0-memory if and only if $\{N_t;\ t \geqslant t_0\}$ is a Markov process.†

Proof. Suppose first that $\{N_t;\ t \geqslant t_0\}$ is a Markov process. As the events generated by the random variables $\{N_t;\ w_1, w_2, \ldots, w_{N_t}\}$ are the same as those generated by the counting paths $\{N_\sigma;\ t_0 \leqslant \sigma < t\}$, we have

$$\begin{aligned}\Pr[N_{t,t+\Delta t} = 1 | N_t;\ w_1, w_2, \ldots, w_{N_t}] &= \Pr(N_{t,t+\Delta t} = 1 | N_\sigma;\ t_0 \leqslant \sigma < t)\\ &= \Pr[N_{t,t+\Delta t} = 1 | N_t], \qquad (5.35)\end{aligned}$$

where the last equality follows because of the Markov hypothesis for N. Thus,

$$\begin{aligned}&\lim_{\Delta t \downarrow 0} (\Delta t)^{-1} \Pr[N_{t,t+\Delta t} = 1 | N_t;\ w_1, w_2, \ldots, w_{N_t}]\\ &\qquad = \lim_{\Delta t \downarrow 0} (\Delta t)^{-1} \Pr[N_{t,t+\Delta t} = 1 | N_t] \triangleq \mu_t(N_t), \qquad (5.36)\end{aligned}$$

and the point process is 0-memory.

†Markov processes are defined in Section 4.2.2 of Chapter 4.

Conversely, suppose that the point process is 0-memory so that (5.36) holds. We need to show that for any $s \in [t_0, t)$ there holds

$$\Pr(N_t = n | N_\sigma, t_0 \leqslant \sigma < s) = \Pr[N_t = n | N_s]. \tag{5.37}$$

We establish (5.37) by showing first that the left and right sides satisfy the same linear differential-difference equation and then appealing to the uniqueness of the solution for such equations. Dealing with the left side, we note for $t_0 \leqslant s \leqslant t$ that

$$\begin{aligned}
&\lim_{\Delta t \downarrow 0} (\Delta t)^{-1} \Pr(N_{t,t+\Delta t} \geqslant 1 | N_t, N_\sigma;\ t_0 \leqslant \sigma < s) \\
&\qquad = \lim_{\Delta t \downarrow 0} (\Delta t)^{-1} E\{\Pr(N_{t,t+\Delta t} \geqslant 1 | N_\sigma;\ t_0 \leqslant \sigma < t) | N_t, N_\sigma;\ t_0 \leqslant \sigma < s\} \\
&\qquad = E\left\{ \lim_{\Delta t \downarrow 0} (\Delta t)^{-1} \Pr(N_{t,t+\Delta t} \geqslant 1 | N_\sigma, t_0 \leqslant \sigma < t) | N_t, N_\sigma;\ t_0 \leqslant \sigma < s \right\} \\
&\qquad = E\{ \mu_t(N_t) | N_t, N_\sigma;\ t_0 \leqslant \sigma < s\} = \mu_t(N_t),
\end{aligned} \tag{5.38}$$

where the interchange of the limit and expectation can be justified using the generalized bounded convergence theorem, Theorem 1.4.1, in the same manner as in the proof of Lemma 5.2.1. It also follows from (5.38) and the conditional orderliness of $\{N_t;\ t \geqslant t_0\}$ that

$$\lim_{\Delta t \downarrow 0} (\Delta t)^{-1} \Pr(N_{t,t+\Delta t} = 1 | N_t, N_\sigma;\ t_0 \leqslant \sigma < s) = \mu_t(N_t). \tag{5.39}$$

Thus, we conclude that

$$\Pr(N_{t,t+\Delta t} = 0 | N_t = n, N_\sigma;\ t_0 \leqslant \sigma < s) = 1 - \mu_t(n)\Delta t + o(\Delta t), \tag{5.40}$$

$$\Pr(N_{t,t+\Delta t} = 1 | N_t = n, N_\sigma;\ t_0 \leqslant \sigma < s) = \mu_t(n)\Delta t + o(\Delta t), \tag{5.41}$$

and

$$\Pr(N_{t,t+\Delta t} > 1 | N_t = n, N_\sigma;\ t_0 \leqslant \sigma < s) = o(\Delta t). \tag{5.42}$$

These conditional, incremental occurrence probabilities can be used in the manner of deriving (5.10) and (5.11) to obtain

$$\frac{\partial \Pr(N_t = 0 | N_\sigma;\ t_0 \leqslant \sigma < s)}{\partial t} = -\mu_t(0) \Pr(N_t = 0 | N_\sigma;\ t_0 \leqslant \sigma < s) \tag{5.43}$$

for $n=0$ and

$$\frac{\partial \Pr(N_t=n|N_\sigma;\ t_0\leqslant\sigma<s)}{\partial t}=-\mu_t(n)\Pr(N_t=n|N_\sigma;\ t_0\leqslant\sigma<s)$$
$$+\mu_t(n-1)\Pr(N_t=n-1|N_\sigma;\ t_0\leqslant\sigma<s) \quad (5.44)$$

for $n\geqslant 1$.

Dealing now with the right side of (5.37), we note that

$$\Pr(N_t=n|N_s)=E\{\Pr(N_t=n|N_\sigma;\ t_0\leqslant\sigma<s)|N_s\}. \quad (5.45)$$

Thus, by taking the conditional expectation of both sides of (5.43) and (5.44) given N_s, we deduce that

$$\frac{\partial \Pr(N_t=0|N_s)}{\partial t}=-\mu_t(0)\Pr(N_t=0|N_s) \quad (5.46)$$

and, for $n\geqslant 1$,

$$\frac{\partial \Pr(N_t=n|N_s)}{\partial t}=-\mu_t(n)\Pr(N_t=n|N_s)+\mu_t(n-1)\Pr(N_t=n-1|N_s). \quad (5.47)$$

At $t=s$, we have the condition

$$\Pr(N_s=n|N_\sigma;\ t_0\leqslant\sigma<s)=\Pr(N_s=n|N_s).$$

Therefore, the left and right sides of (5.37) satisfy the same linear differential-difference equation and boundary condition at $t=s$ for each n. On the basis of the uniqueness of the solution for such equations, we therefore conclude that (5.37) holds, and the theorem is established.

It is evident for a 0-memory self-exciting point process that the count-conditional intensity $\lambda_t(n)$ of (5.7) equals $\mu_t(n)$ of (5.36) for $n=0,1,2,\cdots$. Examination of Theorems 5.2.1 and 5.3.1 indicates that the associated counting process $\{N_t;\ t\geqslant t_0\}$ is properly termed a *Markov birth process* with birth rate $\mu_t(n)$. It is possible for N to be unstable for certain birthrates that depend nonlinearly on the population size. The instability manifests itself in the form of a population *explosion* in which N_t, the population size at time t, becomes infinite for t finite. This phenomenon occurs in the following example.

Example 5.3.1 (A Population Explosion). Let $\{N_t;\ t\geqslant t_0\}$ be a Markov birth process with a birth rate $\{\mu_t(N_t);\ t\geqslant 0\}$ that depends quadratically on the population size as

$$\mu_t(N_t) = (N_t + 1)^2$$

for $N_t = 0, 1, 2, \cdots$. Let $P_n(t) = \Pr[N_t = n]$ denote the probability that the population is of size n at time t, and assume that $P_n(0) = 1$ for $n = 0$ and $P_n(0) = 0$ for $n \geqslant 1$. From (5.10) and (5.11), we see that $P_n(t)$ satisfies the differential equations

$$\frac{\partial P_0(t)}{\partial t} = -P_0(t), \qquad P_0(0) = 1 \tag{5.48}$$

for $n = 0$ and

$$\frac{\partial P_n(t)}{\partial t} = -(n+1)^2 P_n(t) + n^2 P_{n-1}(t), \qquad P_n(0) = 0 \tag{5.49}$$

for $n \geqslant 1$. These equations can be solved by means of Laplace transforms. For this, let $\tilde{P}_n(s)$ be the Laplace transform of $P_n(t)$. By transforming (5.48) and (5.49), we obtain

$$s\tilde{P}_0(s) - 1 = -\tilde{P}_0(s) \tag{5.50}$$

and

$$s\tilde{P}_n(s) = -(n+1)^2 \tilde{P}_n(s) + n^2 \tilde{P}_{n-1}(s), \tag{5.51}$$

respectively. It follows, therefore, that

$$\tilde{P}_n(s) = \frac{1}{(n+1)^2} \prod_{m=1}^{n+1} \frac{m^2}{s+m^2} \tag{5.52}$$

for $n = 0, 1, \ldots$. Let us now define $S_K(t)$ as the probability that the population is at most of size K at time t,

$$S_K(t) \triangleq \Pr[N_t < K+1] = \sum_{n=0}^{K} P_n(t). \tag{5.53}$$

By summing (5.48) and (5.49), it is readily seen that $S_K(t)$ satisfies the differential equation

$$\frac{\partial S_K(t)}{\partial t} = -(K+1)^2 P_K(t), \qquad S_K(0) = 1. \tag{5.54}$$

Thus, the Laplace transform $\tilde{S}_K(s)$ of $S_K(t)$ is given by

$$\tilde{S}_K(s) = \frac{1}{s} - \frac{1}{s}(K+1)^2\tilde{P}_K(s)$$

$$= \frac{1}{s} - \frac{1}{s}\prod_{m=1}^{K+1}\frac{m^2}{s+m^2}, \tag{5.55}$$

where we have used (5.52) to obtain the second equality. By expanding the product term in (5.55) in a partial fraction expansion, we have the alternative expression

$$\tilde{S}_K(s) = \sum_{i=1}^{K+1} c_i(K)[s+i^2]^{-1} \tag{5.56}$$

where

$$c_i(K) = \lim_{s\to -i^2}(s+i^2)\frac{1}{s}\prod_{m=1}^{K+1}\frac{m^2}{s+m^2}$$

$$= -\prod_{\substack{m=1\\ m\neq i}}^{K+1}\frac{m^2}{m^2-i^2}. \tag{5.57}$$

In the limit as $K\to\infty$, it can be verified that $c_i(\infty)=2(-1)^{i+1}$. Thus, we conclude upon inverse Laplace transforming (5.57) and letting $K\to\infty$ that

$$\Pr[N_t<\infty] = -2\sum_{i=1}^{\infty}(-1)^i\exp(-i^2t). \tag{5.58}$$

This expression is related to the theta function, which is tabulated. A plot of this probability is shown in Figure 5.7. It is seen that the probability the population is of finite size rapidly approaches zero with increasing time. Conversely, the probability of a population explosion rapidly approaches unity.

Population explosions can be prevented by suitably restricting the birth rate, as indicated in the following theorem.

Theorem 5.3.2 (Conditions for Nonexploding Birth Processes). Let $\{N_t;\ t\geqslant t_0\}$ be a Markov birth process with birth rate $\{\mu_t(N_t);\ t\geqslant t_0\}$. If the

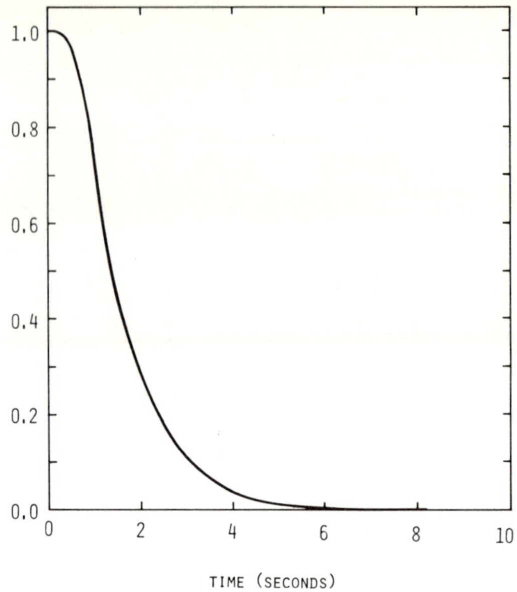

Figure 5.7. Probability of a population of finite size, $\Pr(N_t < \infty)$.

sum

$$\sum_{n=0}^{\infty} \left[\sup_{t_0 \leqslant \sigma < t} \mu_\sigma(n) \right]^{-1} \tag{5.59}$$

is infinite, then for all finite $t \geqslant t_0$, there holds

$$\Pr[N_t < \infty] = \sum_{n=0}^{\infty} \Pr[N_t = n] = 1. \tag{5.60}$$

That is, there is no population explosion with probability one.

Proof. Our proof follows that in W. Feller [7, p. 452] for the homogeneous situation when μ does not depend on t. As in Example 5.3.1, let $P_n(t) = \Pr[N_t = n]$ and set

$$S_K(t) = \Pr[N_t < K+1] = \sum_{n=0}^{K} P_n(t). \tag{5.61}$$

Clearly, $1 - S_K(t)$ is a bounded, nonincreasing function of K; consequently, the limit

$$\xi_t = \lim_{K\to\infty} [1 - S_K(t)] \tag{5.62}$$

exists. Summing (5.10) and (5.11) over $n=0,1,\ldots,K$ shows that

$$\frac{\partial S_K(t)}{\partial t} = -\mu_t(K)P_K(t), \qquad S_K(t_0)=1. \tag{5.63}$$

Hence,

$$1 - S_K(t) = \int_{t_0}^{t} \mu_\sigma(K)P_K(\sigma)\,d\sigma \tag{5.64}$$

for all $K \geqslant 0$. As $\xi_t \leqslant 1 - S_n(t)$, we have

$$\xi_t \leqslant \int_{t_0}^{t} \mu_\sigma(n)P_n(\sigma)\,d\sigma \leqslant \left[\sup_{t_0\leqslant\sigma<t} \mu_\sigma(n)\right]\int_{t_0}^{t} P_n(\sigma)\,d\sigma. \tag{5.65}$$

After multiplying by $[\sup\mu(n)]^{-1}$, summing on $n=0,1,\ldots,K$, and using $S_k(\sigma)\leqslant 1$ for $t_0 \leqslant \sigma < t$, we have

$$\xi_t \sum_{n=0}^{K}\left[\sup_{t_0\leqslant\sigma<t} \mu_\sigma(n)\right]^{-1} \leqslant \int_{t_0}^{t} S_K(\sigma)\,d\sigma \leqslant t - t_0. \tag{5.66}$$

Consequently, if the sum (5.59) diverges, we conclude for all finite $t \geqslant t_0$ that $\xi_t = 0$, and in view of (5.62), this implies (5.60). The theorem is proved.

Two final remarks about Theorem 5.3.2 can be made. The first is that the condition (5.59) is also necessary as well as sufficient for there to be no explosions when the birth rate is independent of time; see Problem 5.3.1. As the number of points in $[t_0, t)$ for a general self-exciting point process has the same distribution as if it were a Markov birth process with birth rate equal to the count conditional intensity $\hat{\lambda}_t(N_t)$, it follows from Theorem 5.3.2 that a sufficient condition for the nonexistence of explosions in a general self-exciting point process is that the count-conditional intensity $\hat{\lambda}_t(n)$ in (5.7) be such that the sum

$$\sum_{n=0}^{\infty}\left[\sup_{t_0\leqslant\sigma<t} \hat{\lambda}_\sigma(n)\right]^{-1} \tag{5.67}$$

is infinite for all finite $t \geqslant t_0$.

We refer to A. Bharucha-Reid [1, Chapter 2], W. Feller [7, p. 448], and I. Gikhman and A. Skorokhod [10, Chapter 7] for further discussion of Markov birth processes.

Characterization of m-Memory Point Processes for $m>0$

Self-exciting point processes having memory greater than zero are characterized in the following theorem.

Theorem 5.3.3 (Characterization of m-Memory Self-Exciting Point Processes). A self-exciting point process is m-memory for $m=1,2,\ldots$ if and only if its occurrence times $\{w_n;\ =1,2,\ldots\}$ form an mth-order Markov sequence.

Proof. Suppose first that the occurrence times do form an mth-order Markov sequence. Then the survival probability for the $(n+1)$st point, where $n \geqslant m$, satisfies

$$\mathcal{P}_{w_{n+1}|w_1,w_2,\ldots,w_n}(t|W_1,W_2,\ldots,W_n)=\mathcal{P}_{w_{n+1}|w_{n-m+1},\ldots,w_n}(t|W_{n-m+1},\ldots,W_n). \tag{5.68}$$

It follows from (5.17) that μ of (5.2) depends only on the m most recent occurrence times, and, therefore, the self-exciting point process is m-memory. Conversely, suppose now that the point process is m-memory. We have by differentiating (5.15) that the conditional probability density function for the $(n+1)$st waiting time is given in general by

$$p_{w_{n+1}|w_1,w_2,\ldots,w_n}(t|W_1,W_2,\ldots,W_n)$$

$$=\mu_t(n;\ W_1,W_2,\ldots,W_n)\exp\left[-\int_{W_n}^{t}\mu_\sigma(n;\ W_1,W_2,\ldots,W_n)\,d\sigma\right]. \tag{5.69}$$

Thus, for an m-memory process, we have for $n \geqslant m$

$$p_{w_{n+1}|w_1,w_2,\ldots,w_n}(t|W_1,W_2,\ldots,W_n)$$

$$=\mu_t(n;\ W_{n-m+1},\ldots,W_n)\exp\left[-\int_{W_n}^{t}\mu_\sigma(n;\ W_{n-m+1},\ldots,W_n)\,d\sigma\right]. \tag{5.70}$$

In view of this and the relation

$$p_{w_{n+1}|w_{n-m+1},\ldots,w_n}(t|W_{n-m+1},\ldots,W_n)$$
$$=\int\cdots\int p_{w_{n+1}|w_1,w_2,\ldots,w_n}(t|W_1,W_2,\ldots,W_n)$$
$$\times p_{w_1,\ldots,w_{n-m}|w_{n-m+1},\ldots,w_n}(W_1,\ldots,W_{n-m}|W_{n-m+1},\ldots,W_n)\,dW_1\,dW_2\cdots dW_{n-m},$$

it is trivially obvious that

$$p_{w_{n+1}|w_1,w_2,\ldots,w_n}(t|W_1,W_2,\ldots,W_n)=p_{w_{n+1}|w_{n-m+1},\ldots,w_n}(t|W_{n-m+1},\ldots,W_n)$$

for $n \geqslant m$. Consequently, the occurrence times $\{w_n\}$ form an mth-order Markov sequence. Theorem 5.3.3 is proved.

The remainder of our discussion of m-memory self-exciting point processes will be for the particular case $m=1$ and $t_0=0$. The future evolution of a 1-memory point process beyond a time t depends at most on t, the number of points N_t that have occurred up to time t, and the most recent occurrence time w_{N_t}. It is often hypothesized in applications that this dependence is not on t and w_{N_t} separately but only on their difference $t-w_{N_t}$; that is, for many physical situations it is only the elasped time following an occurrence point that matters and not the absolute times t and w_{N_t}. These homogeneous, 1-memory, self-exciting point processes are characterized in the following theorem and its corollary.

Theorem 5.3.4 (Homogeneous 1-Memory Point Processes). A self-exciting process is 1-memory with

$$\mu_t(N_t;\,w_{N_t})=f(N_t;\,t-w_{N_t}) \tag{5.71}$$

for some function f if and only if the interarrival times $t_1=w_1, t_2=w_2-w_1,\ldots,t_n=w_n-w_{n-1},\ldots$ form a sequence of statistically independent random variables.

Proof. Let us suppose first that (5.71) holds. The exceedence probability for the nth interarrival time can be expressed in terms of the survival probability for the nth point as

$$\Pr[t_n \geqslant T_n|t_1=T_1,t_2=T_2,\ldots,t_{n-1}=T_{n-1}]$$
$$=\mathcal{P}_{w_n|w_1,w_2,\ldots,w_{n-1}}(T_1+T_2+\cdots+T_n|T_1,T_1+T_2,\ldots,T_1+T_2+\cdots+T_{n-1}). \tag{5.72}$$

But (5.71) implies that the process is 1-memory and, consequently, the occurrence times form a first-order Markov sequence by Theorem 5.3.3. Thus,

$$\begin{aligned}\Pr[t_n \geqslant T_n | t_1 = T_1, t_2 = T_2, \ldots, t_{n-1} = T_{n-1}] \\ &= \mathcal{P}_{w_n|w_{n-1}}(T_1 + T_2 + \cdots + T_n | T_1 + T_2 + \cdots + T_{n-1}) \qquad (5.73) \\ &= \exp\left[-\int_{T_1+T_2+\cdots+T_{n-1}}^{T_1+T_2+\cdots+T_n} \mu_\sigma(n-1;\, T_1 + T_2 + \cdots + T_{n-1})\, d\sigma \right],\end{aligned}$$

where the last equality follows from (5.70). By incorporating (5.71), we have that

$$\Pr[t_n \geqslant T_n | t_1 = T_1, t_2 = T_2, \ldots, t_{n-1} = T_{n-1}] = \exp\left[-\int_0^{T_n} f(n-1;\, \sigma)\, d\sigma \right]. \qquad (5.74)$$

It is trivially obvious from this and the relation

$$\begin{aligned}\Pr[t_n \geqslant T_n] = \int \cdots \int \Pr[t_n \geqslant T_n | t_1 = T_1, t_2 = T_2, \ldots, t_{n-1} = T_{n-1}] \\ \times p_{t_1, t_2, \ldots, t_{n-1}}(T_1, T_2, \ldots, T_{n-1})\, dT_1\, dT_2 \cdots dT_{n-1}\end{aligned}$$

that

$$\Pr[t_n \geqslant T_n | t_1 = T_1, t_2 = T_2, \ldots, t_{n-1} = T_{n-1}] = \Pr[t_n \geqslant T_n]. \qquad (5.75)$$

Thus, (5.71) implies that the interarrival times form an independent sequence. Suppose next that the interarrival times do form an independent sequence. We need to show that (5.71) holds. Equation (5.72) and the independence of the interarrival times imply that the conditional survival probability for the $(n+1)$st point satisfies

$$\mathcal{P}_{w_{n+1}|w_1, w_2, \ldots, w_n}(t | W_1, W_2, \ldots, W_n) = \Pr[t_{n+1} \geqslant t - W_n]. \qquad (5.76)$$

By (5.17), we then have

$$\mu_t(n;\, W_1, W_2, \ldots, W_n) = -\frac{\partial \ln \Pr[t_{n+1} \geqslant t - W_n]}{\partial t}. \qquad (5.77)$$

The right side depends only on the nth occurrence time W_n and, therefore, the process is 1-memory. Furthermore, the dependence on t and W_n is only through the difference $t-W_n$ and, therefore, (5.71) holds with

$$f(n;\, t-W_n) = -\frac{\partial \ln \Pr[t_{n+1} \geqslant t-W_n]}{\partial t}. \tag{5.78}$$

This establishes the theorem.

Renewal processes form a special but widely used class of homogeneous, 1-memory self-exciting point processes. These are defined as follows:

Renewal Processes. A self-exciting point process with independent, identically distributed interarrivals is termed an *ordinary renewal process*. If all interarrivals but the first are identically distributed, it is termed a *modified renewal process*. A modified renewal process in which the first interarrival time has the probability density

$$p_{t_1}(T) = \frac{1-P_{t_n}(T)}{E(t_n)}, \tag{5.79}$$

where $P_{t_n}(T)$ is the common probability distribution function for the nth interarrival time ($n>1$), is termed an *equilibrium renewal process*.

Renewal processes are characterized in the following corollary to Theorem 5.3.4.

Corollary to Theorem 5.3.4. A self-exciting point process is a modified renewal process if and only if μ of (5.2) has the form

$$\mu_t(N_t;\, w_1, w_2, \ldots, w_{N_t}) = h(t-w_{N_t}) \tag{5.80}$$

for some function h. It is an ordinary renewal process if and only if in addition to (5.80), $\mu_t(0) = h(t)$.

Proof. If (5.80) holds the process is 1-memory, and by Theorem 5.3.4 it has independent interarrivals. By (5.14), the exceedence probability for the first interarrival $t_1 = w_1$ is given by

$$\Pr(t_1 \geqslant T) = \exp\left(-\int_0^T \mu_\sigma(0)\, d\sigma\right).$$

By (5.74) and (5.75), the exceedence probability for the nth interarrival t_n, for $n>1$, is given by

$$\Pr(t_n \geqslant T) = \exp\left(-\int_0^T h(\sigma)\,d\sigma\right). \tag{5.81}$$

Thus, interarrivals for $n>1$ are identically distributed, and (5.80) implies that the process is a modified renewal process. It is an ordinary renewal process if in addition $\mu_t(0)=h(t)$. Conversely, suppose that the point process is a modified renewal process with the first interarrival t_1 having a probability density function $p_{t_1}(T)$ and all subsequent interarrival times $t_2, t_3, \ldots$ identically distributed as a random variable t having a probability density function $p_t(T)$. The survival probability for the first point is given by

$$\mathcal{P}_{w_1}(T) = \Pr[t_1 \geqslant T] = \int_T^\infty p_{t_1}(\alpha)\,d\alpha.$$

Hence, from (5.14)

$$\mu_T(0) = \frac{p_{t_1}(T)}{1-P_{t_1}(T)} \tag{5.82}$$

where $P_{t_1}(T) = \int_0^T p_{t_1}(\alpha)\,d\alpha$ is the probability distribution function for t_1. Similarly, the survival probability for the $(n+1)$st point, $n>1$, is given by

$$\begin{aligned} &\mathcal{P}_{w_{n+1}|w_1,w_2,\ldots,w_n}(T \mid W_1, W_2, \ldots, W_n) \\ &\qquad = \Pr[t_{n+1} \geqslant T - W_n] \\ &\qquad = \int_{T-W_n}^\infty p_t(\alpha)\,d\alpha. \end{aligned}$$

Hence, from (5.17)

$$\mu_T(n;\, W_1, W_2, \ldots, W_n) = \frac{p_t(T-W_n)}{1-P_t(T-W_n)} \tag{5.83}$$

where $P_t(T)$ is the distribution function for t. Thus, $\mu_T(n;\, W_1, W_2, \ldots, W_n)$ has the form in (5.80) with

$$h(T) = \frac{p_t(T)}{1-P_t(T)}. \tag{5.84}$$

Furthermore, $\mu_T(0)=h(T)$ if $p_{t_1}(T)=p_t(T)$. Thus, the corollary to Theorem 5.3.4 is proved.

The function $h(T)$ in (5.84) is sometimes referred to as the *hazard function*. We see from (5.80) and the definition of μ in (5.2) that to order Δt, the quantity $h(t-\tau)\Delta t$ is the conditional probability the nth point occurs in the infinitesimal interval $[t,t+\Delta t)$ given that the $(n-1)$st point occurs at time τ. An easy calculation shows that the hazard function for a homogeneous Poisson process is a constant independent of T. Hazard functions are often estimated in applications in order to make a judgment about the appropriateness of employing a Poisson model; significant departures from a constant hazard may preclude the use of a homogeneous Poisson model and suggest employing a more complicated renewal model. The estimated hazard function can be used to form an estimate of the interarrival density needed in this renewal model. Some measured hazard functions are shown in Figure 5.8. These are reported by J. Evans [6] in his studies of low-frequency atmospheric noise recorded at the output of a wideband radio receiver; see Figure 4.2. The intervals between lightning-induced bursts were measured to form three sample hazard functions at three different noise power levels. It is seen that the measured hazard departs from a constant characteristic of a homogeneous Poisson process for interarrival times less than about 0.2 seconds; the probability for small intervals between lightning discharges is higher than would be predicted using a Poisson model. On the other hand, the hazard is nearly constant for interarrivals greater than about 0.2 seconds. Therefore, a Poisson model might be adequate in situations where large interarrivals are of most significance.

The parameter estimation and hypothesis testing results in Examples 5.2.3 and 5.2.5 for general self-exciting point processes apply, of course, in the special case of limited-memory point processes and, in particular, for renewal processes. An ordinary renewal process is characterized completely by the common distribution of its interarrival times. It is not surprising, therefore, that the expressions required for parameter estimation and hypothesis testing for observed renewal processes can be put in a form involving only this distribution. To see how, suppose $\{N_t;\ t\geqslant 0\}$ is the counting process associated with an ordinary renewal process having interarrival times with the probability density and distribution functions $p_t(T)$ and $P_t(T)=\int_0^T p_t(\sigma)\,d\sigma$, respectively. The hazard function is then

$$h(T)=\frac{p_t(T)}{1-P_t(T)}. \tag{5.85}$$

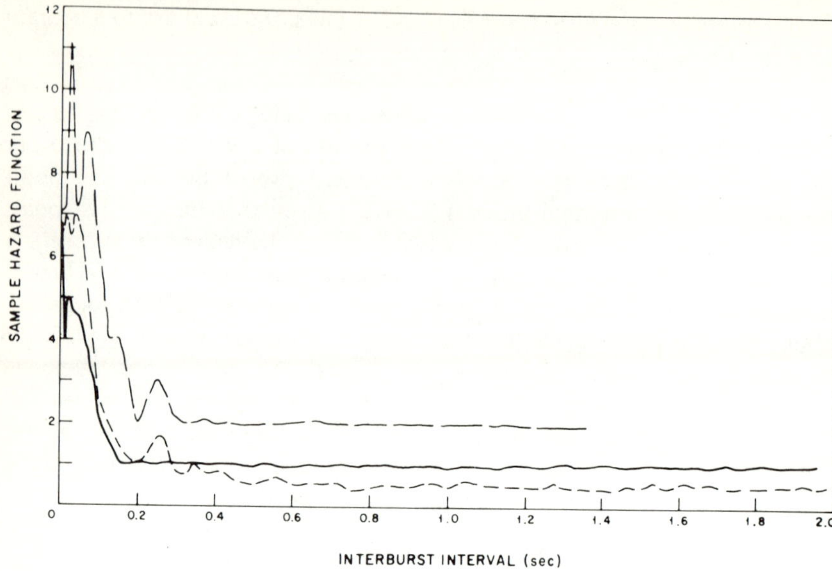

Figure 5.8. Measured hazard functions for interarrivals in lightning discharges. From Evans [6].

We note that

$$\int_{\alpha}^{\beta} h(\sigma)\,d\sigma = \ln\left(\frac{1-P_t(\alpha)}{1-P_t(\beta)}\right). \tag{5.86}$$

A possible intensity process $\{\lambda_t;\ t \geqslant 0\}$ defined in (5.3) is shown in Figure 5.9 for a particular realization of N. By using (5.29), (5.86), and some manipulations, it is seen that the sample-function density for an ordinary renewal process is given by

$$p[\{N_\sigma;\ t_0 \leqslant \sigma < T\}] = \begin{cases} [1-P_t(T)], & N_T = 0 \\ [1-P_t(T)]h(W_1), & N_T = 1 \\ [1-P_t(T)]h(W_1)h(W_2-W_1)\cdots h(W_n-W_{n-1}), & \\ & N_T = n \geqslant 2. \end{cases} \tag{5.87}$$

As $h(T)$ can be expressed in terms of $P_t(T)$, so can the sample-function density. If the interarrival distribution depends on a set of parameters $\mathbf{X}$ to be estimated from an observed path of N on $[0, T)$, then the maximum

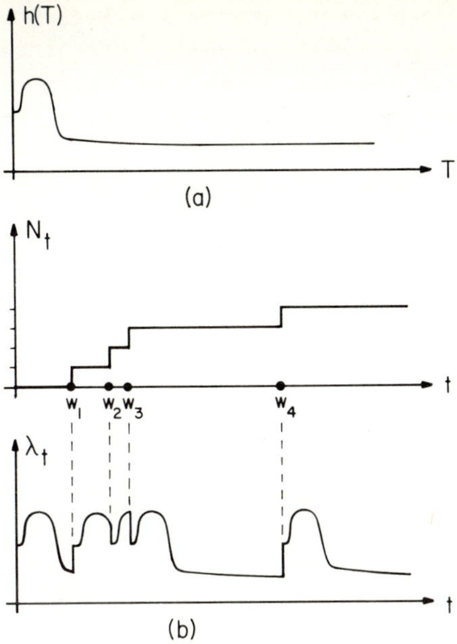

Figure 5.9. (a) A hazard function, and (b) an associated counting and intensity process.

likelihood estimate of **X** is the value of **X** that maximizes

$$l(\mathbf{X}) = \begin{cases} \ln\{[1 - P_t(T:\mathbf{X})]\}, & N_T = 0 \\ \ln\{[1 - P_t(T:\mathbf{X})]h(W_1:\mathbf{X})\}, & N_T = 1 \\ \ln\{[1 - P_t(T:\mathbf{X})]h(W_1:\mathbf{X})h(W_2 - W_1:\mathbf{X})\cdots h(W_n - W_{n-1}:\mathbf{X})\}, & \\ & N_T = n \geq 2 \end{cases} \quad (5.88)$$

subject to any constraints on the possible values of **X**. Here, the notation $(\cdot:\mathbf{X})$ indicates the dependence of the distribution and hazard functions on **X**. If one of two possible interarrival distributions $P_t^{(1)}(T)$ or $P_t^{(2)}(T)$ govern the observed counting path, then a sufficient statistic for deciding which is given by

$$\Lambda = \begin{cases} [1 - P_t^{(2)}(T)][1 - P_t^{(1)}(T)]^{-1}, & N_T = 0 \\ [1 - P_t^{(2)}(T)]h^{(2)}(W_1)\{[1 - P_t^{(1)}(T)]h^{(1)}(W_1)\}^{-1}, & N_T = 1 \\ [1 - P_t^{(2)}(T)]h^{(2)}(W_1)h^{(2)}(W_2 - W_1)\cdots h^{(2)}(W_n - W_{n-1}) & \\ \times\{[1 - P_t^{(1)}(T)]h^{(1)}(W_1)h^{(1)}(W_2 - W_1)\cdots h^{(1)}(W_n - W_{n-1})\}^{-1}, & \\ & N_T = n \geq 2. \end{cases} \quad (5.89)$$

Further discussion of renewal processes is given by W. Smith [23], E. Parzen [17, Chapter 5], and D. Cox and H. Miller [3, Chapter 9].

Summary. Self-exciting point processes in which the future evolution depends on only a limited number of past occurrences have been characterized in this section. An m-memory process depends at most on the m most recent occurrence times. The important conclusions to be drawn from the section are:

1. A necessary and sufficient condition for a self-exciting point process to be a 0-memory process is that its associated counting process be a Markov birth process.
2. A necessary and sufficient condition for a self-exciting point process to be an m-memory process for $m \geqslant 1$ is that its occurrence time sequence form an mth-order Markov sequence.
3. A renewal process is a particular type of 1-memory process that is temporally homogeneous. Thus, 1-memory self-exciting point processes may be regarded as a generalization of the renewal process without the restriction of temporal homogeneity.

REFERENCES

1. A. T. Bharucha-Reid, *Elements of the Theory of Markov Processes and Their Applications*, McGraw-Hill, New York, 1960.
2. E. Cinlar, "Superposition of Point Processes," in *Stochastic Point Processes Statistical Analysis, Theory, and Applications*, Ed. P.A.W. Lewis, Wiley, New York, 1972.
3. D. R. Cox and H. D. Miller, *The Theory of Stochastic Processes*, Wiley, New York, 1965.
4. D. J. Daley and D. Vere-Jones, "A Summary of the Theory of Point Processes," in *Stochastic Point Processes, Statistical Analysis, Theory, and Applications*, Ed. P.A.W. Lewis, Wiley, New York, 1972.
5. W. B. Davenport, Jr. and W. L. Root, *An Introduction to the Theory of Random Signals and Noise*, McGraw-Hill, New York, 1958.
6. J. Evans. "Preliminary Analysis of ELF Noise," Tech. Note 1969-18, M.I.T. Lincoln Laboratory, Lexington, Mass., March 1969.
7. W. Feller, *An Introduction to Probability Theory and Its Applications: Volume* 1, Wiley, New York, 1950.
8. P. M. Fishman, "Statistical Inference for Space-Time Point Processes," D.Sc. Thesis, Sever Institute of Technology, Washington University, St. Louis, Missouri, June 1974.
9. P. Franken, "A Refinement of the Limit Theorem for the Superposition of Independent Renewal Processes," *Teor. Veroyatnost. i Primen.*, **8**, 320–328 (1963).
10. I. I. Gikhman and A. V. Skorokhod, *Introduction to the Theory of Random Processes*, Saunders, Philadelphia, 1965.

11. B. Grigelionis, "On the Convergence of Sums of Random Step Processes to a Poisson Process," *Teor. Veroyatnost. i Primen.*, **8**, 189–194 (1963).

12. P. I. Kuznetsov and R. L. Stratonovich, "A Note on the Mathematical Theory of Correlated Random Points," *Izv. Akad. Nauk SSSR, Ser. Mat.*, **20**, 167 (1956). Also in *Nonlinear Transformations of Random Processes*; Pergamon, New York, 1965.

13. O. Macchi, "Stochastic Point Processes and Multicoincidences," *IEEE Trans. Information Theory*, **IT-17**, No. 1, 2–7 (Jan. 1971).

14. O. Macchi, "Processus Ponctuels et Coincidences," D.Sc. These, Centre D'Orsay, Univ. de Paris Sud, Jan. 1972.

15. J. A. McFadden, "On the Lengths of Intervals in a Stationary Point Process," *J. Roy. Stat. Soc., Ser. B*, **24**, 364–382 (1962).

16. J. E. Moyal, "The General Theory of Stochastic Population Processes," *Acta Math.*, **108**, 1–31 (1962).

17. E. Parzen, *Stochastic Processes*, Holden-Day, San Francisco, 1962.

18. A. Ramakrishnan, "Stochastic Processes Relating to Particles Distributed in Continuous Infinity of States," *Proc. Camb. Phil. Soc.*, **46**, 595–602 (1950).

19. A. Ramakrishnan, "Stochastic Theory of Evolutionary Processes (1937–71)," in *Stochastic Point Processes, Statistical Analysis, Theory and Applications*, Ed. P.A.W. Lewis, Wiley, New York, 1972.

20. S. O. Rice, "Mathematical Analysis of Random Noise," *Bell Syst. Tech. J.*, **23–24**, 1–162 (1944).

21. I. Rubin, "Reduced Memory Information Processors for Point Processes," Proceedings of the Tenth Annual Allerton Conference on Circuit and System Theory, Univ. of Ill., Urbana, pp. 303–312, 1972.

22. I. Rubin, "Regular Point Processes and Their Detection," *IEEE Trans. Information Theory*, **IT-18**, No. 8, 547–557 (Sept. 1972).

23. W. L. Smith, "Renewal Theory and Its Ramifications," *J. Roy. Stat. Soc., Ser. B*, **20**, 243–302 (1958).

24. S. K. Srinivasan and R. Vasudevan, *Introduction to Random Differential Equations and Their Applications*, Elsevier, New York, 1971.

25. H. Wold, "Sur les Processus Stationnaires Ponctuels," in *Le Calcul des Probabilités et Applications*, CNRS, Paris, 1949, pp. 75–86.

PROBLEMS

5.2.1. Complete the proof of Theorem 5.2.1.

5.2.2. Let $\{N_t;\ t \geqslant t_0\}$ be a self-exciting point process satisfying the hypotheses of Theorem 5.2.1. Establish the following expressions for the conditional probability of there being no points on the interval $[s,t)$ given the number and occurrence times for those on $[t_0,s)$:

$$\Pr[N_{s,t}=0|N_s]=\exp\left[-\int_s^t \mu_\sigma(0)\,d\sigma\right]$$

for $N_s = 0$ and

$$\Pr[N_{s,t}=0|N_s;\, w_1,w_2,\ldots,w_{N_s}] = \exp\left[-\int_s^t \hat{\mu}_\sigma(N_s;\, w_1,w_2,\ldots,w_{N_s})\,d\sigma\right]$$

for $N_s \geqslant 1$, where we define

$$\hat{\mu}_\sigma(N_s;\, w_1,w_2,\ldots,w_{N_s}) = E\left[\mu_\sigma(N_\sigma;\, w_1,w_2,\ldots,w_{N_\sigma})|N_s;\, w_1,w_2,\ldots,w_{N_s}\right]$$

for $s \leqslant \sigma < t$.

5.2.3 (Counting Moments for a Self-Exciting Point Process). Show that on the interval $[t_0,t)$, the kth moment of the number of points for a self-exciting point process satisfies

$$\frac{dE(N_t^k)}{dt} = \sum_{j=1}^{k} \binom{k}{j} E\left[N_t^{k-j}\hat{\lambda}_t(N_t)\right], \qquad E\left(N_{t_0}^k\right) = 0$$

for $k=1,2,\ldots,$where $\hat{\lambda}_t(N_t)$ is the count-conditional intensity of the process as given in (5.7). Deduce, therefore, that the mean number of points is given by

$$E(N_t) = \int_{t_0}^t E(\lambda_\sigma)\,d\sigma.$$

[Hint: Multiply (5.11) by n^k and sum on n.]

5.2.4. Deduce from Problem 5.2.3 that

$$E(N_t^2) = \int_{t_0}^t E(\lambda_\sigma)\,d\sigma + 2\int_{t_0}^t E(N_\sigma \lambda_\sigma)\,d\sigma.$$

Hence, conclude that if

$$\int_{t_0}^t E[(N_\sigma+1)\lambda_\sigma]\,d\sigma < \infty$$

then $E(N_t^2) < \infty$. Show that for a Poisson process, this condition is equivalent to $\int_{t_0}^t \lambda_\sigma\,\mathrm{d}\sigma < \infty$.

5.2.5. Let $\{N_t;\ t \geqslant t_0\}$ be a self-exciting counting process such that

$\mu_t(0)=\alpha_t$ for $N_t=0$ and $\mu_t(N_t;\ w_1,w_2,\ldots,w_{N_t})=(N_t+1)\alpha_t$ for $N_t\geqslant 1$. Thus, $\{N_t;\ t\geqslant 0\}$ is an inhomogeneous, linear birth-process.

a. Show that

$$\Pr[N_t=n]=\Pr[N_t=0]\Pr^n[N_t>0]$$

[Hint: Use (5.12), (5.13), and induction.]

b. Let $m_t=E(N_t)$ be the expected number of points on $[t_0,t)$. Determine m_t by using the result of Problem 5.2.3, and verify your answer using the result in part (a).

5.2.6 (Product Densities). *Product densities* are quantities that completely characterize a self-exciting point process. Let $\{N_t;\ t\geqslant t_0\}$ be the counting process associated with a self-exciting point process having $\{\lambda_t\ ;\ t\geqslant t_0\}$ as its intensity process. The product density of degree n is defined by

$$f_n(t_1,t_2,\ldots,t_n)=\lim_{\max(\Delta t_i)\downarrow 0} E\left\{\prod_{i=1}^{n}(\Delta t_i)^{-1}\Delta N_{t_i}\right\}$$

for $t_0\leqslant t_1<t_2<\cdots<t_n$ and $n=1,2,\ldots,$ where we define $\Delta N_{t_i}=N_{t_i+\Delta t_i}-N_{t_i}$. Because N is orderly, an entirely equivalent definition for f is

$$f_n(t_1,t_2,\ldots,t_n)=\lim_{\max(\Delta t_i)\downarrow 0}\left(\prod_{i=1}^{n}(\Delta t_i)^{-1}\right)\Pr\left[\Delta N_{t_1}=1,\Delta N_{t_2}=1,\ldots,\Delta N_{t_n}=1\right].$$

Thus, $f_n(t_1,t_2,\ldots,t_n)dt_1\,dt_2\cdots dt_n$ is the n-fold correlation or n-fold probability between points occurring on infinitesimal intervals $[t_1,t_1+dt_1),[t_2,t_2+dt_2),\ldots,[t_n,t_n+dt_n)$ regardless of all other point occurrences.

a. Establish the following relation between the product densities and the intensity process:

$$f_n(t_1,t_2,\ldots,t_n)=\begin{cases}E(\lambda_{t_1}), & n=1\\ E(\lambda_{t_1})\displaystyle\prod_{i=2}^{n}E(\lambda_{t_i}|t_1,t_2,\ldots,t_{i-1}), & n\geqslant 2\end{cases}$$

where $E(\lambda_{t_n}|t_1,t_2,\ldots,t_{n-1})$ is the expectation of λ_t at time $t=t_n$ given that $t_1,t_2,\ldots,$ and t_{n-1} are occurrence times.

Hint: use the fact that

$$\begin{aligned}&\Pr\left[\Delta N_{t_n}=1|\Delta N_{t_1}=1,\ldots,\Delta N_{t_{n-1}}=1\right]\\ &\qquad=E\left(\Pr\left[\Delta N_{t_n}=1|\{N_\sigma;\ t_0\leqslant\sigma<t_n\}\right]|\Delta N_{t_1}=1,\ldots,\Delta N_{t_{n-1}}=1\right).\end{aligned}$$

b. The probability of a single occurrence point in the infinitesimal interval $[t,t+dt)$ is $f_1(t)dt$ to order dt. In view of this, explain why the integral $\int_{t_0}^{t} f_1(t)\,dt$ does not equal the probability of one occurrence in $[t_0,T)$ but, rather, equals the total expected number of occurrences in this interval.

c. Determine the product densities for an inhomogeneous Poisson process with intensity function λ_t, $t \geqslant t_0$.

5.2.7 (Product Densities, Continued). The sample-function density for a self-exciting point process can be expressed in terms of all its product densities. As a result, the product densities provide a complete statistical characterization of the point process. We now indicate the steps used by P. Kuznetsov and R. Stratonovich [12] to establish this.

Consider a partition of $[t_0,T]$ as $t_0 \leqslant t_1 < t_2 < \cdots < t_k = T$ and a realization $\{N_\sigma;\ t_0 \leqslant \sigma < T\}$ of N on this interval. If $N_T = n \geqslant 1$, denote the n occurrence times in $[t_0,T)$ by $W_1, W_2, \ldots, W_n$, and take the partition sufficiently fine that these times fall in different intervals:

$$t_{k_1} \leqslant W_1 < t_{k_1+1}, t_{k_2} \leqslant W_2 < t_{k_2+1}, \ldots, t_{k_n} \leqslant W_n < t_{k_n+1}.$$

If $N_T = n = 0$, there are no occurrence times in $[t_0,T)$.

Let $\xi_k = N_{t_{k+1}} - N_{t_k}$. Then for $n \geqslant 1$, $\xi_{k_i} = 1$ for $i = 1,2,\ldots,n$ and $\xi_k = 0$ for $k = 0,1,\ldots,K-1$ $(k \neq k_1,k_2,\ldots,k_n)$. For $n = 0$, $\xi_k = 0$ for $k = 0,1,\ldots,K-1$. Denote the $M = K-n$ ξ_k's that are zero by $\gamma_1,\gamma_2,\ldots,\gamma_M$. The sample function density for the counting path $\{N_\sigma;\ t_0 \leqslant \sigma < T\}$ is given formally by the limit

$$p[\{N_\sigma;\ t_0 \leqslant \sigma < T\}] = \begin{cases} \lim \mathcal{P}_0, & n = 0 \\ \lim \left[\prod_{i=1}^{n} (\Delta t_{k_i})^{-1} \right] \mathcal{P}_n, & n \geqslant 1 \end{cases}$$

where the limit is taken as $K \to \infty$ and $\max(\Delta t_i) \downarrow 0$, $\Delta t_i = t_{i+1} - t_i$, and where we define

$$\mathcal{P}_n = \Pr(A_n, \gamma_1 = 0, \gamma_2 = 0, \ldots, \gamma_M = 0)$$

in which A_n is the event

$$A_n = \begin{cases} \Omega, & n = 0 \\ \bigcap_{i=1}^{n} \{\xi_{k_i} = 1\}, & n \geqslant 1 \end{cases}$$

where Ω is the certain event. The objective of what follows is to express $\mathcal{P}_n$

in terms only of A_n and events of the form $\{\gamma_j=1\}$.

a. From the orderliness of N, we have

$$\Pr(A_n,\gamma_1=0)=\Pr(A_n)-\Pr(A_n,\gamma_1=1).$$

Show that

$$\Pr(A_n,\gamma_1=0,\gamma_2=0)=\Pr(A_n)-\sum_{i=1}^{2}\Pr(A_n,\gamma_i=1)$$
$$+\Pr(A_n,\gamma_1=1,\gamma_2=1)$$

and that

$$\Pr(A_n,\gamma_1=0,\gamma_2=0,\gamma_3=0)=\Pr(A_n)-\sum_{i=1}^{3}\Pr(A_n,\gamma_i=1)$$
$$+\sum\sum_{1\leqslant i_1<i_2\leqslant 3}\Pr(A_n,\gamma_{i_1}=1,\gamma_{i_2}=1)$$
$$-\Pr(A_n,\gamma_1=1,\gamma_2=1,\gamma_3=1).$$

Deduce in general that

$$\mathcal{P}_n=\Pr(A_n)$$
$$+\sum_{m=1}^{M}(-1)^m\left[\sum\sum_{i\leqslant i_1<i_2<\cdots<i_m\leqslant M}\cdots\sum\Pr(A_n,\gamma_{i_1}=1,\gamma_{i_2}=1,\ldots,\gamma_{i_m}=1)\right].$$

b. By using the expression for $\mathcal{P}_n$ in part (a), introducing the product densities, and passing to the limit, show that for $N_T=n=0$,

$$p[\{N_\sigma;\ t_0\leqslant\sigma<T\}]$$
$$=1+\sum_{m=1}^{\infty}(-1)^m\int\int_{t_0\leqslant\sigma_1<\sigma_2<\cdots<\sigma_m\leqslant T}\cdots\int f_m(\sigma_1,\sigma_2,\ldots,\sigma_m)\,d\sigma_1\,d\sigma_2\cdots d\sigma_m.$$

And for $N_T=n\geqslant 1$, show that

$$p[\{N_\sigma;\ t_0\leqslant\sigma<T\}]=f_n(W_1,W_2,\ldots,W_n)$$
$$+\sum_{m=1}^{\infty}(-1)^m\int\int_{t_0\leqslant\sigma_1<\sigma_2<\cdots<\sigma_m\leqslant T}\cdots\int f_{n+m}(\alpha_1,\alpha_2,\ldots,\alpha_{n+m})\,d\sigma_1\,d\sigma_2\cdots d\sigma_m,$$

where α_i is the ith term in the ordering of $W_1,W_2,\ldots,W_n,\sigma_1,\sigma_2,\ldots,\sigma_m$; that is,

$$\alpha_1 = \text{minimum of } W_1, \ldots, W_n, \sigma_1, \ldots, \sigma_m$$
$$\alpha_2 = \text{minimum of } W_1, \ldots, W_n, \sigma_1, \ldots, \sigma_m \text{ (excluding } \alpha_1)$$
$$\vdots$$
$$\alpha_{n+m} = \text{maximum of } W_1, \ldots, W_n, \sigma_1, \ldots, \sigma_m.$$

c. Use the result of part (b) to show that the sample-function density for an inhomogeneous Poisson process with intensity function λ_t is given by

$$p[\{N_\sigma;\, t_0 \leqslant \sigma < T\}] = \begin{cases} \exp\left[-\int_{t_0}^{T} \lambda_\sigma \, d\sigma\right], & N_T = n = 0 \\ \left(\prod_{i=1}^{n} \lambda_{W_i}\right) \exp\left[-\int_{t_0}^{T} \lambda_\sigma \, d\sigma\right], & N_T = n \geqslant 1. \end{cases}$$

5.2.8. Let $\{N_t;\, t \geqslant t_0\}$ be a self-exciting counting process with intensity process $\{\lambda_t;\, t \geqslant t_0\}$. Let $\{b_t;\, t \geqslant t_0\}$ be a random process that is at most causally dependent on N; in particular, b_t is statistically independent of any future increment $N_u - N_t$ of N, where $u \geqslant t$. Suppose, further, that for almost every realization of N and for all $\sigma \in [t_0, t)$ that b_σ is bounded and piecewise continuous; b is left continuous at any discontinuities. Then from Theorem 4.2.1, the counting integral

$$I_t = \int_{t_0}^{t} b_\sigma \, dN_\sigma$$

exists and has the evaluation

$$I_t = \sum_{i=1}^{N_t} b_{w_i}.$$

By paralleling the derivation of (4.60), show that

$$E(I_t) = \int_{t_0}^{t} E(b_\sigma \lambda_\sigma) \, d\sigma.$$

5.2.9. The entropy $H(t_0, T)$ associated with the paths of a self-exciting counting process on $[t_0, T)$ is defined by

$$H(t_0, T) = -E(\ln p[\{N_\sigma;\, t_0 \leqslant \sigma < T\}]),$$

where p is the sample-function density. Use the result in Problem 5.2.8 to show that if $\ln\lambda$ is bounded on $[t_0, T)$, then

$$H(t_0,T)=\int_{t_0}^{T}E[\lambda_\sigma(1-\ln\lambda_\sigma)]\,d\sigma.$$

5.2.10 (Moments of the Sample-Function Density). Let $\{\zeta_t;\ t\geqslant t_0\}$ be defined by

$$\zeta_t=-\int_{t_0}^{t}\lambda_\sigma\,d\sigma+\int_{t_0}^{t}\ln\lambda_\sigma\,dN_\sigma,$$

where $\{N_t;\ t\geqslant t_0\}$ is a self-exciting point process and $\{\lambda_t;\ t\geqslant t_0\}$ the associated intensity process. Recall from (5.29) that $\phi(\zeta_t)=\exp(\zeta_t)$ is the sample-function density for N on $[t_0,t)$.

a. Show that

$$\phi^k(\zeta_t)=1-k\int_{t_0}^{t}\phi^k(\zeta_\sigma)\lambda_\sigma\,d\sigma+\int_{t_0}^{t}\phi^k(\zeta_\sigma)[\lambda_\sigma^k-1]\,dN_\sigma.$$

b. Conclude from (a) and Problem 5.2.8 that

$$E[\phi^k(\zeta_t)]=1-\int_{t_0}^{t}E[\phi^k(\zeta_\sigma)\{\lambda_\sigma^{k+1}-(k+1)\lambda_\sigma\}]\,d\sigma.$$

5.3.1. Show that if the birth rate $\mu_t(n)$ for a Markov birth process is such that the sum

$$\sum_{n=0}^{\infty}\Big[\inf_{t_0\leqslant\sigma<\infty}\mu_\sigma(n)\Big]^{-1}$$

converges, then $\Pr[N_t<\infty]<1$ and a population explosion can occur with nonzero probability. Use this result to show that the condition of Theorem 5.3.2 is necessary as well as sufficient for there to be no explosions when the birth rate is independent of time.

5.3.2. Prove that an m-memory self-exciting point process has the function $\mu_t(N_t;\ w_1,w_2,\ldots,w_{N_t})$ of the form

$$\mu_t(N_t;\ w_1,w_2,\ldots,w_{N_t})=f_t(w_{N_t-m+1},\ldots,w_{N_t})$$

for all $N_t \geqslant m$ if and only if the sequence of occurrence times $w_1, w_2, \ldots$ forms an mth-order Markov sequence. Here, f is independent of both N_t and the $N_t - m$ earlier times.

5.3.3. Let $\{N_t;\ t \geqslant 0\}$ be an ordinary renewal counting process, $m_t = E(N_t)$, and $p_t(T)$ the common interarrival-time probability density.

a. Show that

$$m_t = \sum_{n=1}^{\infty} P_{w_n}(t),$$

where $P_{w_n}(t)$ is the probability distribution function for the nth occurrence time. [Hint: First establish that $P_r(N_t = n) = P_{w_n}(t) - P_{w_{n+1}}(t)$.]

b. Let

$$M(s) = \int_0^{\infty} m_t \exp(-st)\,dt \quad \text{and} \quad \Phi(s) = \int_0^{\infty} p_t(T) \exp(-sT)\,dT$$

be the Laplace transforms of m_t and $p_t(T)$, respectively. Show that

$$sM(s) = \Phi(s)[1 - \Phi(s)]^{-1}.$$

Conclude from this that the mean-value function m_t of a renewal process provides a complete statistical characterization because the interarrival density does.

5.3.4. Suppose the process N in Problem 5.3.3 is a modified renewal process.

a. Show that

$$sM(s) = \Phi_1(s)[1 - \Phi(s)]^{-1},$$

where $\Phi_1(s)$ is the Laplace transform for the first interarrival density and $\Phi(s)$ is the transform for the subsequent interarrival densities.

b. Assuming it applies, use the final value theorem of Laplace transforms to show for a modified renewal process that

$$\lim_{t \to \infty} \left[m_t - \frac{t}{\mu} \right] = \frac{E(t_1^2)}{2\mu^2} - \frac{E(t_1)}{\mu},$$

where $\mu = E(t_n)$ is the mean interarrival time for $n \geqslant 2$. Thus, conclude for this process that

$$m_t = \frac{t}{\mu} + \frac{E(t_1^2)}{2\mu^2} - \frac{E(t_1)}{\mu} + O(t^{-1})$$

where $O(t^{-1}) \to 0$ as $t \to \infty$; for large t, $m_t \approx t/\mu$.

c. Show for an equilibrium renewal process that $sM(s) = (s\mu)^{-1}$. Hence, conclude for all $t \geqslant 0$ that $m_t = t/\mu$.

5.3.5. Pulses arrive at a scintillation detector as a homogeneous Poisson process with intensity λ. The detector is nonparalyzable and has a fixed dead time of duration τ.

a. Show the output is either a modified or ordinary renewal process corresponding to whether the detector is initially unlocked or locked in a dead time.

b. Determine the hazard function for the output point process if the detector is initially locked. Sketch the intensity process for a typical output counting path.

c. Suppose an output counting path is observed on the interval $[0, T)$. Use (5.88) to determine an expression for the maximum likelihood estimate for λ in terms of this path, and compare your answer to that obtained in Example 5.2.4.

5.3.6 (Paralyzable Counter). Suppose events arrive at a counter as a homogeneous Poisson process with a constant intensity λ. Each incident event causes the counter to be inoperative for a duration τ. Thus, an incident event arriving during an inoperative period extends the inoperative period. Let $\{t_n\}$ be the sequence of interarrival times for the *registered events*. Here, t_1 is exponentially distributed with parameter λ. Denote the probability density for t_n, $n \geqslant 2$, by $p_{t_n}(t)$ and the Laplace transform of $p_{t_n}(t)$ by $\Phi(s) = E[\exp(-st)]$.

a. Let $\{N_t;\ t \geqslant 0\}$ and $\{M_t;\ t \geqslant 0\}$ be the counting processes for the incident and registered events, respectively. And let $m_t = E(M_t)$. By using

$$E[M_{t+\Delta t} - M_t] = \Pr(M_{t,t+\Delta t} = 1) + o(\Delta t)$$

$$= \Pr(N_{t-\tau,t} = 0, N_{t,t+\Delta t} = 1) + o(\Delta t),$$

show

$$\frac{dm_t}{dt}=\begin{cases}\lambda e^{-\lambda\tau}, & t\geqslant\tau\\ \lambda e^{-\lambda t}, & t<\tau.\end{cases}$$

b. Using this result and part (a) of Problem 5.3.4, show

$$\Phi(s)=\frac{\lambda e^{-\tau(s+\lambda)}}{s+\lambda e^{-\tau(s+\lambda)}}.$$

c. Evaluate the mean $E(t_n)$ and variance $\operatorname{var}(t_n)$ for the nth interarrival time t_n for $n\geqslant 2$.

5.3.7. Let $m_k(t)=E(N_t^k)$, where $\{N_t;\ t\geqslant 0\}$ is an ordinary renewal process. Show that

$$m_2(t)=m_1(t)+2\int_0^t m_1(t-\sigma)\dot{m}_1(\sigma)\,d\sigma$$

where $\dot{m}_1(t)=dm_1(t)/dt$.

DOUBLY STOCHASTIC POISSON PROCESSES

CHAPTER SIX

6.1 INTRODUCTION

The self-exciting point processes of the previous chapter were obtained from the Poisson process by allowing its intensity to become causally dependent on the point process itself. In this way, the intensity is transformed into a random process having paths known exactly given the point process. Point processes in this chapter are also obtained by randomizing the intensity of the Poisson process. Here, however, the intensity is not randomized through self-excitation by the point process itself but, rather, by an "outside" process so that the paths of the resulting intensity process are not known exactly given the point process.

The point processes of interest in this chapter are conditionally Poisson processes given the intensity process. We shall term them *doubly stochastic Poisson processes*, a terminology introduced by D. Cox [6], who first studied them.[†] To be specific, let $\{\mathbf{x}_t; t \geqslant t_0\}$ be a left-continuous, vector-valued stochastic process that is the "outside" process influencing the evolution of a point process whose associated counting process is $\{N_t; t \geqslant t_0\}$. A doubly stochastic Poisson process is defined as follows.

Doubly Stochastic Poisson process. $\{N_t; t \geqslant t_0\}$ is a doubly stochastic Poisson process with intensity process $\{\lambda_t(\mathbf{x}_t); t \geqslant t_0\}$ if for almost every given path of the process $\{\mathbf{x}_t; t \geqslant t_0\}, N$ is a Poisson process with intensity function $\lambda_t(\mathbf{x}_t)$. In other words, $\{N_t; t \geqslant t_0\}$ is conditionally a Poisson process with intensity function $\lambda_t(\mathbf{x}_t)$ given $\{\mathbf{x}_t; t \geqslant t_0\}$.

The process $\{\mathbf{x}_t; t \geqslant t_0\}$ arises in many applications we shall encounter as the quantity that conveys desired information, and for this reason we call it the *information process*. The simplest examples occur when the information process is a finite collection of random variables independent of time. An information process of this simple type is encountered in such diverse applications as nuclear medicine, optical detection, and auditory electrophysiology when the parameters of interest are modeled as random variables in contrast to Examples 2.4.2, 2.4.3, and 2.4.4, respectively, where they are modeled as nonrandom parameters; the new aspect in this chapter is that the parameters are assumed to have a known probability distribution. More generally, there are only relatively weak restrictions on the time dependence and statistics of the information process; to get our most explicit results, we will assume it to be either a vector-valued Gaussian process or a vector-valued Markovian process.

[†]This terminology is not used universally; *conditional, compound*, and *mixed* Poisson processes are also popular, the latter especially when the intensity process is a random variable.

There appear to be numerous physical situations for which doubly stochastic Poisson processes are reasonable mathematical models. Several specific illustrations are developed in the chapter. These are predominantly, but not exclusively, in the area of optical-detection and optical-communication systems where the applicability of the model may be motivated as follows.

Example 6.1.1 (Optical Detection). If the optical field incident on the photodetector of Example 2.4.3 changes randomly in time, the resulting photoelectron stream forms a doubly stochastic Poisson process to an accurate approximation because it is well modeled by a Poisson process conditional on the field. The optical field may be randomly varying because of fluctuations in the intensity of an incoherent light source, as with a stellar object, scattered solar radiation, and an incandescent light bulb, or because of fluctuations in the properties of the medium in which the optical wave propagates, as when atmospheric turbulence causes random changes in the refractive index of the atmosphere.

Example 6.1.2 (Optical Phase Tracking). If the reference phase $\theta_r(t)$ in Example 4.2.3 changes randomly in time, the resulting photoelectron stream is accurately modeled as a doubly stochastic Poisson process because it is well modeled as a Poisson process conditional on the reference phase. Randomness in $\theta_r(t)$ may be intentional, as when phase or frequency modulation are used to communicate an analog information signal, or unintentional, which occurs when the reference oscillator is unstable and has a randomly drifting phase.

The chapter may be divided conceptually into two broad problem areas. The first category is what we term the *forward problem*. This is concerned with the determination of the statistics of a doubly stochastic Poisson process. Thus, in Sections 6.2 and 6.3, we study the counting and time statistics, respectively. A fundamental observation in Section 6.2 is that a doubly stochastic Poisson process is a particular form of self-exciting point process studied in Chapter 5. Certain statistical properties of the doubly stochastic Poisson process are an immediate consequence of this observation. Most importantly, it leads in Section 6.3 to a characterization of the sample-function density of a doubly stochastic Poisson process. The form of the sample-function density is remarkably similar to that of a Poisson process, and its use in decision and estimation problems for observed doubly stochastic Poisson processes parallels that in earlier chapters for other processes. For example, the likelihood ratio, which in principle solves the binary hypothesis testing problem, may be expressed in terms of the sample-function density as in (2.110). The expression for the likelihood ratio is developed in this way in Section 6.4, where we also introduce the important philosophy associated with the *separation theorem of detection*. This principle indicates that the detector for doubly stochastic Poisson processes is that in Figure 2.8 for Poisson processes with the only modification being that certain estimates for the random intensities on each hypothesis are used as if they were the known intensities, $\lambda_t^{(0)}$ and $\lambda_t^{(1)}$, of the Poisson process.

We term the second broad problem area of the chapter the *inverse problem*. This is concerned with the determination of the conditional statistics of the information process given observations of the doubly stochastic Poisson process it influences. Motivation for study of the inverse problem arises from the fact that while a point process may be observed, it is often not this process itself that is of primary interest. Rather, what is desired is an underlying mechanism, here the information process, that can only be observed indirectly through its influence on the point process. For instance, in Example 6.1.2, the observer of the photoelectron stream desires to know the underlying and not directly observable reference phase $\theta_r(t)$ that influences the incident optical field. The conditional statistics for the information process $\mathbf{x}_t$ at time t are characterized in Section 6.5, and this characterization is used to develop estimates for $\mathbf{x}_t$ in terms of the available observations. The mean-square performance of estimators when the information process is normally distributed is also developed. Similar results are obtained in Section 6.7 for the more general situation where $\mathbf{x}_{t-\tau}$, for $\tau \geqslant 0$, is to be estimated in terms of observations of the point process up to time t; here, the time for the estimate lags the observation time by τ units.

In Section 6.6, we briefly extend the results of Sections 6.2 to 6.5 to situations where more than one doubly stochastic Poisson process is observed simultaneously.

6.2 COUNTING STATISTICS FOR DOUBLY STOCHASTIC POISSON PROCESSES

Let $\{N_t; t \geqslant t_0\}$ be a doubly stochastic Poisson process with intensity process $\{\lambda_t(\mathbf{x}_t); t \geqslant t_0\}$, where $\{\mathbf{x}_t; t \geqslant t_0\}$ is the underlying information process. In this section, we want to determine the statistics for N_t, the number of points occurring on $[t_0, t)$. The statistics of interest include various moments for N_t as well as the probability that N_t takes on a particular value n. Two methods for the determination of these statistics are described, the *method of conditioning* and the *method of self-exciting processes*. The first uses the property that $\{N_t; t \geqslant t_0\}$ is conditionally a Poisson counting process given the information process, and the second uses the property, yet to be established, that $\{N_t; t \geqslant t_0\}$ is a self-exciting counting process. Two observations are perhaps worthy of mention before these methods are described. The first is that explicit, closed-form expressions for the statistics of N_t are difficult to obtain even though the problem to be solved is generally straightforward conceptually. The second observation is that the conditional expectation

$$\hat{\lambda}_t \triangleq E[\lambda_t(\mathbf{x}_t)|N_\sigma; t_0 \leqslant \sigma < t]$$

will be seen to play an important role in this and subsequent sections; this conditional expectation of the intensity process will be investigated in some depth in Section 6.5.

Counting Statistics by the Method of Conditioning

The counting statistics for a doubly stochastic Poisson process can be determined in principle by conditioning on the information process, using the fact that thus conditioned $\{N_t; t \geqslant t_0\}$ is a Poisson counting process, and then removing the conditioning by taking the expectation over the paths of the information process. Thus, for example, the probability that the number of points occurring on $[t_0, t)$ is n is given by

$$\Pr(N_t = n) = E[\Pr(N_t = n|\mathbf{x}_\sigma; \sigma \geqslant t_0)]$$

$$= E\left[(n!)^{-1}\left(\int_{t_0}^t \lambda_\sigma(\mathbf{x}_\sigma)\,d\sigma\right)^n \exp\left(-\int_{t_0}^t \lambda_\sigma(\mathbf{x}_\sigma)\,d\sigma\right)\right]. \tag{6.1}$$

The expectation on the right is generally difficult, if not impossible, to evaluate. Some situations where the evaluation is possible are given in the following two examples and in the problems.

Example 6.2.1 (Randomly Scaled Inhomogeneous Poisson Process). Suppose the information process is simply a nonnegative random variable x and that for $t \geqslant t_0, \lambda_t(x) = x\nu_t$ where ν_t is a deterministic function of time. Thus, $\{N_t; t \geqslant t_0\}$ can be viewed as an inhomogeneous Poisson counting process having an intensity function ν_t that is scaled randomly by x. From (6.1), we have that

$$\Pr(N_t = n) = (n!)^{-1}\left(\int_{t_0}^t \nu_\sigma\,d\sigma\right)^n \int_0^\infty X^n \exp\left(-X\int_{t_0}^t \nu_\sigma\,d\sigma\right) dP_x(X), \tag{6.2}$$

where $P_x(X)=\Pr(x \leqslant X)$ is the probability distribution function for x. As an example, suppose x is a continuous random variable having the gamma distribution

$$p_x(X)=\frac{dP_x(X)}{dX}=\begin{cases}\dfrac{\alpha^{\beta+1}}{\Gamma(\beta+1)}X^{\beta}\exp(-\alpha X), & X\geqslant 0\\ 0, & X<0\end{cases} \tag{6.3}$$

where $\Gamma(\cdot)$ denotes the gamma function. Evaluation of (6.2) results in

$$\Pr(N_t=n)=\frac{\Gamma(n+\beta+1)}{\Gamma(n+1)\Gamma(\beta+1)}\left[\frac{\alpha}{\alpha+\int_{t_0}^{t}\nu_\sigma\,d\sigma}\right]^{\beta+1}\left[\frac{\int_{t_0}^{t}\nu_\sigma\,d\sigma}{\alpha+\int_{t_0}^{t}\nu_\sigma\,d\sigma}\right]^{n} \tag{6.4}$$

for $n=0,1,2,\ldots$. This is the negative binomial distribution, and the associated process $\{N_t; t\geqslant t_0\}$ is termed an *inhomogeneous Polya process*. Equation (6.2) can, of course, be evaluated for other distributions; some examples are given in the problems. If it exists, the moment generating function for x is given by

$$M_x(s)=E[\exp(sx)]=\int_0^{\infty}\exp(sX)\,dP_x(X), \tag{6.5}$$

and we have

$$M_x^{(n)}(s)=\frac{d^nM_x(s)}{ds^n}=\int_0^{\infty}X^n\exp(sX)\,dP_x(X). \tag{6.6}$$

Thus, if the moment generating for x exists, we have from (6.2) that

$$\Pr(N_t=n)=(n!)^{-1}\left(\int_{t_0}^{t}\nu_\sigma\,d\sigma\right)^{n}M_x^{(n)}\left(-\int_{t_0}^{t}\nu_\sigma\,d\sigma\right). \tag{6.7}$$

Example 6.2.2 (Photocount Statistics for Gaussian Light). In this example, we examine the counting statistics at the output of an ideal photodetector when the optical field incident on it is a narrowband Gaussian process. Such an incident field occurs in optical communications when a known signal field is observed in the presence of a background radiation such as scattered sunlight (see E. Hoversten, R. Harger, and S. Halme [19]). It can also occur when the signal field propagates through a scattering medium such as a cloud or fog (see R. Kennedy [23]) or when an optical radar ranging pulse is reflected from a surface that is rough. The counting statistics are important for determining the error performance when a receiver is used that simply counts the number of photoelectrons generated on a signaling interval and disregards their occurrence times. The model of this example will be used throughout the chapter. As in Example 4.1.9 and Equation (4.22), let $S(t,\vec{r})$ be the complex envelope of a scalar, narrowband, optical field incident on an ideal photodetector having an active area $\mathcal{A}$. We assume here that $\mathcal{A}$ is sufficiently small that any spatial variations in $S(t,\vec{r})$ can be neglected; this restriction can easily be removed (see J. Clark and E. Hoversten [7]). Thus, we write $S(t,\vec{r})$ simply as $S(t,\vec{r})=X_t$, and (4.22) becomes

$$\lambda_t(X_t)=\alpha|X_t|^2, \tag{6.8}$$

where $\alpha=\eta\mathcal{A}/h\nu$. Here, $\{X_t;t\geqslant t_0\}$ is the complex envelope of a narrowband field; that is, the real field is

$$x_t=\sqrt{2}\ \mathrm{Re}[X_t\exp(j2\pi\nu t)]$$

where ν is the optical carrier frequency.

From Example 4.1.9, we know that the output of the ideal photodetector can be modeled conditionally as a Poisson counting process $\{N_t;t\geqslant t_0\}$ given $\{X_t;t\geqslant t_0\}$. We take $\{x_t;t\geqslant t_0\}$ to be a narrowband Gaussian process, the mean $m_t\triangleq E(x_t)$ corresponds to a deterministic signal field, and $b_t\triangleq x_t-m_t$ corresponds to a zero-mean Gaussian field due to background radiation. In correspondance with $x_t=m_t+b_t$, we write the complex envelope as $X_t=M_t+B_t$. In this case, $\{\mathrm{N}_t;t\geqslant t_0\}$ is a doubly stochastic Poisson process with intensity process $\{\lambda_t(X_t);t\geqslant t_0\}$ given by (6.8). It is convenient to expand $\{X_\sigma;t_0\leqslant\sigma<t\}$ in a generalized Fourier series (Karhunen–Loève expansion) as

$$X_\sigma=\sum_{i=1}^{\infty}X_i\varphi_i(\sigma),\qquad t_0\leqslant\sigma<t \tag{6.9}$$

where the coefficients X_i are given by

$$X_i=\int_{t_0}^{t}X_\sigma\varphi_i^*(\sigma)\,d\sigma. \tag{6.10}$$

Here, the asterisk denotes complex conjugation. The coefficients $\{X_i\}$ form a sequence of independent, complex-valued, Gaussian random variables if the orthonormal basis functions in the series (6.9) are the eigenfunctions of the integral equation

$$\lambda_i \varphi_i(\sigma) = \int_{t_0}^{t} K(\sigma,u)\varphi_i(u)\,du, \qquad t_0 \leqslant \sigma < t \tag{6.11}$$

in which $K(\sigma,u) = E(X_\sigma X_u^*) - M_\sigma M_u^*$ is the covariance function for X. The eigenvalue λ_i is the variance of the ith coefficient X_i. The real and imaginary parts of X_i are statistically independent Gaussian random variables with mean $\mathrm{Re}(M_i)$ and $\mathrm{Im}(M_i)$, respectively, and each has variance $\lambda_i/2$, where

$$M_i = E(X_i) = \int_{t_0}^{t} M_\sigma \varphi_i^*(\sigma)\,d\sigma.$$

These and other properties of the expansion (6.9) are discussed by M. Rosenblatt [35, pp. 185–195] and H. L. Van Trees [48, pp. 584–589]. Because of the orthonormality of the eigenfunctions on $[t_0, t)$, substitution of (6.9) into (6.8) shows that

$$\int_{t_0}^{t} \lambda_\sigma(X_\sigma)\,d\sigma = \alpha \sum_{i=1}^{\infty} |X_i|^2. \tag{6.12}$$

We now examine the counting statistics following the procedure of S. Karp and J. Clark [21] and S. Karp, E. O'Neill, and R. Gagliardi [22]; an historical discussion of this important problem is given by C. Mehta [31]. Let $M_{N_t}(ju) = E[\exp(juN_t)]$ be the characteristic function for the number of detected photons at time t. Using (2.35) and the method of conditioning, we have

$$\begin{aligned} M_{N_t}(ju) &= E\left\{ \exp\left[(e^{ju}-1) \int_{t_0}^{t} \lambda_\sigma(X_\sigma)\,d\sigma \right] \right\} \\ &= \prod_{i=1}^{\infty} E\{\exp[\alpha(e^{ju}-1)|X_i|^2]\}, \end{aligned} \tag{6.13}$$

where we have used (6.12) and the independence of the coefficients $\{X_i\}$. After evaluating the expectation in (6.13), we have

$$M_{N_t}(ju) = \prod_{i=1}^{\infty} [1-\alpha\lambda_i(e^{ju}-1)]^{-1} \exp\left\{ \alpha |M_i|^2 (e^{ju}-1)[1-\alpha\lambda_i(e^{ju}-1)]^{-1} \right\}. \tag{6.14}$$

Each factor in the product is the characteristic function of a Laguerre-distributed random variable; call this nonnegative integer-valued variable $N_t(i)$. By inverse Fourier transformation (see B. Lachs [25]), there results

$$\Pr[N_t(i)=n_i]=\frac{(\alpha\lambda_i)^{n_i}}{(1+\alpha\lambda_i)^{n_i+1}}\exp\left(-\frac{\alpha|M_i|^2}{1+\alpha\lambda_i}\right)$$

$$\times L_{n_i}\left[\frac{|M_i|^2}{\lambda_i(1+\alpha\lambda_i)}\right], \tag{6.15}$$

where $L_n(x)$ is the Laguerre polynomial

$$L_n(x)\triangleq\frac{(n!)^{-1}\exp(x)d^n[x^n\exp(-x)]}{dx^n}$$

$$=\sum_{k=0}^{n}\binom{n}{k}(k!)^{-1}(-x)^k.$$

It follows that the total number N_t of photoelectrons on $[t_0,t)$ can be represented as the sum $\Sigma_{i=0}^{\infty}N_t(i)$ of independent Laguerre random variables. Two limiting situations are of interest. In the limit as $\lambda_i\to 0$ for all i, we have from (6.14) that

$$M_{N_t}(ju)=\exp\left\{(e^{ju}-1)\alpha\sum_{i=1}^{\infty}|M_i|^2\right\}. \tag{6.16}$$

By Parseval's theorem,

$$\sum_{i=1}^{\infty}|M_i|^2=\int_{t_0}^{t}|M_\sigma|^2\,d\sigma.$$

Consequently, in the limit as the background radiation, $B_t=X_t-M_t$, tends to zero, the photocount N_t tends to a Poisson-distributed random variable with parameter $\alpha\int_{t_0}^{t}|M_\sigma|^2d\sigma$. On the other hand, for any component with $M_i=0$, we have

$$\Pr[N_t(i)=n_i]=\frac{(\alpha\lambda_i)^{n_i}}{(1+\alpha\lambda_i)^{n_i+1}}, \tag{6.17}$$

which is the Bose–Einstein distribution; thus, any coordinate in the series (6.9) that has no deterministic part contributes a Bose–Einstein variate to the total photocount.

Counting Statistics by the Method of Self-Exciting Point Processes

Doubly stochastic Poisson processes are self-exciting point processes. This important characterization is established in the following theorem and will be used throughout the chapter. The implication, of course, is that all the results of Chapter 5 may be directly applied in the study of doubly stochastic Poisson processes.

Theorem 6.2.1 (Characterization of a Doubly Stochastic Poisson Process as a Self-Exciting Point Process). Let $\{N_t; t \geqslant t_0\}$ be a doubly stochastic Poisson process with intensity process $\{\lambda_t(\mathbf{x}_t); t \geqslant t_0\}$, and suppose that $E[\lambda_t(\mathbf{x}_t)] < \infty$ for all $t \geqslant t_0$. Then $\{N_t; t \geqslant t_0\}$ is a self-exciting counting process with intensity process $\{\hat{\lambda}_t; t \geqslant t_0\}$, where

$$\hat{\lambda}_t = E[\lambda_t(\mathbf{x}_t) | N_\sigma; t_0 \leqslant \sigma < t] = \begin{cases} E[\lambda_t(\mathbf{x}_t) | N_t], & N_t = 0 \\ E[\lambda_t(\mathbf{x}_t) | N_t; w_1, w_2, \ldots, w_{N_t}], & N_t \geqslant 1 \end{cases}. \tag{6.18}$$

Proof. We need to evaluate the functions $\mu_t(0)$ and $\mu_t(N_t; w_1, w_2, \cdots, w_{N_t})$ of (5.2). Suppose, first, that $N_t \geqslant 1$. Let $\{\rho_n\}$ be a sequence of positive numbers converging to zero as n tends to infinity, and define

$$f_n = \rho_n^{-1} \Pr[N_{t,t+\rho_n} = 1 | N_t; w_1, w_2, \ldots, w_{N_t}, \mathbf{x}_\sigma; \sigma \geqslant t_0].$$

Then $E(f_n) < \infty$. Since $\{N_t; t \geqslant t_0\}$ is conditionally a Poisson process given $\{\mathbf{x}_\sigma; \sigma \geqslant t_0\}$, we have that

$$f_n = \rho_n^{-1} \int_t^{t+\rho_n} \lambda_\sigma(\mathbf{x}_\sigma)\, d\sigma \exp\left[-\int_t^{t+\rho_n} \lambda_\sigma(\mathbf{x}_\sigma)\, d\sigma \right]. \tag{6.19}$$

Consequently, the sequence $\{f_n\}$ converges to $\lambda_t(\mathbf{x}_t)$ with probability one as n tends to infinity. Also, define

$$g_n = \rho_n^{-1} \int_t^{t+\rho_n} \lambda_\sigma(\mathbf{x}_\sigma)\, d\sigma.$$

Then $E(g_n)<\infty$ because $E[\lambda_\sigma(\mathbf{x}_\sigma)]<\infty$ by assuumption. The sequence $\{g_n\}$ converges to $\lambda_t(\mathbf{x}_t)$ with probability one as n tends to infinity, $f_n \leqslant g_n$ for each n, and $E[\lambda_t(\mathbf{x}_t)]=\lim_{n\to\infty} E(g_n)$. Thus, the sequences $\{f_n\}$ and $\{g_n\}$ satisfy the conditions of the generalized bounded convergence theorem, Theorem 1.4.1, and we have that

$$\begin{aligned}E&\left[\lambda_t(\mathbf{x}_t)|N_t;w_1,w_2,\ldots,w_{N_t}\right]\\&=\lim_{n\to\infty} E(f_n|N_t;w_1,w_2,\ldots,w_{N_t})\\&=\lim_{n\to\infty}\rho_n^{-1}\Pr[N_{t,t+\rho_n}=1|N_t;w_1,w_2,\ldots,w_{N_t}].\end{aligned}\tag{6.20}$$

The limit on the right side in (6.20) is by definition $\mu_t(N_t;w_1,w_2,\ldots,w_{N_t})$. Consequently, (6.18) holds for $N_t \geqslant 1$. The proof for $N_t=0$ is analogous when f_n is defined to be

$$f_n=\rho_n^{-1}\Pr[N_{t,t+\rho_n}=1|N_t=0,\mathbf{x}_\sigma;\sigma\geqslant t_0].$$

This completes the proof of Theorem 6.2.1.

Theorem 6.2.1 indicates the basic role of the conditional expectation $\hat{\lambda}_t$ of the intensity $\lambda_t(\mathbf{x}_t)$ of the doubly stochastic Poisson process given the entire past $\{N_\sigma;t_0\leqslant\sigma<t\}$ of the point process up to time t. We will encounter this conditional expectation repeatedly throughout the chapter. It is characterized and studied in depth in Section 6.5. In particular, we show that $\hat{\lambda}_t$ may be interpreted as an estimate of $\lambda_t(\mathbf{x}_t)$ in terms of $\{N_\sigma;t_0\leqslant\sigma<t\}$ that has the smallest mean-square error $E\{[\lambda_t(\mathbf{x}_t)-\hat{\lambda}_t]^2\}$ of any other function of $\{N_\sigma;t_0\leqslant\sigma<t\}$.

It is reasonably straightforward to extend the doubly stochastic Poisson process model to that of a doubly stochastic self-exciting point process by incorporating a possible dependence in $\lambda_t(\mathbf{x}_t)$ on the count N_t and the occurrence times $w_1,w_2,\ldots,w_{N_t}$. In this way, doubly stochastic renewal processes, Markov birth processes, and more general self-exciting processes can be accomodated. In most of our results of the chapter, this extension is accomplished by replacing the estimate $\hat{\lambda}_t$ by the estimate $\hat{\lambda}_t(N_t;w_1,w_2,\ldots,w_{Nt})$ where

$$\hat{\lambda}_t(N_t;w_1,w_2,\ldots,w_{N_t})=E\left\{\lambda_t(\mathbf{x}_t,N_t;w_1,w_2,\ldots,w_{N_t})|N_t;w_1,w_2,\ldots,w_{N_t}\right\}.\tag{6.21}$$

This extension is contained as a special case of the doubly stochastic, marked point processes considered in Chapter 7.

As a consequence of Theorems 6.2.1 and 5.2.1, the counting statistics for a doubly stochastic Poisson process satisfy the following differential-difference equations:

$$\frac{\partial \Pr(N_t=0)}{\partial t} = -\hat{\lambda}_t(0)\Pr(N_t=0), \qquad \Pr(N_{t_0}=0)=1 \tag{6.22}$$

and, for $n \geqslant 1$,

$$\frac{\partial \Pr(N_t=n)}{\partial t} = -\hat{\lambda}_t(n)\Pr(N_t=n)+\hat{\lambda}_t(n-1)\Pr(N_t=n-1),$$

$$\Pr(N_{t_0}=n)=0, \tag{6.23}$$

where $\hat{\lambda}_t(n)$ is the count-conditional intensity

$$\hat{\lambda}_t(n)=E\big(\hat{\lambda}_t|N_t=n\big)=E[\lambda_t(\mathbf{x}_t)|N_t=n]. \tag{6.24}$$

This indicates that the counting statistics for a doubly-stochastic Poisson process are identical to those of a pure Markov birth process with a population-dependent birthrate equal for each n to the estimate $\hat{\lambda}_t(n)$. Equations (6.22) and (6.23) can be solved recursively for the counting probabilities $\Pr(N_t=0), \Pr(N_t=1)$, $\Pr(N_t=2),\ldots$ and so forth. For this purpose, we note from (6.22) that

$$\Pr(N_t=0)=\exp\left(-\int_{t_0}^{t}\hat{\lambda}_\sigma(0)\,d\sigma\right) \tag{6.25}$$

and from (6.23)

$$\Pr(N_t=n)=\int_{t_0}^{t}\hat{\lambda}_\tau(n-1)\Pr[N_\tau=n-1]\exp\left(-\int_{t_0}^{t}\hat{\lambda}_\sigma(n)\,d\sigma\right)d\tau \tag{6.26}$$

for $n=1,2,\ldots$.

The estimate $\hat{\lambda}_t(n)$ is, in general, difficult if not impossible to determine, so (6.22)–(6.26) must be viewed as characterizations rather than formulas for the counting statistics. The following is an example where $\hat{\lambda}_t(n)$ can be evaluated explicitly and the counting statistics determined.

Example 6.2.3. Consider again the randomly scaled inhomogeneous Poisson process of Example 6.2.1. To use (6.25) and (6.26), we need to determine the estimate $\hat{\lambda}_t(N_t) = E(xv_t|N_t)$. By using Bayes' rule, we obtain

$$\hat{\lambda}_t(N_t) = \frac{v_t \int_0^\infty X^{N_t+1} \exp\left(-X\int_{t_0}^t v_\sigma \, d\sigma\right) dP_x(X)}{\int_0^\infty X^{N_t} \exp\left(-X\int_{t_0}^t v_\sigma \, d\sigma\right) dP_x(X)}. \tag{6.27}$$

This determines $\hat{\lambda}$ for any given distribution of x, and the counting statistics can be determined, in principle, from (6.25) and (6.26). If the moment generating function for x exists, (6.27) can be expressed alternatively by using (6.6) as

$$\hat{\lambda}_t(N_t) = \frac{v_t M_x^{(N_t+1)}\left(-\int_{t_0}^t v_\sigma \, d\sigma\right)}{M_x^{(N_t)}\left(-\int_{t_0}^t v_\sigma \, d\sigma\right)}. \tag{6.28}$$

For a specific example, suppose x has the gamma distribution in (6.3), so that $\{N_t; t \geqslant t_0\}$ is an inhomogeneous Polya process. Then an easy calculation shows that $M_x(s) = [\alpha/(\alpha - s)]^{\beta+1}$. It then follows from (6.28) that

$$\hat{\lambda}_t(N_t) = (N_t + \beta + 1)\left(\alpha + \int_{t_0}^t v_\sigma \, d\sigma\right)^{-1}. \tag{6.29}$$

Thus, the count-conditional intensity for an inhomogeneous Polya process is linearly dependent on the count. We leave it as an exercise for the reader to derive the counting statistics in (6.4) for this process from (6.25), (6.26), and (6.29).

Limit Theorems for Counting Statistics

Because of the analytical intractability in evaluating the expectations in (6.1) or (6.24), it is in general impossible to determine an explicit expression for the counting probability $\Pr(N_t = n)$ for t finite. It is of some importance, therefore, that this probability can be more completely characterized asymptotically for large t. These characterizations can be summarized briefly as follows. Let $\Lambda_t = \int_{t_0}^t \lambda_\sigma(\mathbf{x}_\sigma) \, d\sigma$, and define centered and normalized (i.e., zero-mean and unit-variance) versions of N_t and Λ_t

by

$$N_t^{cn} = \frac{N_t - m_N(t)}{\sigma_N(t)} \quad \text{and} \quad \Lambda_t^{cn} = \frac{\Lambda_t - m_\Lambda(t)}{\sigma_\Lambda(t)}, \tag{6.30}$$

where $m_N(t)$ and $m_\Lambda(t)$ are the mean-value functions for N_t and Λ_t, and $\sigma_N^2(t)$ and $\sigma_\Lambda^2(t)$ are the corresponding variances. Suppose N_t^{cn} and Λ_t^{cn} approach random variables η and ξ in distribution, respectively, as t approaches infinity, and let g be a zero-mean, unit-variance Gaussian random variable independent of ξ. Then, depending on the value of c, where $c = \lim_{t\to\infty}\sigma_\Lambda^2(t)/m_\Lambda(t)$, η can be characterized as $\eta = g$ for $c=0$, $\eta = g/\sqrt{1+c} + \xi/\sqrt{1+c^{-1}}$ for $0<c<\infty$, and $\eta=\xi$ for $c=\infty$. These results are established in the following theorems due independently to J. Grandell [16] and R. Serfozo [40]. The proofs we use parallel those of Grandell for the situation when λ is a stationary process.

Theorem 6.2.2 (Limit Theorem for Counting Statistics). Suppose $m_\Lambda(t) \to \infty$ and $\sigma_\Lambda^2(t)/m_\Lambda(t) \to c$ as $t\to\infty$, where c is a constant such that $0<c<\infty$. If Λ_t^{cn} converges in distribution to ξ as $t\to\infty$, then N_t^{cn} converges in distribution to $g(1+c)^{-1/2} + \xi(1+c^{-1})^{-1/2}$ as $t\to\infty$, where g is a zero-mean, unit-variance Gaussian random variable independent of ξ.

The condition that $m_\Lambda(t)$ tends to infinity as t tends to infinity implies that the expected number of points in the infinite interval $[t_0, \infty)$ is infinite. This condition precludes the possibility that $\lambda_t(\mathbf{x}_t)$ be zero for all $t \geqslant T$ for some finite time T.

Proof. Throughout the proof, we use the relations established in Problem 6.2.5 that

$$m_N(t) = m_\Lambda(t), \tag{6.31}$$

$$\sigma_N^2(t) = \sigma_\Lambda^2(t) + m_\Lambda(t). \tag{6.32}$$

We have to prove that

$$\lim_{t\to\infty} E[\exp(juN_t^{cn})] = M_\xi\left[ju(1+c^{-1})^{-1/2}\right]\exp\left[-\tfrac{1}{2}u^2(1+c)^{-1}\right], \tag{6.33}$$

where $M_\xi(ju) = E[\exp(ju\xi)]$ is the characteristic function for ξ. By the method of conditioning, and with some minor manipulating, we deduce that

$$E[\exp(juN_t^{cn})] = \exp\left[-\frac{jum_N(t)}{\sigma_N(t)}\right]E\left\{\exp\left[\Lambda_t(e^{ju/\sigma_N(t)}-1)\right]\right\}$$

$$= \exp\left[-\frac{jum_N(t)}{\sigma_N(t)} + m_\Lambda(t)(e^{ju/\sigma_N(t)}-1)\right]$$

$$\times E\left\{\exp\left[\Lambda_t^{cn}\sigma_\Lambda(t)(e^{ju/\sigma_N(t)}-1)\right]\right\}. \tag{6.34}$$

Using (6.31) and (6.32), we have

$$\lim_{t\to\infty}\exp\left[\frac{-jum_N(t)}{\sigma_N(t)} + m_\Lambda(t)(e^{ju/\sigma_N(t)}-1)\right] = \exp\left[-\tfrac{1}{2}u^2(1+c^{-1})^{-1}\right]. \tag{6.35}$$

Provided the interchange of the limit and expectation is justified, we also have

$$\lim_{t\to\infty} E\left\{\exp\left[\Lambda_t^{cn}\sigma_\Lambda(t)(e^{ju/\sigma_N(t)}-1)\right]\right\}$$

$$= E\left\{\lim_{t\to\infty}\exp\left[\Lambda_t^{cn}\sigma_\Lambda(t)(e^{ju/\sigma_N(t)}-1)\right]\right\}$$

$$= E\left\{\exp\left[ju(1+c^{-1})^{-1/2}\xi\right]\right\}. \tag{6.36}$$

Thus, the limit of (6.34) as $t\to\infty$ yields (6.33) and establishes the theorem. The interchange of the limit and expectation is justified by the bounded convergence theorem, Theorem 1.4.1, if we can demonstrate that $\exp[\Lambda_t^{cn}\sigma_\Lambda(t)(e^{ju/\sigma_N(t)}-1)]$ is bounded in magnitude independently of t. This we do as follows. Since $\Lambda_t \geqslant 0$ and $\cos(u/\sigma_N(t))-1 \leqslant 0$, we have

$$\left|\exp\left[\Lambda_t^{cn}\sigma_\Lambda(t)(e^{ju/\sigma_N(t)}-1)\right]\right| \leqslant \exp\left\{-m_\Lambda(t)\left[\cos\left(\frac{u}{\sigma_N(t)}\right)-1\right]\right\}$$

$$\leqslant \exp\left\{\frac{\frac{1}{2}u^2 m_\Lambda(t)}{\sigma_N^2(t)}\right\} \leqslant \exp\left\{\tfrac{1}{2}u^2\right\} < \infty, \tag{6.37}$$

where we have used the inequalities $1-\cos x \leqslant \frac{1}{2}x^2$ and, from (6.32), $m_\Lambda(t)/\sigma_N^2(t) \leqslant 1$. The theorem is proved.

Theorem 6.2.3 (Limit Theorem for Counting Statistics). Suppose $m_\Lambda(t) \to \infty$ and $\sigma_\Lambda^2(t)/m_\Lambda(t) \to 0$ as $t\to\infty$. Then N_t^{cn} converges in distribution to g as $t\to\infty$, where g is a zero-mean, unit-variance Gaussian random variable.

Proof. The proof is the same as that for Theorem 6.2.2 except that the limits in (6.35) and (6.36) are replaced by

$$\lim_{t\to\infty} \exp\left[\frac{-jum_N(t)}{\sigma_N(t)} + m_\Lambda(t)\left(e^{ju/\sigma_N(t)}-1\right)\right] = \exp\left[-\tfrac{1}{2}u^2\right] \quad (6.38a)$$

and

$$\lim_{t\to\infty} E\left\{\exp\left[\Lambda_t^{cn}\sigma_\Lambda(t)\left(e^{ju/\sigma_N(t)}-1\right)\right]\right\} = 1, \quad (6.38b)$$

respectively.

Theorem 6.2.4 (Limit Theorem for Counting Statistics). Suppose $m_\Lambda(t) \to \infty$ and $\sigma_\Lambda^2(t)/m_\Lambda(t) \to \infty$ as $t\to\infty$. If Λ_t^{cn} converges in distribution to ξ as $t\to\infty$, then N_t^{cn} also converges in distribution to ξ as $t\to\infty$.

Proof. The proof is the same as that for Theorem 6.2.2 except that the limits in (6.35) and (6.36) are replaced by

$$\lim_{t\to\infty} \exp\left[\frac{-jum_N(t)}{\sigma_N(t)} + m_\Lambda(t)\left(e^{ju/\sigma_N(t)}-1\right)\right] = 1 \quad (6.39a)$$

and

$$\lim_{t\to\infty} E\left\{\exp\left[\Lambda_t^{cn}\sigma_\Lambda(t)\left(e^{ju/\sigma_N(t)}-1\right)\right]\right\} = E\left\{\exp\left[ju\xi\right]\right\}, \quad (6.39b)$$

respectively.

Example 6.2.4 (Poisson Process). Suppose $\{N_t; t \geqslant t_0\}$ is a Poisson process with intensity ν_t for $t \geqslant t_0$. Then $m_\Lambda(t) = \Lambda_t = \int_{t_0}^t \nu_\sigma \, d\sigma$ and $\sigma_\Lambda^2(t) = 0$ for all $t \geqslant t_0$. If $\Lambda_\infty = \infty$, then according to Theorem 6.2.3, there holds

$$\lim_{t\to\infty} \Pr\left[\Lambda_t^{-1/2}(N_t - \Lambda_t) \leqslant Y\right] = \Phi(Y), \quad (6.40)$$

where

$$\Phi(Y) = \int_{-\infty}^{Y} (2\pi)^{-1/2} \exp\left(-\tfrac{1}{2}u^2\right) du.$$

Table 6.1 contains values of $\Pr[\Lambda_t^{-1/2}(N_t-\Lambda_t)\leqslant Y]$ and of $\Phi(Y)$ for a homogeneous Poisson process with $\nu_t=1$ and for selected values of Y and t. It is seen for $Y=0$ and $Y=1$ that $\Pr[\Lambda_t^{-1/2}(N_t-\Lambda_t)\leqslant Y]-\Phi(Y)$ tends to zero as $t\to\infty$. The implication of Theorem 6.2.3 is that this trend to zero occurs for any choice of Y.

Table 6.1

	$\Pr[N_t\leqslant t^{1/2}Y+t]$					$\Phi(Y)$
Y	$t=1$	$t=4$	$t=9$	$t=16$	$t=25$	$t=\infty$
0	0.736	0.629	0.587	0.566	0.553	0.5
1	0.920	0.890	0.876	0.868	0.863	0.841

Example 6.2.5. Suppose $\{N_t; t\geqslant t_0\}$ is a doubly stochastic Poisson process with intensity process $\{\lambda_t(\mathbf{x}_t); t\geqslant t_0\}$. Assume for each $t\geqslant t_0$ that $\Lambda_t=\int_{t_0}^t\lambda_\sigma(\mathbf{x}_\sigma)\,d\sigma$ is a gamma-distributed random variable.

$$p_{\Lambda_t}(X)=\frac{\alpha^{\beta t+1}}{\Gamma(\beta t+1)}X^{\beta t}\exp(-\alpha X),\qquad X\geqslant 0. \tag{6.41}$$

Then $m_\Lambda(t)=\alpha^{-1}(\beta t+1)$ and $\sigma_\Lambda^2(t)=\alpha^{-2}(\beta t+1)$. Hence, the constant c in Theorem 6.2.2 equals α^{-1}. It is straightforward to verify that Λ_t^{cn} converges in distribution to zero as t tends to infinity. If $0<\alpha<\infty$, then according to Theorem 6.2.2, N_t^{cn} converges in distribution to a normal random variable with zero mean and variance $\alpha(1+\alpha)^{-1}$.

By using the method of conditioning, we find that $E(N_t)=\alpha^{-1}(\beta t+1)$, $\sigma_N^2(t)=\alpha^{-2}(1+\alpha)(\beta t+1)$, and

$$\Pr[N_t=n]=\frac{\Gamma(\beta t+n+1)}{\Gamma(n+1)\Gamma(\beta t+1)(1+\alpha)^n}\left[\frac{\alpha}{1+\alpha}\right]^{\beta t+1} \tag{6.42}$$

for $n=0, 1, 2,\ldots$. Thus,

$$\lim_{t\to\infty}\Pr\left[N_t\leqslant\alpha^{-1}(1+\alpha)^{1/2}(\beta t+1)^{1/2}Y+\alpha^{-1}(\beta t+1)\right]$$

$$=\Phi\left[(1+\alpha)^{1/2}\alpha^{-1/2}Y\right], \tag{6.43}$$

where $\Phi(Y)$ denotes the standard Gaussian distribution function as defined in Example 6.2.4. Table 6.2 contains values of the probability on the left side of (6.43) and its limit on the right side of (6.43) for $\alpha=\beta=1$, $Y=0$, and selected values of t. The approach of the probability to its limit as t increases is evident. Theorem 6.2.2 implies that this trend transpires regardless of the values of α, β, and Y.

Table 6.2

	$\Pr\{N_t \leqslant [2(t+1)]^{1/2}Y+(t+1)\}$					$\Phi(\sqrt{2}\,Y)$
Y	$t=1$	$t=2$	$t=4$	$t=8$	$t=16$	$t=\infty$
0	0.688	0.656	0.623	0.593	0.568	0.5

Summary. Counting statistics for a doubly stochastic Poisson process may be determined by the methods of conditioning and of self-exciting point processes. The method of conditioning uses the property that N_T is conditionally Poisson distributed with parameter $\int_{t_0}^{T}\lambda_\sigma(\mathbf{x}_\sigma)\,d\sigma$ given $\{\mathbf{x}_\sigma;\sigma \geqslant t_0\}$. The unconditional statistics for N_T are then found by averaging the Poisson distribution with respect to the statistics for $\{\mathbf{x}_\sigma;\sigma\geqslant t_0\}$. The last step is often so complicated analytically that explicit expressions for $\Pr[N_T=n]$ cannot be obtained. The method of self-exciting point processes uses the property that a doubly stochastic Poisson process is a particular self-exciting point process. Recursion equations developed in Chapter 5 for $\Pr[N_T=n]$ may then be applied. However, this procedure requires an expression for the count-conditional intensity $\hat{\lambda}_t(n)$, which is generally difficult to obtain. Theorems 6.2.2–6.2.4 contain conditions under which the distribution for N_T can be approximated asymptotically for large T.

6.3 TIME STATISTICS FOR DOUBLY STOCHASTIC POISSON PROCESSES

Various expressions associated with the statistics of the occurrence times of a doubly stochastic Poisson process are derived in this section. The quantities of interest are the *survival probability*, the *joint occurrence density*, and the *sample-function density*. Of these, the sample-function density will be the most important in what follows. Its use in decision and estimation problems for doubly-stochastic Poisson processes will be identical to that in Chapters 2 and 3 for Poisson and compound Poisson processes and to that in Chapter 5 for self-exciting point processes. Moreover, we will find in Section 6.5 that the sample-function density plays an important role in the filtering problem for observed doubly stochastic Poisson processes. As with the counting statistics of Section 6.2, two methods can be used for determining the time statistics, the *method of conditioning* and the *method of self-exciting processes*.

Time Statistics by the Method of Conditioning

The joint-occurrence density and the sample-function density for a doubly stochastic Poisson process $\{N_t; t \geqslant t_0\}$ can be determined in principle by conditioning on the information process $\{\mathbf{x}_t; t \geqslant t_0\}$, using the fact that thus conditioned $\{N_t; t \geqslant t_0\}$ is a Poisson counting process, and then removing the conditioning by evaluating an expectation with respect to the statistics of the information process. The last step is where the method of conditioning often fails because of analytic intractability. Even so, there are important models for which the expectation can be accomplished, and two examples are given below.

Let $p_{\mathbf{w}}^{(n)}(\mathbf{W})$ be the joint occurrence density for the first n occurrence times $w_1, w_2, \ldots, w_n$, and let $p[\{N_\sigma; t_0 \leqslant \sigma < T\}]$ be the sample-function density for the doubly stochastic Poisson process $\{N_t; t \geqslant t_0\}$ with intensity process $\{\lambda_t(\mathbf{x}_t); t \geqslant t_0\}$. Then using (2.39), (2.51), and the first two steps of the method of conditioning, we have for the joint-occurrence density that

$$p_{\mathbf{w}}^{(n)}(\mathbf{W}) = E\left\{\prod_{i=1}^{n} \lambda_{W_i}(\mathbf{x}_{W_i}) \exp\left[-\int_{t_0}^{W_n} \lambda_\sigma(\mathbf{x}_\sigma)\, d\sigma\right]\right\} \tag{6.44}$$

and for the sample-function density that

$$p[\{N_\sigma; t_0 \leqslant \sigma < T\}] = E\left\{\exp\left[-\int_{t_0}^{T} \lambda_\sigma(\mathbf{x}_\sigma)\, d\sigma + \int_{t_0}^{T} \ln \lambda_\sigma(\mathbf{x}_\sigma)\, dN_\sigma\right]\right\}. \tag{6.45}$$

The expectation in (6.45) is with respect to the statistics of the information process for a fixed path $\{N_\sigma; t_0 \leqslant \sigma < T\}$ of the counting process.

Example 6.3.1 (Randomly Scaled Inhomogeneous Poisson Process). As in Examples 6.2.1 and 6.2.3, let $\{N_t; t \geqslant t_0\}$ be an inhomogeneous Poisson counting process having an intensity $\{\nu_t; t \geqslant t_0\}$ that is randomly scaled by a nonnegative random variable x. The result is a doubly stochastic Poisson process with intensity process $\{x\nu_t; t \geqslant t_0\}$. From (6.45), the sample-function density is

$$p[\{N_\sigma; t_0 \leqslant \sigma < T\}] = E\left\{x^{N_T} \exp\left(-x\int_{t_0}^{T} \nu_\sigma \, d\sigma\right)\right\} \exp\left(\int_{t_0}^{T} \ln \nu_\sigma \, dN_\sigma\right). \tag{6.46}$$

The expectation can be evaluated in principle given the distribution for x. If the moment generating function $M_x(s) = E[\exp(sx)]$ exists, (6.46) can be written in terms of it as

$$p[\{N_\sigma; t_0 \leqslant \sigma < T\}] = M_x^{(N_T)}\left(-\int_{t_0}^{T} \nu_\sigma \, d\sigma\right) \exp\left(\int_{t_0}^{T} \ln \nu_\sigma \, dN_\sigma\right), \tag{6.47}$$

where $M_x^{(n)}(s) = d^n M_x(s)/ds^n$. When x has the gamma distribution in (6.3) so that $\{N_t; t \geqslant t_0\}$ is the inhomogeneous Polya process, we have from (6.46) that

$$p[\{N_\sigma; t_0 \leqslant \sigma < T\}] = \frac{\Gamma(N_T + \beta + 1)}{\Gamma(\beta + 1)} \left[\frac{\alpha}{\alpha + \int_{t_0}^{T} \nu_\sigma \, d\sigma}\right]^{\beta+1}$$

$$\times \left[\frac{1}{\alpha + \int_{t_0}^{T} \nu_\sigma \, d\sigma}\right]^{N_T} \exp\left(\int_{t_0}^{T} \ln \nu_\sigma \, dN_\sigma\right). \tag{6.48}$$

The particular selection of $\beta = 0$ corresponds to an exponential scaling of ν_t; that is, $p_x(X) = \alpha \exp(-\alpha X)$ for $X \geqslant 0$. From (6.48), the sample-function density for exponential scaling is

$$p[\{N_\sigma; t_0 \leqslant \sigma < T\}]$$

$$= \Gamma(N_T + 1)\alpha\left[\alpha + \int_{t_0}^{T} \nu_\sigma \, d\sigma\right]^{-(N_T+1)} \exp\left(\int_{t_0}^{T} \ln \nu_\sigma \, dN_\sigma\right). \tag{6.49}$$

By setting the derivative of (6.49) with respect to α equal to zero, it is seen that the maximum likelihood estimate $\hat{\alpha}_{\text{ML}}(\{N_\sigma; t_0 \leqslant \sigma < T\})$ of α, the parameter of the exponential distribution of x, given an observed counting path is

$$\hat{\alpha}_{\text{ML}}(\{N_\sigma; t_0 \leqslant \sigma < T\}) = N_T^{-1} \int_{t_0}^{T} \nu_\sigma \, d\sigma. \tag{6.50}$$

This estimate depends only on the total number of points on $[t_0, T)$ and not on the times of their occurrence. Furthermore, if $\nu_t = 1$, the maximum likelihood estimate of α is simply the observed average count rate, $\hat{\alpha}_{ML} = (T - t_0)/N_T$.

Example 6.3.2 (Sample-Function Density for Gaussian Light). In Example 6.2.2, we examined the photocounting statistics for the response of an ideal photodetector to a scalar optical field modeled as a narrowband Gaussian process. This model is widely used in the theories of optical coherence, optical communications, and optical radar. It is of some importance, therefore, that the sample-function density can be determined thus characterizing the time statistics of the photoelectron process at the output of the detector. In this example, we use the method of conditioning to evaluate the sample-function density for the Gaussian-light model of Example 6.2.2. The procedure is straightforward conceptually, but there are some rather tedious manipulations involved. Let the photoelectron counting process $\{N_t; t \geqslant t_0\}$ be a doubly stochastic Poisson counting process with intensity process $\{\lambda_t(X_t) = \alpha|X_t|^2;\ t \geqslant t_0\}$, where $\{X_t; t \geqslant t_0\}$ is the complex envelope of a narrowband Gaussian process $\{x_t; t \geqslant t_0\}$ representing incoherent light. Here, we assume that X_t has zero mean. Let us recall from Example 6.2.2 that for $t_0 \leqslant t < T$, X_t can be written as

$$X_t = \sum_{i=1}^{\infty} X_i \varphi_i(t), \qquad t_0 \leqslant t < T \tag{6.51}$$

where $\{\varphi_i(t)\}$ are the eigenfunctions of the covariance function $K(t,u) = E[X_t X_u^*]$, and where the coefficients $\{X_i\}$ form a sequence of independent, complex-valued, Gaussian random variables such that $\mathrm{Re}(X_i)$ and $\mathrm{Im}(X_i)$ are independent, zero-mean Gaussian random variables with equal variance $\lambda_i/2$. The eigenvalues $\{\lambda_i\}$ and eigenfunctions $\{\varphi_i(t)\}$ satisfy the integral equation (6.11). Define $D_{\mathfrak{F}}(\alpha) = \prod_{i=1}^{\infty}(1 + \alpha\lambda_i)$; this function of the eigenvalues is called the *Fredholm determinant* of $K(t,u)$. Finally, define $h(t,u)$ in terms of the eigenfunctions and eigenvalues of $K(t,u)$ by

$$h(t,u) = \sum_{i=1}^{\infty} \frac{\lambda_i}{1 + \alpha\lambda_i} \varphi_i(t)\varphi_i^*(u), \qquad (t,u) \in [t_0, T) \times [t_0, T). \tag{6.52}$$

An equivalent specification for $h(t,u)$ is that it should be the solution to the integral equation

$$h(t,u) + \alpha \int_{t_0}^{T} K(t,v)h(v,u)\,dv = K(t,u), \qquad (t,u) \in [t_0, T) \times [t_0, T). \tag{6.53}$$

This integral equation has been studied extensively in the context of estimation theory for Gaussian processes; see T. Kailath [24], A. Baggeroer [3. Chapter 6], and H. Van Trees [47, Chapter 6].

By using the method of conditioning, O. Macchi [28, 29] has established that the sample-function density associated with a counting path $\{N_\sigma; t_0 \leqslant \sigma < T\}$ with $N_T = n$ and occurrence times $W_1, W_2, \ldots, W_n$ is given by

$$p[\{N_\sigma; t_0 \leqslant \sigma < T\}] = \begin{cases} D_{\mathcal{F}}^{-1}(\alpha), & N_T = n = 0 \\ \alpha^n D_{\mathcal{F}}^{-1}(\alpha) \sum_{\eta \in \mathcal{P}_n} h(W_1, W_{\eta_1}) h(W_2, W_{\eta_2}) \cdots h(W_n, W_{\eta_n}), & N_T = n \geqslant 1 \end{cases} \tag{6.54}$$

where $\mathcal{P}_n$ is the set of all the $n!$ permutations of the integers $1, 2, \ldots, n$. Simply to illustrate the notation, suppose $n = 2$; then $\mathcal{P}_n = \{(1,2), (2,1)\}$ and (6.54) becomes

$$p[\{N_\sigma; t_0 \leqslant \sigma < T\}] = \alpha^2 D_{\mathcal{F}}^{-1}(\alpha)[h(W_1, W_1) h(W_2, W_2) + h(W_1, W_2) h(W_2, W_1)]. \tag{6.55}$$

Macchi's proof of (6.54) by the method of conditioning proceeds as follows. By incorporating (6.12) into (6.45), we need to evaluate the expectation in

$$p[\{N_\sigma; t_0 \leqslant \sigma < T\}] = E\left[\exp\left(-\alpha \sum_{i=1}^{\infty} |X_i|^2 + \int_{t_0}^{T} \ln \alpha |X_\sigma|^2 dN_\sigma\right)\right]. \tag{6.56}$$

If $N_T = 0$, the counting integral is zero, and using the independence of the X_i, we have

$$p[\{N_\sigma; t_0 \leqslant \sigma < T\}] = \prod_{i=1}^{\infty} E[\exp(-\alpha |X_i|^2)].$$

But $u_i \triangleq |X_i|^2 = \mathrm{Re}^2(X_i) + \mathrm{Im}^2(X_i)$ has the exponential distribution $p_{u_i}(U_i) = \lambda_i^{-1} \exp(-\lambda_i^{-1} U_i)$; consequently, $E[\exp(-\alpha |X_i|^2)] = (1 + \alpha\lambda_i)^{-1}$. This proves (6.54) for $N_T = n = 0$. Now take $N_T = n \geqslant 1$. Then (6.56) becomes

$$p[\{N_\sigma; t_0 \leqslant \sigma < T\}]$$

$$= \alpha^n E\left\{|X(W_1)|^2 |X(W_2)|^2 \cdots |X(W_n)|^2 \exp\left(-\alpha \sum_{i=1}^{\infty} |X_i|^2\right)\right\}$$

$$= \alpha^n \sum_{\substack{j_1, j_2, \ldots, j_n = 1 \\ k_1, k_2, \ldots, k_n = 1}}^{\infty} a(\mathbf{j}, \mathbf{k}) \prod_{l=1}^{n} \varphi_{j_l}(W_l) \varphi_{k_l}^*(W_l), \tag{6.57}$$

where we have used (6.51) to get the second equality and where

$$a(\mathbf{j}, \mathbf{k}) \triangleq E\left[X_{j_1} X_{j_2} \cdots X_{j_n} X_{k_1}^* X_{k_2}^* \cdots X_{k_n}^* \exp\left(-\alpha \sum_{i=1}^{\infty} |X_i|^2\right)\right]. \tag{6.58}$$

Let p_m and q_m be the number of j_i's and k_i's equal to m, respectively; there holds

$$\sum_{m=1}^{\infty} p_m = \sum_{m=1}^{\infty} q_m = n.$$

Because of the independence of the coefficients $\{X_i\}$, we have

$$a(\mathbf{j}, \mathbf{k}) = \prod_{m=1}^{\infty} E\left[X_m^{p_m} X_m^{*q_m} \exp(-\alpha |X_m|^2)\right]. \tag{6.59}$$

The moment factorization property of Gaussian random variables and the fact that $E(X_i^2) = 0$ imply that $a(\mathbf{j}, \mathbf{k}) = 0$ if p_m does not equal q_m. This and the exponential distribution for $|X_i|^2$ yield

$$a(\mathbf{j}, \mathbf{k}) = \begin{cases} 0, & p_m \neq q_m \\ \displaystyle\prod_{m=1}^{\infty} \frac{m!}{1 + \alpha\lambda_m} \left(\frac{m}{1 + \alpha\lambda_m}\right)^{p_m}, & p_m = q_m. \end{cases} \tag{6.60}$$

The condition $p_m = q_m$ requires that the set of integers $k_1, k_2, \ldots, k_n$ be a permutation of the set of integers $j_1, j_2, \ldots, j_n$; denote this restriction on k_1,

$k_2,\ldots,k_n$ by $\mathbf{k}\in\mathcal{P}(\mathbf{j})$. Then (6.57) becomes

$$p[\{N_\sigma;t_0\leqslant\sigma<T\}]$$

$$=\alpha^n D_{\mathcal{F}}^{-1}(\alpha)\sum_{j_1,j_2,\ldots,j_n=1}^{n}\left[\prod_{m=1}^{\infty}\left(\frac{\lambda_m}{1+\alpha\lambda_m}\right)^{p_m}\prod_{l=1}^{\infty}\varphi_{j_l}(W_l)\right]f(j_1,j_2,\ldots,j_n),\tag{6.61}$$

where

$$f(j_1,j_2,\ldots,j_n)$$

$$\triangleq\prod_{m=1}^{\infty}(p_m!)\sum_{\substack{k_1,k_2,\ldots,k_n=1\\ \mathbf{k}\in\mathcal{P}(\mathbf{j})}}^{n}[\varphi_{k_1}^*(W_1)\varphi_{k_2}^*(W_2)\cdots\varphi_{k_n}^*(W_n)]$$

$$=\sum_{\eta\in\mathcal{P}_n}\varphi_{j_1}^*(W_{\eta_1})\varphi_{j_2}^*(W_{\eta_2})\cdots\varphi_{j_n}^*(W_{\eta_n}),\tag{6.62}$$

where the last equality can be verified by direct comparison with the defined expression for $f(j_1,j_2,\ldots,j_n)$. Thus,

$$p[\{N_\sigma;t_0\leqslant\sigma<T\}]$$

$$=\alpha^n D_{\mathcal{F}}^{-1}(\alpha)\sum_{\eta\in\mathcal{P}_n}\sum_{j_1,j_2,\ldots,j_n=1}^{n}\left[\prod_{m=1}^{\infty}\left(\frac{\lambda_m}{1+\alpha\lambda_m}\right)^{p_m}\prod_{l=1}^{n}\varphi_{j_l}(W_l)\varphi_{j_l}^*(W_{\eta_l})\right]$$

$$=\alpha^n D_{\mathcal{F}}^{-1}(\alpha)\sum_{\eta\in\mathcal{P}_n}h(W_1,W_{\eta_1})h(W_2,W_{\eta_2})\cdots h(W_n,W_{\eta_n}).\tag{6.63}$$

The last equality is obtained using (6.52). This establishes the asserted form (6.54) for the sample-function density of Gaussian light.

The sample-function density for zero-mean, Gaussian light can be evaluated using (6.54) whenever the Fredholm determinant $D_{\mathcal{F}}(\alpha)$ and the function $h(t,u)$ can be evaluated. As both $D_{\mathcal{F}}(\alpha)$ and $h(t,u)$ are determined by the eigenvalues $\{\lambda_i\}$ and eigenfunctions $\{\varphi_i(t)\}$ of the covariance function $K(t,u)$ of the Gaussian process, the sample-function density can be evaluated whenever (6.11) can be solved; $h(t,u)$ may also be determined by solving (6.53). Indicated below are four situations where $D_{\mathfrak{F}}(\alpha)$ and $h(t,u)$ can be evaluated or closely approximated.

a. Stationary Process, Long Observation Interval. It is often reasonable to assume Gaussian light to be a stationary process. Then $K(t,u)=K(t-u,0)$. If, in addition, the observation interval $[t_0,T)$ is long in comparison to the time over which the Gaussian process has significant correlation, then $D_{\mathfrak{F}}(\alpha)$ and $h(t,u)$ can be specified in terms of the power-density spectrum $S(f)$ defined by

$$S(f)=\int_{-\infty}^{\infty} K(\tau,0)\exp(-j2\pi f\tau)\,d\tau.$$

As discussed by H. Van Trees [48, pp. 101 and 150], there holds

$$\ln D_{\mathfrak{F}}(\alpha)\approx(T-t_0)\int_{-\infty}^{\infty}\ln[1+\alpha S(f)]\,df$$

and

$$h(t,u)\approx\int_{-\infty}^{\infty}H(f)\exp(j2\pi f(t-u))\,df,$$

where

$$H(f)=S(f)[1+\alpha S(f)]^{-1}.$$

b. Separable Covariance. The covariance function $K(t,u)$ is termed separable when it can be expressed in the form

$$K(t,u)=\sum_{i=1}^{M}\lambda_i\varphi_i(t)\varphi_i^*(u),$$

where $\varphi_1(t)$, $\varphi_2(t),\ldots,$and $\varphi_M(t)$ form an orthonormal set and M is a finite number. It follows easily that

$$D_{\mathfrak{F}}(\alpha)=\prod_{i=1}^{M}(1+\alpha\lambda_i)$$

and that

$$h(t,u)=\sum_{i=1}^{M}\frac{\lambda_i}{1+\alpha\lambda_i}\varphi_i(t)\varphi_i^*(u).$$

Separable covariance functions arise in optical communication and radar systems that employ time or frequency diversity. H. Van Trees [48, pp. 122–130] may be consulted for several specific examples.

c. Finite-Dimensional Covariance. The covariance function $K(t,u)$ is termed finite dimensional when it can be expressed in the form

$$K(t,u)=\begin{cases}\mathbf{a}'(t)\mathbf{b}^*(u), & t\geqslant u\\ \mathbf{b}'(t)\mathbf{a}^*(u), & t\leqslant u\end{cases}$$

for some finite-dimensional vector functions $\mathbf{a}(t)$ and $\mathbf{b}(t)$. A Gaussian process with a finite-dimensional covariance can be realized with a linear dynamical system (see Example 4.2.2) excited by a broadband Gaussian process; this realization problem is discussed by B. Anderson, J. Moore, and S. Loo [1]. A. Baggeroer [3, Chapter 6] and H. Van Trees [48, p. 371] discuss the determination of $h(t,u)$ and $D_{\mathfrak{F}}(\alpha)$ for this situation.

d. Low-Energy Coherence. The expected energy in the ith mode of X_t is from (6.51)

$$E\left(\int_{t_0}^{T}|X_i\varphi_i(t)|^2\,dt\right)=\lambda_i. \tag{6.64}$$

Thus, the condition $\alpha\lambda_i\ll 1$ may be interpreted as requiring the expected energy in the ith mode to be much smaller than α^{-1}. Following H. Van Trees [48, p. 131], we use the term low-energy coherence for the situation when $\alpha\lambda_i\ll 1$ for all i. Note that this does not require that the total expected energy $\Sigma_{i=1}^{\infty}\lambda_i$ be small. Under the low-energy coherence condition, we have to a close approximation that $D_{\mathfrak{F}}(\alpha)\approx 1$ and $h(t,u)\approx K(t,u)$.

From its definition in (2.49), the sample-function density may be written as

$$p[\{N_\sigma;t_0\leqslant\sigma<T\}]=\begin{cases}\Pr(N_T=0), & N_T=n=0\\ p_{\mathbf{w}}(\mathbf{W}|N_T=n)\Pr(N_T=n), & N_T=n\geqslant 1.\end{cases} \tag{6.65}$$

Thus, the counting statistics, $\Pr(N_T=n)$ for $n=1,2,\ldots,$ can be determined from the integration

$$\Pr(N_T=n)=\underset{t_0\leqslant W_1<W_2\cdots<W_n<T}{\int\int\cdots\int} p[\{N_\sigma;t_0\leqslant\sigma<T\}]\,dW_1\,dW_2\cdots dW_n. \tag{6.66}$$

For example, the photo-count statistics for zero-mean Gaussian light are seen from (6.66) and (6.54) to be

$$\Pr(N_T=n)$$

$$=\alpha^n D_{\mathfrak{F}}^{-1}(\alpha)\sum_{\eta\in\mathcal{P}_n}\int\int_{t_0\leq W_1<\cdots<W_n<T}\cdots\int\prod_{i=1}^{n}h(W_i,W_{\eta_i})\,dW_1\,dW_2\cdots dW_n \tag{6.67}$$

for $n=1,2,\ldots$.

Time Statistics by the Method of Self-Exciting Point Processes

In Theorem 6.2.1, we proved that a doubly stochastic Poisson process $\{N_t;t\geq t_0\}$ with intensity process $\{\lambda_t(\mathbf{x}_t);t\geq t_0\}$ is a self-exciting process with an intensity process $\{\hat{\lambda}_t;t\geq t_0\}$ defined in (6.18). Consequently, the time statistics for a doubly-stochastic Poisson process can be expressed in terms of those developed in Section 5.2 of Chapter 5 for a self-exciting process.

By combining (5.15) and (6.18), we have that the *conditional survival probability* for the $(n+1)$st point of a doubly stochastic Poisson process is given for $t\geq W_n$ and $n\geq 1$ by

$$\mathcal{P}_{w_{n+1}|w_1,w_2,\ldots,w_n}(t|W_1,W_2,\ldots,W_n)=\exp\left(-\int_{W_n}^{t}\hat{\lambda}_\sigma\,d\sigma\right), \tag{6.68}$$

where

$$\hat{\lambda}_\sigma=E[\lambda_\sigma(\mathbf{x}_\sigma)|N_\sigma=n;W_1,W_2,\ldots,W_n].$$

By combining (5.22), (5.23), and (6.18), we have that the *joint occurrence density* for the first n occurrence times of a doubly stochastic Poisson process is given by

$$p_{\mathbf{w}}^{(n)}(\mathbf{W})=\exp\left(-\int_{t_0}^{W_n}\hat{\lambda}_\sigma\,d\sigma+\int_{t_0}^{W_n}\ln\hat{\lambda}_\sigma\,dN_\sigma\right), \tag{6.69}$$

where

$$\hat{\lambda}_\sigma=\begin{cases}E[\lambda_\sigma(\mathbf{x}_\sigma)|N_\sigma=0], & \text{for } N_\sigma=0\\ E[\lambda_\sigma(\mathbf{x}_\sigma)|N_\sigma=n;W_1,W_2,\ldots,W_n], & \text{for } N_\sigma=n\geq 1.\end{cases} \tag{6.70}$$

We have used the counting integrals of Chapter 4 in writing (6.69). Finally, by combining (5.29) and (6.18), the *sample-function density* for a doubly stochastic Poisson process is given by

$$p[\{N_\sigma; t_0 \leqslant \sigma < T\}] = \exp\left(-\int_{t_0}^{T} \hat{\lambda}_\sigma \, d\sigma + \int_{t_0}^{T} \ln \hat{\lambda}_\sigma \, dN_\sigma\right), \tag{6.71}$$

where $\hat{\lambda}_\sigma$ is the same as in (6.70).

The estimate $\hat{\lambda}_t$ appearing in (6.68), (6.69), and (6.71) is in general difficult, if not impossible, to determine. Consequently, these expressions for the time statistics of a doubly stochastic Poisson process must be viewed as representations rather than explicit formulas. In Section 6.4, we will investigate in some detail the problem of determining $\hat{\lambda}_t$.

Summary. Time statistics for a doubly stochastic Poisson process may be determined by the methods of conditioning and of self-exciting point processes. The method of conditioning uses the property that $\{N_t; t \geqslant t_0\}$ is conditionally a Poisson process given $\{\mathbf{x}_t; t \geqslant t_0\}$, and the method of self-exciting point processes uses the property that a doubly stochastic Poisson process is a particular form of self-exciting point process. The most important quantity developed in the section is the sample-function density for a doubly stochastic Poisson process. This quantity has the same form as the sample-function density for a Poisson process with an intensity $\hat{\lambda}_t = E[\lambda_t(\mathbf{x}_t) | N_\sigma; t_0 \leqslant \sigma < t]$.

6.4 HYPOTHESIS TESTING FOR DOUBLY STOCHASTIC POISSON PROCESSES

The problem of hypothesis testing frequently arises in applications involving the doubly stochastic Poisson process model. For example, this is the central problem for the design of efficient receiver processing to be performed subsequent to photoelectron conversion in optical communication systems used to relay digital data and in optical radar systems whenever random effects influence the optical signals used. Our most important result in this section is a fundamental *separation theorem of detection* indicating that the hypothesis testing problem for an observed doubly stochastic Poisson process can be solved by first solving two related problems separately and independently and then combining their solutions in a natural fashion. The importance of this separation theorem of detection is the insight it provides into the structure of the receiver that achieves minimum error probability or minimum Bayes' risk.

We consider only the binary hypothesis testing problem, which can be described as follows. Let $\{N_t; t \geqslant t_0\}$ be a doubly stochastic Poisson process with an intensity process that is either $\{\lambda_t^{(0)}(\mathbf{x}_t);\ t \geqslant t_0\}$ or $\{\lambda_t^{(1)}(\mathbf{x}_t);\ t \geqslant t_0\}$. The problem is to decide which given an observed counting record $\{N_\sigma; t_0 \leqslant \sigma < T\}$ on the interval $[t_0, T)$. From Section 2.5 of Chapter 2, we know that the rule for making this decision can be stated in terms of the likelihood ratio Λ defined by

$$\Lambda = \frac{p^{(1)}[\{N_\sigma; t_0 \leqslant \sigma < T\}]}{p^{(0)}[\{N_\sigma; t_0 \leqslant \sigma < T\}]}, \tag{6.72}$$

where $p^{(0)}[\{N_\sigma; t_0 \leqslant \sigma < T\}]$ and $p^{(1)}[\{N_\sigma; t_0 \leqslant \sigma < T\}]$ denote the sample-function densities associated with the intensities $\lambda_t^{(0)}$ and $\lambda_t^{(1)}$. The rule is: Accept $\lambda_t^{(1)}$ if $\Lambda \geqslant \eta$ and accept $\lambda_t^{(0)}$ if $\Lambda < \eta$, where Λ is evaluated for the particular observed counting record and where η is a nonnegative constant that does not depend on the observed data but only on the prior probabilities of $\lambda_t^{(1)}$ and $\lambda_t^{(0)}$ and the decision performance criterion.

The importance of the sample-function density in problems of hypothesis testing for doubly stochastic Poisson processes is evident from these remarks. The likelihood ratio Λ and, therefore, the decision rule can be specified explicitly whenever the sample-function densities associated with both $\lambda_t^{(0)}$ and $\lambda_t^{(1)}$ can be specified explicitly. Unfortunately, this is only rarely the case; two notable exceptions are those of Examples 6.3.1 and 6.3.2 for a randomly scaled inhomogeneous Poisson process and for Gaussian light. Fundamental insight into the structure of the likelihood ratio in more general situations is provided by the representation of the sample-function density in (6.71). By incorporating this representation into Λ, we see that the likelihood-ratio test becomes

$$\ln\Lambda = -\int_{t_0}^{T}\left(\hat{\lambda}_\sigma^{(1)} - \hat{\lambda}_\sigma^{(0)}\right)d\sigma + \int_{t_0}^{T}\ln\left(\frac{\hat{\lambda}_\sigma^{(1)}}{\hat{\lambda}_\sigma^{(0)}}\right)dN_\sigma \underset{\lambda^{(0)}}{\overset{\lambda^{(1)}}{\gtrless}} \ln\eta, \tag{6.73}$$

where $\hat{\lambda}_\sigma^{(i)}$ for $i=0$ and $i=1$ is that function of the observed counting record defined by

$$\hat{\lambda}_\sigma^{(i)} = \begin{cases} E[\lambda_\sigma^{(i)}(\mathbf{x}_\sigma)|H_i, N_\sigma], & N_\sigma = 0 \\ E[\lambda_\sigma^{(i)}(\mathbf{x}_\sigma)|H_i, N_\sigma = k; W_1, W_2, \ldots, W_k], & N_\sigma = k \geqslant 1. \end{cases} \tag{6.74}$$

Here, H_i denotes the hypothesis that $\lambda^{(i)}$ is the intensity governing the evolution of N. The interpretation of $\hat{\lambda}^{(i)}$ is that it is the minimum mean square-error estimate of $\lambda_\sigma^{(i)}$ in terms of the observed counting record *if* $\lambda^{(i)}$ is, in fact, the underlying intensity; otherwise, $\hat{\lambda}^{(i)}$ is a well-defined function of the counting record but has no interpretation as an estimate. In other words, $\hat{\lambda}_\sigma^{(i)}$ is designed according to (6.74) as if H_i is the true hypothesis. But, of course, H_i may or may not be true; $\hat{\lambda}_\sigma^{(i)}$ has an interpretation as an estimate if it is, but not otherwise. For this reason we term $\hat{\lambda}_\sigma^{(1)}$ and $\hat{\lambda}_\sigma^{(0)}$ in (6.73) *pseudo estimates* of $\lambda_\sigma^{(1)}$ and $\lambda_\sigma^{(0)}$, respectively.

Further interpretation of (6.73) is possible. We note from it that the form of the hypothesis test for doubly stochastic Poisson processes is remarkably similar to that in (2.112) for inhomogeneous Poisson processes, the only difference being that pseudo estimates of the intensities are to be incorporated in the test in exactly the same way as if the intensities themselves were deterministic and known. This indicates a fundamental structural decomposition in the detector or data processing used to realize the hypthesis test in (6.73). Namely, the detector consists of two operations on the observed counting record, operations that can be realized independently and then combined. The first is the operation to generate the pseudo estimates and the second operation is to incorporate these into the same detector used to implement the hypothesis test for inhomogeneous Poisson processes. We term this decomposition implied by (6.73) the *separation theorem of detection* for doubly-stochastic Poisson processes. The general structure of the processing suggested by the separation theorem of detection is shown schematically in Figure 6.1. The only modification of the processing indicated in Figure 2.8 for the Poisson process is that the function generator is replaced by an estimator whose output depends upon the observed point process.

Two additional observations associated with the separation theorem of detection are of sufficient significance to be mentioned here. The fact that the pseudo estimates depend only causally on the observed counting record means that the detector can be updated continuously in time as the doubly stochastic Poisson process is observed. This is important for situations where the hypothesis test must be realized in real time and for implementing sequential tests. The second observation is that for doubly stochastic Poisson processes, the problem of causally estimating the intensity from observed counting records is the essential problem to be solved for the testing of hypotheses. This estimation problem is investigated in the next section. For some limited number of situations, the estimates required for (6.73) can be specified explicitly. More generally, we are only able to specify reasonable approximations to the required estimates. The separation theorem provides a natural motivation to incorporate these approximate estimates, or any others selected for ease of implementation, into the detector structure in the same way as if they were exact estimates.

Example 6.4.1 (Optical Radar). Perhaps the simplest situation occurs in the optical-radar context when the intensity process $\lambda_t^{(0)}$ corresponding to

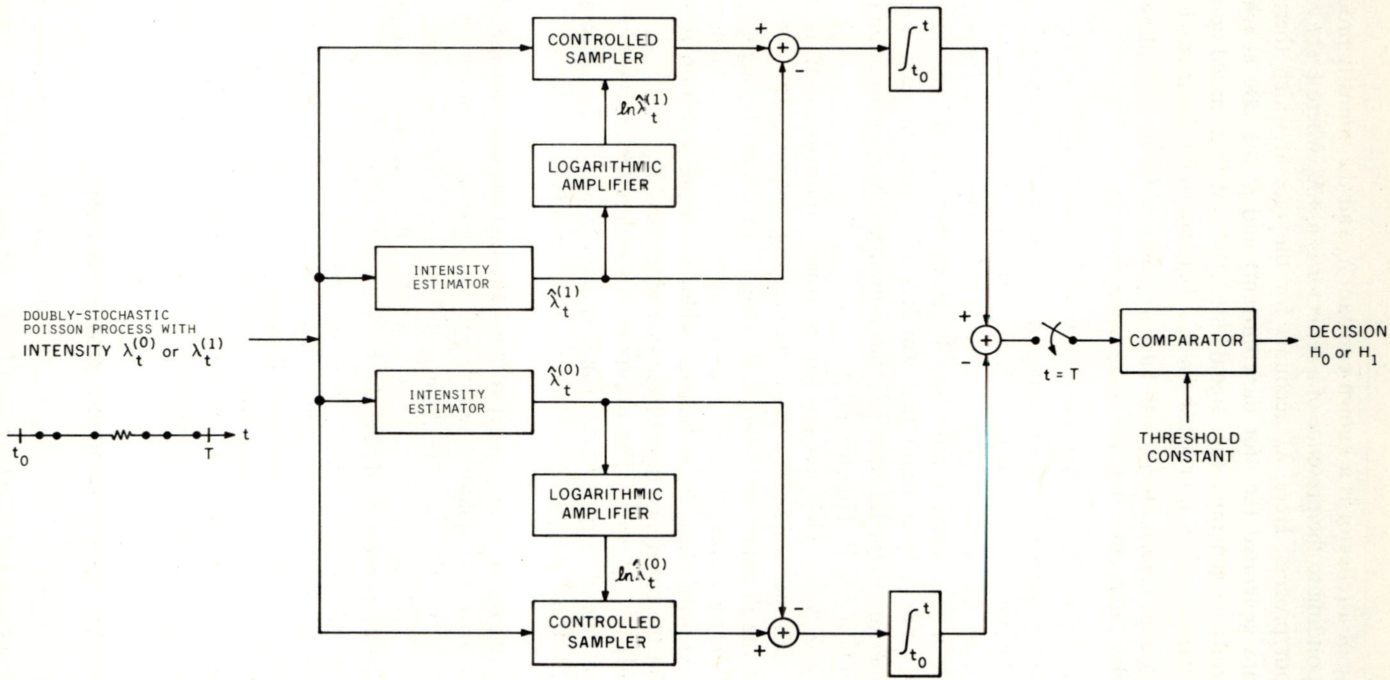

Figure 6.1. Optimum test for observations of a doubly-stochastic Poisson process.

the absence of a reflector is a known constant λ_0 and the intensity process $\lambda_t^{(1)}$ corresponding to the presence of a reflector is $\mu_t + \lambda_0$, where $\{\mu_t; t \geq t_0\}$ is a random process. Here, λ_0 accounts for the rate at which thermo-electrons are generated (i.e., for dark current) and μ_t the rate at which photoelectrons are generated (i.e., signal current). If the transmitted optical radar pulse is a narrowband, deterministic signal $s_t = \sqrt{2}\ \mathrm{Re}[S_t \exp(j2\pi\nu t)]$, then μ_t can have a variety of forms depending on the particular radar environment. Some examples are:

a. $\mu_t = \alpha|ZS_t|^2$. This occurs for a stationary, rough reflector. For flat (i.e., time-independent) Rayleigh fading, Z is a zero-mean complex-valued Gaussian random variable; the mean is nonzero for flat Ricean fading. For propagation of the optical signal in clear air turbulence, Z is frequently taken to be a lognormal variate (see E. Hoversten, R. Harger, and S. Halme [19]). For a laser beam that wanders in a random walk over a receiving aperture in the far field, $|Z|^2$ has a modified beta distribution (see R. Esposito [10] and P. Titterton [45]).

b. $\mu_t = \alpha|Z_t S_t|^2$. This occurs for a rough reflector that is not stationary; its aspect or radar cross section changes significantly over the duration of the illumination. Nonflat Rayleigh and Ricean fading correspond to Z_t being a complex-valued Gaussian process. Nonflat lognormal fading occurs with clear air turbulence when the signal duration is long compared to the coherence time of the fading process Z_t.

c. $\mu_t = \alpha|\int_U S_{t-u} Z_u \, du|^2$. This occurs for a rough reflector whose physical size is sufficiently large to cause range spreading (see H. Van Trees [48, Chapter 12]). Here, $\{Z_u; u \in U\}$ is a complex-valued Gaussian process.

In these examples, α is a nonegative constant that depends on the photodetector's area and quantum efficiency. Background radiation in the form of incoherent light can also be accommodated in these examples; for instance, (a) becomes $\mu_t = \alpha|ZS_t + B_t|^2$, where $\{B_t; t \geq t_0\}$ is a complex-valued Gaussian process modeling the background radiation.

The radar detection problem is to decide whether or not a reflector is present given an observed output of a photodetector. Assuming this output to be modeled as a doubly stochastic Poisson process and using (6.73), we see that the receiver processing to be performed subsequent to photoelectron conversion is that shown schematically in Figure 6.2. The procedure for designing the intensity estimator is developed in the next section.

6.5 FILTERING FOR DOUBLY STOCHASTIC POISSON PROCESSES

Estimation problems for observed doubly stochastic Poisson processes are of basic theoretical and practical importance. As we have already seen in

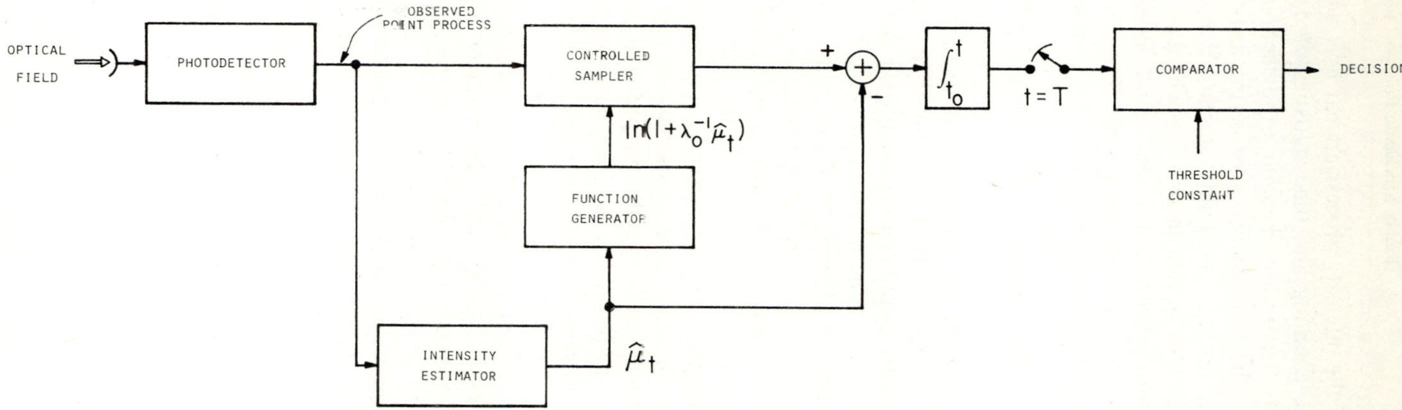

Figure 6.2. Optical radar receiver.

Sections 6.2 and 6.3, estimates of the intensity process appear as a natural component in expressions for the counting and time statistics of doubly stochastic Poisson processes. The separation theorem of detection described in Section 6.4 brings out the fundamental role of intensity estimates in problems of hypothesis testing for doubly stochastic Poisson processes. The estimation problem also occurs in practice whenever it is desired to determine the information process that modulates the intensity.

Our emphasis in this section is on a particular form of estimation problem commonly referred to as the *filtering problem*. This problem is described as follows. Let $\{N_t; t \geqslant t_0\}$ be a doubly stochastic Poisson process with intensity process $\{\lambda_t(\mathbf{x}_t); t \geqslant t_0\}$, where $\{\mathbf{x}_t; t \geqslant t_0\}$ is the underlying information process. We suppose that N is observed on the interval $[t_0, t)$ and that the entire counting path $\{N_\sigma; t_0 \leqslant \sigma < t\}$ is available for processing to form estimates. The endpoint time t is allowed to increase from t_0 corresponding to a real-time parameter as additional data are accumulated. We desire to estimate the value at time t of some specified function of the information process; for example, this function could be the intensity process $\lambda_t(\mathbf{x}_t)$ or the information process $\mathbf{x}_t$ itself. An estimate is sought that evolves in time as new data become available; all the data accumulated up to any time can be used to form the estimate at that time. This causal dependence of the estimate on the observed counting record implies that there is a potentiality for generating the estimate in real time as data arrive. It is important to note that in the filtering problem studied in this section, we are not allowing the possibility of returning to update or improve any previous estimate after new data are obtained. This type of estimation, termed smoothing, is examined in Section 6.7.

There are three main topics in the section. The first is the linear filtering problem. In this problem, treated in Section 6.5.1, the estimate is constrained to be a linear function of the counting record. The motivation for this constraint is that it leads to estimators that are relatively easy to realize in practice with either hardware or numerical algorithms. The disadvantage is that for many applications of interest, linearly constrained estimates are not sufficiently accurate. Thus, the second topic, treated in Section 6.5.2, is the filtering problem without the linearity constraint. Here, the major difficulty is analytic intractability, but we do manage to make substantial progress in specifying suboptimal estimators that are shown to work well in certain situations even at very low count rates. The final topic of the section, treated in Section 6.5.3, is the mean-square performance of estimators.

6.5.1 Linear Minimum Mean Square-Error Filtering

Let $\{N_t;\ t \geq t_0\}$ be a doubly stochastic Poisson process with intensity process $\{\lambda_t;\ t \geq t_0\}$. We seek an estimate of the intensity at time t in terms of an observed counting path on the interval $[t_0, t)$. The estimator is constrained to be linear; that is, the estimate λ_t^{*l} of λ_t must have the form

$$\lambda_t^{*l} = a_t + \int_{t_0}^{t} h(t,u)\, dN_u \tag{6.75}$$

for some deterministic function a_t and impulse response function $h(t,u)$. As an example, a possible linear estimator occurs for the selection

$$a_t = 0 \quad \text{and} \quad h(t,u) = \begin{cases} \dfrac{1}{\tau}, & t-\tau \leq u < t \\ 0, & \text{otherwise.} \end{cases} \tag{6.76}$$

For this choice,

$$\lambda_t^{*l} = \frac{1}{\tau}\int_{t-\tau}^{t} dN_u = \frac{1}{\tau}(N_t - N_{t-\tau}). \tag{6.77}$$

This is a moving-average, histogram estimate of the intensity process. Such an estimate is widely used because it has the advantage of requiring almost no knowledge about the intensity process beyond that needed to select the averaging time τ. It is common practice to sample the moving-average estimate at discrete times, for instance at integral multiples of τ, and then to least-squares curve fit an assumed time function to the sampled values.

It is at least intuitively evident that estimators having performance superior to the moving average estimator can be designed if more knowledge about the statistics of the intensity process is available. Such statistical knowledge might be available in practice because of having measured it experimentally or because of having a sufficient understanding of the physical mechanisms involved to postulate it. The estimator we consider next requires a knowledge of the first- and second-order statistics of the intensity process for its implementation.

Let $\hat{\lambda}_t^l$ denote the linear estimate λ_t^{*l} that results in (6.75) when both a_t and $h(t,u)$ are selected to minimize the mean-square error $E[(\lambda_t - \lambda_t^{*l})^2]$. This linear, minimum mean square-error estimate is given in the following theorem due independently to J. Grandell [16, 17], O. Macchi [28, 29], and J. Clark [8]. We use Grandell's proof.

Theorem 6.5.1 (Optimum Linear Estimator). Let λ_t^{*l} be a linear estimate of λ_t and $E[(\lambda_t - \lambda_t^{*l})^2]$ the corresponding mean-square error. Denote the estimate λ_t^{*l} that minimizes this mean-square error by $\hat{\lambda}_t^l$. Then

$$\hat{\lambda}_t^l = E(\lambda_t) + \int_{t_0}^t h_0(t,u)[dN_u - E(\lambda_u)\,du], \tag{6.78}$$

where the optimum impulse response $h_0(t,u)$ is the solution to the integral equation

$$h_0(t,u)E(\lambda_u) + \int_{t_0}^t h_0(t,\sigma)K_\lambda(\sigma,u)\,d\sigma = K_\lambda(t,u), \qquad t_0 \leqslant u \leqslant t. \tag{6.79}$$

Here, K_λ is the covariance function of the intensity process. Furthermore, the resulting minimum mean-square error is given by

$$E\left[\left(\lambda_t - \hat{\lambda}_t\right)^2\right] = h_0(t,t)E(\lambda_t). \tag{6.80}$$

Proof. Let λ_t^{*l} be an arbitrary linear estimate of λ_t in terms of the counting record $\{N_\sigma;\ t_0 \leqslant \sigma < t\}$, and suppose $h_0(t,u)$ satisfies (6.79). Write λ_t^{*l} as

$$\lambda_t^{*l} = \alpha_t + \int_{t_0}^t [h_0(t,u) + g(t,u)][dN_u - E(\lambda_u)\,du]. \tag{6.81}$$

By using the assertions in Problem 6.2.6, we then deduce that

$$\begin{aligned}
E\left[\left(\lambda_t - \lambda_t^{*l}\right)^2\right] = E\Bigg[\Bigg\{&(\lambda_t - E(\lambda_t)) - (\alpha_t - E(\lambda_t)) \\
&- \int_{t_0}^t [h_0(t,u) + g(t,u)][dN_u - E(\lambda_u)\,du]\Bigg\}^2\Bigg] \\
= K_\lambda(t,t) &+ (\alpha_t - E(\lambda_t))^2 + \int_{t_0}^t E(\lambda_u)[h_0(t,u) + g(t,u)]^2\,du \\
&+ \int_{t_0}^t\int_{t_0}^t [h_0(t,u) + g(t,u)]K_\lambda(u,v)[h_0(t,v) + g(t,v)]\,du\,dv \\
&- 2\int_{t_0}^t [h_0(t,u) + g(t,u)]K_\lambda(t,u)\,du.
\end{aligned}$$

Manipulation and use of (6.79) then shows that

$$E\left[(\lambda_t - \lambda_t^{*l})^2\right] = K_\lambda(t,t) + (\alpha_t - E(\lambda_t))^2 - \int_{t_0}^{t} h_0(t,u) K_\lambda(t,u)\,du$$

$$+ \int_{t_0}^{t}\int_{t_0}^{t} g(t,u) K_\lambda(u,v) g(t,v)\,du\,dv$$

$$+ \int_{t_0}^{t} E(\lambda_u) g^2(t,u)\,du. \tag{6.82}$$

Because $E(\lambda_t) \geqslant 0$ and K_λ is nonnegative definite, we have

$$\int_{t_0}^{t} E(\lambda_u) g^2(t,u)\,du \geqslant 0$$

and

$$\int_{t_0}^{t}\int_{t_0}^{t} g(t,u) K_\lambda(u,v) g(t,v)\,du\,dv \geqslant 0.$$

Consequently, $E[(\lambda_t - \lambda_t^{*l})^2]$ is minimized by setting $\alpha_t = E(\lambda_t)$ and $g(t,u) = 0$. It follows from (6.81) that $\hat{\lambda}_t^l$ is given by (6.78). We have also that

$$E\left[(\lambda_t - \hat{\lambda}_t^l)^2\right] = K_\lambda(t,t) - \int_{t_0}^{t} h_0(t,u) K_\lambda(t,u)\,du. \tag{6.83}$$

Equation (6.80) follows from this and (6.79), and the theorem is proven.

The linear estimator that minimizes the mean square-error is shown schematically in Figure 6.3. The impulse response of the optimum linear filter $h_0(t,u)$ satisfies the integral equation (6.79). The mean-square performance of this estimator is given in terms of the optimum impulse response in (6.80). In order to implement this estimator and evaluate its performance, we need to know the mean-value function and covariance function of the intensity process.

It is perhaps worth emphasizing that only the mean and covariance functions are needed to design the estimator. Any additional statistical or structural information that may be available about the intensity process is not used and can be disregarded. Thus, for instance, while we have required that the mean of the intensity process be nonnegative for (6.82), we have not made use of the property that the intensity process itself is nonnegative. Two intensity processes with strikingly different sample paths

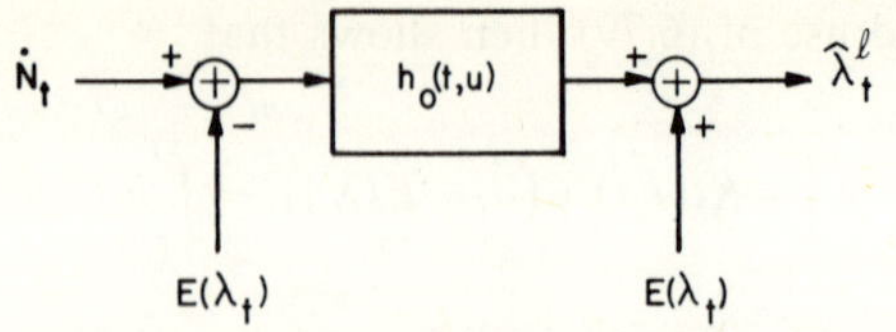

Figure 6.3. Optimum linear intensity estimator.

will lead to the same linear intensity estimator if their mean and covariance functions are identical.

The integral equation (6.79) has been studied extensively in connection with the linear filtering problem for observations that contain additive noise (see, for example, H. Van Trees [48, Chapter 6]). The additive model that leads to the same integral equation is the following. Let $\{\lambda_t;\ t \geqslant t_0\}$ be the intensity process defined above, and suppose observations of the additive form

$$r_t = \lambda_t + [E(\lambda_t)]^{1/2} w_t, \qquad t \geqslant t_0 \tag{6.84}$$

are available, where $\{w_t;\ t \geqslant t_0\}$ is a zero-mean, white noise process with unit intensity and independent of λ. Then the linear estimator that minimizes the mean-square error in estimating λ_t in terms of an observed path $\{r_\sigma;\ t_0 \leqslant \sigma < t\}$ is that shown in Figure 6.3 with $h_0(t,u)$ specified by (6.79).

This last remark is very important. It implies that all the highly developed, linear-filtering theory for the additive observation model can be applied directly to the doubly stochastic Poisson process observation model. Perhaps the most notable aspect of this theory because of its computational advantage is the Kalman–Bucy formulation developed over the past several years. As extensive treatments of this theory are readily available elsewhere (see, for example, E. Wong [50, Chapter 3], R. Bucy and P. Joseph [5, Chapter 4], H. Van Trees [47, Chapter 6], or A. Jazwinski [20, Chapter 7]), it is not our intention to rederive this formulation here, but rather, we want simply to indicate how the Kalman–Bucy filtering algorithm can be used with doubly stochastic Poisson process observations. We first need to postulate a linear dynamical model for the intensity process. Such a model might be postulated from the physical mechanisms generating the intensity process, but this is not necessary because, as we have emphasized above, the model need only match the mean and covariance function of the intensity process, so that even a Gaussian model can be used as long as it has the correct mean and covariance function; systematic techniques are available for finding linear dynamical realizations for specified covariance functions (see B. Anderson, J. Moore, and S. Loo [1]).

Let us suppose then that the following linear dynamical model is available for the intensity process:

$$dx_t = A_t x_t\,dt + b_t\,du_t, \qquad x_{t_0} = x_0$$

$$\lambda_t = E(\lambda_t) + c_t' x_t, \tag{6.85}$$

where $\mathbf{x}$ is an n-dimensional state vector; $\mathbf{A}_t$ is a known $n \times n$ matrix and $\mathbf{b}_t$ and $\mathbf{c}_t$ are known n-dimensional vectors; $\mathbf{x}_0$ is an initial random variable with zero mean and known covariance $\boldsymbol{\Sigma}_0$; and $\{u_t;\ t \geqslant t_0\}$ is a zero-mean random process with stationary orthogonal increments. We normalize u so that $E(u_s u_t) = \min(s,t)$. It now follows from (6.84), (6.85), and the Kalman–Bucy algorithm that $\hat{\lambda}_t^l$ is given by the following equations:

$$d\hat{x}_t^l = A_t \hat{x}_t^l\,dt + \Sigma_t c_t E^{-1}(\lambda_t)\left[dN_t - E(\lambda_t)\,dt - c_t'\hat{x}_t^l\,dt\right],$$

$$\hat{x}_{t_0}^l = \mathbf{0}, \tag{6.86a}$$

$$\frac{d\Sigma_t}{dt} = A_t\Sigma_t + \Sigma_t A_t' + b_t b_t' - \Sigma_t c_t E^{-1}(\lambda_t) c_t' \Sigma_t,$$

$$\Sigma_{t_0} = \Sigma_0, \tag{6.86b}$$

$$\hat{\lambda}_t^l = E(\lambda_t) + c_t'\hat{x}_t^l. \tag{6.86c}$$

Furthermore, the mean-square error in estimating λ_t using this linear estimator is given by

$$E\left[\left(\lambda_t - \hat{\lambda}_t^l\right)^2\right] = c_t'\Sigma_t c_t. \tag{6.87}$$

Thus, the solution to the nonlinear, matrix, differential equation (6.86b) determines both the structure of the estimator through (6.86a) and its performance through (6.87). In some few examples, it is possible to solve (6.86a) and (6.86b) analytically to get an explicit expression for the estimate $\hat{\lambda}_t^l$ in terms of the observed counting path. More important, however, is the view of (6.86a) and (6.86b) as defining numerical updating algorithms that indicate how the estimate $\hat{\lambda}_t^l$ is to be propagated forward in time as data are accumulated. Equation (6.86b) is a Riccati equation and has been extensively studied; several numerical algorithms are available for its solution. The Riccati equation can either be solved on line as data are accumulated or in advance because (6.86b) does not depend on observed data; on-line evaluation may be desired when only limited data storage is available.

We now give some examples illustrating the use of these equations.

Example 6.5.1 (Bayesian Regression). Suppose the intensity process has the form

$$\lambda_t = \sum_{i=1}^{M} \alpha_i c_{t,i} = \mathbf{c}_t' \boldsymbol{\alpha} \tag{6.88}$$

for $t \geqslant t_0$, where each regression coefficient α_i is a random variable with known mean $E(\alpha_i)$, and each regression function $c_{t,i}$ is known. Denote the covariance matrix for the coefficient vector $\boldsymbol{\alpha}$ by $\boldsymbol{\Sigma}_0$; we assume this matrix is known and positive definite. The regression functions could be a sequence of exponentials with known decay rates, a sequence of orthonormal functions, a sequence of powers of t, and so on. The problem is to observe the doubly stochastic Poisson process with the intensity in (6.88) and estimate the regression coefficients $\alpha_1, \alpha_2, \ldots, \alpha_M$. The linear minimum mean square-error estimator can be determined by modeling the intensity in the form of (6.85). For this purpose, let us define $\mathbf{x}_t$ by $\mathbf{x}_t = \boldsymbol{\alpha} - E(\boldsymbol{\alpha})$. Then (6.85) becomes

$$d\mathbf{x}_t = \mathbf{0}, \qquad \mathbf{x}_{t_0} = \mathbf{x}_0$$

$$\lambda_t = \mathbf{c}_t' E(\boldsymbol{\alpha}) + \mathbf{c}_t' \mathbf{x}_t, \tag{6.89}$$

where $\mathbf{x}_0$ is a zero mean random variable with covariance $\boldsymbol{\Sigma}_0$. The linear estimator of the coefficients is obtained by specializing (6.86) for the model of (6.89). The result is

$$d\hat{\mathbf{x}}_t^l = \boldsymbol{\Sigma}_t \mathbf{c}_t E^{-1}(\lambda_t)[dN_t - E(\lambda_t)dt - \mathbf{c}_t' \hat{\mathbf{x}}_t^l dt], \qquad \hat{\mathbf{x}}_{t_0}^l = \mathbf{0} \tag{6.90a}$$

$$\frac{d\boldsymbol{\Sigma}_t}{dt} = -\boldsymbol{\Sigma}_t \mathbf{c}_t E^{-1}(\lambda_t) \mathbf{c}_t' \boldsymbol{\Sigma}_t, \qquad \boldsymbol{\Sigma}_{t_0} = \boldsymbol{\Sigma}_0 \tag{6.90b}$$

where $E(\lambda_t) = \mathbf{c}_t' E(\boldsymbol{\alpha})$. The solution to (6.90b) is

$$\boldsymbol{\Sigma}_t = \left[\mathbf{I} + \boldsymbol{\Sigma}_0 \int_{t_0}^{t} \mathbf{c}_\sigma E^{-1}(\lambda_\sigma) \mathbf{c}_\sigma' \, d\sigma \right]^{-1} \boldsymbol{\Sigma}_0. \tag{6.91}$$

The estimate of the regression coefficients is given by $\hat{\boldsymbol{\alpha}}_t^l = \hat{\mathbf{x}}_t^l + E(\boldsymbol{\alpha})$ and of the intensity by $\hat{\lambda}_t^l = \mathbf{c}_t' \hat{\boldsymbol{\alpha}}_t^l$. A schematic diagram for the estimator is shown in Figure 6.4.

Example 6.5.2 (Subcarrier Amplitude Modulation in Direct-Detection Optical Communication Systems). Suppose an optical communication system is used to relay an analog information process y_t to a remote user.

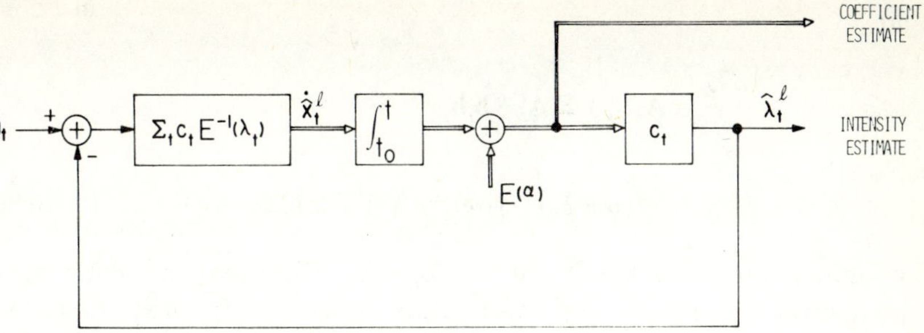

Figure 6.4. Linear estimator for regression coefficients.

For this purpose, y_t amplitude modulates a subcarrier $\cos(\omega t)$ of frequency ω. This modulated subcarrier is then used to modulate the light intensity of an optical source. The term $|S(t,\vec{r})|^2$ appearing in Equation (4.22) can be taken to be (see W. Pratt [33, p. 178])

$$|S(t,\vec{r})|^2 = P(1 + my_t \cos\omega t) \tag{6.92}$$

where P is a nonnegative constant and m is the subcarrier modulation index. The response of an ideal photodector can be modeled as a doubly stochastic Poisson process with intensity process $\{\lambda_t;\ t \geqslant t_0\}$, where

$$\lambda_t = \alpha(1 + my_t \cos\omega t) + \lambda_0 \tag{6.93}$$

in which α is a nonnegative constant that depends on the detector area and quantum efficiency and λ_0 accounts for the thermoelectron count rate. We assume that the information process $\{y_t;\ t \geqslant t_0\}$ has a finite-dimensional state-variable realization as

$$\begin{aligned} d\mathbf{x}_t &= \mathbf{A}_t \mathbf{x}_t\, dt + \mathbf{b}_t\, du_t, \qquad \mathbf{x}_{t_0} = \mathbf{x}_0 \\ y_t &= \mathbf{h}_t' \mathbf{x}_t, \end{aligned} \tag{6.94}$$

where $\{u_t;\ t \geqslant t_0\}$ is a zero mean random process with orthogonal increments and normalized so that $E(u_s u_t) = \min(s,t)$. The initial state $\mathbf{x}_0$ has zero mean and covariance $\mathbf{\Sigma}_0$. It follows that the intensity process has the representation in (6.85) with $E(\lambda_t) = \alpha + \lambda_0$ and $\mathbf{c}_t = (\alpha m \cos\omega t)\mathbf{h}_t$. Equations (6.86a) and (6.86b) become

$$\begin{aligned} d\hat{\mathbf{x}}_t^l &= \mathbf{A}_t \hat{\mathbf{x}}_t^l\, dt + (\alpha + \lambda_0)^{-1} (\alpha m \cos\omega t) \mathbf{\Sigma}_t \mathbf{h}_t \\ &\quad \times \left\{ dN_t - \left[\alpha(1 + m\mathbf{h}_t' \hat{\mathbf{x}}_t^l \cos\omega t) + \lambda_0\right] dt \right\} \end{aligned} \tag{6.95}$$

and

$$\frac{d\mathbf{\Sigma}_t}{dt} = \mathbf{A}_t\mathbf{\Sigma}_t + \mathbf{\Sigma}_t\mathbf{A}_t' + \mathbf{b}_t\mathbf{b}_t'$$

$$-(\alpha+\lambda_0)^{-1}(\alpha m \cos\omega t)^2\mathbf{\Sigma}_t\mathbf{h}_t\mathbf{h}_t'\mathbf{\Sigma}_t. \tag{6.96}$$

The initial conditions are $\hat{\mathbf{x}}_{t_0}^l = \mathbf{0}$ and $\mathbf{\Sigma}_{t_0} = \mathbf{\Sigma}_0$. The linear minimum mean square-error estimate of the information process is $\hat{y}_t^l = \mathbf{h}_t'\hat{\mathbf{x}}_t^l$. Equations (6.95) and (6.96) define the structure of the estimator in general, and the corresponding demodulator is shown schematically in Figure 6.5*a*. Further simplification is possible in practice because the information process normally has a bandwidth much smaller than the subcarrier frequency ω. If this condition holds, it is common practice to eliminate terms in (6.95) and (6.96) that vary at twice the subcarrier frequency since they contribute only negligibly. The resulting simplified demodulator is shown in Figure 6.5*b*. The corresponding minimum mean-square error in estimating the information process is

$$E\left[(y_t - \hat{y}_t^l)^2\right] = \mathbf{h}_t'\mathbf{\Sigma}_t\mathbf{h}_t. \tag{6.97}$$

Example 6.5.3 (On–Off Modulated Light). Suppose a light source is turned on and off by a random telegraph wave. The rate μ_t at which photoelectrons are generated in a photodetector will then follow the random telegraph wave; a typical sample function is shown in Figure 6.6. This photocount intensity has the following structure. At time t_0, the intensity μ_{t_0} has the value 0 or α with equal probability. For $t > t_0$, the intensity alternates between the values 0 and α, switching at times corresponding to the occurrence times of a homogeneous Poisson point process with constant intensity ν. Thus, for $t \geqslant t_0$, μ_t can be expressed as

$$\mu_t = \left(\frac{\alpha}{2}\right) + z\left(\frac{\alpha}{2}\right)(-1)^{M_t}, \tag{6.98}$$

where z is a discrete random variable having the value minus one or plus one with equal probability, and where $\{M_t;\ t \geqslant t_0\}$ is a homogeneous Poisson process with intensity ν and independent of z. Let us suppose that the photodetector also generates thermoelectrons at a rate λ_0. Then the detector output is a doubly-stochastic Poisson process with intensity pro-

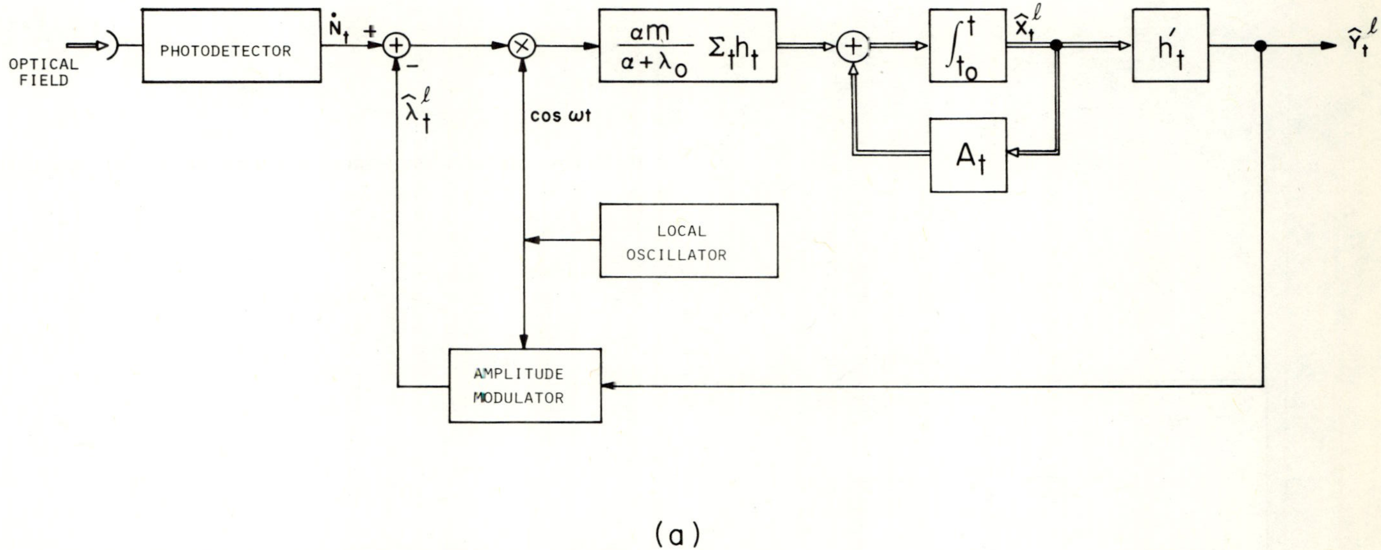

Figure 6.5. (a) Demodulator for subcarrier amplitude modulation.

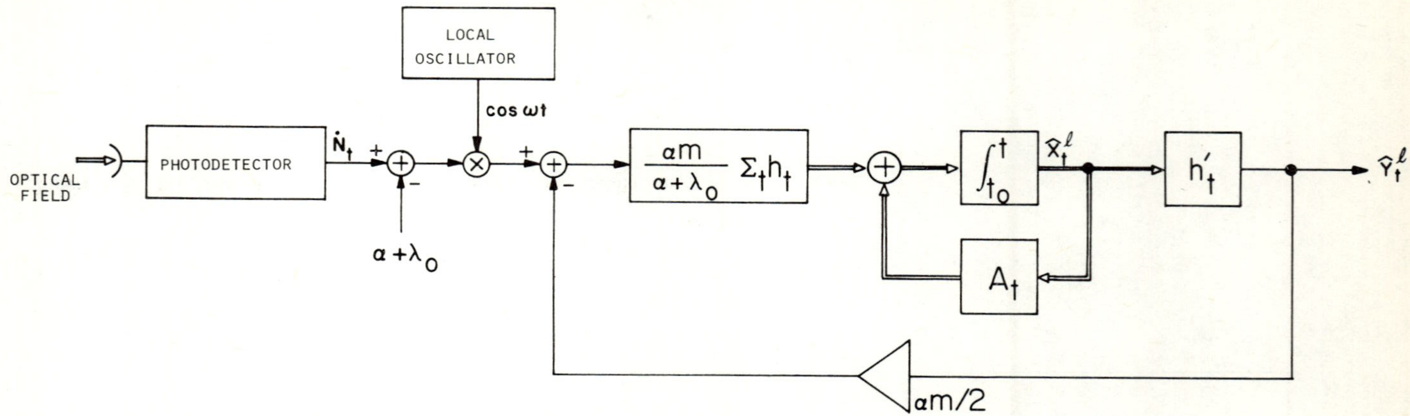

(b)

Figure 6.5. (b) Demodulator with double-frequency terms suppressed.

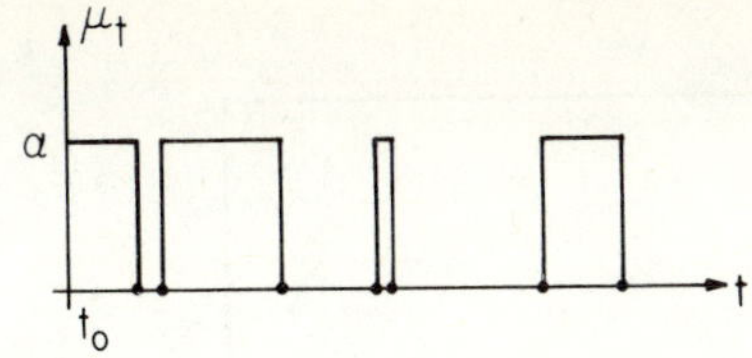

Figure 6.6. Photocount intensity for light on–off modulated by a random telegraph wave.

cess $\{\lambda_t;\ t\geqslant t_0\}$, where

$$\lambda_t=\mu_t+\lambda_0=\left(\left(\frac{\alpha}{2}\right)+\lambda_0\right)+z\left(\frac{\alpha}{2}\right)(-1)^{M_t}. \tag{6.99}$$

We need the mean-value and covariance functions for λ. It is easily seen that $E(\lambda_t)=(\alpha/2)+\lambda_0$. The covariance function is for $t_0\leqslant t<u$

$$\begin{aligned}K_\lambda(t,u)&=E[(\lambda_t-E(\lambda_t))(\lambda_u-E(\lambda_u))]\\&=E(z^2)\left(\frac{\alpha}{2}\right)^2E\left[(-1)^{M_t+M_u}\right]\\&=\left(\frac{\alpha}{2}\right)^2\exp[-2\nu(u-t)],\qquad t_0\leqslant t<u.\end{aligned}$$

In evaluating the last expectation, we have used

$$E\left[(-1)^{M_t+M_u}\right]=E\left[(-1)^{M_u-M_t}(-1)^{2M_t}\right]=E\left[(-1)^{M_u-M_t}\right].$$

As $K_\lambda(t,u)=K_\lambda(u,t)$, we have for all $(t,u)\geqslant t_0$,

$$K_\lambda(t,u)=\left(\frac{\alpha}{2}\right)^2\exp[-2\nu|t-u|]. \tag{6.100}$$

We now want to determine the linear minimum mean square-error estimator of the signal intensity μ_t in terms of the photodetector output. There are two ways to proceed. The first is to solve the integral equation (6.79) for the optimum impulse response and the second is to formulate a linear dynamical model for λ and use (6.86). We choose to illustrate the latter approach. The model does not have to reproduce the sample-function structure of the intensity process in (6.99). Rather, it only has to reproduce the mean-value and covariance function of λ. A one-dimensional state variable model that accomplishes this is

$$\begin{aligned}&dx_t=-2\nu x_t\,dt+du_t, &&x_{t_0}=x_0\\&\mu_t=\left(\frac{\alpha}{2}\right)+\alpha\sqrt{\nu}\,x_t, &&\lambda_t=\mu_t+\lambda_0\end{aligned} \tag{6.101}$$

where x_0 is a zero-mean random variable with variance $(4\nu)^{-1}$, and where $\{u_t;\ t\geqslant t_0\}$ is a zero-mean random process with stationary, orthogonal

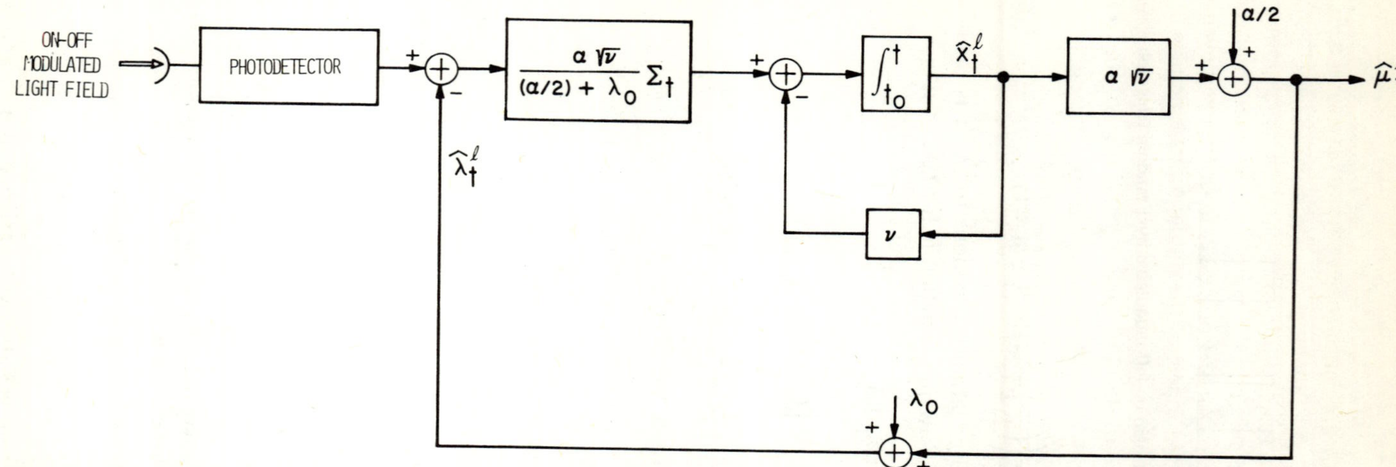

Figure 6.7. Linear estimator for light on–off modulated by a random telegraph wave.

increments, independent of x_0, and normalized so that $E(u_s u_t) = \min(s,t)$. We reemphasize that the sample functions for the process $\{\lambda_t;\ t \geqslant t_0\}$ in (6.101) may differ markedly from those of a random telegraph wave. Indeed, it is consistent with our requirements here to select u to be a Wiener process in which case the model process λ in (6.101) will even take on negative values. The essential aspect of the model that we require is that the mean and covariance functions for the intensity process in (6.99) and its model in (6.101) are identical. The linear estimator of the signal intensity process is now seen from (6.86) to be given by the equations

$$d\hat{x}_t^l = -2\nu\hat{x}_t^l\,dt + \frac{\alpha\sqrt{\nu}}{(\alpha/2)+\lambda_0}\Sigma_t\left(dN_t - \hat{\lambda}_t^l\,dt\right), \qquad \hat{x}_{t_0}^l = 0 \qquad (6.102a)$$

$$\frac{d\Sigma_t}{dt} = -4\nu\Sigma_t + 1 - \frac{\alpha^2\nu}{(\alpha/2)+\lambda_0}\Sigma_t^2, \qquad \Sigma_{t_0} = (4\nu)^{-1} \qquad (6.102b)$$

$$\hat{\lambda}_t^l = \hat{\mu}_t^l + \lambda_0 = \left(\frac{\alpha}{2}\right) + \alpha\sqrt{\nu}\,\hat{x}_t^l + \lambda_0. \qquad (6.102c)$$

The estimator is shown in schematic form in Figure 6.7.

We have simulated the processes of this model for the parameters $\nu = 2.0$, $\alpha = 20.0$, and $\lambda_0 = 0.1$. The results are shown in Figure 6.8. The process $\{\mu_t;\ t \geqslant 0\}$ realized in the simulation had an initial value of 20.0 at $t = 0$, and transitions between the on and off state occurred at times 0.49, 1.81, 4.49, 8.70, 9.45, 9.80, and 9.85 seconds. During the ten second interval shown, a total of 72 points occurred in the observed doubly-stochastic Poisson process. The estimate $\{\hat{\mu}_t^l;\ t \geqslant 0\}$ obtained by implementing (6.102) numerically and processing the observed points is shown in Figure 6.8.

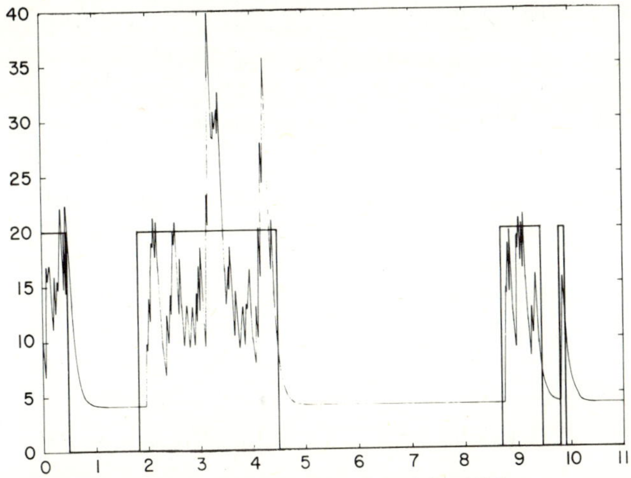

Figure 6.8. Linear estimation of on–off modulated light.

6.5.2 Nonlinear Minimum Mean-Square Error Filtering

We now examine the filtering problem for doubly stochastic Poisson processes when the linearity constraint imposed in the previous subsection is removed. There are two important reasons for removing this constraint. The first is that for many applications, the linearly constrained estimators do not perform as well as simple alternatives designed without this constraint.The second reason is that linearly constrained, minimum mean square-error estimators do not hold the fundamental theoretical position of minimum mean square-error estimators without the constraint. This is evident from our investigations in earlier sections of the counting and time statistics of doubly stochastic Poisson processes; the estimators required in the representations developed are only infrequently linear.

We will adopt the following notations throughout the subsection:

i. The inner, or dot, product of two n-dimensional vectors $\mathbf{v}_1$ and $\mathbf{v}_2$ will be denoted by $\langle \mathbf{v}_1, \mathbf{v}_2 \rangle$, where

$$\langle \mathbf{v}_1, \mathbf{v}_2 \rangle = \sum_{i=1}^{n} v_{1i} v_{2i}. \tag{6.103}$$

ii. The length of a vector $\mathbf{v}$ will be denoted by $|\mathbf{v}|$, where

$$|\mathbf{v}| = \langle \mathbf{v}, \mathbf{v} \rangle^{1/2}. \tag{6.104}$$

iii. Let $\mathbf{h}_t(\mathbf{x}_t)$ be a possibly time-dependent vector-valued function of the information process and such that $E(|\mathbf{h}_t(\mathbf{x}_t)|^2) < \infty$. An arbitrary filtering estimate of $\mathbf{h}_t(\mathbf{x}_t)$ in terms of an observed counting path $\{N_\sigma;\ t_0 \leqslant \sigma < t\}$ on $[t_0, t)$ will be denoted by $\mathbf{h}_t^*$. The conditional expectation of $\mathbf{h}_t(\mathbf{x}_t)$ given the observed counting path will be denoted by $\hat{\mathbf{h}}_t$; thus

$$\hat{\mathbf{h}}_t = E[\mathbf{h}_t(\mathbf{x}_t) | N_\sigma;\ t_0 \leqslant \sigma < t]. \tag{6.105}$$

A well-known relationship between $\mathbf{h}_t^*$ and $\hat{\mathbf{h}}_t$ is given in the following assertion.

Assertion 6.5.1 (The Conditional Mean is Optimum). Let $\{N_t;\ t \geqslant t_0\}$ be a doubly stochastic Poisson process with intensity process $\{\lambda_t(\mathbf{x}_t);\ t \geqslant t_0\}$, where $\{\mathbf{x}_t;\ t \geqslant t_0\}$ is an information process. The estimate $\mathbf{h}_t^*$ of $\mathbf{h}_t(\mathbf{x}_t)$ that minimizes the mean-square error $E(|\mathbf{h}_t(\mathbf{x}_t) - \mathbf{h}_t^*|^2)$ in terms of an observed counting path $\{N_\sigma;\ t_0 \leqslant \sigma < t\}$ is the conditional mean $\hat{\mathbf{h}}_t$.

This assertion is easily proven. By expanding the mean-square error and using the relation

$$\begin{aligned} E(\langle \mathbf{h}_t(\mathbf{x}_t), \mathbf{h}_t^* \rangle) &= E[E(\langle \mathbf{h}_t(\mathbf{x}_t), \mathbf{h}_t^* \rangle | N_\sigma;\ t_0 \leqslant \sigma < t)] \\ &= E(\langle \hat{\mathbf{h}}_t, \mathbf{h}_t^* \rangle), \end{aligned}$$

we have for an arbitrary estimate $\mathbf{h}_t^*$ that

$$\begin{aligned} E(|\mathbf{h}_t(\mathbf{x}_t)-\mathbf{h}_t^*|^2) & \\ &= E(|\mathbf{h}_t(\mathbf{x}_t)|^2)-2E(\langle \mathbf{h}_t(\mathbf{x}_t),\mathbf{h}_t^*\rangle)+E(|\mathbf{h}_t^*|^2) \\ &= E(|\mathbf{h}_t(\mathbf{x}_t)|^2)-2E(\langle \hat{\mathbf{h}}_t,\mathbf{h}_t^*\rangle)+E(|\mathbf{h}_t^*|^2) \\ &= E(|\mathbf{h}_t(\mathbf{x}_t)|^2)+E(|\hat{\mathbf{h}}_t-\mathbf{h}_t^*|^2)-E(|\hat{\mathbf{h}}_t|^2). \end{aligned}$$

Only the second term on the right is influenced by the selection of $\mathbf{h}_t^*$, and this term is nonnegative. It follows that the mean square-error is smallest for the choice $\mathbf{h}_t^*=\hat{\mathbf{h}}_t$, which establishes the assertion. Furthermore, we have that the minimum mean-square error is given by

$$\min_{\mathbf{h}^*} E(|\mathbf{h}_t(\mathbf{x}_t)-\mathbf{h}_t^*|^2)=E(|\mathbf{h}_t(\mathbf{x}_t)|^2)-E(|\hat{\mathbf{h}}_t|^2). \tag{6.106}$$

The nonlinear filtering problem is this. A doubly stochastic Poisson process $\{N_t;\ t\geqslant t_0\}$ having intensity process $\{\lambda_t(\mathbf{x}_t);\ t\geqslant t_0\}$ is observed on the interval $[t_0,t)$. Based on these observations we want to estimate at time t some specified function $\mathbf{h}_t(\mathbf{x}_t)$ of the information process. For instance, this function might be the intensity $\lambda_t(\mathbf{x}_t)$ or the information process $\mathbf{x}_t$. There is no linearity constraint on the estimator or data processor. The problem is to determine the structure of the estimator for some performance criterion.

The performance criterion we adopt is minimum mean-square error, so we know from the above assertion that the estimator must implement the conditional mean of $\mathbf{h}_t(\mathbf{x}_t)$ given the observed counting record. This conditional mean can be evaluated and the nonlinear filtering problem thus solved if we know the posterior statistics of the information process at any time given the observations up to that time.

Alternative performance criteria are useful in certain applications even though they do not seem to hold the same fundamental position as the minimum mean-square error criterion. An estimate that minimizes the expectation of the magnitude of the error is sometimes adopted. With this criterion, large errors are weighted less heavily than with the minimum mean-square error criterion; conversely, small errors are weighted somewhat more heavily. It can be shown that the estimate h_t^* that minimizes $E(|h_t(\mathbf{x}_t)-h_t^*|)$ in terms of $\{N_\sigma;\ t_0\leqslant\sigma<t\}$ is the *conditional median* of $h_t(\mathbf{x}_t)$ given $\{N_\sigma;\ t_0\leqslant\sigma<t\}$; see Problem 6.5.4. A uniform error weighting is also sometimes adopted. With this choice, it is desired to minimize $E[C_\epsilon(|\mathbf{h}_t(\mathbf{x}_t)-\mathbf{h}_t^*|)]$,

$$C_\epsilon(\alpha)=\begin{cases} 0, & 0\leqslant\alpha<\epsilon \\ 1, & \epsilon\leqslant\alpha. \end{cases}$$

It can be shown that in the limit as ϵ tends to zero, the estimate $\mathbf{h}_t^*$ that minimizes $E[C_\epsilon(|\mathbf{h}_t(\mathbf{x}_t)-\mathbf{h}_t^*|)]$ in terms of $\{N_\sigma;\ t_0 \leqslant \sigma < t\}$ is the *conditional mode* of $\mathbf{h}_t(\mathbf{x}_t)$ given $\{N_\sigma;\ t_0 \leqslant \sigma < t\}$. This estimate is termed the *maximum a posteriori probability estimate* by H. Van Trees [47, Chapter 2]. The maximum a posteriori probability estimate may be viewed as a natural generalization of the maximum likelihood estimate studied in Section 2.4 of Chapter 2 when the quantity to be estimated has a known prior probability distribution.

As with the conditional-mean estimate, both the conditional-median and conditional-mode estimates are, in principle, determined by the posterior statistics of the information process at any time given observations of the point process up to that time. With this motivation, we now study these posterior statistics.

Posterior Statistics by the Characteristic Function Method. We shall assume in what follows that the intensity $\lambda_t(\mathbf{x}_t)$ for $t \geqslant t_0$ is a positive function of an n-vector Markov process $\{\mathbf{x}_t;\ t \geqslant t_0\}$. These processes are defined in Section 4.2.2 where a particular example, the Poisson driven Markov process, is studied in detail. In this subsection, we seek a characterization of the posterior statistics of $\mathbf{x}_t$ given $\{N_\sigma;\ t_0 \leqslant \sigma < t\}$ for each $t \geqslant t_0$. We begin by characterizing the prior statistics of $\mathbf{x}_t$ for each $t \geqslant t_0$. The characterization of the posterior statistics will be a natural generalization of this.

The time evolution of the prior statistics of a Markovian information process can be described by a differential equation for the characteristic function $M_t(j\mathbf{v})$ for $\mathbf{x}_t$, where

$$M_t(j\mathbf{v}) = E\{\exp[j\langle \mathbf{v}, \mathbf{x}_t\rangle]\}. \tag{6.107}$$

For the purpose of deriving this differential equation, we need to assume there exists a nonnegative function $g_t(\mathbf{v},\mathbf{x}_t)$ such that $E[|\,g_t(\mathbf{v},\mathbf{x}_t)|] < \infty$ and such that for all $\Delta t > 0$ there holds

$$(\Delta t)^{-1}|E\{(\exp[j\langle \mathbf{v}, \Delta\mathbf{x}_t\rangle]-1)|\mathbf{x}_t\}| \leqslant g_t(\mathbf{v},\mathbf{x}_t), \tag{6.108}$$

where we define $\Delta\mathbf{x}_t = \mathbf{x}_{t+\Delta t}-\mathbf{x}_t$. Denote by $\psi_t(\mathbf{v}|\mathbf{x}_t)$ the limit of $(\Delta t)^{-1} \times E\{(\exp[j\langle \mathbf{v}, \Delta\mathbf{x}_t\rangle]-1)|\mathbf{x}_t\}$ as Δt tends to zero. Thus,

$$\psi_t(\mathbf{v}|\mathbf{x}_t) = \lim_{\Delta t \downarrow 0}(\Delta t)^{-1}E\{(\exp[j\langle \mathbf{v}, \Delta\mathbf{x}_t\rangle]-1)|\mathbf{x}_t\}. \tag{6.109}$$

The function $\psi_t(\mathbf{v}|\mathbf{x}_t)$ is termed the *characteristic form for the differential generator* of the Markov process $\{\mathbf{x}_t;\ t \geqslant t_0\}$. It and the distribution of the initial condition $\mathbf{x}_{t_0}$ completely characterize $\{\mathbf{x}_t;\ t \geqslant t_0\}$ statistically, just as we have seen in Section 4.2.2.

From these assumptions, it can be shown that the time evolution of the unconditional characteristic function of $\mathbf{x}_t$ is described for $t \geqslant t_0$ by

$$\frac{\partial M_t(j\mathbf{v})}{\partial t} = E\left\{\exp[j\langle \mathbf{v}, \mathbf{x}_t\rangle]\psi_t(\mathbf{v}|\mathbf{x}_t)\right\},$$
$$M_{t_0}(j\mathbf{v}) = E\left\{\exp[j\langle \mathbf{v}, \mathbf{x}_{t_0}\rangle]\right\}. \tag{6.110}$$

To see this, note first that

$$\begin{aligned} M_{t+\Delta t}(j\mathbf{v}) &= E\{\exp[j\langle \mathbf{v}, \mathbf{x}_{t+\Delta t}\rangle]\} \\ &= E\{\exp[j\langle \mathbf{v}, \mathbf{x}_t\rangle]\exp[j\langle \mathbf{v}, \Delta\mathbf{x}_t\rangle]\} \\ &= E\{\exp[j\langle \mathbf{v}, \mathbf{x}_t\rangle]E(\exp[j\langle \mathbf{v}, \Delta\mathbf{x}_t\rangle]|\mathbf{x}_t)\}. \end{aligned}$$

Thus,

$$\begin{aligned} &(\Delta t)^{-1}[M_{t+\Delta t}(j\mathbf{v}) - M_t(j\mathbf{v})] \\ &\quad = E\left\{\exp[j\langle \mathbf{v}, \mathbf{x}_t\rangle](\Delta t)^{-1}E((\exp[j\langle \mathbf{v}, \Delta\mathbf{x}_t\rangle]-1)|\mathbf{x}_t)\right\}. \end{aligned}$$

Taking the limit as $\Delta t\downarrow 0$ results in the desired expression (6.110). The interchange of limit and expectation is justified by assumption (6.108) and the bounded convergence theorem, Theorem 1.4.1.

Some examples that illustrate the variety of information processes accomodated by the model are the following:

1 *Random Variables.* In this case, $\mathbf{x}_t = \mathbf{x}_{t_0}$ for all $t \geqslant t_0$. It is easily seen from (6.109) that the function $\psi_t(\mathbf{v}|\mathbf{x}_t)$ is zero. Substitution of this into (6.110) shows that for all $t \geqslant t_0$,

$$M_t(j\mathbf{v}) = M_{t_0}(j\mathbf{v}).$$

2. *Poisson Driven Markov Processes.* These processes were studied in Chapter 4. In this case, $\mathbf{x}_t$ satisfies (4.92), and the function $\psi_t(\mathbf{v}|\mathbf{x}_t)$ is given in (4.100).

3. *Markov Diffusion Processes.* These processes are the continuous analogs of the Poisson driven Markov processes studied in Chapter 4. They differ in the sense that the excitation of the differential equation (4.39) is a continuous, independent-increment process rather than the discontinuous, independent-increment, counting process M. Here, we suppose $\{\mathbf{x}_t;\ t \geqslant t_0\}$

is the solution to a stochastic differential equation of the form

$$d\mathbf{x}_t = \mathbf{a}_t(\mathbf{x}_t)dt + \mathbf{B}_t(\mathbf{x}_t)d\mathbf{w}_t, \qquad \mathbf{x}_{t_0} = \mathbf{x}_0 \tag{6.111}$$

in which $\{\mathbf{w}_t;\ t \geqslant t_0\}$ is a vector of independent Wiener processes such that $E(\mathbf{w}_t\mathbf{w}_u') = \mathbf{I}\min(t,u)$, and where the initial condition $\mathbf{x}_0$ is a random variable independent of $\{\mathbf{w}_t;\ t \geqslant t_0\}$. Under conditions on $\mathbf{a}_t(\mathbf{x}_t)$ and $\mathbf{B}_t(\mathbf{x}_t)$ similar to those of Theorem 4.2.3, the solution to (6.111) exists and is unique; see E. Wong [50, Chapter 4] for explicit details. Furthermore, the solution forms a Markov process, which has as a special case stationary Gaussian processes having finite mean square value and rational power-density spectra. The characteristic form for the differential generator $\psi_t(\mathbf{v}|\mathbf{x}_t)$ for $\mathbf{x}_t$ is given by (see Problem 6.5.5)

$$\psi_t(\mathbf{v}|\mathbf{x}_t) = j\langle \mathbf{v}, \mathbf{a}_t(\mathbf{x}_t)\rangle - \tfrac{1}{2}\langle \mathbf{v}, \mathbf{B}_t(\mathbf{x}_t)\mathbf{B}_t'(\mathbf{x}_t)\mathbf{v}\rangle. \tag{6.112}$$

Substitution of this expression into (6.110) and inverse Fourier transformation shows that the probability density function $p_{\mathbf{x}_t}(\mathbf{X};t) \triangleq p_t(\mathbf{X})$ for $\mathbf{x}_t$, assuming it exists, satisfies the partial differential equation

$$\frac{\partial p_t(\mathbf{X})}{\partial t} = -\sum_{i=1}^{n} \frac{\partial\left[a_t^{(i)}(\mathbf{X})p_t(\mathbf{X})\right]}{\partial X_i} + \frac{1}{2}\sum_{i=1}^{n}\sum_{j=1}^{n} \frac{\partial^2\left[(\mathbf{B}_t(\mathbf{X})\mathbf{B}_t'(\mathbf{X}))_{ij}p_t(\mathbf{X})\right]}{\partial X_i \partial X_j} \tag{6.113}$$

$$p_{t_0}(\mathbf{X}) = p_{\mathbf{x}_0}(\mathbf{X}),$$

where n is the dimension of $\mathbf{x}$ and $a^{(i)}$ and X_i are the ith components of $\mathbf{a}$ and $\mathbf{X}$, respectively. This equation is called the *Fokker–Planck* or *forward Kolmogorov equation*. It describes the time evolution of the unconditional or unobserved statistics of $\mathbf{x}$. The differential operator

$$L(\cdot) = -\sum_{i=1}^{n} \frac{\partial\left[a_t^{(i)}(\mathbf{X})(\cdot)\right]}{\partial X_i} + \frac{1}{2}\sum_{i=1}^{n}\sum_{j=1}^{n} \frac{\partial^2\left[(\mathbf{B}_t(\mathbf{X})\mathbf{B}_t'(\mathbf{X}))_{ij}(\cdot)\right]}{\partial X_i \partial X_j} \tag{6.114}$$

is called the *forward Kolmogorov differential operator*.

4. *Continuous-Time Markov Jump Processes.* Denote by $\mathcal{Z}$ the denumerable state space associated with the continuous-time Markov jump process $\{x_t;\ t \geqslant t_0\}$. The sample paths of x are piecewise constant with randomly occurring jumps between the states ζ_i of $\mathcal{Z}$. Define $a_{ik}(t)$ by

$$a_{ik}(t) = \begin{cases} \lim\limits_{\Delta t \downarrow 0} (\Delta t)^{-1} \Pr(x_{t+\Delta t} = \zeta_k | x_t = \zeta_i), & \text{for } i \neq k \\ -\sum\limits_{j(j \neq i)} a_{ij}(t), & \text{for } i = k. \end{cases} \tag{6.115}$$

The function $a_{ik}(t)$ for $i \neq k$ is the instantaneous rate or intensity of jumps from state ζ_i to state ζ_k. The sum of $a_{ij}(t)$ on j for $j \neq i$ is simply the total instantaneous rate of jumps from state ζ_i. These Markov jump processes include as special cases the following: (i) Poisson counting processes when

$$a_{ik}(t) = \begin{cases} \mu_t, & k = i+1 \\ -\mu_t, & k = i \\ 0, & \text{otherwise} \end{cases} \tag{6.116}$$

and $\mathcal{Z}$ is the set of nonnegative integers; (ii) Markov birth processes when

$$a_{ik}(t) = \begin{cases} \mu_t(i), & k = i+1 \\ -\mu_t(i), & k = i \\ 0, & \text{otherwise} \end{cases} \tag{6.117}$$

and $\mathcal{Z}$ is the set of nonnegative integers; and (iii) Markov birth-death processes when

$$a_{ik}(t) = \begin{cases} \mu_t(i), & k = i+1 \\ \nu_t(i), & k = i-1 \\ -\mu_t(i) - \nu_t(i), & k = i \\ 0, & \text{otherwise} \end{cases} \tag{6.118}$$

and $\mathcal{Z}$ is the set of integers. The characteristic form for the differential generator $\psi_t(v|x_t)$ is given by (see Problem 6.5.6)

$$\psi_t(v|x_t = \zeta_i) = \exp(-jv\zeta_i) \sum_k a_{ik}(t) \exp(jv\zeta_k). \tag{6.119}$$

Let $\pi_i(t) = \Pr(x_t = \zeta_i)$ be the probability that x_t is in state ζ_i at time t. Then substitution of (6.119) in (6.110) and inverse Fourier transformation shows that these unconditional state probabilities for x evolve according to the differential equation

$$\frac{d\pi_i(t)}{dt} = \sum_k a_{ki}(t)\pi_k(t),$$

$$\pi_i(t_0) = \Pr(x_0 = \zeta_i). \tag{6.120}$$

This is known as the *forward Kolmogorov equation.*

These examples illustrate the variety of information processes accommodated by the Markov model. The time evolution of the prior statistics of the information process is described by the differential equation (6.110) for the unconditional characteristic function. We now want to obtain an analogous description for the posterior statistics. In particular, we want a differential equation for the conditional characteristic function of $\mathbf{x}_t$ given $\{N_\sigma;\ t_0 \leqslant \sigma < t\}$. Denote this conditional characteristic function by

$$\hat{M}_t(j\mathbf{v}) = E\{\exp[j\langle \mathbf{v}, \mathbf{x}_t\rangle] | N_\sigma;\ t_0 \leqslant \sigma < t\}. \tag{6.121}$$

Notice that the conditional expectation can be interpreted as a minimum mean square-error estimate of $\exp[j\langle \mathbf{v}, \mathbf{x}_t\rangle]$, which accounts for our notation. The main result is in the following theorem.

Theorem 6.5.2 (Evolution of Posterior Statistics). Let $\{N_t;\ t \geqslant t_0\}$ be a doubly stochastic Poisson process with a positive intensity process $\{\lambda_t(\mathbf{x}_t);\ t \geqslant t_0\}$. Assume that the information process $\{\mathbf{x}_t;\ t \geqslant t_0\}$ is a stochastically continuous Markov process and that there exists a function $g_t(\mathbf{v}, \mathbf{x}_t)$ such that (6.108) holds. Denote the characteristic form for the differential generator of $\mathbf{x}$ by $\psi_t(\mathbf{v}|\mathbf{x}_t)$, where this function is given in (6.109). Then the conditional characteristic function $\hat{M}_t(j\mathbf{v})$ of $\mathbf{x}_t$ satisfies the stochastic differential equation

$$\begin{aligned} d\hat{M}_t(j\mathbf{v}) &= \hat{E}\{\exp[j\langle \mathbf{v}, \mathbf{x}_t\rangle]\psi_t(\mathbf{v}|\mathbf{x}_t)\}\,dt \\ &\quad + \hat{E}\left\{\exp[j\langle \mathbf{v}, \mathbf{x}_t\rangle]\left(\lambda_t(\mathbf{x}_t) - \hat{\lambda}_t\right)\right\}\hat{\lambda}_t^{-1}\left(dN_t - \hat{\lambda}_t\,dt\right) \qquad (6.122) \\ \hat{M}_{t_0}(j\mathbf{v}) &= E\left\{\exp[j\langle \mathbf{v}, \mathbf{x}_{t_0}\rangle]\right\} \end{aligned}$$

where $(\hat{\cdot})$ and $\hat{E}\{(\cdot)\}$ both denote the conditional expectation of $(\cdot)$ given $\{N_\sigma;\ t_0 \leqslant \sigma < t\}$.

Proof. In the proof, we use the notation $\mathcal{N}_t$ for the path $\{N_\sigma;\ t_0 \leqslant \sigma < t\}$.

Corresponding to an infinitesimal time increment $[t, t+\Delta t)$, there is an increment $\Delta N_t = N_{t+\Delta t} - N_t$ in the counting path and an increment

$$\Delta \hat{M}_t(j\mathbf{v}) = \hat{M}_{t+\Delta t}(j\mathbf{v}) - \hat{M}_t(j\mathbf{v})$$

in the conditional characteristic function for $\mathbf{x}$, where

$$\begin{aligned}\hat{M}_{t+\Delta t}(j\mathbf{v}) &= E\{\exp[j\langle \mathbf{v}, \mathbf{x}_t\rangle]\exp[j\langle \mathbf{v}, \Delta\mathbf{x}_t\rangle] \mid \mathfrak{N}_{t+\Delta t}\} \\ &= \int \exp[j\langle \mathbf{v}, \mathbf{X}\rangle]\exp[j\langle \mathbf{v}, \mathbf{Y}\rangle]\, dP(\mathbf{X}, \mathbf{Y} \mid \mathfrak{N}_{t+\Delta t}), \end{aligned} \qquad (6.123)$$

where $\Delta\mathbf{x}_t = \mathbf{x}_{t+\Delta t} - \mathbf{x}_t$ is the increment in the information process and $P(\mathbf{X}, \mathbf{Y} \mid \mathfrak{N}_{t+\Delta t})$ is the joint conditional probability distribution function for $\mathbf{x}_t$ and $\Delta\mathbf{x}_t$ given the path $\mathfrak{N}_{t+\Delta t}$. To evaluate $\hat{M}_{t+\Delta t}(j\mathbf{v})$, let us define

$$\hat{M}^{\Delta}_{t+\Delta t}(j\mathbf{v}) = \int \exp[j\langle \mathbf{v}, \mathbf{X}\rangle]\exp[j\langle \mathbf{v}, \mathbf{Y}\rangle]\, dP(\mathbf{X}, \mathbf{Y} \mid \mathfrak{N}_t, \Delta N_t), \qquad (6.124)$$

where $P(\mathbf{X}, \mathbf{Y} \mid \mathfrak{N}_t, \Delta N_t)$ is the joint distribution function for $\mathbf{x}_t$ and $\Delta\mathbf{x}_t$ given the path $\mathfrak{N}_t$ and the increment ΔN_t. We first evaluate $\hat{M}^{\Delta}_{t+\Delta t}$ and then get $\hat{M}_{t+\Delta t}$ from this evaluation by using the fact that $\hat{M}^{\Delta}_{t+\Delta t}$ approaches $\hat{M}_{t+\Delta t}$ with probability one as Δt tends to zero; this holds because $(\mathfrak{N}_t, \Delta N_t) \rightarrow \mathfrak{N}_{t+\Delta t}$ as $\Delta t \downarrow 0$, which implies by the Martingale convergence theorem (see W. Feller [11, p. 236]) that $P(\mathbf{X}, \mathbf{Y} \mid \mathfrak{N}_t, \Delta N_t)$ approaches $P(\mathbf{X}, \mathbf{Y} \mid \mathfrak{N}_{t+\Delta t})$ with probability one as $\Delta t \downarrow 0$. We evaluate $\hat{M}^{\Delta}_{t+\Delta t}$ for fixed ΔN_t; namely, for $\Delta N_t = 0$, $\Delta N_t = 1$, and $\Delta N_t = m$ where $m > 1$. For $\Delta N_t = 0$, we have

$$\hat{M}^{\Delta}_{t+\Delta t}(j\mathbf{v}) = \int \exp[j\langle \mathbf{v}, \mathbf{X}\rangle]\exp[j\langle \mathbf{v}, \mathbf{Y}\rangle]\, dP(\mathbf{X}, \mathbf{Y} \mid \mathfrak{N}_t, \Delta N_t = 0).$$

The finite difference

$$\delta P(\mathbf{X}, \mathbf{Y} \mid \mathfrak{N}_t, \Delta N_t = 0) \triangleq \Pr(\mathbf{x}_t \in (\mathbf{X} - \delta\mathbf{X}, \mathbf{X}], \Delta\mathbf{x}_t \in (\mathbf{Y} - \delta\mathbf{Y}, \mathbf{Y}] \mid \mathfrak{N}_t, \Delta N_t = 0)$$

can be evaluated using Bayes' rule as

$$\delta P(\mathbf{X}, \mathbf{Y} \mid \mathfrak{N}_t, \Delta N_t = 0) = \Pr(\Delta N_t = 0 \mid \mathbf{x}_t \in (\mathbf{X} - \delta\mathbf{X}, \mathbf{X}], \Delta\mathbf{x}_t \in (\mathbf{Y} - \delta\mathbf{Y}, \mathbf{Y}], \mathfrak{N}_t)$$
$$\times \frac{\delta P(\mathbf{X}, \mathbf{Y} \mid \mathfrak{N}_t)}{\Pr(\Delta N_t = 0 \mid \mathfrak{N}_t)}$$

where as $\delta\mathbf{X}\to\mathbf{0}$ and $\delta\mathbf{Y}\to\mathbf{0}$

$$\Pr(\Delta N_t=0|\mathbf{x}_t\in(\mathbf{X}-\delta\mathbf{X},\mathbf{X}],\Delta\mathbf{x}_t\in(\mathbf{Y}-\delta\mathbf{Y},\mathbf{Y}],\mathcal{N}_t)$$
$$=1-\lambda_t(\mathbf{X})\Delta t+o(\Delta t)$$

and where

$$\Pr(\Delta N_t=0|\mathcal{N}_t)=E\{\Pr(\Delta N_t=0|\mathbf{x}_t,\mathcal{N}_t)|\mathcal{N}_t\}$$
$$=1-\hat{\lambda}_t\Delta t+o(\Delta t) \tag{6.125}$$

where $o(\Delta t)/\Delta t\to 0$ as $\Delta t\downarrow 0$. By putting these results together, we obtain for $\Delta N_t=0$ that

$$\hat{M}^{\Delta}_{t+\Delta t}(j\mathbf{v})$$
$$=E\{\exp[j\langle\mathbf{v},\mathbf{x}_t\rangle]\exp[j\langle\mathbf{v},\Delta\mathbf{x}_t\rangle]$$
$$\times(1-\lambda_t(\mathbf{x}_t)\Delta t)|\mathcal{N}_t\}(1-\hat{\lambda}_t\Delta t)^{-1}+o(\Delta t)$$
$$=E\Big\{\exp[j\langle\mathbf{v},\mathbf{x}_t\rangle]\exp[j\langle\mathbf{v},\Delta\mathbf{x}_t\rangle]$$
$$\times\big(1-\lambda_t(\mathbf{x}_t)\Delta t+\hat{\lambda}_t\Delta t\big)|\mathcal{N}_t\Big\}+o(\Delta t). \tag{6.126}$$

Similarly, by using this same procedure, we obtain for $\Delta N_t=1$ that

$$\hat{M}^{\Delta}_{t+\Delta t}(j\mathbf{v})=E\{\exp[j\langle\mathbf{v},\mathbf{x}_t\rangle]\exp[j\langle\mathbf{v},\Delta\mathbf{x}_t\rangle]\lambda_t(\mathbf{x}_t)|\mathcal{N}_t\}\hat{\lambda}_t^{-1}+o(\Delta t) \tag{6.127}$$

and for $\Delta N_t=m$, $m>1$, that

$$\hat{M}^{\Delta}_{t+\Delta t}(j\mathbf{v})=o(\Delta t). \tag{6.128}$$

It then follows that for ΔN_t variable, we have

$$\hat{M}^{\Delta}_{t+\Delta t}(j\mathbf{v})$$
$$=E\Big\{\exp[j\langle\mathbf{v},\mathbf{x}_t\rangle]\exp[j\langle\mathbf{v},\Delta\mathbf{x}_t\rangle]\big(1-\lambda_t(\mathbf{x}_t)\Delta t+\hat{\lambda}_t\Delta t\big)|\mathcal{N}_t\Big\}\delta_{0,\Delta N_t}$$
$$+E\{\exp[j\langle\mathbf{v},\mathbf{x}_t\rangle]\exp[j\langle\mathbf{v},\Delta\mathbf{x}_t\rangle]\lambda_t(\mathbf{x}_t)|\mathcal{N}_t\}\hat{\lambda}_t^{-1}\delta_{1,\Delta N_t}+o(\Delta t). \tag{6.129}$$

Here, δ_{ij} is the Kronecker delta function. Because $\delta_{0,\Delta N_t}$ approaches $1-\Delta N_t$

and $\delta_{1,\Delta N_t}$ approaches ΔN_t with probability one as $\Delta t \downarrow 0$, we can write

$$\hat{M}^{\Delta}_{t+\Delta t}(j\mathbf{v}) = E\{\exp[j\langle \mathbf{v}, \mathbf{x}_t\rangle]\exp[j\langle \mathbf{v}, \Delta\mathbf{x}_t\rangle] | \mathfrak{N}_t\}$$

$$+ E\left\{\exp[j\langle \mathbf{v}, \mathbf{x}_t\rangle]\exp[j\langle \mathbf{v}, \Delta\mathbf{x}_t\rangle]\big(\lambda_t(\mathbf{x}_t) - \hat{\lambda}_t\big) | \mathfrak{N}_t\right\}\hat{\lambda}_t^{-1}\big(\Delta N_t - \hat{\lambda}_t \Delta t\big) + o(\Delta t). \tag{6.130}$$

By substracting the conditional characteristic function for $\mathbf{x}_t$ given $\mathfrak{N}_t$, we therefore have

$$\hat{M}^{\Delta}_{t+\Delta t}(j\mathbf{v}) - \hat{M}_t(j\mathbf{v})$$

$$= E\left\{\exp[j\langle \mathbf{v}, \mathbf{x}_t\rangle](\Delta t)^{-1} E(\exp[j\langle \mathbf{v}, \Delta\mathbf{x}_t\rangle] - 1 | \mathbf{x}_t, \mathfrak{N}_t) | \mathfrak{N}_t\right\}\Delta t$$

$$+ E\left\{\exp[j\langle \mathbf{v}, \mathbf{x}_t\rangle]\exp[j\langle \mathbf{v}, \Delta\mathbf{x}_t\rangle]\big(\lambda_t(\mathbf{x}_t) - \hat{\lambda}_t\big) | \mathfrak{N}_t\right\}\hat{\lambda}_t^{-1}\big(\Delta N_t - \hat{\lambda}_t \Delta t\big)$$

$$+ o(\Delta t). \tag{6.131}$$

However,

$$E(\exp[j\langle \mathbf{v}, \Delta\mathbf{x}_t\rangle] - 1 | \mathbf{x}_t, \mathfrak{N}_t) = E\{\exp[j\langle \mathbf{v}, \Delta\mathbf{x}_t\rangle] - 1 | \mathbf{x}_t\}$$

because $\{\mathbf{x}_t;\ t \geqslant t_0\}$ is a Markov process. Thus, we can now establish the theorem as follows:

$$d\hat{M}_t(j\mathbf{v}) = \lim_{\Delta t \downarrow 0}\left[\hat{M}_{t+\Delta t}(j\mathbf{v}) - \hat{M}_t(j\mathbf{v})\right]$$

$$= \lim_{\Delta t \downarrow 0}\left[\hat{M}^{\Delta}_{t+\Delta t}(j\mathbf{v}) - \hat{M}_t(j\mathbf{v})\right]$$

$$= E\{\exp[j\langle \mathbf{v}, \mathbf{x}_t\rangle]\psi_t(\mathbf{v}|\mathbf{x}_t) | \mathfrak{N}_t\}\, dt$$

$$+ E\left\{\exp[j\langle \mathbf{v}, \mathbf{x}_t\rangle]\left[\lambda_t(\mathbf{x}_t) - \hat{\lambda}_t\right] | \mathfrak{N}_t\right\}\hat{\lambda}_t^{-1}\big(dN_t - \hat{\lambda}_t\, dt\big), \tag{6.132}$$

where the interchange of limit and expectation required for the first term in the last equality is justified by assumption (6.108) and the bounded convergence theorem, Theorem 1.4.1. The second term follows from the stochastic continuity of **x**. This ends the proof of Theorem 6.5.2.

Equation (6.122) is one of our basic results for solving the nonlinear filtering problem for observed doubly stochastic Poisson processes. The procedure for implementing an estimate of some specified function $\mathbf{h}_t(\mathbf{x}_t)$ of $\mathbf{x}_t$ is clearly very complicated in general. First, it is necessary to solve (6.122) for $\hat{M}_t(j\mathbf{v})$ for the given observed counting path $\{N_\sigma;\ t_0 \leqslant \sigma < t\}$. Then this characteristic function must be inverse Fourier transformed to determine the posterior probability density or distribution function for $\mathbf{x}_t$ given $\{N_\sigma;\ t_0 \leqslant \sigma < t\}$. Finally, the desired estimate must be determined by evaluating the expectation

$$\hat{\mathbf{h}}_t = E[\mathbf{h}_t(\mathbf{x}_t) | N_\sigma;\ t_0 \leqslant \sigma < t].$$

These steps are generally intractable analytically so that numerical techniques will be required. As we shall see, the equations are well suited for this.

We now give several examples that illustrate the use of (6.122) in solving the filtering problem for Markovian information processes.

Example 6.5.4 (Estimation of Random Variables). Suppose that the information process is simply a time-independent vector $\mathbf{x}$ of random variables with prior probability density function $p_{\mathbf{x}}(\mathbf{X})$. In this case, $\psi_t(\mathbf{v}|\mathbf{x}_t)$ of (6.109) is zero. If we let $p_t(\mathbf{X}|N_\sigma;\ t_0 \leqslant \sigma < t)$ denote the conditional probability density function for $\mathbf{x}_t$ given $\{N_\sigma;\ t_0 \leqslant \sigma < t\}$, and if we inverse Fourier transform (6.122), then we obtain

$$dp_t(\mathbf{X}|N_\sigma;\ t_0 \leqslant \sigma < t) = p_t(\mathbf{X}|N_\sigma;\ t_0 \leqslant \sigma < t) \times \left[\lambda_t(\mathbf{X}) - \hat{\lambda}_t\right]\hat{\lambda}_t^{-1}\left(dN_t - \hat{\lambda}_t\,dt\right),$$

$$p_{t_0}(\mathbf{X}|N_{t_0}) = p_{\mathbf{x}}(\mathbf{X}), \tag{6.133}$$

where

$$\hat{\lambda}_t = E[\lambda_t(\mathbf{x})|N_\sigma;\ t \leqslant \sigma < t] = \int \cdots \int \lambda_t(\mathbf{X}) p_t(\mathbf{X}|N_\sigma;\ t_0 \leqslant \sigma < t)\, dX_1 \cdots dX_n. \tag{6.134}$$

This equation provides three lines of departure for solving the problem of estimating a specified function of $\mathbf{x}$ given the observed counting record. The first is that of solving the equation for the posterior density. This can be accomplished, as will be seen momentarily. However, in more general situations where the information process is time dependent, we will be unable to pursue this line because of analytic intractability. The second line is that of numerical solutions using (6.133) as an updating algorithm for the posterior density. This idea will be described in more detail after we examine the analytic solution of the equation. The third line is to use (6.133) to obtain estimates that are feasible to generate in practice but which are only reasonable approximations to the minimum mean square-error estimate. This line is developed later in Section 6.5.3.

Our procedure for solving (6.133) analytically makes use of the differential rule of Theorem 4.2.2, a rule that will be used frequently in the following sections. Define $\phi_t(\mathbf{X}) = \ln p_t(\mathbf{X}|N_\sigma;\ t_0 \leqslant \sigma < t)$. Then from (6.133) and the differential rule of Theorem 4.2.2 [see particularly (4.72) and (4.73)], we have

$$d\phi_t(\mathbf{X}) = -\left(\lambda_t(\mathbf{X}) - \hat{\lambda}_t\right)dt + \ln\left(\lambda_t(\mathbf{X})\hat{\lambda}_t^{-1}\right)dN_t$$
$$\phi_{t_0}(\mathbf{X}) = \ln p_{\mathbf{x}}(\mathbf{X}) \tag{6.135}$$

Consequently, we see that (6.133) is equivalent to

$$p_t(\mathbf{X}|N_\sigma;\ t_0 \leqslant \sigma < t) = p_{\mathbf{x}}(\mathbf{X}) \exp\left[-\int_{t_0}^{t} \left(\lambda_\sigma(\mathbf{X}) - \hat{\lambda}_\sigma\right) d\sigma \right.$$
$$\left. + \int_{t_0}^{t} \ln\left(\lambda_\sigma(\mathbf{X})\hat{\lambda}_\sigma^{-1}\right) dN_\sigma \right]. \tag{6.136}$$

As $\hat{\lambda}_t$ requires an integration with respect to the posterior density $p_t(\mathbf{X}|N_\sigma;\ t_0 \leqslant \sigma < t)$, equation (6.136) is a nonlinear integral equation for the posterior density. We have traded the nonlinear differential-integral equation (6.133) for the nonlinear integral equation (6.136). The integral equation can be solved easily using the expressions we have developed in Section 6.3 for the sample-function density. In particular, by equating (6.45) and (6.71), we have the relation

$$E\left\{ \exp\left[-\int_{t_0}^{t} \lambda_\sigma(\mathbf{x})\, d\sigma + \int_{t_0}^{t} \ln\lambda_\sigma(\mathbf{x})\, dN_\sigma \right] \right\}$$
$$= \exp\left[-\int_{t_0}^{t} \hat{\lambda}_\sigma d\sigma + \int_{t_0}^{t} \ln\hat{\lambda}_\sigma\, dN_\sigma \right]. \tag{6.137}$$

Substitution of this expression into (6.136) yields the desired solution for the posterior density:

$$p_t(\mathbf{X}|N_\sigma;\ t_0 \leqslant \sigma < t) = \frac{\exp[H_t(\mathbf{X})]\, p_{\mathbf{x}}(\mathbf{X})}{E\{\exp[H_t(\mathbf{x})]\}}, \tag{6.138}$$

where we define

$$H_t(\mathbf{x}) = -\int_{t_0}^{t} \lambda_\sigma(\mathbf{x})\, d\sigma + \int_{t_0}^{t} \ln \lambda_\sigma(\mathbf{x})\, dN_\sigma. \tag{6.139}$$

Equation (6.138) in principle solves the nonlinear filtering problem for information processes that are simply time-independent random variables. From Assertion 6.5.1 and (6.138), we have that the minimum mean square-error estimate of a specified function $\mathbf{h}_t(\mathbf{x})$ of $\mathbf{x}$ is given by

$$\begin{aligned}\hat{\mathbf{h}}_t &= E[\mathbf{h}_t(\mathbf{x})|N_\sigma;\ t_0 \leqslant \sigma < t]\\ &= \frac{E\{\mathbf{h}_t(\mathbf{x})\exp[H_t(\mathbf{x})]\}}{E\{\exp[H_t(\mathbf{x})]\}}.\end{aligned} \tag{6.140}$$

In a limited number of situations, the required integrations with respect to the prior statistics of $\mathbf{x}$ can be performed analytically to give a closed-form expression for $\hat{\mathbf{h}}_t$ in terms of the counting path. Such an example is given in Problem 6.5.10. More generally, however, numerical evaluation of the integrals is required. The procedure for determining $\hat{\mathbf{h}}_t$ numerically is first to form $H_t(\mathbf{X})$ from the observed counting path $\{N_\sigma;\ t_0 \leqslant \sigma < t\}$ according to (6.139). Then the two integrals in (6.140) must be evaluated. These integrations can, at least in principle, be done by multidimensional Gaussian quadratures or multidimensional Monte Carlo techniques (see S. Haber [18]). The advantage of this approach is that the resulting estimate is exact, the only source of error being in the numerical implementation. However, for the dimension of $\mathbf{x}$ greater than one or two, as is often the case in applications, the numerical integrations may require a substantial computational capability, certainly more than that available with a small computer. Moreover, a large amount of computer memory will be required for storage of the posterior density.

Equation (6.133) can also be viewed as defining an updating algorithm according to which the prior density $p_{\mathbf{x}}(\mathbf{X})$ is propagated forward in time to form the posterior density as data are accumulated. For this interpretation, it is convenient to rewrite (6.133) in the finite-difference form

$$\begin{aligned}&p_{t+\Delta t}(\mathbf{X}|N_\sigma;\ t_0 \leqslant \sigma < t+\Delta t)\\ &\quad= p_t(\mathbf{X}|N_\sigma;\ t_0 \leqslant \sigma < t)\left\{1 + \left(\lambda_t(\mathbf{X}) - \hat{\lambda}_t\right)\hat{\lambda}_t^{-1}\left(\Delta N_t - \hat{\lambda}_t \Delta t\right)\right\} + o(\Delta t).\end{aligned} \tag{6.141}$$

For Δt sufficiently small, the term $o(\Delta t)$ can be disregarded and $\Delta N_t = N_{t+\Delta t} - N_t$ will be either zero or one according to the nonoccurrence or occurrence of a point in $[t, t+\Delta t)$. The posterior density at time $t+\Delta t$ can be evaluated by knowing the density at time t and whether or not a point occurs in $[t, t+\Delta t)$; this evaluation requires an integration with respect to the density at t in order to determine $\hat{\lambda}_t$. The estimate $\hat{\mathbf{h}}_t$ of a specified function $\mathbf{h}_t(\mathbf{x})$ of $\mathbf{x}$ can be determined at each stage of the updating process by performing the integration

$$\hat{\mathbf{h}}_t = \int \cdots \int \mathbf{h}_t(\mathbf{X}) p_t(\mathbf{X}|N_\sigma;\ t_0 \leqslant \sigma < t)\, dX_1 \cdots dX_n$$

with respect to the updated density. Some simulation results illustrating this updating idea are summarized in Figures 6.9 and 6.10. The results in Figure 6.9*a* and 6.9*b* were obtained by assuming that $\lambda_t(x) = x$, where x had a uniform prior probability density between 1.0 and 2.0. The value of x used in the simulation was selected at random with a random number generator. The numerically generated time evolution for $p_t(x|N_\sigma;\ t_0 \leqslant \sigma < t)$ is shown as well as for the minimum mean square-error estimate of $\lambda_t(x) = x$.

The results in Figures 6.10*a*–6.10*c* were obtained by assuming that $\lambda_t(x) = 100\exp(-xt)$ for $t \geqslant 0$ where x had a uniform prior probability density between 0.5 and 1.0. As before, the value of x used in the simulation was selected at random with a random number generator. The numerically generated time evolution for $p_t(x|N_\sigma;\ t_0 \leqslant \sigma < t)$ is shown as well as for $\hat{\lambda}_t$ and $\hat{x}_t$. The results shown in both these simulation examples are typical of those obtained with other randomly selected values for x.

The view of (6.133) as defining an updating algorithm for generating estimates numerically clearly does not obviate the need for the same multidimensional integrations and large storage requirements we encountered with the use of the analytic expression (6.140). This motivates our later study in Section 6.5.3 of estimates $\mathbf{h}_t^*$ that only approximate the exact minimum mean square-error estimate $\hat{\mathbf{h}}_t$.

Example 6.5.5 (Estimation of a Poisson Driven Markov Process). Suppose that the information process $\{\mathbf{x}_t;\ t \geqslant t_0\}$ is a Poisson driven Markov process satisfying (4.38) with $\mathbf{h}_t(\mathbf{x}_t, \mathbf{u}) = \mathbf{b}_t(\mathbf{u})$ not a function of $\mathbf{x}_t$. The function $\psi_t(\mathbf{v}|\mathbf{x}_t)$ required in (6.122) is given in (4.100). Substitution of this function into (6.122) and inverse Fourier transformation shows that the posterior probability density for $\mathbf{x}_t$ given $\{N_\sigma;\ t_0 \leqslant \sigma < t\}$ satisfies

$$\begin{aligned} &dp_t(\mathbf{X}|N_\sigma;\ t_0 \leqslant \sigma < t) \\ &\quad = -\sum_{i=1}^{n} \frac{\partial\left[a_t^{(i)}(\mathbf{X}) p_t(\mathbf{X}|N_\sigma;\ t_0 \leqslant \sigma < t)\right]}{\partial X_i\, dt} \\ &\qquad + \lambda_t \int_{\mathfrak{U}} \left[p_t(\mathbf{X} - \mathbf{b}_t(\mathbf{U})|N_\sigma;\ t_0 \leqslant \sigma < t) - p_t(\mathbf{X}|N_\sigma;\ t_0 \leqslant \sigma < t)\right] dP_{\mathbf{u}}(\mathbf{U})\, dt \\ &\qquad + p_t(\mathbf{X}|N_\sigma;\ t_0 \leqslant \sigma < t)\left[\lambda_t(\mathbf{X}) - \hat{\lambda}_t\right]\hat{\lambda}_t^{-1}\left(dN_t - \hat{\lambda}_t\, dt\right) \end{aligned}$$

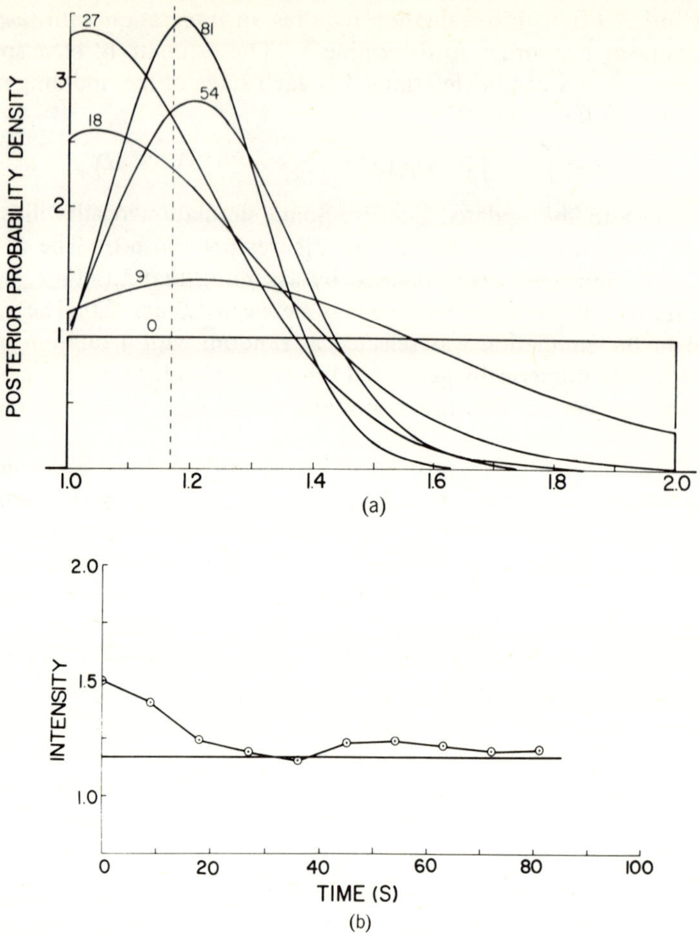

Figure 6.9. (a) Time evolution for the posterior probability density function of x when $\lambda_t(x)=x$. The true value of x generated at random, uniformly between 1.0 and 2.0, is indicated by the dashed line at $x=1.170$. The density is shown for $t=0$, 9, 18, 27, 54, and 81 seconds. All densities are zero outside the interval [1,2]. (b) Time evolution for the minimum mean square-error estimate of $\lambda_t(x)=x$. The solid curve is the true value of $\lambda_t(x)=1.170$, and the data points are the minimum mean square-error estimates of this value taken at 9.0 second intervals.

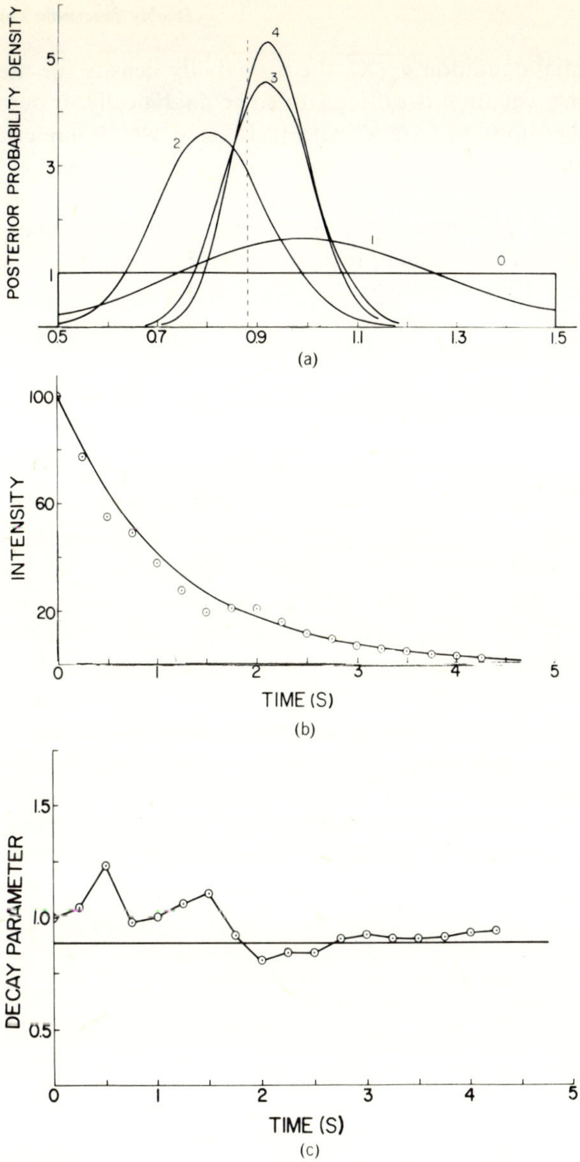

Figure 6.10. (a) Time evolution for the posterior probability density of x when $\lambda_t(x) = 100\exp(-xt)$. The true value of x generated at random, uniformly between 0.5 and 1.5, equals 0.8822. The posterior density is shown for $t=0$, 1, 2, 3, and 4 seconds. All densities are zero outside the interval [0.5, 1.5]. (b) Time evolution for the minimum mean square-error estimate for $\lambda_t(x)=100\exp[-xt]$. The solid curve is the true value of $\lambda_t(x)=100\exp[-0.8822t]$, and the data points are the minimum mean square-error estimates of this curve taken at 0.25-second intervals. (c) Time evolution for the minimum mean square-error estimate of x. The solid curve is the true parameter value of 0.8822 and the data points are the minimum mean square-error estimates of this value at 0.25-second intervals.

with the initial condition $p_{\mathbf{x}_0}(\mathbf{X})$, the probability density for the initial state $\mathbf{x}_0$. While this equation is difficult to solve analytically, it does provide an updating algorithm, in parallel with (6.141), by which numerical solutions are possible.

Example 6.5.6 (Estimation of a Markov Diffusion Process). If $\{\mathbf{x}_t; t \geqslant t_0\}$ is a Markov diffusion process satisfying (6.111), the function $\psi_t(\mathbf{v}|\mathbf{x}_t)$ required in (6.122) is given in (6.112). Substitution of this function into (6.122) and inverse Fourier transformation shows that the posterior probability density for $\mathbf{x}_t$ given $\{N_\sigma;\ t_0 \leqslant \sigma < t\}$ satisfies

$$dp_t(\mathbf{X}|N_\sigma;\ t_0 \leqslant \sigma < t) = L[p_t(\mathbf{X}|N_\sigma;\ t_0 \leqslant \sigma < t)]\,dt$$

$$+ p_t(\mathbf{X}|N_\sigma;\ t_0 \leqslant \sigma < t)\big[\lambda_t(\mathbf{X}) - \hat{\lambda}_t\big]\hat{\lambda}_t^{-1}\big(dN_t - \hat{\lambda}_t\,dt\big), \tag{6.142}$$

where $L(\cdot)$ is the forward Kolmogorov differential operator defined in (6.114). This equation is the counterpart for observed doubly-stochastic Poisson processes of the Kushner–Stratonovich equation encountered in the additive white Gaussian noise observation model (see A. Jazwinski [20], p. 178). Again, the equation is difficult to solve analytically, but it does provide an updating algorithm analogous to (6.141) by which numerical solutions are possible.

Example 6.5.7 (Conditional Moments for a Markov Diffusion Process). The results of this example will be important in Section 6.5.3.

Evolution equations for the conditional moments of the information process can be obtained using (6.122) and the moment generating properties of the conditional characteristic function. In particular, assuming they exist, the first two conditional moments of $\mathbf{x}_t$ given $\{N_\sigma;\ t_0 \leqslant \sigma < t\}$ are given in terms of the conditional characteristic function $\hat{M}_t(j\mathbf{v})$ by the relations

$$\hat{\mathbf{x}}_t = \left.\frac{j^{-1}\partial \hat{M}_t(j\mathbf{v})}{\partial \mathbf{v}}\right|_{\mathbf{v}=\mathbf{0}} \tag{6.143}$$

and

$$\hat{E}\{\mathbf{x}_t\mathbf{x}_t'\} = \left.\frac{j^{-2}\partial^2 \hat{M}_t(j\mathbf{v})}{\partial \mathbf{v}^2}\right|_{\mathbf{v}=\mathbf{0}} \tag{6.144}$$

Here, $\partial(\cdot)/\partial\mathbf{v}$ denotes the gradient vector of $(\cdot)$ with respect to $\mathbf{v}$, a vector whose ith component is $\partial(\cdot)/\partial v_i$. And $\partial^2(\cdot)/\partial \mathbf{v}^2$ denotes a symmetric matrix whose i, jth element is $\partial^2(\cdot)/\partial v_i \partial v_j$.

We illustrate this idea when the information process is the Markov diffusion process of the preceding example. Recall that $\psi_t(\mathbf{v}|\mathbf{x}_t)$ is given in (6.112). We will need the relations

$$\frac{\partial \exp[j\langle \mathbf{v}, \mathbf{x}_t \rangle]\psi_t(\mathbf{v}|\mathbf{x}_t)}{\partial \mathbf{v}}$$
$$= (j\mathbf{x}_t\psi_t(\mathbf{v}|\mathbf{x}_t) + j\mathbf{a}_t(\mathbf{x}_t) - \mathbf{B}_t(\mathbf{x}_t)\mathbf{B}_t'(\mathbf{x}_t)\mathbf{v})\exp[j\langle \mathbf{v}, \mathbf{x}_t \rangle] \quad (6.145)$$

and

$$\frac{\partial^2 \exp[j\langle \mathbf{v}, \mathbf{x}_t \rangle]\psi_t(\mathbf{v}|\mathbf{x}_t)}{\partial \mathbf{v}^2}$$
$$= (j^2\mathbf{x}_t\mathbf{x}_t'\psi_t(\mathbf{v}|\mathbf{x}_t) + j^2\mathbf{a}_t(\mathbf{x}_t)\mathbf{x}_t' + j^2\mathbf{x}_t\mathbf{a}_t'(\mathbf{x}_t) - \mathbf{B}_t(\mathbf{x}_t)\mathbf{B}_t'(\mathbf{x}_t)$$
$$- j\mathbf{B}_t(\mathbf{x}_t)\mathbf{B}_t'(\mathbf{x}_t)\mathbf{v}\mathbf{x}_t' - j\mathbf{x}_t\mathbf{v}'\mathbf{B}_t(\mathbf{x}_t)\mathbf{B}_t'(\mathbf{x}_t))\exp[j\langle \mathbf{v}, \mathbf{x}_t \rangle]. \quad (6.146)$$

These relations can be verified by direct evaluation.

By taking the derivative of both sides of (6.122) with respect to $\mathbf{v}$, using (6.145), setting $\mathbf{v}=\mathbf{0}$, and using (6.143), the equation we obtain for the conditional mean is

$$d\hat{\mathbf{x}}_t = \hat{\mathbf{a}}_t\,dt + \hat{E}\left\{\mathbf{x}_t\left[\lambda_t(\mathbf{x}_t) - \hat{\lambda}_t\right]\right\}\hat{\lambda}_t^{-1}\left(dN_t - \hat{\lambda}_t\,dt\right),$$
$$\hat{\mathbf{x}}_{t_0} = E(\mathbf{x}_0). \quad (6.147)$$

Similarly, by taking the second derivative with respect to $\mathbf{v}$, using (6.146), setting $\mathbf{v}=\mathbf{0}$, and using (6.144), the equation we obtain for the conditional second moment matrix is

$$d\hat{E}\left\{\mathbf{x}_t\mathbf{x}_t'\right\} = \hat{E}\left\{\mathbf{a}_t(\mathbf{x}_t)\mathbf{x}_t' + \mathbf{x}_t\mathbf{a}_t'(\mathbf{x}_t) + \mathbf{B}_t(\mathbf{x}_t)\mathbf{B}_t'(\mathbf{x}_t)\right\}dt$$
$$+ \hat{E}\left\{\mathbf{x}_t\mathbf{x}_t'\left[\lambda_t(\mathbf{x}_t) - \hat{\lambda}_t\right\}\hat{\lambda}_t^{-1}\left(dN_t - \hat{\lambda}_t\,dt\right),$$
$$\hat{E}\left\{\mathbf{x}_{t_0}\mathbf{x}_{t_0}'\right\} = E(\mathbf{x}_0\mathbf{x}_0'). \quad (6.148)$$

These equations for the conditional first and second moments of $\mathbf{x}_t$ may not appear particularly useful because of the unknown and difficult to determine expectations involved on the right-hand sides. Nevertheless, we shall find them a natural point of departure in Section 6.5.3 for formulating reasonable approximations to the minimum mean square-error estimate of $\mathbf{x}_t$. For this later prupose, it is convenient to carry (6.148) somewhat further. In particular, we will need an equation for the conditional error-covariance matrix $\hat{\boldsymbol{\Sigma}}_t$ defined by

$$\hat{\boldsymbol{\Sigma}}_t = E[(\mathbf{x}_t - \hat{\mathbf{x}}_t)(\mathbf{x}_t - \hat{\mathbf{x}}_t)' | N_\sigma;\ t_0 \leqslant \sigma < t] = \hat{E}\{\mathbf{x}_t\mathbf{x}_t'\} - \hat{\mathbf{x}}_t\hat{\mathbf{x}}_t' \qquad (6.149)$$

We have an expression for the differential $d\hat{E}\{\mathbf{x}\mathbf{x}'\}$ in (6.148). Thus, to get an expression for $d\hat{\boldsymbol{\Sigma}}$, we need to investigate the differential $d\hat{\mathbf{x}}\hat{\mathbf{x}}'$. We do this by using (6.147) and the differential rule of Theorem 4.2.2. Let $\mathbf{e}_i$ denote a unit vector with ith element equal to one and all other elements equal to zero. Also, let $\varphi_{ij}(\hat{\mathbf{x}}_t) = \mathbf{e}_i'\hat{\mathbf{x}}_t\hat{\mathbf{x}}_t'\mathbf{e}_j$. Then

$$\frac{\partial \varphi_{ij}(\hat{\mathbf{x}}_t)}{\partial \hat{\mathbf{x}}_t} = (\mathbf{e}_i\mathbf{e}_j' + \mathbf{e}_j\mathbf{e}_i')\hat{\mathbf{x}}_t.$$

From (6.147) and the differential rule, we deduce that

$$\begin{aligned} d\varphi_{ij}(\hat{\mathbf{x}}_t) = {} & \Big\langle (\mathbf{e}_i\mathbf{e}_j' + \mathbf{e}_j\mathbf{e}_i')\hat{\mathbf{x}}_t, \hat{\mathbf{a}}_t - \hat{E}\big\{\mathbf{x}_t\big[\lambda_t(\mathbf{x}_t) - \hat{\lambda}_t\big]\big\}\Big\rangle\, dt \\ & + \mathbf{e}_i'\hat{E}\big\{(\mathbf{x}_t\hat{\mathbf{x}}_t' + \hat{\mathbf{x}}_t\mathbf{x}_t')\big[\lambda_t(x_t) - \hat{\lambda}_t\big]\big\}\mathbf{e}_j\hat{\lambda}_t^{-1}dN_t \\ & + \mathbf{e}_i'\hat{E}\big\{\mathbf{x}_t\big[\lambda_t(\mathbf{x}_t) - \hat{\lambda}_t\big]\big\}\hat{E}\big\{\mathbf{x}_t'\big[\lambda_t(\mathbf{x}_t) - \hat{\lambda}_t\big]\big\}\mathbf{e}_j\hat{\lambda}_t^{-2}dN_t. \end{aligned} \qquad (6.150)$$

Consequently,

$$\begin{aligned} d\hat{\mathbf{x}}_t\hat{\mathbf{x}}_t' = {} & (\hat{\mathbf{a}}_t\hat{\mathbf{x}}_t' + \hat{\mathbf{x}}_t\hat{\mathbf{a}}_t')dt \\ & + \hat{E}\big\{(\mathbf{x}_t\hat{\mathbf{x}}_t' + \hat{\mathbf{x}}_t\mathbf{x}_t')\big[\lambda_t(\mathbf{x}_t) - \hat{\lambda}_t\big]\big\}\hat{\lambda}_t^{-1}\big(dN_t - \hat{\lambda}_t dt\big) \\ & + \hat{E}\big\{\mathbf{x}_t\big[\lambda_t(\mathbf{x}_t) - \hat{\lambda}_t\big]\big\}\hat{E}\big\{\mathbf{x}_t'\big[\lambda_t(\mathbf{x}_t) - \hat{\lambda}_t\big]\big\}\hat{\lambda}_t^{-2}dN_t. \end{aligned} \qquad (6.151)$$

Our desired equation for the conditional error-covariance matrix is now obtained from (6.148), (6.151), and the relation

$$d\hat{\boldsymbol{\Sigma}}_t = d\hat{E}\{\mathbf{x}_t\mathbf{x}_t'\} - d\hat{\mathbf{x}}_t\hat{\mathbf{x}}_t'.$$

The result is

$$d\hat{\Sigma}_t = \hat{E}\left\{\mathbf{a}_t(\mathbf{x}_t)\hat{\mathbf{x}}_t' - \hat{\mathbf{a}}_t\hat{\mathbf{x}}_t' + \hat{\mathbf{x}}_t\mathbf{a}'(\mathbf{x}_t) - \hat{\mathbf{x}}_t\hat{\mathbf{a}}_t' + \mathbf{B}_t(\mathbf{x}_t)\mathbf{B}_t'(\mathbf{x}_t)\right\}dt$$

$$+\hat{E}\left\{(\mathbf{x}_t\mathbf{x}_t' - \mathbf{x}_t\hat{\mathbf{x}}_t' - \hat{\mathbf{x}}_t\mathbf{x}_t')\left[\lambda_t(\mathbf{x}_t) - \hat{\lambda}_t\right]\right\}\hat{\lambda}_t^{-1}\left(dN_t - \hat{\lambda}_t\,dt\right)$$

$$-\hat{E}\left\{\mathbf{x}_t\left[\lambda_t(\mathbf{x}_t) - \hat{\lambda}_t\right]\right\}\hat{E}\left\{\mathbf{x}_t'\left[\lambda_t(\mathbf{x}_t) - \hat{\lambda}_t\right]\right\}\hat{\lambda}_t^{-2}dN_t,$$

$$\hat{\Sigma}_{t_0} = E(\mathbf{x}_0\mathbf{x}_0') - E(\mathbf{x}_0)E(\mathbf{x}_0'). \tag{6.152}$$

We summarize these results in the following theorem. Equations (6.147) and (6.152) have been modified slightly in the theorem for later use in Section 6.5.3.

Theorem 6.5.3 (Conditional Mean and Error Covariance for a Markov Diffusion Process). Let $\{N_t;\ t \geqslant t_0\}$ be a doubly-stochastic Poisson process with a positive intensity process $\{\lambda_t(\mathbf{x}_t);\ t \geqslant t_0\}$. Suppose the information process $\{\mathbf{x}_t;\ t \geqslant t_0\}$ is a Markov diffusion satisfying (6.111). Suppose, further, that the conditional first and second moments for $\mathbf{x}_t$ given $\{N_\sigma;\ t_0 \leqslant \sigma < t\}$ exist. Then the conditional mean $\hat{\mathbf{x}}_t$, or minimum mean square-error estimate, of $\mathbf{x}_t$ and the conditional error-covariance matrix $\hat{\Sigma}_t$ satisfy

$$d\hat{\mathbf{x}}_t = \hat{\mathbf{a}}_t\,dt + \hat{E}\left\{(\mathbf{x}_t - \hat{\mathbf{x}}_t)\left[\lambda_t(\mathbf{x}_t) - \hat{\lambda}_t\right]\right\}\hat{\lambda}_t^{-1}\left(dN_t - \hat{\lambda}_t\,dt\right),$$

$$\hat{\mathbf{x}}_{t_0} = E(\mathbf{x}_0), \tag{6.153}$$

and

$$d\hat{\Sigma}_t = \hat{E}\left\{[\mathbf{a}_t(\mathbf{x}_t) - \hat{\mathbf{a}}_t][\mathbf{x}_t - \hat{\mathbf{x}}_t]' + [\mathbf{x}_t - \hat{\mathbf{x}}_t][\mathbf{a}_t(\mathbf{x}_t) - \hat{\mathbf{a}}_t]' + \mathbf{B}_t(\mathbf{x}_t)\mathbf{B}_t'(\mathbf{x}_t)\right\}dt$$

$$+\hat{E}\left\{(\mathbf{x}_t - \hat{\mathbf{x}}_t)(\mathbf{x}_t - \hat{\mathbf{x}}_t)'\left[\lambda_t(\mathbf{x}_t) - \hat{\lambda}_t\right]\right\}\hat{\lambda}_t^{-1}\left(dN_t - \hat{\lambda}_t\,dt\right)$$

$$-\hat{E}\left\{(\mathbf{x}_t - \hat{\mathbf{x}}_t)\left[\lambda_t(\mathbf{x}_t) - \hat{\lambda}_t\right]\right\}\hat{E}\left\{(\mathbf{x}_t - \hat{\mathbf{x}}_t)'\left[\lambda_t(\mathbf{x}_t) - \hat{\lambda}_t\right]\right\}\hat{\lambda}_t^{-2}dN_t,$$

$$\hat{\Sigma}_{t_0} = E(\mathbf{x}_0\mathbf{x}_0') - E(\mathbf{x}_0)E(\mathbf{x}_0'). \tag{6.154}$$

Example 6.5.8 (Estimation of a Markov Jump Process). Suppose that the information process is a continuous-time Markov jump process with states $\{\zeta_i\}$. Then the function $\psi_t(v|x_t)$ required in (6.122) is given in (6.119). Substitution of this function into (6.122) shows that the conditional state occupancy probabilities $\hat{\pi}_i(t) \triangleq \Pr(x_t=\zeta_i|N_\sigma;\ t_0\leqslant\sigma<t)$ given $\{N_\sigma;\ t_0\leqslant\sigma<t\}$ satisfy

$$d\hat{\pi}_i(t)=\sum_k a_{ki}(t)\hat{\pi}_k(t)dt+\hat{\pi}_i(t)\big[\lambda_t(\zeta_i)-\hat{\lambda}_t\big]\hat{\lambda}_t^{-1}\big(dN_t-\hat{\lambda}_t dt\big),$$

$$\hat{\pi}_i(t_0)=\Pr(x_0=\zeta_i), \tag{6.155}$$

where

$$\hat{\lambda}_t=\sum_i \lambda_t(\zeta_i)\hat{\pi}_i(t) \tag{6.156}$$

and where the instantaneous transition rates $\{a_{ik}(t)\}$ are defined in (6.115). Equation (6.155) provides an updating algorithm by which the conditional state occupancy probabilities of the information process can be generated numerically. This algorithm has the following recursive interpretation. Suppose a point is observed to occur at time s, where $s\geqslant t_0$, and no points occur on the interval (s,t). Then from (6.155) we have for $t>s$ that

$$\frac{d\hat{\pi}_i(t)}{dt}=\sum_k a_{ki}(t)\hat{\pi}_k(t)-\hat{\pi}_i(t)\big[\lambda_t(\zeta_i)-\hat{\lambda}_t\big], \tag{6.157}$$

$$\hat{\pi}_i(s+)=\hat{\pi}_i(s-)\lambda_{s-}(\zeta_i)\hat{\lambda}_{s-}^{-1}. \tag{6.158}$$

Equations (6.157) and (6.158) can be used recursively to determine the state occupancy probabilities. Thus, starting at time t_0, (6.157) is used to determine $\hat{\pi}_i(t)$ for t up to the first occurrence time w_1. Then (6.158) is used to determine $\hat{\pi}_i(w_1+)$ from $\hat{\pi}_i(w_1-)$, and (6.157) is used again for t up to the second occurrence time w_2. Then (6.158) is used again, and so forth. M. Rudemo [37] shows that the solution to (6.157) can be obtained by solving an appropriate set of linear differential equations.

We now use (6.157) and (6.158) to reexamine the model of Example 6.5.3.

Example 6.5.9 (On–Off Modulated Light, Continued). The information process $\{\mu_t;\ t\geqslant t_0\}$ of Example 6.5.3 is a two-state Markov jump process. The state space is $\mathfrak{Z}=\{\zeta_0=0,\zeta_1=\alpha\}$. The transition rates from state ζ_0 to state ζ_1 and from state ζ_1 to state ζ_0 both equal ν. Thus, the instantaneous intensities $a_{ik}(t)$ of (6.115) are given by

$$\begin{bmatrix} a_{00}(t) & a_{01}(t) \\ a_{10}(t) & a_{11}(t) \end{bmatrix} = \begin{bmatrix} -\nu & \nu \\ \nu & -\nu \end{bmatrix}. \tag{6.159}$$

The problem of estimating the information process from an observed realization of the point process it influences can be solved using (6.157) and (6.158). For this purpose, we note that $\hat{\pi}_0(t) = 1 - \hat{\pi}_1(t)$ and that

$$\begin{aligned} \hat{\lambda}_t &= \lambda_t(0)\hat{\pi}_0(t) + \lambda_t(\alpha)\hat{\pi}_1(t) \\ &= \lambda_0(1 - \hat{\pi}_1(t)) + (\alpha + \lambda_0)\hat{\pi}_1(t) = \lambda_0 + \alpha\hat{\pi}_1(t). \end{aligned}$$

From (6.157), the equation for the conditional occupancy probability $\hat{\pi}_1(t)$ of state $\zeta_1 = \alpha$ between observed points becomes

$$\frac{d\hat{\pi}_1(t)}{dt} = \nu - (2\nu + \alpha)\hat{\pi}_1(t) + \alpha\hat{\pi}_1^2(t). \tag{6.160}$$

The solution to this equation is

$$\hat{\pi}_1(t) = \frac{(\beta - \gamma) - (\beta + \gamma)\xi(s)\exp[-2\alpha\gamma(t-s)]}{1 - \xi(s)\exp[-2\alpha\gamma(t-s)]}, \tag{6.161}$$

where we have defined

$$\beta = \tfrac{1}{2}\left(1 + \frac{2\nu}{\alpha}\right),$$

$$\gamma = \tfrac{1}{2}\sqrt{1 + \left(\frac{2\nu}{\alpha}\right)^2},$$

$$\xi(s) = [\hat{\pi}_1(s+) - (\beta - \gamma)][\hat{\pi}_1(s+) - (\beta + \gamma)]^{-1}.$$

From (6.158), the update of $\hat{\pi}_1(t)$ at an occurrence time s is

$$\hat{\pi}_1(s+) = \hat{\pi}_1(s-)[\lambda_0 + \alpha][\lambda_0 + \alpha\hat{\pi}_1(s-)]^{-1}. \tag{6.162}$$

We have simulated the processes of this model for the parameters $\nu = 2.0$, $\alpha = 20.0$, and $\lambda_0 = 0.1$. The results are shown in Figure 6.11. The process

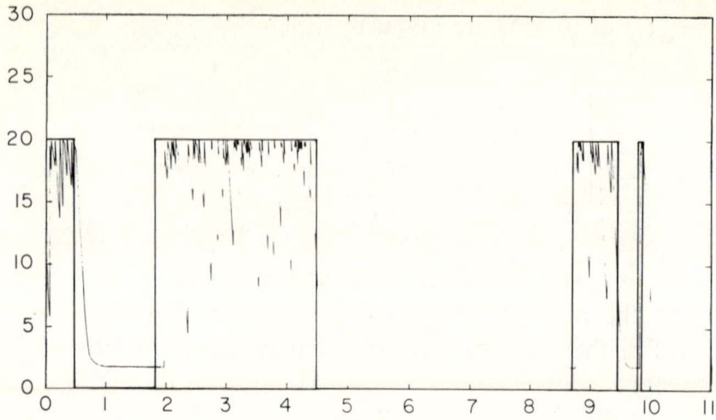

Figure 6.11. Nonlinear estimation of on–off modulated light.

$\{\mu_t;\ t \geqslant 0\}$ is the same as that obtained in the simulation described in Example 6.5.3 as is the observed point process. The estimate $\{\hat{\mu}_t;\ t \geqslant 0\}$ obtained by implementing (6.160) and (6.162) to process the observed points is shown in Figure 6.11. The estimate $\{\hat{\mu}_t;\ t \geqslant 0\}$ may be compared to the suboptimal, linear estimate $\{\hat{\mu}_t^l;\ t \geqslant 0\}$ shown in Figure 6.8; $\{\hat{\mu}_t^l;\ t \geqslant 0\}$ was generated using (6.102) to process the same observed data.

6.5.3 SUBOPTIMAL NONLINEAR FILTERING

In Section 6.5.2, we have developed a general framework for causally estimating information processes that influence the observed points of a doubly stochastic Poisson process. The equations we have given for the minimum mean square-error estimates are, in general, too complicated for analytic solution, and they even require a substantial computational capability for numerical solution. For this reason, we are motivated to develop suboptimal estimators that not only approximate the minimum mean square-error estimators but also have the advantage of a less extravagant computational requirement for their implementation. The estimators we obtain are nearly optimal when the estimation error is sufficiently small.

To be explicit, we will assume throughout the section that the information process $\{\mathbf{x}_t;\ t \geqslant t_0\}$ is a Markov diffusion satisfying the stochastic differential equation

$$d\mathbf{x}_t = \mathbf{a}_t(\mathbf{x}_t)\,dt + \mathbf{B}_t\,d\mathbf{w}_t, \tag{6.163}$$

$$\mathbf{x}_{t_0} = \mathbf{x}_0,$$

in which $\{\mathbf{w}_t;\ t \geqslant t_0\}$ is a vector of independent Wiener processes such that $E(\mathbf{w}_t\mathbf{w}'_u) = \mathbf{I}\min(t,u)$, and where the initial condition $\mathbf{x}_0$ is a random variable independent of $\{\mathbf{w}_t;\ t \geqslant t_0\}$. This is the model of (6.111) but with the assumption that $\mathbf{B}$ is not a function of $\mathbf{x}$. We make this assumption here only for simplicity of notation; the more general situation when $\mathbf{B}$ depends on $\mathbf{x}$ can be included by a straightforward modification of our procedure.

The choice of $\mathbf{x}$ as a Markov diffusion satisfying (6.163) includes as a special case time-independent random variables, nonstationary Gauss–Markov processes, and stationary Gaussian processes with rational power-density spectra. Consequently, by selecting a particular Markov diffusion and intensity process $\{\lambda_t(\mathbf{x}_t);\ t \geqslant t_0\}$, many physical phenomena encountered in applications can be accomodated. We will illustrate this with applications from radiation physics and optical communications.

The procedure we adopt in developing suboptimal estimators parallels that used several times with other models (A. Jazwinski [20, Chapter 9], D. Snyder [42, Chapter 4]) and can be described briefly as follows. We start with equations (6.153) and (6.154) for the exact conditional mean $\hat{\mathbf{x}}_t$ and conditional error covariance $\hat{\boldsymbol{\Sigma}}_t$ for $\mathbf{x}_t$ given $\{N_\sigma;\ t_0 \leqslant \sigma < t\}$. The nonlinear functions $\mathbf{a}_t(\mathbf{x}_t)$ and $\lambda_t(\mathbf{x}_t)$ are expanded in a Taylor series about the estimate $\hat{\mathbf{x}}_t$. This results in an infinite series in moments of the error $\mathbf{x}_t - \hat{\mathbf{x}}_t$ on the right-hand sides of (6.153) and (6.154). The Gaussian moment-factorization property is used on these error moments; that is, higher-order moments are written in terms of second-order moments as if the error were normally distributed. The infinite series are truncated at the "most significant" terms on the assumption that the error moments are sufficiently small. Just what these most significant terms are will depend on the particular functions $\mathbf{a}$ and λ involved; consequently, the usefulness of the particular truncation we employ must be verified in individual applications. Our choice of truncation will be shown to be reasonable for certain situations encountered in radiation physics and optical communications.

The result of the manipulations described above is a pair of coupled equations defining the suboptimal estimator. The first equation is

$$d\mathbf{x}_t^* = \mathbf{a}_t(\mathbf{x}_t^*)\,dt + \tfrac{1}{2}\sum_i \sum_j \sigma_{ij}^* \frac{\partial^2 \mathbf{a}_t(\mathbf{x}_t^*)}{\partial x_i^* \partial x_j^*}\,dt$$

$$+\boldsymbol{\Sigma}_t^*\left[\frac{\partial \lambda_t(\mathbf{x}_t^*)}{\partial \mathbf{x}_t^*}\right]\lambda_t^{-1}(\mathbf{x}_t^*)\left[dN_t - \lambda_t(\mathbf{x}_t^*)\,dt\right], \qquad (6.164)$$

$$\mathbf{x}_{t_0}^* = E(\mathbf{x}_0),$$

where $\boldsymbol{\Sigma}_t^*$ is a symmetric matrix that satisfies the second equation, σ_{ij}^* is the i,j-element of $\boldsymbol{\Sigma}_t^*$, and $\partial\lambda(\mathbf{x})/\partial\mathbf{x}$ denotes the gradient vector having as its ith element $\partial\lambda(\mathbf{x})/\partial x_i$. The second equation is

$$d\boldsymbol{\Sigma}_t^* = \left[\frac{\partial \mathbf{a}_t(\mathbf{x}_t^*)}{\partial \mathbf{x}_t^*}\right]'\boldsymbol{\Sigma}_t^*\,dt + \boldsymbol{\Sigma}_t^*\left[\frac{\partial \mathbf{a}_t(\mathbf{x}_t^*)}{\partial \mathbf{x}_t^*}\right]dt$$

$$+\mathbf{B}_t\mathbf{B}_t'\,dt - \boldsymbol{\Sigma}_t^*\left[\frac{\partial^2\lambda_t(\mathbf{x}_t^*)}{\partial \mathbf{x}_t^{*2}}\right]\boldsymbol{\Sigma}_t^*\,dt$$

$$+\boldsymbol{\Sigma}_t^*\left[\frac{\partial^2 \ln\lambda_t(\mathbf{x}_t^*)}{\partial \mathbf{x}_t^{*2}}\right]\boldsymbol{\Sigma}_t^*\,dN_t, \qquad (6.165)$$

$$E_{t_0}^* = \operatorname{cov}(\mathbf{x}_0) = E(\mathbf{x}_0\mathbf{x}_0') - E(\mathbf{x}_0)E(\mathbf{x}_0'),$$

where $\partial\mathbf{a}(\mathbf{x})/\partial\mathbf{x}$ denotes the Jacobian matrix having as its i-row j-column element $\partial a_j(\mathbf{x})/\partial x_i$ and where $\partial^2\lambda(\mathbf{x})/\partial\mathbf{x}^2 = \partial[\partial\lambda(\mathbf{x})/\partial\mathbf{x}]/\partial\mathbf{x}$.

We shall refer to (6.164) as the *processor equation* and to (6.165) as the *variance equation*. It is seen that in general the processor and variance equations are coupled and both depend on the observed doubly stochastic Poisson process N. The processor equation can be viewed as defining an updating algorithm according to which the suboptimal estimate $\mathbf{x}_{t+dt}^*$ at time $t+dt$ following an incremental observation is expressed in terms of the estimate $\mathbf{x}_t^*$ at time t preceding the observation and the data dN_t observed on the interval $[t,t+dt)$. Here, dN_t is either one or zero depending upon whether or not a point occurs in $[t,t+dt)$. The only unknown quantity in the updating is the matrix $\boldsymbol{\Sigma}_t^*$, and this is specified by the variance equation, which can similarly be viewed as defining an updating algorithm for $\boldsymbol{\Sigma}_t^*$.

As we indicated in Example 6.5.6, the evolution equation for the posterior density $p_t(\mathbf{X}|N_\sigma;\ t_0 \leqslant \sigma < t)$ in (6.142) can be viewed as defining an updating algorithm for estimating $\mathbf{x}_t$. This procedure has the advantage compared to (6.164) and (6.165) that no approximations are required beyond those needed for the numerical implementation, so the estimates obtained are optimal. On the other hand, the advantages of the suboptimal estimator defined by (6.164) and (6.165) in comparison to this exact procedure are: (i) no integrations are required; (ii) a large memory for storing the up-to-date version of $p_t(\mathbf{X}|N_\sigma;\ t_0 \leqslant \sigma < t)$ is not required; and (iii) only the prior mean and covariance of $\mathbf{x}$ are required, as opposed to the prior distribution needed in the exact analysis. Because of the symmetry of $\boldsymbol{\Sigma}^*$, the number of memory locations required for the storage of $\mathbf{x}^*$ and $\boldsymbol{\Sigma}^*$ is $n(n+3)/2$, where n is the dimension of $\mathbf{x}$.

Derivation of the Processor Equation (6.164). We assume that $\mathbf{a}_t(\mathbf{x}_t)$ and $\lambda_t(\mathbf{x}_t)$ have Taylor series about the estimate $\hat{\mathbf{x}}_t$ as

$$\mathbf{a}_t(\mathbf{x}_t) = \mathbf{a}_t(\hat{\mathbf{x}}_t) + \left[\frac{\partial \mathbf{a}_t(\hat{\mathbf{x}}_t)}{\partial \hat{\mathbf{x}}_t}\right]'(\mathbf{x}_t - \hat{\mathbf{x}}_t) + \tfrac{1}{2}\sum_i \sum_j (x_i - \hat{x}_i)(x_j - \hat{x}_j)\frac{\partial^2 \mathbf{a}_t(\hat{\mathbf{x}}_t)}{\partial \hat{x}_i \partial \hat{x}_j} + \cdots \tag{6.166}$$

and

$$\lambda_t(\mathbf{x}_t) = \lambda_t(\hat{\mathbf{x}}_t) + \left[\frac{\partial \lambda_t(\hat{\mathbf{x}}_t)}{\partial \hat{\mathbf{x}}_t}\right]'(\mathbf{x}_t - \hat{\mathbf{x}}_t) + \tfrac{1}{2}\sum_i \sum_j (x_i - \hat{x}_i)(x_j - \hat{x}_j)\frac{\partial^2 \lambda_t(\hat{\mathbf{x}}_t)}{\partial \hat{x}_i \partial \hat{x}_j} + \cdots \tag{6.167}$$

where $\partial \mathbf{v}(\mathbf{x})/\partial \mathbf{x}$ denotes the Jacobian matrix associated with the vector $\mathbf{v}(\mathbf{x})$; the i-row, j-column element of $\partial \mathbf{v}(\mathbf{x})/\partial \mathbf{x}$ is $\partial v_j/\partial x_i$.

Upon substituting these expansions into (6.153), we obtain

$$d\hat{\mathbf{x}}_t = \mathbf{a}_t(\hat{\mathbf{x}}_t)dt + \tfrac{1}{2}\sum_i \sum_j \hat{\sigma}_{ij}\frac{\partial^2 \mathbf{a}_t(\hat{\mathbf{x}}_t)}{\partial \hat{x}_i \partial \hat{x}_j}dt + \hat{\boldsymbol{\Sigma}}_t\left[\frac{\partial \lambda_t(\hat{\mathbf{x}}_t)}{\partial \hat{\mathbf{x}}_t}\right]\hat{\lambda}_t^{-1}\left[dN_t - \hat{\lambda}_t dt\right] + o(\mathbf{x}_t - \hat{\mathbf{x}}_t), \tag{6.168}$$

in which $\hat{\sigma}_{ij}$ is the i,j element of the conditional error covariance matrix

$$\hat{\mathbf{\Sigma}}_t = E[(\mathbf{x}_t - \hat{\mathbf{x}}_t)(\mathbf{x}_t - \hat{\mathbf{x}}_t)' | N_\sigma;\ t_0 \leqslant \sigma < t],$$

and where $o(\mathbf{x}_t - \hat{\mathbf{x}}_t)$ denotes terms of order three or more in error moments. No approximations have yet been introduced; $\hat{\mathbf{x}}_t$ is the optimal minimum mean square-error estimate of $\mathbf{x}_t$ given $\{N_\sigma;\ t_0 \leqslant \sigma < t\}$.

We now introduce the following three approximations:

a. replace $\hat{\lambda}_t$ by $\lambda_t(\hat{\mathbf{x}}_t)$,
b. disregard all the higher than second-order error-moment terms in $o(\mathbf{x}_t - \hat{\mathbf{x}}_t)$,
c. replace $\hat{\mathbf{\Sigma}}_t$ by $\mathbf{\Sigma}_t^*$, which is defined below.

It is evident from (6.176) that approximation (a) is consistent with the disregarding of second-order terms of approximation (b). These approximations are ad hoc and must be examined for specific functions $\mathbf{a}_t$ and λ_t encountered in applications, but they appear reasonable if the estimation error is sufficiently small. The result of these approximations is the processor equation (6.164) for the suboptimal estimate $\mathbf{x}_t^*$.

Derivation of the Variance Equation (6.165). By using the series (6.166) and (6.167), we have that

$$\hat{E}\{[\mathbf{a}_t(\mathbf{x}_t) - \hat{\mathbf{a}}_t](\mathbf{x}_t - \hat{\mathbf{x}}_t)'\} = \left[\frac{\partial \mathbf{a}_t(\hat{\mathbf{x}}_t)}{\partial \hat{\mathbf{x}}_t}\right]' \hat{\mathbf{\Sigma}}_t + \cdots,$$

$$\hat{E}\left\{\left[\lambda_t(\mathbf{x}_t) - \hat{\lambda}_t\right](\mathbf{x}_t - \hat{\mathbf{x}}_t)'\right\} = \left[\frac{\partial \lambda_t(\hat{\mathbf{x}}_t)}{\partial \hat{\mathbf{x}}_t}\right]' \hat{\mathbf{\Sigma}}_t + \cdots,$$

and

$$\hat{E}\left\{\left[\lambda_t(\mathbf{x}_t) - \hat{\lambda}_t\right](\mathbf{x}_t - \hat{\mathbf{x}}_t)(\mathbf{x}_t - \hat{\mathbf{x}}_t)'\right\}$$

$$= \sum_i \hat{E}\{(x_i - \hat{x}_i)(\mathbf{x}_t - \hat{\mathbf{x}}_t)(\mathbf{x}_t - \hat{\mathbf{x}}_t)'\}\left[\frac{\partial \lambda_t(\hat{\mathbf{x}}_t)}{\partial \hat{x}_i}\right]$$

$$+ 1/2 \sum_i \sum_j \hat{E}\{(x_i - \hat{x}_i)(x_j - \hat{x}_j)(\mathbf{x}_t - \hat{\mathbf{x}}_t)(\mathbf{x}_t - \hat{\mathbf{x}}_t)(\mathbf{x}_t - \hat{\mathbf{x}}_t)'\}\left[\frac{\partial^2 \lambda_t(\hat{\mathbf{x}}_t)}{\partial \hat{x}_i \partial \hat{x}_j}\right]$$

$$+ \cdots.$$

We now substitute these expressions into the right side of (6.154) and make the following approximations:

a. replace $\hat{\lambda}_t$ by $\lambda_t(\hat{\mathbf{x}}_t)$,
b. retain only the leading terms indicated in the above expansions,
c. disregard third-order moments of the error $\mathbf{x}_t - \hat{\mathbf{x}}_t$,
d. factor fourth-order moments of the error $\mathbf{x}_t - \hat{\mathbf{x}}_t$ in products of second-order moments as if the error were normally distributed,
e. replace $\hat{\mathbf{x}}_t$ by $\mathbf{x}_t^*$, which is defined above.

These approximations lead us to define $\boldsymbol{\Sigma}_t^*$ by the equation

$$d\boldsymbol{\Sigma}_t^* = \left[\frac{\partial \mathbf{a}_t(\mathbf{x}_t^*)}{\partial \mathbf{x}_t^*}\right]'\boldsymbol{\Sigma}_t^*\,dt + \boldsymbol{\Sigma}_t^*\left[\frac{\partial \mathbf{a}_t(\mathbf{x}_t^*)}{\partial \mathbf{x}_t^*}\right]dt + \mathbf{B}_t\mathbf{B}_t'\,dt$$

$$+\boldsymbol{\Sigma}_t^*\left[\frac{\partial^2 \lambda_t(\mathbf{x}_t^*)}{\partial \mathbf{x}_t^{*2}}\right]\boldsymbol{\Sigma}_t^*\lambda_t^{-1}(\mathbf{x}_t^*)[dN_t - \lambda_t(\mathbf{x}_t^*)\,dt]$$

$$-\boldsymbol{\Sigma}_t^*\left[\frac{\partial \lambda_t(\mathbf{x}_t^*)}{\partial \mathbf{x}_t^*}\right]\left[\frac{\partial \lambda_t(\mathbf{x}_t^*)}{\partial \mathbf{x}_t^*}\right]'\boldsymbol{\Sigma}_t^*\lambda_t^{-2}(\mathbf{x}_t^*)\,dN_t.$$

The variance equation (6.165) results from a straightforward rearrangement of this equation and the use of the relation

$$\frac{\partial^2 \ln \lambda_t(\mathbf{x})}{\partial \mathbf{x}^2} = \left[\frac{\partial \lambda_t(\mathbf{x})}{\partial \mathbf{x}}\right]\left[\frac{\partial \lambda_t(\mathbf{x})}{\partial \mathbf{x}}\right]'\lambda_t^{-2}(\mathbf{x})$$

$$+\left[\frac{\partial^2 \lambda_t(\mathbf{x})}{\partial \mathbf{x}^2}\right]\lambda_t^{-1}(\mathbf{x}).$$

We take the processor equation (6.164) and the variance equation (6.165) as our definition of the suboptimal estimator of the state $\mathbf{x}_t$ in terms of the observed counting path $\{N_\sigma; t_0 \leqslant \sigma < t\}$. For emphasis, we mention again that the utility of this definition depends on the particular form of $\mathbf{a}_t(\mathbf{x}_t)$ and $\lambda_t(\mathbf{x}_t)$ and must be investigated in individual applications. The definition we have adopted appears reasonable when the estimation error is small.

We now examine several applications. The results indicate the potential usefulness of the approximations introduced above.

Example 6.5.10 (One-Parameter Exponential Intensity). To illustrate the use of the processor and variance equations as defining an updating algorithm, let us consider a one-parameter intensity process of the form $\lambda_t(x) = A\exp(-xt)$ for $t \geqslant 0$, where A is a known positive constant and x is a nonnegative random variable with known mean $\bar{x}_0$ and variance σ_0^2. In this case (6.164) and (6.165) become:

$$\begin{aligned} dx_t^* &= At\Sigma_t^* \exp(-x_t^*)\,dt - t\Sigma_t^*\,dN_t, \qquad x_0^* = \bar{x}_0 \\ d\Sigma_t^* &= -At^2\Sigma_t^{*2}\exp(-x_t^*)\,dt, \qquad \Sigma_0^* = \sigma_0^2. \end{aligned} \tag{6.169}$$

We have simulated a doubly stochastic Poisson process with the intensity process $\lambda_t(x) = A\exp(-xt)$. The value $A = 100$ was used, and the prior probability density for x was taken as uniform on the interval [0.5, 1.5]; thus, $\bar{x}_0 = 1.0$ and $\sigma_0^2 = \frac{1}{12}$. The optimal and suboptimal estimates of x and $\lambda_t(x)$ are shown in Figures 6.12 and 6.13, respectively, for three parameter

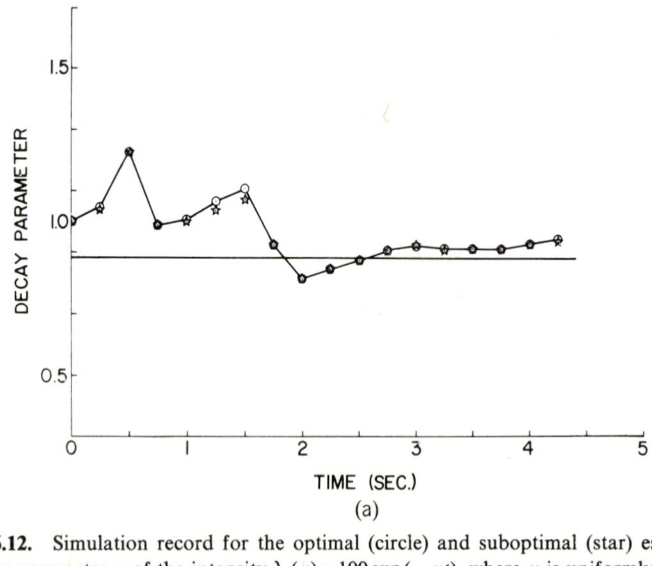

(a)

Figure 6.12. Simulation record for the optimal (circle) and suboptimal (star) estimates for the decay parameter x of the intensity $\lambda_t(x) = 100\exp(-xt)$, where x is uniformly distributed on [0.5, 1.5]. The values for x are 0.8822, 0.538, and 1.429 in (a), (b), and (c), respectively. At $t = 4.25$ seconds, the total number of counts is 100 and 144 in (a) and (b), respectively. At $t = 3.00$ seconds, the total number of counts is 59 in (c).

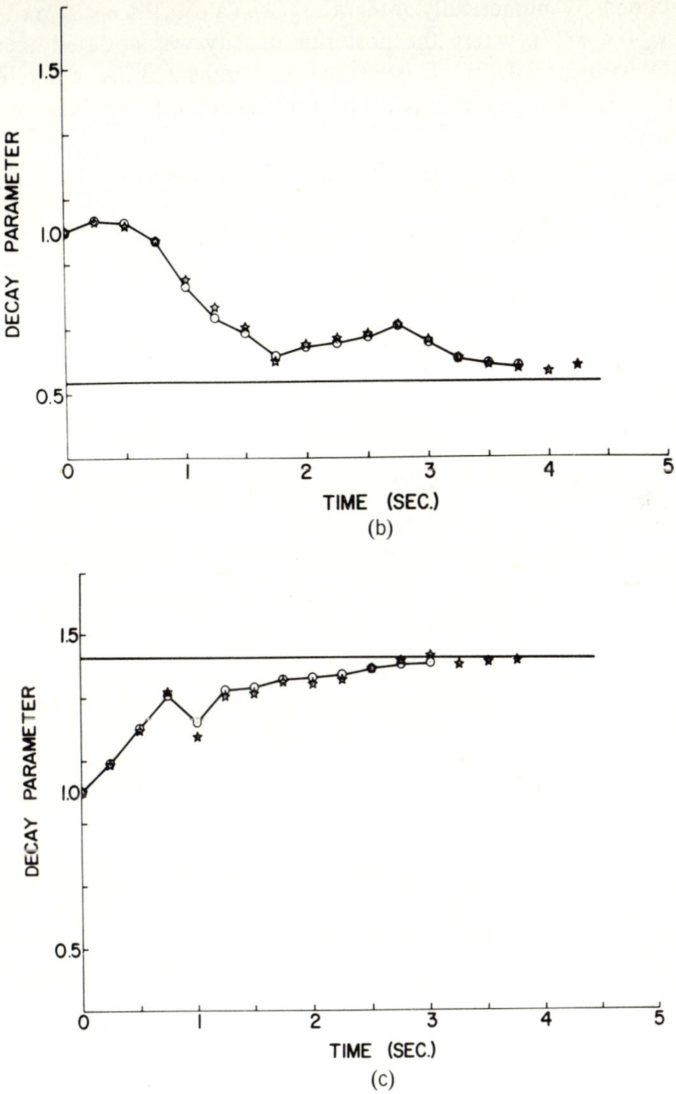

Figure 6.12. (*continued*)

values. The procedure by which the data shown in these figures were obtained consisted of using a random number generator to select a value for x, generating an inhomogeneous Poisson process with the appropriate decaying-exponential intensity, and then generating the estimates from the realization of the Poisson process. The optimal estimate was generated using the procedure described in Example 6.5.4; specifically, the estimates were obtained by numerically integrating $Xp_t(X|N_\sigma;0\leqslant\sigma<t)$ and $\lambda_t(X)\times p_t(X|N_\sigma;0\leqslant\sigma<t)$, where the posterior density was updated according to (6.141) with $\Delta t=0.001$. This estimate is indicated by a circle. The suboptimal estimate was generated by the following finite-difference version of (6.169):

$$x^*_{t+\Delta t}=x^*_t+100t\Sigma^*_t\exp(-x^*_t t)\Delta t+t\Sigma^*_t\Delta N_t$$

$$\Sigma^*_{t+\Delta t}=\Sigma^*_t-100t^2\Sigma^{*2}_t\exp(-x^*_t t)\,\Delta t$$

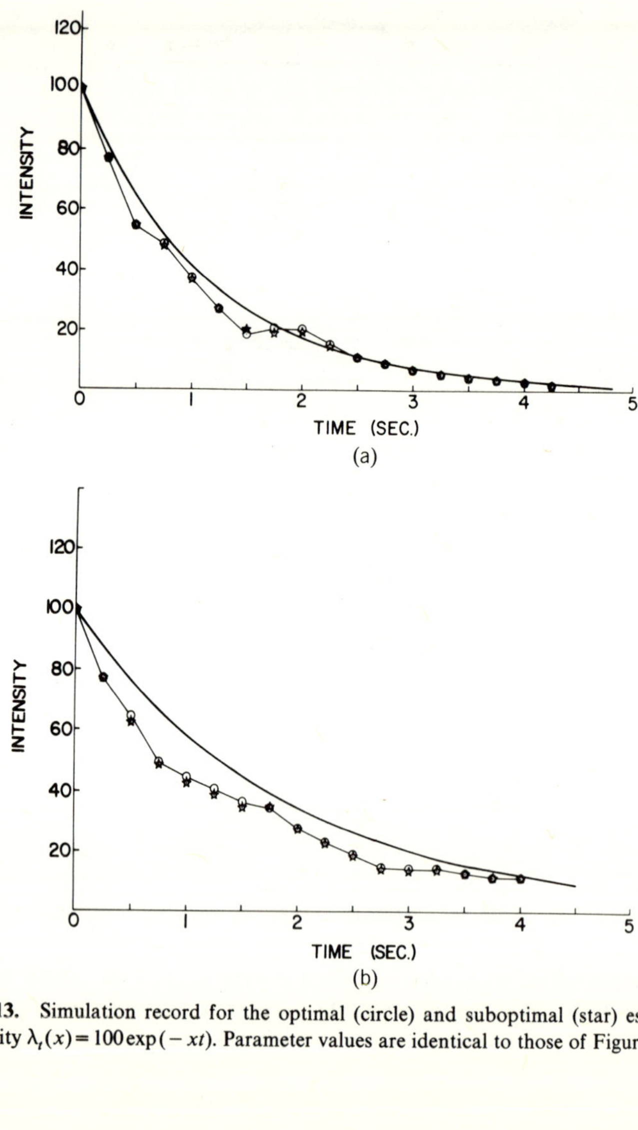

Figure 6.13. Simulation record for the optimal (circle) and suboptimal (star) estimates for the intensity $\lambda_t(x)=100\exp(-xt)$. Parameter values are identical to those of Figure 6.12.

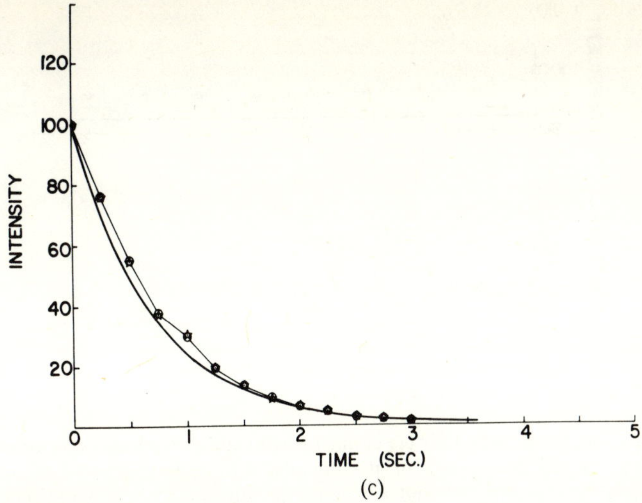

Figure 6.13. *(continued)*

with $\Delta t = 0.001$. This estimate is indicated by an asterisk. The agreement between the optimal and suboptimal estimates is striking in view of the small number of accumulated counts.

Example 6.5.11 (Two-Parameter Exponential Intensity). For this example, we consider a two-parameter intensity process of the form $\lambda_t(x_1, x_2) = x_1 \exp(-x_2 t), t \geqslant 0$, where x_1 and x_2 are uncorrelated, nonnegative random variables with known means $\bar{x}_1$ and $\bar{x}_2$ and variances σ_1^2 and σ_2^2. In this case, (6.164) and (6.165) become

$$d\mathbf{x}_t^* = -\Sigma_t^*[1 \quad -tx_1^*]' \exp(-x_2^* t)\, dt + \Sigma_t^*[x_1^{*-1} \quad -t]'\, dN_t,$$

$$\mathbf{x}_0^* = [\bar{x}_1 \bar{x}_2], \tag{6.170a}$$

$$d\Sigma_t^* = -\Sigma_t^* \begin{bmatrix} 0 & -t \\ -t & t^2 x_1^* \end{bmatrix} \Sigma_t^* \exp(-x_2^* t)\, dt$$

$$-\Sigma_t^* \begin{bmatrix} x_1^{*-2} & 0 \\ 0 & 0 \end{bmatrix} \Sigma_t^*\, dN_t,$$

$$\Sigma_0^* = \begin{bmatrix} \sigma_0^2 & 0 \\ 0 & \sigma_1^2 \end{bmatrix}. \tag{6.170b}$$

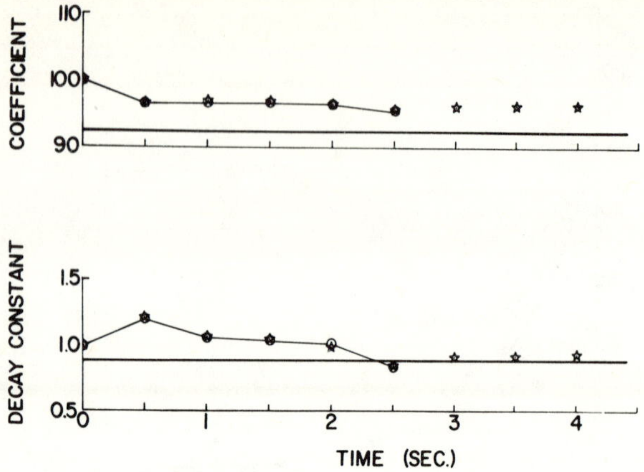

Figure 6.14. Simulation record for the optimal (circle) and suboptimal (star) estimates of the coefficient and decay parameters of the intensity $\lambda_t(x_1,x_2)=x_1\exp(-x_2t)$, where x_1 is uniformly distributed on [90, 100] and x_2 is uniformly distributed on [0.5, 1.5]. The values for x_1 and x_2 are 92.19 and 0.8822, respectively. At $t=4.00$ seconds, the total number of counts is 94.

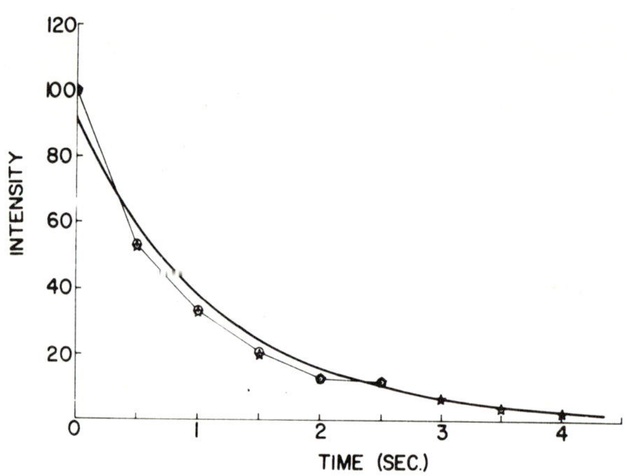

Figure 6.15. Simulation record for the optimal (circle) and suboptimal (star) estimate for the intensity $\lambda_t(x_1,x_2)=x_1\exp(-x_2t)$. Parameter values are identical to those of Figure 6.14.

A doubly stochastic Poisson process having this two-parameter intensity process was simulated, and the optimal and suboptimal estimates of x_1, x_2, and $\lambda_t(x_1,x_2)$ were generated using the procedure described in Example 6.5.10. The results are shown in Figures 6.14 and 6.15. The optimal and suboptimal estimates are indicated by a circle and a star, respectively. Both x_1 and x_2 were assumed to have a uniform prior probability density function on the intervals (90, 110) and (0.5, 1.5), respectively; thus $\bar{x}_{10}$

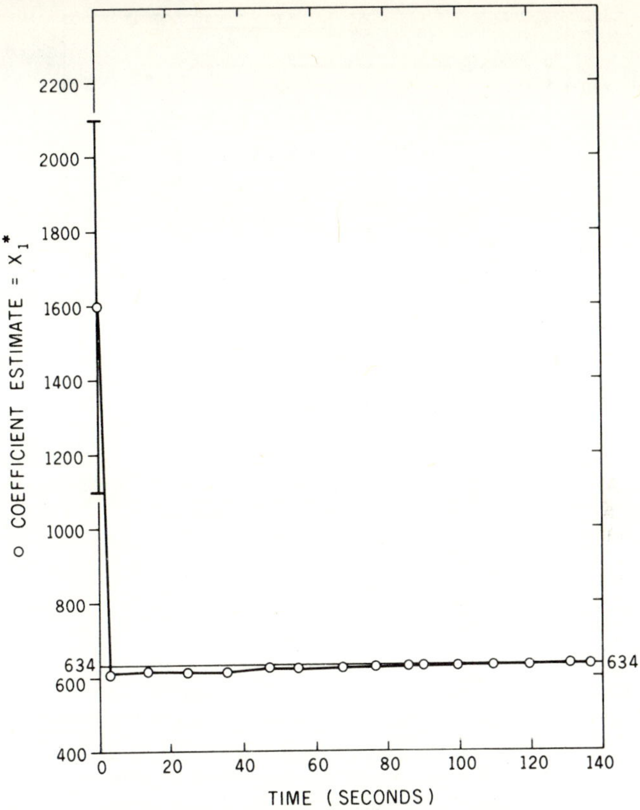

Figure 6.16. Time evolution of the suboptimal MMSE estimate of x_1 of the intensity $\lambda_t(x_1,x_2) = x_1 \exp(-x_2 t)$ for monitored oxygen-15.

$= 100$, $\bar{x}_{20} = 1.0$, $\sigma_{10}{}^2 = 100/3$, and $\sigma_{20}{}^2 = 1/12$. The important aspect of this simulation is not the accuracy of the estimates but, rather, the close correspondence between the optimal and suboptimal estimates even at the low count-rates of the simulation.

Oxygen-15 has a radioactive half-life of about 2 minutes. Physical principles suggest that the radioactivity of such an isotope disappears exponentially. With this in mind, we placed a quantity of oxygen-15 labeled water in the field of view of a scintillation detector and observed the times of occurrence of detected annihilation photons for about 8 minutes.† We assumed that the detected events formed a doubly-stochastic

†Oxygen – 15 is a cyclotron-produced isotope that is used to form a variety of radioactive pharmaceutical tracers used in nuclear medicine and biological and physiological research. The data described were collected in the Radiation Sciences Division of the Mallinckrodt Institute of Radiology, Washington University School of Medicine, St. Louis, Missouri.

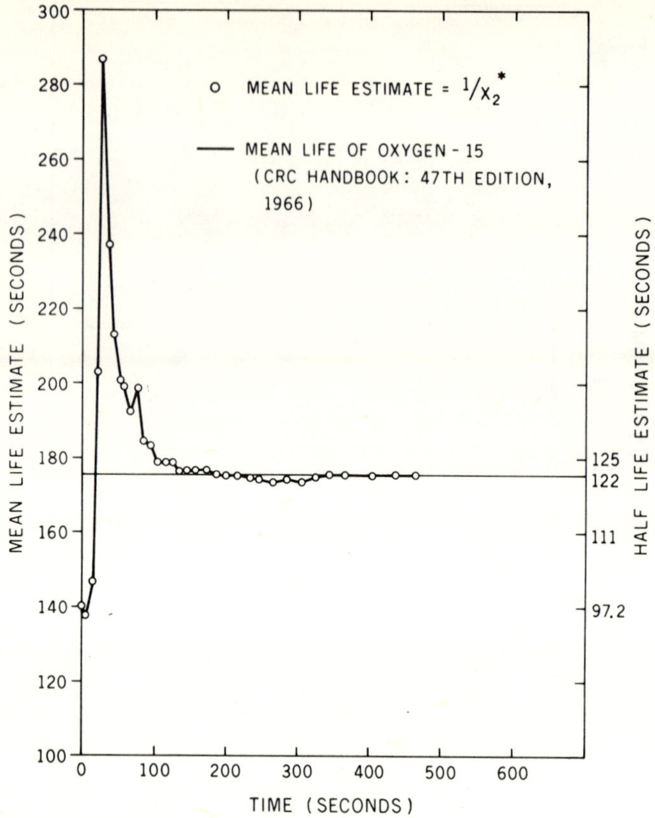

Figure 6.17. Time evolution of the suboptimal MMSE estimate of x_2 of the intensity $\lambda_t(x_1, x_2) = x_1 \exp(-x_2 t)$ for monitored oxygen-15.

Poisson process with the two-parameter exponential intensity $\lambda_t(x_1,x_2) = x_1 \exp(-x_2 t)$, where x_1 depends on the quantity of labeled water and x_2 on the physical properties of the labeling isotope. Estimates of x_1 and x_2 were generated according to (6.170) with the initial conditions $(x_1^*, x_2^*)_0 = (1600, 0.7143 \times 10^{-2})$ and $\Sigma_0^* = \text{diag}(2500, 0.4 \times 10^{-5})$. The estimates are shown in Figures 6.16 and 6.17.

Example 6.5.12 (Subcarrier Angle Modulation). A common technique for transmitting analog information at optical frequencies is to modulate the light intensity of the optical carrier with an angle-modulated subcarrier. This was discussed in Example 4.2.3 of Chapter 4. The output of an ideal photodetector at the receiver can be taken as a doubly stochastic

Poisson process with an intensity process given by

$$\lambda_t(\mathbf{x}_t) = \left(\frac{\eta \mathcal{A} P}{h\nu}\right)[1 + \alpha \cos(\omega t + \theta_t)] + \lambda_0 \tag{6.171}$$

where η is the quantum efficiency of the detector, $\mathcal{A}$ is the detector area, P is the received signal power per unit area, h is Planck's constant, ν is the optical carrier frequency, α is the subcarrier modulation index and satisfies $|\alpha| \leqslant 1$, and λ_0 is the thermoelectron generation rate in the detector. The angle-modulation process $\{\theta_t; t \geqslant t_0\}$ is assumed here to be a Gaussian process defined by $\theta_t = \mathbf{c}_t' \mathbf{x}_t$, where $\{\mathbf{x}_t; t \geqslant t_0\}$ is a Gaussian diffusion given by (6.163) when $\mathbf{a}_t(\mathbf{x}_t) = \mathbf{A}_t \mathbf{x}_t$ and $\mathbf{x}_0$ is a normal random variable with mean zero and covariance $\mathbf{\Sigma}_0$.

By selection of $\mathbf{c}_t$ and $\mathbf{A}_t$, this model includes as special cases the following commonly used angle-modulation formats:

a. *Phase Modulation*. For phase modulation, $\theta_t = \beta y_t$ where β is a constant, called the modulation index, and y_t is the information process. Here, we take $y_t = \mathbf{h}_t' \mathbf{x}_t$ where $\mathbf{x}_t$ satisfies (6.163) when $\mathbf{a}_t(\mathbf{x}_t) = \mathbf{A}_t \mathbf{x}_t$ and $\mathbf{x}_0$ is normally distributed. In this case, $\mathbf{c}_t = \beta \mathbf{h}_t$.

b. *Frequency Modulation*. For frequency modulation, $\theta_t = d_f \int_{t_0}^{t} y_\sigma \, d\sigma$ where d_f is a constant, called the frequency deviation factor, and y_t is the information process. Again, we take $y_t = \mathbf{h}_t' \mathbf{x}_t$ where $\mathbf{x}_t$ satisfies (6.163) when $\mathbf{a}_t(\mathbf{x}_t) = \mathbf{A}_t \mathbf{x}_t$ and $\mathbf{x}_0$ is normally distributed. To accommodate frequency modulation in our model, we define an additional state $x_{0t} = \int_{t_0}^{t} y_\sigma \, d\sigma$ and construct an expanded state vector $\tilde{\mathbf{x}}$ of dimension $n+1$, where n is the dimension of $\mathbf{x}$,

$$\tilde{\mathbf{x}}_t = [x_{0t} \quad x_{1t} \quad \cdots \quad x_{nt}]'. \tag{6.172}$$

Then, $\tilde{\mathbf{x}}_t$ satisfies

$$d\tilde{\mathbf{x}}_t = \tilde{\mathbf{A}}_t \tilde{\mathbf{x}}_t + \tilde{\mathbf{B}}_t \, d\mathbf{w}_t. \tag{6.173}$$

Here,

$$\tilde{\mathbf{A}}_t \triangleq \left[\begin{array}{c|c} 0 & \mathbf{h}_t' \\ \hline \mathbf{0} & \mathbf{A}_t \end{array}\right], \qquad \tilde{\mathbf{B}}_t \triangleq \left[\begin{array}{c} \mathbf{0}' \\ \hline \mathbf{B}_t \end{array}\right].$$

In terms of $\tilde{\mathbf{x}}$, the phase θ_t can be written as $\theta_t = \mathbf{c}_t' \tilde{\mathbf{x}}_t$, where $\mathbf{c}_t$ is an

$(n+1)$-dimensional vector of the form

$$\mathbf{c}_t = [d_f \quad 0 \quad 0 \quad \cdots \quad 0]'.$$

c. *Preemphasized Frequency Modulation.* For preemphasized frequency modulation,

$$\theta_t = \int_{t_0}^{t} g(t,u) y(u)\, du$$

for some specified preemphasis-filter impulse-response $g(t,u)$. If $g(t,u)$ has a state variable realization of dimension m, then preemphasized frequency-modulation can be accommodated in our model by defining an additional state variable $x_{0t} = \theta_t$ and an expanded state vector $\tilde{\mathbf{x}}$ of dimension $1+m+n$ having as elements x_{0t}, $\mathbf{x}_t$, and the state vector associated with the preemphasis filter.

Simply to economize on the number of parameters in our estimation equations, we now rewrite the intensity of (6.171) in the form

$$\lambda_t(\mathbf{x}_t) = G[1 + m\cos(\omega t + \theta_t)], \tag{6.174}$$

where

$$G = \frac{\eta \mathscr{Q} P}{h\nu} + \lambda_0 \quad \text{and} \quad m = \alpha\left(\frac{\eta \mathscr{Q} P}{h\nu}\right) G^{-1}.$$

As we have seen, this intensity includes as special cases commonly used angle-modulation formats, and the problem of estimating $\mathbf{x}_t$ given $\{N_\sigma; t_0 \leqslant \sigma < t\}$ corresponds to the problem of designing the receiver processing that should be performed on the output of the photodetector to recover information processes that modulate the light incident on the detector. For the intensity (6.174), the processor equation (6.164) becomes

$$d\mathbf{x}_t^* = \mathbf{A}_t \mathbf{x}_t^*\, dt$$

$$- \frac{m \sin(\omega t + \theta_t^*)}{1 + m\cos(\omega t + \theta_t^*)} \boldsymbol{\Sigma}_t^* \mathbf{c}_t \{ dN_t - G[1 + m\cos(\omega t + \theta_t^*)]\, dt \}, \tag{6.175}$$

$$\mathbf{x}_{t_0}^* = E(\mathbf{x}_0) = \mathbf{0},$$

where $\theta_t^* = \mathbf{c}_t'\mathbf{x}_t^*$. The variance equation (6.165) becomes

$$d\boldsymbol{\Sigma}_t^* = \mathbf{A}_t\boldsymbol{\Sigma}_t^*\,dt + \boldsymbol{\Sigma}_t^*\mathbf{A}_t'\,dt + \mathbf{B}_t\mathbf{B}_t'\,dt$$

$$+ Gm\cos(\omega t + \theta_t^*)\boldsymbol{\Sigma}_t^*\mathbf{c}_t\mathbf{c}_t'\boldsymbol{\Sigma}_t^*\,dt$$

$$- \frac{m[m + \cos(\omega t + \theta_t^*)]}{[1 + m\cos(\omega t + \theta_t^*)]^2}\boldsymbol{\Sigma}_t^*\mathbf{c}_t\mathbf{c}_t'\boldsymbol{\Sigma}_t^*\,dN_t \tag{6.176}$$

$$\boldsymbol{\Sigma}_{t_0}^* = \operatorname{cov}(\mathbf{x}_0) = \boldsymbol{\Sigma}_0.$$

Equations (6.175) and (6.176) define the suboptimal estimate of $\mathbf{x}_t$ and the phase θ_t. However, if ω is sufficiently large, as is typically the case in practice, additional approximations that appear reasonable can be made. These result in estimation equations having the distinct advantage that the variance equation is uncoupled from both the suboptimal state estimate and the data process N_t; the equation is a Riccati equation of the type occurring in Kalman–Bucy filtering, and which we encountered previously (6.86) in the linear filtering problem. Furthermore, the estimation equations suggest the use of a tanlock loop (L. Robinson [34]) operating on the output of the photodetector to estimate θ_t.

We assume for the additional approximations that ω is sufficiently large that the elements of $\mathbf{A}_t$, $\mathbf{B}_t$, $\mathbf{c}_t$, $\mathbf{x}_t$, $\mathbf{x}_t^*$, and $\boldsymbol{\Sigma}_t^*$ are slowly varying in comparison to $\sin(\omega t + \theta_t^*)$ and $\cos(\omega t + \theta_t^*)$. It then appears reasonable to neglect the subtractive term $G[1 + m\cos(\omega t + \theta_t^*)]$ in (6.175) because of the implied time integration of the product of the slowly varying function $Gm\boldsymbol{\Sigma}_t^*\mathbf{c}_t$ and $\sin(\omega t + \theta_t^*)$. We arbitrarily replace dN_t in (6.176) by $G[1 + m\cos(\omega + \theta_t^*)]\,dt$. Our motivation for doing so is based on the following conditional expectation and approximation that is consistent with our suboptimal estimator,

$$E(dN_t | N_\sigma; t_0 \leqslant \sigma < t) = \hat{\lambda}_t\,dt \approx \lambda_t(\mathbf{x}_t^*)\,dt$$

$$= G[1 + m\cos(\omega t + \theta_t^*)]\,dt.$$

With this approximation, (6.176) becomes

$$d\boldsymbol{\Sigma}_t^* = \mathbf{A}_t\boldsymbol{\Sigma}_t^*\,dt + \boldsymbol{\Sigma}_t^*\mathbf{A}_t'\,dt + \mathbf{B}_t\mathbf{B}_t'\,dt$$

$$- \frac{Gm^2\sin^2(\omega t + \theta_t^*)}{1 + m\cos(\omega t + \theta_t^*)}\boldsymbol{\Sigma}_t^*\mathbf{c}_t\mathbf{c}_t'\boldsymbol{\Sigma}_t^*\,dt. \tag{6.177}$$

Let us now examine the fourth term on the right. The function

$$f(\phi)=\frac{\sin^2(\phi)}{1+m\cos(\phi)} \tag{6.178}$$

is an even periodic function of ϕ. Consequently, f has a Fourier cosine series of the form

$$f(\phi)=a_0+a_1\cos(\phi)+a_2\cos(2\phi)+\cdots,$$

where it can be verified that a_0 is given by

$$a_0=\left[1-(1-m^2)^{1/2}\right]m^{-2}.$$

If we let $\phi=\omega t+\theta_t^*$ and substitute this series into the fourth term, then it appears reasonable to neglect all the harmonic components involving $\cos(n\omega t+n\theta_t^*)$, $n\geqslant 1$, and retain only a_0 because of the implied time integration.

The result of these approximations is the following pair of equations for the suboptimal estimate of the phase θ_t^*:

$$\begin{aligned} d\mathbf{x}_t^* &= \mathbf{A}_t\mathbf{x}_t^*\,dt-m\sin(\omega t+\theta_t^*) \\ &\quad\times[1+m\cos(\omega t+\theta_t^*)]^{-1}\boldsymbol{\Sigma}_t^*\mathbf{c}_t\,dN_t, \\ \mathbf{x}_{t_0}^* &= \mathbf{0} \end{aligned} \tag{6.179}$$

and

$$\begin{aligned} \frac{d\boldsymbol{\Sigma}_t^*}{dt} &= \mathbf{A}_t\boldsymbol{\Sigma}_t^*+\boldsymbol{\Sigma}_t^*\mathbf{A}_t'+\mathbf{B}_t\mathbf{B}_t'-G\left[1-(1-m^2)^{1/2}\right]\boldsymbol{\Sigma}_t^*\mathbf{c}_t\mathbf{c}_t'\boldsymbol{\Sigma}_t^*, \\ \boldsymbol{\Sigma}_{t_0}^* &= \boldsymbol{\Sigma}_0 \end{aligned} \tag{6.180}$$

where $\theta_t^*=\mathbf{c}_t'\mathbf{x}_t^*$. Equation (6.179) suggests the use of a form of tanlock loop to estimate θ_t, and (6.180) is a Riccati equation. Because (6.180) is uncoupled from both $\mathbf{x}_t^*$ and N_t, the matrix $\boldsymbol{\Sigma}_t^*$ can be precomputed. We will see in Section 6.5.4 that $\boldsymbol{\Sigma}_t^*$ lower bounds the performance in estimating $\mathbf{x}_t$ and θ_t.

The simulation results shown in Figure 6.18 were obtained for the intensity $\lambda_t(\theta)=G[1+m\cos(\omega t+\theta)]$ for the special case of a constant phase parameter uniformly distributed on $[-\pi,\pi]$. For this problem, the

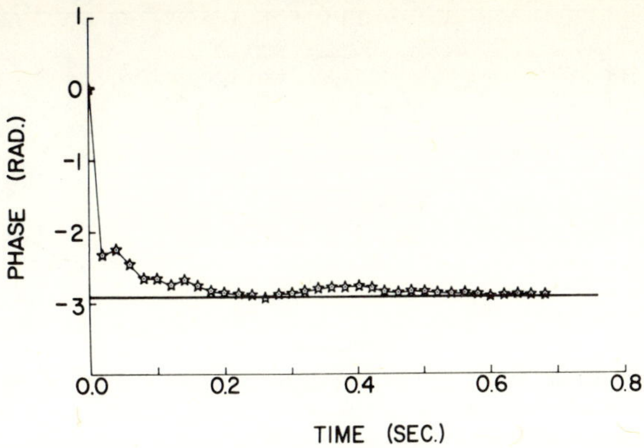

Figure 6.18. Simulation record for the suboptimal (star) estimate of θ for the intensity $\lambda_t(\theta)=G(1+m\cos(\omega_0 t+\theta))$, where $G=500$, $m=0.75$, $\omega_0=100(2\pi)$, and θ is uniformly distributed on $(-\pi,\pi)$. The value of θ is -2.905 rad. The total number of counts at $t=0.6$ seconds is 267.

model for $\{\theta_t; t\geqslant 0\}$ is

$$dx_t=0,\qquad E(x_0)=0,\qquad \operatorname{cov}(x_0)=\tfrac{1}{3}\pi^2,\qquad \theta_t=x_t.$$

The processor and variance equations, (6.179) and (6.180), become

$$dx_t^*=-m\sin(\omega t+\theta_t^*)[1+m\cos(\omega t+\theta_t^*)]^{-1}\Sigma_t^*\,dN_t,$$
$$x_0^*=0 \tag{6.181}$$

and

$$\frac{d\Sigma_t^*}{dt}=-G\left[1-(1-m^2)^{1/2}\right]\Sigma_t^{*2},\qquad \Sigma_0^*=\tfrac{1}{3}\pi^2 \tag{6.182}$$

where $\theta_t^*=x_t^*$. The solution to (6.182) is

$$\Sigma_t^*=\left\{\left(\frac{3}{\pi^2}\right)+G\left[1-(1-m^2)^{1/2}\right]t\right\}^{-1}. \tag{6.183}$$

The procedure used in the simulation consisted of selecting a value of θ by using a random number generator, then generating a Poisson process with an intensity $\lambda_t(\theta)$ for the value of θ obtained, and then generating the suboptimal estimate θ_t^* by using (6.181) as an updating algorithm.

The simulation results shown in Figure 6.19 were obtained for the intensity $\lambda_t(\theta_t)=G[1+m\cos(\omega t+\theta_t)]$ where $\{\theta_t; t\geqslant 0\}$ is a Wiener process

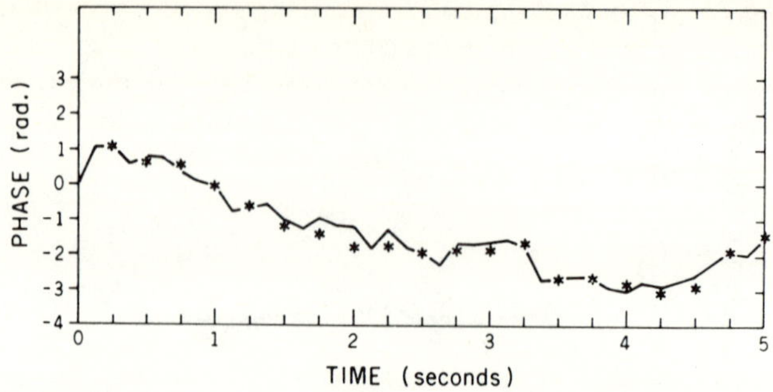

Figure 6.19. Simulation record for the suboptimal estimate (star) of θ_t for the intensity $\lambda_t(\theta_t)=G[1+m\cos(\omega t+\theta_t)]$, where $G=1000$, $m=0.99$, and $\{\theta_t; t\geqslant 0\}$ is a Wiener process with coherence time parameter $\tau_c=1$.

with parameter $1/\tau_c$. The parameter τ_c may be interpreted as the coherence time of an unstable oscillator; it is the time at which one radian rms of phase drift is accumulated. The problem of estimating θ_t is equivalent to the problem of establishing phase or time synchronization at a direct-detection receiver when the pilot-tone oscillator used to generate the subcarrier is unstable. For this situation, the model for $\{\theta_t; t\geqslant 0\}$ is

$$dx_t=\tau_c^{-1/2}\,dw_t, \qquad x_0=0, \qquad \theta_t=x_t$$

where $\{w_t;\ t\geqslant 0\}$ is a unit-parameter Wiener process. The processor equation, (6.179), is the same as when θ_t is a constant independent of time; namely, (6.181). The variance equation, (6.180), becomes

$$d\Sigma_t^*=\tau_c^{-1}-G\left[1-(1-m^2)^{1/2}\right]\Sigma_t^{*2}, \qquad \Sigma_0^*=0. \tag{6.184}$$

The solution to (6.184) is

$$\Sigma_t^*=\left\{\tau_c G\left[1-(1-m^2)^{1/2}\right]\right\}^{-1/2}\tanh\left[\left(\tau_c^{-1}G\left[1-(1-m^2)^{1/2}\right]\right)^{1/2}t\right]. \tag{6.185}$$

The procedure used in the simulation consisted of generating a sample function of the phase process θ, then generating a Poisson process with intensity $\lambda_t(\theta_t)$ for the sample function obtained, and then generating the suboptimal estimate of θ by using (6.181) as an updating algorithm.

Example 6.5.13 (Binary Detection in the Presence of Lognormal Fading). The binary detection problem in the context of optical radar was discussed in Example 6.4.1. We reexamine this problem here for the situation when μ_t, the photoelectron generation rate, has the following form, which corresponds to time-dependent lognormal fading,

$$\mu_t = \alpha |S_t|^2 \exp(y_t),$$

where $\{y_t; t \geqslant t_0\}$ is a Gaussian random process with the representation

$$d\mathbf{x}_t = \mathbf{A}_t \mathbf{x}_t\, dt + \mathbf{B}_t\, d\mathbf{w}_t, \qquad \mathbf{x}_{t_0} = \mathbf{x}_0, \qquad y_t = \mathbf{c}_t' \mathbf{x}_t$$

in which $\mathbf{x}_0$ is a zero-mean Gaussian variable with covariance $\boldsymbol{\Sigma}_0$ and $\{\mathbf{w}_t; t \geqslant t_0\}$ is a Weiner process with normalized covariance $E(\mathbf{w}_t \mathbf{w}_u') = \mathbf{I} \min(t,u)$.

The optimal detector for deciding whether or not a reflector is present is considered in Example 6.4.1 and is shown in Figure 6.2. This detector requires the optimal minimum mean square-error estimator of μ_t. However, as we have seen, this estimator is difficult to realize. Thus, there is a strong motivation to use a suboptimal detector. A reasonable suboptimal detector is that obtained by using the suboptimal estimator $\mu_t^* = \alpha |S_t|^2 \exp(y_t^*)$, where $y_t^* = \mathbf{c}_t' \mathbf{x}_t^*$, in the detector structure of Figure 6.2 in exactly the same way as if μ_t^* were the optimal estimate $\hat{\mu}_t$. The suboptimal estimator is specified by the processor and variance equations, (6.164) and (6.165), on the assumption that the intensity is that corresponding to the presence of a reflector; that is, assuming $\lambda_t(\mathbf{x}_t) = \alpha |S_t|^2 \exp(\mathbf{c}_t' \mathbf{x}_t) + \lambda_0$. The processor equation becomes

$$d\mathbf{x}_t^* = \mathbf{A}_t \mathbf{x}_t^*\, dt + \mu_t^* \boldsymbol{\Sigma}_t^* \mathbf{c}_t \lambda_t^{*-1} [dN_t - \lambda_t^*\, dt], \qquad \mathbf{x}_{t_0}^* = \mathbf{0}$$

where $\lambda_t^* = \mu_t^* + \lambda_0$ and where $\boldsymbol{\Sigma}_t^*$ satisfies the variance equation, which we will not write out explicitly because it does not simplify appreciably.

6.5.4 Nonlinear Filtering Performance for Gaussian Information Processes

The structure of optimal and suboptimal minimum mean square-error estimators has been investigated in Sections 6.5.2 and 6.5.3. We now investigate the mean-square performance of these estimators. Unfortunately, only a limited amount can be said about this performance because in general the direct evaluation of the mean-square error presents insurmountable analytic difficulties. For this reason, we restrict our attention to Gaussian information processes and establish a lower bound on the mean-square error for any estimate casually derived from the observed doubly stochastic Poisson process. We use the bound to study the performance of direct-detection optical communication systems that employ subcarrier angle modulation. The main result is contained in the following theorem due to D. Snyder and I. Rhodes [41].

Theorem 6.5.4 (Performance Lower Bound). Let $\{N_t; t_0 \leqslant t \leqslant T\}$ be a doubly stochastic Poisson process with intensity process $\{\lambda_t(\mathbf{x}_t); t_0 \leqslant t \leqslant T\}$, where $\{\mathbf{x}_t; t_0 \leqslant t \leqslant T\}$ is a Gaussian process with zero mean and autocovariance $\mathbf{K}_{t,u} = E(\mathbf{x}_t \mathbf{x}_u')$. Let $\mathbf{x}_t^*$ denote any causal estimate of $\mathbf{x}_t$ in terms of past data $\{N_\sigma; t_0 \leqslant \sigma < t\}$. The particular estimate $\mathbf{x}_t^* = E(\mathbf{x}_t | N_\sigma; t_0 \leqslant \sigma < t)$ will be denoted by $\hat{\mathbf{x}}_t$; this is the minimum mean square-error estimate of $\mathbf{x}_t$. Then for all t such that $t_0 \leqslant t \leqslant T$, there holds

$$E[(\mathbf{x}_t - \mathbf{x}_t^*)(\mathbf{x}_t - \mathbf{x}_t^*)'] \geqslant E[(\mathbf{x}_t - \hat{\mathbf{x}}_t)(\mathbf{x}_t - \hat{\mathbf{x}}_t)']$$

$$\geqslant \boldsymbol{\Sigma}_t \triangleq \lim_{u \uparrow t} \boldsymbol{\Gamma}_{t,u}, \tag{6.186}$$

where $\boldsymbol{\Gamma}$ satisfies the integral equation

$$\boldsymbol{\Gamma}_{t,u} + \int_{t_0}^{t} \boldsymbol{\Gamma}_{t,\tau} \mathbf{P}_\tau \mathbf{K}_{\tau,u}\, d\tau = \mathbf{K}_{t,u}, \qquad t_0 \leqslant u \leqslant t \leqslant T \tag{6.187}$$

and where $\mathbf{P}$ is a precomputable matrix defined by

$$\mathbf{P}_t = E\left\{\lambda_t^{-1}(\mathbf{x}_t)\left[\frac{\partial \lambda_t(\mathbf{x}_t)}{\partial \mathbf{x}_t}\right]\left[\frac{\partial \lambda_t(\mathbf{x}_t)}{\partial \mathbf{x}_t}\right]'\right\}. \tag{6.188}$$

The matrix inequality $\mathbf{M} \geqslant \mathbf{N}$ appearing in (6.186) means that $\mathbf{M} - \mathbf{N}$ is nonnegative definite; for any vector $\mathbf{v}$, there holds $\mathbf{v}'\mathbf{M}\mathbf{v} \geqslant \mathbf{v}'\mathbf{N}\mathbf{v}$. Thus, for example, if $y_t = \mathbf{c}_t'\mathbf{x}_t$, then (6.186) implies

$$E\left[(y_t - y_t^*)^2\right] \geqslant E\left[(y_t - \hat{y}_t)^2\right] \geqslant \mathbf{c}_t' \boldsymbol{\Sigma}_t \mathbf{c}_t \tag{6.189}$$

where $y_t^* = \mathbf{c}_t'\mathbf{x}_t^*$ and $\hat{y}_t = \mathbf{c}_t'\hat{\mathbf{x}}_t$.

We make three important remarks before proving the theorem.

Remark 1. An interpretation of $\boldsymbol{\Sigma}_t$ can be given in terms of an appropriate linear filtering problem involving additive Gaussian noise observations. If $\mathbf{P}_t$ is factored as $\mathbf{C}_t'\mathbf{C}_t$, then postmultiplication of (6.187) by $\mathbf{C}_u'$ shows $\mathbf{\Gamma}_{t,u}\mathbf{C}_u'$ is the impulse response matrix of the causal linear, least-squares filter for estimating $\mathbf{x}_t$ in terms of linear observations of the additive form $\{d\mathbf{z}_\sigma = \mathbf{C}_\sigma \mathbf{x}_\sigma d\sigma + d\mathbf{w}_\sigma; t_0 \leq \sigma \leq t\}$, where $\mathbf{w}$ is a vector Wiener process normalized so $E(\mathbf{w}_t\mathbf{w}_u') = \mathbf{I}\min(t,u)$. The lower bound matrix $\mathbf{\Sigma}_t$ of (6.186) is the covariance of the corresponding estimation error.

Remark 2. When $\mathbf{x}$ is a finite-dimensional Gauss–Markov process, the lower bound matrix $\mathbf{\Sigma}_t$ of (6.186) can be determined by solving a matrix Riccati equation. Suppose $\mathbf{x}$ has a finite-dimensional state representation as

$$d\mathbf{x}_t = \mathbf{A}_t\mathbf{x}_t\,dt + \mathbf{B}_t\,d\mathbf{w}_t, \qquad \mathbf{x}_{t_0} = \mathbf{x}_0 \tag{6.190}$$

where $\{\mathbf{w}_t; t \geq t_0\}$ is a vector Wiener process normalized so that $E(\mathbf{w}_t\mathbf{w}_u') = \mathbf{I}\min(t,u)$, and $\mathbf{x}_0$ is normally distributed with zero mean, covariance $\mathbf{\Sigma}_0$, and independent of $\{\mathbf{w}_t; t \geq t_0\}$. Then, the lower bound matrix $\mathbf{\Sigma}_t$ satisfies

$$\frac{d\mathbf{\Sigma}_t}{dt} = \mathbf{A}_t\mathbf{\Sigma}_t + \mathbf{\Sigma}_t\mathbf{A}_t' + \mathbf{B}_t\mathbf{B}_t' - \mathbf{\Sigma}_t\mathbf{P}_t\mathbf{\Sigma}_t, \qquad \mathbf{\Sigma}_{t_0} = \mathbf{\Sigma}_0. \tag{6.191}$$

Remark 3. An asymptotically efficient estimate is an estimate for which equality in (6.186) is achieved asymptotically as some parameter, typically a measure of the signal-to-noise ratio, increases. Specifically, we will say that a causal estimate $\mathbf{x}_t^*$ of $\mathbf{x}_t$ is asymptotically efficient with a parameter α if for any $\epsilon > 0$, there exists a real number a such that for all $\alpha \geq a$, we have

$$\mathbf{\Sigma}_t \leq E[(\mathbf{x}_t - \mathbf{x}_t^*)(\mathbf{x}_t - \mathbf{x}_t^*)'] \leq \mathbf{\Sigma}_t + \epsilon\mathbf{I}. \tag{6.192}$$

We are unable to give conditions on λ that guarantee the existence of an asymptotically efficient estimate. An example for which the suboptimal estimators of Section 6.5.3 appear to be asymptotically efficient is given below for subcarrier angle modulation.

Proof of Theorem 6.5.4. The first inequality in (6.186) is trivial because the conditional mean $\hat{\mathbf{x}}_t$ is the minimum mean square-error estimate of $\mathbf{x}_t$. To prove the second inequality, we consider an approximation

$$\mathbf{x}_t^{(M)} \triangleq \sum_{m=1}^{M} \chi_m \psi_t^{(m)}, \qquad t_0 \leq t \leq T \tag{6.193}$$

to $\mathbf{x}$ over the interval $[t_0, T)$ motivated by the Karhunen–Loeve representation (H. Van Trees [47, p. 200]) in which $\mathbf{x}_t^{(M)}$ converges in the mean-square sense to $\mathbf{x}_t$ as $M \to \infty$. The coefficients $\{\chi_m\}$ are the Fourier coefficients of the representation and $\{\psi_t^{(m)}; t_0 \leqslant t \leqslant T\}$ are the eigenvectors associated with the integral equation

$$\lambda_m \psi_t^{(m)} = \int_{t_0}^{T} \mathbf{K}_{t,\tau} \psi_\tau^{(m)} \, d\tau, \qquad t_0 \leqslant t \leqslant T \tag{6.194}$$

in which $\mathbf{K}$ is the autocovariance for $\mathbf{x}$. The coefficients $\{\chi_m\}$ are independent, zero-mean, Gaussian variables with variances $\{\lambda_m\}$, where λ_i is the eigenvalue associated with the eigenvector $\psi_t^{(i)}$.

Denote an observed counting path $\mathfrak{N}_t = \{N_\sigma; t_0 \leqslant \sigma < t\}$ when $\mathbf{x}$ is replaced by its M-term approximation $\mathbf{x}^{(M)}$ by $\mathfrak{N}_t^{(M)}$. Let $p_\chi(\mathbf{X} | \mathfrak{N}_t^{(M)})$ be the posterior probability density of the coefficient vector χ given $\mathfrak{N}_t^{(M)}$. Let $\mathbf{J}_t^{(M)}$ be the matrix with k,ith element

$$[\mathbf{J}_t^{(M)}]_{k,i} = -E\left\{ \frac{\partial^2 \ln p_\chi(\chi | \mathfrak{N}_t^{(M)})}{\partial \chi_k \, \partial \chi_i} \right\}. \tag{6.195}$$

Finally, define $\mathbf{S}_t^{(M)} = [\mathbf{J}_t^{(M)}]^{-1}$. Then

$$E[(\chi - \hat{\chi}_t)(\chi - \hat{\chi}_t)'] \geqslant \mathbf{S}_t^{(M)}, \tag{6.196}$$

where $\hat{\chi}_t = E(\chi | \mathfrak{N}_t^{(M)})$ is the causal minimum mean-square error of χ in terms of $\mathfrak{N}_t^{(M)}$. Equation (6.196) is the Cramér–Rao inequality for random variables (H. Van Trees [47, p. 84]).

We need next to evaluate the matrix $\mathbf{J}_t^{(M)}$ of (6.195). We have from the above definitions that $\{N_t^{(M)}; t_0 \leqslant t \leqslant T\}$ is a doubly-stochastic Poisson process with an intensity process $\{\mu_t(\chi); t_0 \leqslant t \leqslant T\}$, where

$$\mu_t(\chi) = \lambda_t\left(\sum_{m=1}^{M} \chi_m \psi_t^{(m)} \right) = \lambda_t(\mathbf{x}_t^{(M)}). \tag{6.197}$$

Thus, Bayes' rule and the normality of χ imply that

$$p_\mathbf{x}(\mathbf{X} | \mathfrak{N}_t^{(M)}) = C \exp\left\{ -\int_{t_0}^{t} \mu_\sigma(\mathbf{X}) \, d\sigma + \int_{t_0}^{t} \ln \mu_\sigma(\mathbf{X}) \, dN_\sigma^{(M)} - \tfrac{1}{2} \mathbf{X}' \mathbf{\Lambda}^{-1} \mathbf{X} \right\}, \tag{6.198}$$

where $\mathbf{\Lambda} = \operatorname{diag}(\lambda_i)$ and C is a normalization constant independent of $\mathbf{X}$. When this expression for the posterior density of χ is used in (6.195), we obtain the following expression for the k,ith element of $\mathbf{J}_t^{(M)}$ after some manipulations:

$$[\mathbf{J}_t^{(M)}]_{k,i} = \lambda_i^{-1}\delta_{ki} + \int_{t_0}^{t} E\left\{ \mu_\sigma^{-1}(\chi)\left[\frac{\partial\mu_\sigma(\chi)}{\partial\chi_k}\right]\left[\frac{\partial\mu_\sigma(\chi)}{\partial\chi_i}\right]\right\} d\sigma. \quad (6.199)$$

From the chain rule and (6.197), there follows

$$\frac{\partial\mu_\sigma(\chi)}{\partial\chi_i} = \left[\frac{\partial\lambda_\sigma(\mathbf{x}_\sigma^{(M)})}{\partial\mathbf{x}_\sigma^{(M)}}\right]'\psi_\sigma^{(i)}.$$

Substitution of this relation into (6.199) yields the desired expression for $\mathbf{J}_t^{(M)}$,

$$\mathbf{J}_t^{(M)} = \mathbf{\Lambda}^{-1} + \int_{t_0}^{t} \mathbf{\Psi}_\sigma^{(M)\prime}\mathbf{P}_\sigma^{(M)}\mathbf{\Psi}_\sigma^{(M)}\, d\sigma, \quad (6.200)$$

where $\mathbf{\Psi}_t^{(M)}$ is the matrix of eigenvectors

$$\mathbf{\Psi}_t^{(M)} \triangleq [\psi_t^{(1)} \quad \psi_t^{(2)} \quad \cdots \quad \psi_t^{(M)}]$$

and $\mathbf{P}_t^{(M)}$ is defined by

$$\mathbf{P}_t^{(M)} = E\left\{\lambda_t^{-1}(\mathbf{x}_t^{(M)})\left[\frac{\partial\lambda_t(\mathbf{x}_t^{(M)})}{\partial\mathbf{x}_t^{(M)}}\right]\left[\frac{\partial\lambda_t(\mathbf{x}_t^{(M)})}{\partial\mathbf{x}_t^{(M)}}\right]'\right\}. \quad (6.201)$$

We now use (6.196) and (6.200) to complete the proof of the theorem. Let

$$\hat{\mathbf{x}}_t^{(M)} = E(\mathbf{x}_t^{(M)} \mid \mathcal{N}_t^{(M)}) = \sum_{m=1}^{M} \hat{\chi}_t^{(m)}\psi_t^{(m)}$$
$$= \mathbf{\Psi}_t^{(M)}\hat{\chi}_t \quad (6.202)$$

be the causal minimum mean square-error estimate of $\mathbf{x}_t^{(M)}$ in terms of $\mathcal{N}_t^{(M)}$. Using (6.196), we find

$$E[(\mathbf{x}_t^{(M)} - \hat{\mathbf{x}}_t^{(M)})(\mathbf{x}_t^{(M)} - \hat{\mathbf{x}}_t^{(M)})']$$
$$= \mathbf{\Psi}_t^{(M)}E[(\chi - \hat{\chi}_t)(\chi - \hat{\chi}_t)']\mathbf{\Psi}_t^{(M)\prime}$$
$$\geqslant \mathbf{\Psi}_t^{(M)}\mathbf{S}_t^{(M)}\mathbf{\Psi}_t^{(M)\prime} \triangleq \mathbf{\Sigma}_t^{(M)}. \quad (6.203)$$

Define

$$\mathbf{\Gamma}_{t,\tau}^{(M)} = \mathbf{\Psi}_t^{(M)} \mathbf{S}_t^{(M)} \mathbf{\Psi}_\tau^{(M)\prime}, \qquad t_0 \leqslant t \leqslant T. \tag{6.204}$$

Postmultiplying (6.204) by $\mathbf{P}_\tau^{(M)} \psi_\tau^{(M)}$, integrating on $[t_0, T]$, and using (6. 200), and then using $\mathbf{S}_t^{(M)} \mathbf{J}_t^{(M)} = \mathbf{I}$, yields

$$\int_{t_0}^{t} \mathbf{\Gamma}_{t,\tau}^{(M)} \mathbf{P}_\tau^{(M)} \psi_\tau^{(M)} \, d\tau = \mathbf{\Psi}_t^{(M)} \mathbf{S}_t^{(M)} \left[\mathbf{J}_t^{(M)} - \mathbf{\Lambda}^{-1} \right]$$

$$= \mathbf{\Psi}_t^{(M)} \left[\mathbf{I} - \mathbf{S}_t^{(M)} \mathbf{\Lambda}^{-1} \right].$$

Now postmultiply by $\mathbf{\Lambda \Psi}_u^{(M)\prime}$, use (6.204), and the definition $\mathbf{K}_{t,u}^{(M)} = \mathbf{\Psi}_t^{(M)} \mathbf{\Lambda \Psi}_u^{(M)\prime}$ to obtain

$$\mathbf{\Gamma}_{t,u}^{(M)} + \int_{t_0}^{t} \mathbf{\Gamma}_{t,\tau}^{(M)} \mathbf{P}_\tau^{(M)} \mathbf{K}_{\tau,u}^{(M)} \, d\tau = \mathbf{K}_{t,u}^{(M)}, \qquad t_0 \leqslant u \leqslant t \leqslant T. \tag{6.205}$$

An interpretation of (6.205) in parallel to that of Remark 1 for (6.187) is that $\mathbf{\Gamma}_{t,u}^{(M)} \mathbf{C}_u^{(M)\prime}$ is the impulse response of the optimum linear filter for causally estimating $\mathbf{x}_t^{(M)}$ in terms of the additive observations $\{ d\mathbf{z}_\sigma = \mathbf{C}_\sigma^{(M)} \mathbf{x}_\sigma^{(M)} \, d\sigma + d\mathbf{w}_\sigma; t_0 \leqslant \sigma \leqslant t \}$ and $\mathbf{\Sigma}_t^{(M)}$ is the corresponding covariance of the error $\mathbf{x}_t^{(M)} - \hat{\mathbf{x}}_t^{(M)}$, where $\mathbf{P}_t^{(M)} = \mathbf{C}_t^{(M)\prime} \mathbf{C}_t^{(M)}$.

Now as $M \to \infty$, we have

$$\mathbf{P}_t^{(M)} \to \mathbf{P}_t,$$

$$\mathbf{C}_t^{(M)} \mathbf{K}_{t,u}^{(M)} \mathbf{C}_u^{(M)\prime} \to \mathbf{C}_t \mathbf{K}_{t,u} \mathbf{C}_u',$$

$$\mathbf{x}_t^{(M)} - \hat{\mathbf{x}}_t^{(M)} \to \mathbf{x}_t - \hat{\mathbf{x}}_t,$$

and

$$\mathbf{\Gamma}_{t,u}^{(M)} \to \mathbf{\Gamma}_{t,u},$$

where $\mathbf{\Gamma}_{t,u}$ satisfies (6.187). As (6.203) holds for all M, it holds in the limit as $M \to \infty$, and the proof of Theorem 6.5.4 is complete.

We now apply the performance lower bound of Theorem 6.5.4 to direct-detection optical communication systems. As we have seen in earlier examples, the output of an ideal photodetector can be modeled as a doubly stochastic Poisson process. A stochastic intensity process can occur in practice by a variety of unintentional and intentional effects: some unintentional effects are fading of optical signals because of atmospheric turbulence and the absence of phase and timing coherence because of oscillator instabilities; intentional effects arise from modulation of optical carriers by information-bearing signals. The accuracy bounds on attainable performance developed in Theorem 6.5.4 apply to those situations when the stochastic intensity is modulated by one or more Gaussian processes, any one of which may be unintentional or intentional in origin. For brevity, we now examine only intensities that are modulated in phase and frequency by a Gaussian process; the application to alternative situations is then obvious.

Example 6.5.14 (Performance for Subcarrier Phase Modulation). The model for subcarrier phase modulation is given in Example 6.5.12 of the preceding section. The intensity process is

$$\lambda_t(\mathbf{x}_t) = G[1 + m\cos(\omega t + \theta_t)]$$

where $\theta_t = \mathbf{c}_t'\mathbf{x}_t$. Some straightforward calculations result in

$$\lambda_t^{-1}(\mathbf{x}_t)\left[\frac{\partial \lambda_t(\mathbf{x}_t)}{\partial \mathbf{x}_t}\right]\left[\frac{\partial \lambda_t(\mathbf{x}_t)}{\partial \mathbf{x}_t}\right]' = Gm^2 f(\omega t + \theta_t)\mathbf{c}_t\mathbf{c}_t' \tag{6.206}$$

where f is the function defined in (6.178); this function is even and periodic, so it can be written in a Fourier cosine series as

$$f(\varphi) = a_0 + \sum_{n=1}^{\infty} a_n \cos(n\varphi) \tag{6.207}$$

where a_0 is given by

$$a_0 = \left[1 - (1 - m^2)^{1/2}\right] m^{-2}.$$

For (6.187), we need the expectation of (6.206). By using the series for f and assuming that θ_t is a Gaussian random variable with zero mean and variance σ_t^2, we obtain

$$\mathbf{P}_t = \left\{\left[1 - (1 - m^2)^{1/2}\right]G + \sum_{n=1}^{\infty} a_n G\left[\exp\left(-\tfrac{1}{2}n^2\sigma_t^2\right)\right]\cos(n\omega t)\right\}\mathbf{c}_t\mathbf{c}_t'. \tag{6.208}$$

Because of the integration in (6.187) and the implied integration in (6.191), we retain only the first term in (6.208) on the assumption that the subcarrier frequency ω is sufficiently large that the rapid oscillations in the last term average to zero. Thus, for instance, if $\mathbf{x}_t$ has a state representation as in (6.190), then we have that the mean-square error for any causal estimate θ_t^* of $\theta_t = \mathbf{c}_t'\mathbf{x}_t$ satisfies the inequality

$$E\left[(\theta_t - \theta_t^*)^2\right] \geqslant \mathbf{c}_t'\boldsymbol{\Sigma}_t\mathbf{c}_t, \tag{6.209}$$

where $\boldsymbol{\Sigma}$ satisfies the Riccati equation

$$\frac{d\boldsymbol{\Sigma}_t}{dt} = \mathbf{A}_t\boldsymbol{\Sigma}_t + \boldsymbol{\Sigma}_t\mathbf{A}_t' + \mathbf{B}_t\mathbf{B}_t'$$

$$- G\left[1-(1-m^2)^{1/2}\right]\boldsymbol{\Sigma}_t\mathbf{c}_t\mathbf{c}_t'\boldsymbol{\Sigma}_t, \qquad \boldsymbol{\Sigma}_{t_0} = \boldsymbol{\Sigma}_0. \tag{6.210}$$

In Example 6.5.12 we have shown that a tanlock loop can be used to generate an approximation to the causal, minimum mean square-error estimate of the phase process θ_t. The structure of this loop is specified in terms of the solution to a Riccati equation (6.180) that is identical to (6.210). Thus, the solution to (6.180) determines both the structure of this causal estimator as well as a lower bound on its tracking accuracy.

The fact that (6.210) corresponds to the equation for the error covariance matrix of an appropriate Kalman–Bucy filtering problem implies that there are numerous specializations for which an explicit lower bound (6.209) can be given. One such example occurs when $t_0 = -\infty$ and θ_t is a stationary Gaussian process with a rational power-spectral density $S(f)$. Then from (6.210) and the Yovits–Jackson equation (see M. Yovits and J. Jackson [51] or A. Viterbi [49, p. 143]), we deduce that

$$E\left[(\theta_t - \theta_t^*)^2\right] \geqslant \alpha^{-1}\int_{-\infty}^{\infty} \ln\left[1 + \alpha S(f)\right] df, \tag{6.211}$$

where $\alpha \triangleq G[1-(1-m^2)^{1/2}]$.

The simplest nonstationary example occurs when $\{\theta_t; t \geqslant 0\}$ is a Wiener process with parameter $1/\tau_c$. As discussed in Example 6.5.12, the problem of estimating θ_t is equivalent to that of establishing a reference subcarrier when the subcarrier oscillator is unstable and has a coherence time τ_c. It follows easily from (6.210) and (6.185) that the steady-state mean-square error in establishing a coherent phase reference satisfies the inequality

$$\lim_{t\to\infty} E\left[(\theta_t - \theta_t^*)^2\right] \geqslant \left\{\tau_c G\left[1-(1-m^2)^{1/2}\right\}^{-1/2}. \tag{6.212}$$

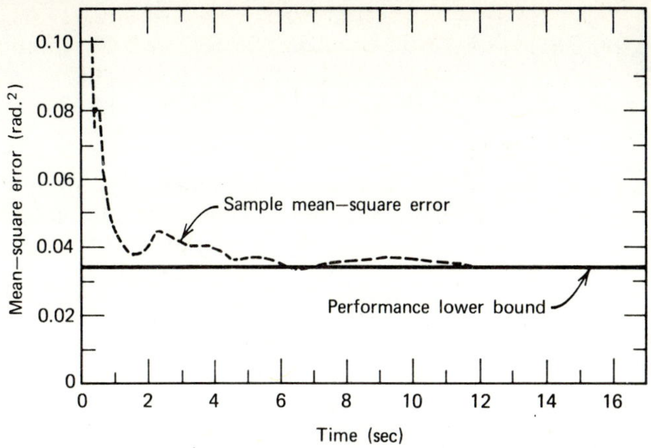

Figure 6.20. Sample mean-square error for the intensity $\lambda_t(\theta_t)=G[1+m\cos(\omega t+\theta_t)]$, where $G=1000$, $m=0.99$, and $\{\theta_t;t\geqslant 0\}$ is a Wiener process with coherence-time parameter $\tau_c=1$.

Shown in Figure 6.20 is the sample mean-square error obtained in the simulation of Figure 6.19. It is seen that the sample mean-square error approaches a constant value with increasing time. The close proximity of this value to the performance lower bound (6.212) suggests that the suboptimal estimator defined by (6.181) and (6.182) is asymptotically efficient.

It is possible to analyze the performance somewhat more completely for the above situation when θ_t is a Wiener process. By using the approximation procedure developed in Example 4.2.12, it has been shown by R. Forrester and D. Snyder [14] that the steady-state probability density $p_\infty(\Phi)$ of the phase-estimation error $\varphi_t \triangleq \theta_t-\theta_t^*$ is approximated by

$$p_\infty(\Phi)\approx C(1+\Gamma\cos\Phi)^{\Lambda\Gamma^{-1}-1}, \qquad |\Phi|\leqslant\pi \tag{6.213}$$

where C is the normalization constant

$$C=\left[\int_{-\pi}^{\pi}(1+\Gamma\cos\Phi)^{\Lambda\Gamma^{-1}-1}\,d\Phi\right]^{-1}$$

and where Λ and Γ are defined by

$$\Lambda=2\left\{\tau_c G\left[1-(1-m^2)^{1/2}\right]\right\}^{1/2}\left[1+(1-m^2)^{-1/2}\right]^{-1}$$

and

$$\Gamma=\left[1-(1-m^2)^{-1/2}\right]\left[1+(1-m^2)^{-1/2}\right]^{-1}.$$

Shown in Figure 6.21 is the sample mean-square error obtained in the same manner as that of Figure 6.20 but for the parameter values $G=300$, $m=0.33$, and $\tau_c=1$. The performance lower bound (6.212) is shown in the figure as is the theoretical performance determined numerically from (6.213) according to $\int_{-\pi}^{\pi}\Phi^2 p_\infty(\Phi)\,d\Phi$. For the parameter values selected here, the sample mean-square error does not appear to approach the lower bound, but it is accurately predicted from (6.213).

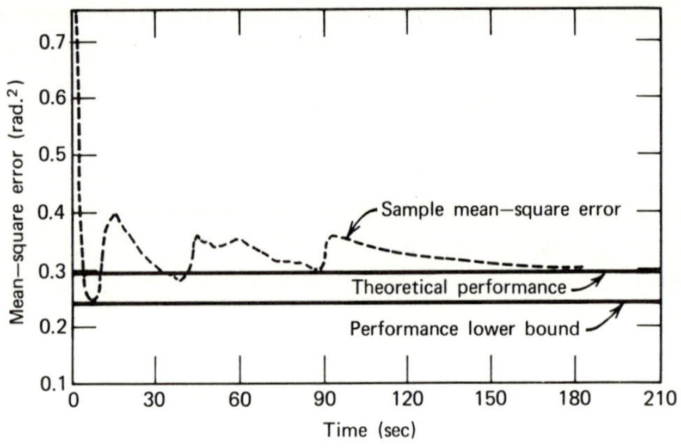

Figure 6.21. Sample mean-square error for the intensity $\lambda_t(\theta_t)=G[1+m\cos(\omega t+\theta_t)]$, where $G=300$, $m=0.33$, and $\{\theta_t; t\geq 0\}$ is a Wiener process with coherence-time parameter $\tau_c=1$.

Example 6.5.15 (Performance for Subcarrier Frequency Modulation). The model for subcarrier frequency modulation is given in Example 6.5.12 of the preceding section. The intensity process is

$$\lambda_t(\tilde{\mathbf{x}}_t)=G[1+m\cos(\omega t+\theta_t)]$$

where $\theta_t=\mathbf{c}_t'\tilde{\mathbf{x}}_t=d_f x_{0t}$. Here, $\tilde{\mathbf{x}}_t$ is an expanded state vector defined in (6.172), d_f is the frequency deviation factor, $x_{0t}=\int_{t_0}^{t} y_\sigma\,d\sigma$ is the integrated message, and $y_t=[0 \quad \mathbf{h}_t']\tilde{\mathbf{x}}_t$. We have

$$\lambda_t^{-1}(\tilde{\mathbf{x}}_t)\left[\frac{\partial\lambda_t(\tilde{\mathbf{x}}_t)}{\partial\tilde{\mathbf{x}}_t}\right]\left[\frac{\partial\lambda_t(\tilde{\mathbf{x}}_t)}{\partial\tilde{\mathbf{x}}_t}\right]'=Gm^2 f(\omega t+\theta_t)\mathbf{c}_t\mathbf{c}_t', \qquad (6.214)$$

where f is defined in (6.178), as with phase modulation. Using the expansion for f in (6.207) and dropping double-frequency terms, as before, we thus conclude from (6.186) that for any causal estimate $\tilde{\mathbf{x}}_t^*$ of $\tilde{\mathbf{x}}_t$ we have

$$E[(\tilde{\mathbf{x}}_t - \tilde{\mathbf{x}}_t^*)(\tilde{\mathbf{x}}_t - \tilde{\mathbf{x}}_t^*)'] \geqslant \tilde{\mathbf{\Sigma}}_t \tag{6.215}$$

where $\tilde{\mathbf{\Sigma}}_t$ satisfies the Riccati equation

$$\frac{d\tilde{\mathbf{\Sigma}}_t}{dt} = \tilde{\mathbf{A}}_t\tilde{\mathbf{\Sigma}}_t + \mathbf{\Sigma}_t\tilde{\mathbf{A}}_t' + \tilde{\mathbf{B}}_t\tilde{\mathbf{B}}_t'$$

$$- G\left[1-(1-m^2)^{1/2}\right]\tilde{\mathbf{\Sigma}}_t\mathbf{c}_t\mathbf{c}_t'\tilde{\mathbf{\Sigma}}_t \tag{6.216}$$

with the initial condition

$$\tilde{\mathbf{\Sigma}}_{t_0} = \left[\begin{array}{c|c} & \mathbf{0}' \\ \hline \mathbf{0} & \mathbf{\Sigma}_0 \end{array}\right].$$

It further follows that the mean-square phase and frequency tracking errors satisfy

$$E\left[(x_{0t} - x_{0t}^*)^2\right] \geqslant [1 \quad 0 \quad 0 \quad \cdots \quad 0]\tilde{\mathbf{\Sigma}}_t[1 \quad 0 \quad 0 \quad \cdots \quad 0]' \tag{6.127}$$

and

$$E\left[(y_t - y_t^*)^2\right] \geqslant [0 \quad \mathbf{h}_t']\tilde{\mathbf{\Sigma}}_t[0 \quad \mathbf{h}_t']', \tag{6.218}$$

respectively, for any causal estimator.

The structure of the suboptimal minimum mean square-error estimator of Section 6.5.3 for the phase x_{0t} and the frequency y_t is determined by the Riccati equation (6.216). Thus, (6.216) determines both the structure and a lower bound on the performance of this suboptimal estimator.

The fact that (6.216) corresponds to the error covariance matrix of an appropriate Kalman–Bucy filtering problem implies that there are numerous specializations for which explicit lower bounds, (6.217) and (6.218), can be given. One such example occurs when $t_0 = -\infty$ and y_t is a stationary Gaussian process with a rational power-spectral density $S(f)$. Then from (6.216) and the results of D. Snyder [42], we deduce that for any causal estimator

$$E\left[(x_{0t} - x_{0t}^*)^2\right] \geqslant \alpha^{-1}I_1 \tag{6.219}$$

and

$$E\left[(y_t - y_t^*)^2\right] \geqslant \left(\frac{1}{3\alpha}\right) I_1^3 + I_2, \tag{6.220}$$

where

$$\alpha \triangleq G d_f^2\left[1 - (1 - m^2)^{1/2}\right]$$

$$I_1 \triangleq \int_{-\infty}^{\infty} \ln[1 + \alpha f^2 S(f)]\, df$$

and

$$I_2 = \int_{-\infty}^{\infty} f^2 \alpha^{-1} \ln[1 + \alpha f^2 S(f)]\, df.$$

Plots of the lower bounds (6.219) and (6.220) are given by D. Snyder [42, pp. 91–94] for the Butterworth class of power-density spectra.

Summary. In this section, we have studied the filtering problem for observed doubly stochastic Poisson processes. Four major topics were discussed: (1) linear filtering, (2) optimal nonlinear filtering, (3) suboptimal nonlinear filtering, and (4) filtering performance.

Estimation of the intensity process was emphasized for the linear filtering problem. Here, the intensity estimator is selected to minimize the mean-square error in estimating the intensity at time t subject to the constraint that the estimate be a linear function of the observed doubly stochastic Poisson process up to time t. The optimum linear estimator is specified in Theorem 6.5.1. The estimator can be realized as a Kalman–Bucy filter for a wide class of intensity processes; in these cases, the estimator can be realized in a straightforward manner numerically or with hardware. Linear estimates of the intensity process are developed in Problem 6.5.1 for histogram data. Linear estimates of desired processes other than the intensity process are developed in Problem 6.5.2.

The linearity constraint results in estimates that are not sufficiently accurate in many applications. Thus, in Section 6.5.2, we removed this constraint and studied the problem of estimating a prescribed function of the information process at time t in terms of the observed doubly-stochastic Poisson process up to time t. We considered Markovian information processes, which encompass a wide variety of useful models. The principal result is in Theorem 6.5.2, which contains a representation for the conditional characteristic function of the information process given the observed doubly stochastic Poisson process. Several examples were given showing how this representation may be used to develop numerical algorithms for generating estimates of the information process. The results in the subsection may be generalized in several ways. The restrictions that the information process be Markovian and that the observed point process be conditionally a Poisson process given the information process are removed in Chapter 7, where we study doubly stochastic self-exciting point processes. For alternative approaches to the filtering problem, we refer to P. Fishman and D. Snyder [12, 13], K. Dunn and I. Rhodes [9], A. Segall [39], P. Bremaud [4], and J. Van Schuppen [46]. P. Fishman develops a general measure-theoretic approach to the filtering problem in which a conditional measure on the information-process probability space given the observed point process is related to the unconditional measure by the sample-function density. K. Dunn develops an approach based on certain measure transformations. A. Segall, P. Bremaud, and J. Van Schuppen develop an approach based on the properties of martingales.

Estimators without the linearity constraint are frequently too complicated to implement without a large computational capability. For this reason, suboptimal estimators are derived in Section 6.5.3. These are obtained by truncating Taylor series expansions of all nonlinear functions of the information process about the estimate of the information process. While the suboptimal estimators that result are easily implemented and nearly optimal for small estimation errors, they must be carefully examined in individual applications where alternative approximations may be more appropriate.

A precomputable lower bound on the mean-square performance in estimating a normal information process was derived in Section 6.5.4. The lower bound applies to an arbitrary estimator including the linear, optimal, and suboptimal estimators derived in Sections 6.5.1, 6.5.2, and 6.5.3, respectively. It was shown that under favorable circumstances, the suboptimal estimators of Section 6.5.3 can nearly achieve the lower bound. Furthermore, the lower bound was seen to be closely related to the design of the suboptimal estimators for certain applications arising in optical communication.

We now generalize these results to situations where more than one doubly stochastic Poisson process is observed simultaneously.

6.6 MULTICHANNEL DOUBLY STOCHASTIC POISSON PROCESSES

The results of the preceding sections can be extended to the situation where more than one doubly stochastic Poisson process is observed and an estimate is desired of an underlying information process that influences them all. The following are several applications that motivate this multichannel extension.

Example 6.6.1 (Nuclear Medicine). Multichannel data are routinely collected in nuclear medicine. This occurs in a variety of procedures and equipment configurations. Three examples are multiple-isotope studies, the gamma-ray camera, and multiprobe detection systems.

a. *Multiple-Isotope Studies* (D. Ross and C. Harris [36, p. 179]). Multiple radioactive isotopes with different energy spectra are introduced simultaneously into an organ and radioactivity is monitored externally with a scintillation detector. As in Figure 3.5, a point process associated with each isotope is obtained by processing the pulses from the scintillation detector with energy discriminators. Ideally, each point process is influenced by only one isotope, but, in practice, the energy spectra overlap so that this cannot be achieved precisely.

b. *Gamma-Ray Camera* (H. Anger [2]; N. Moody et al. [32]). The Anger gamma-ray camera is widely used in nuclear medicine. It consists of an array of 19 photomultipliers that view a single scintillation crystal in the form of a large disk about 28 centimeters in diameter. Radioactivity monitored with the camera results in a multichannel point process with 19 components.

c. *Multiprobe Detection Systems* (M. Ter-Pogossian et al. [44]). Multiprobe detection systems consist of an array of several independent scintillation detectors placed over an organ to monitor radioactivity. Each component of a multichannel point process can be associated with one of the scintillation-detector outputs.

Example 6.6.2 (Direct-Detection Optical Communication Systems with Spatial Diversity). Atmospheric turbulence causes the received light field in an optical communication system to be spatially incoherent (E. Hoversten et al. [19].) This effect can be compensated for to some extent by using an array of photodetectors at the receiver. Each component of a multichannel point process can be associated with the output of a photodetector in the array.

Example 6.6.3 (Intensity Interferometry). In intensity interferometry, light incident on two spatially separated apertures is first detected and then processed to determine desired properties of the optical source, which is typically a stellar object (J. Goodman [15, p. 19].) The outputs of two photomultipliers used to detect the light can be modeled as components of a two-dimensional multichannel point process.

Our model for a multichannel doubly stochastic Poisson process is comprised of K single-channel doubly stochastic Poisson processes that are conditionally mutually independent Poisson processes given the information process. Thus, let $\{N_t^{(k)};\ t \geq t_0\}$ for $k=1, 2,\ldots,K$ be doubly stochastic Poisson processes with corresponding intensity processes $\{\lambda_t^{(k)}(\mathbf{x}_t);\ t \geq t_0\}$ for $k=1, 2,\ldots,K$, where $\{\mathbf{x}_t;t \geq t_0\}$ is an information process. We assume $N^{(1)}$, $N^{(2)},\ldots,$ and $N^{(K)}$ are mutually independent given $\{\mathbf{x}_t;t \geq t_0\}$. We term the vector $\mathbf{N}_t$, where

$$\mathbf{N}_t \triangleq [N_t^{(1)} \quad N_t^{(2)} \quad \cdots \quad N_t^{(K)}]',$$

a multichannel doubly stochastic Poisson process with intensity process $\boldsymbol{\lambda}_t(\mathbf{x}_t)$, where

$$\boldsymbol{\lambda}_t(\mathbf{x}_t) \triangleq [\lambda_t^{(1)}(\mathbf{x}_t) \quad \lambda_t^{(2)}(\mathbf{x}_t) \quad \cdots \quad \lambda_t^{(K)}(\mathbf{x}_t)]'.$$

In the next paragraphs, we indicate how the important results of the preceding sections are modified for multichannel doubly stochastic Poisson processes. Most of these extensions are given without proof because the proofs are straightforward modifications of those for the single-channel model.

Multichannel Differential Rule (Extension of Theorem 4.2.2). Let $\{\zeta_t;\ t \geq t_0, \zeta_t \in R^n\}$ satisfy

$$d\zeta_t = \boldsymbol{\alpha}_t(\zeta_t)dt + \boldsymbol{\beta}_t(\zeta_t)d\mathbf{N}_t \tag{6.221}$$

where $\{\mathbf{N}_t, t \geq t_0\}$ is a multichannel doubly stochastic Poisson process. Here, $\boldsymbol{\beta}$ is an $(n\times K)$-dimensional matrix. Suppose $\varphi_t(\mathbf{X})$ is a real valued function of $\mathbf{X} \in R^n$ with continuous partial derivatives in t and X_1, $X_2,\ldots,X_n$. Then

$$d\varphi_t(\zeta_t) = \left[\frac{\partial \varphi_t(\zeta_t)}{\partial t} + \left\langle \frac{\partial \varphi_t(\zeta_t)}{\partial \zeta_t}, \boldsymbol{\alpha}_t(\zeta_t) \right\rangle \right] dt$$

$$+ \sum_{k=1}^{K} [\varphi_t(\zeta_t + \boldsymbol{\beta}_t(\zeta_t)\mathbf{e}_k) - \varphi_t(\zeta_t)]dN_t^{(k)}, \tag{6.222}$$

where $\mathbf{e}_k$ is a K-dimensional unit vector with a one as its kth component and zero for all others. Equation (6.222) can be proved by paralleling the proof of Theorem 4.2.2; this is left as an exercise; see Problem 6.6.1.

Multichannel Sample-Function Density [Extension of (6.45) and (6.71)]. Let $\{\mathbf{N}_t; t \geqslant t_0\}$ be a multichannel doubly stochastic Poisson process with intensity process $\{\boldsymbol{\lambda}_t(\mathbf{x}_t); t \geqslant t_0\}$, where $\{\mathbf{x}_t; t \geqslant t_0\}$ is an information process. Denote the joint sample-function density of the components $N^{(1)}$, $N^{(2)}, \ldots,$ and $N^{(K)}$ of $\mathbf{N}$ by $p[\{\mathbf{N}_\sigma; t_0 \leqslant \sigma < t\}]$. By using the method of conditioning, the conditional independence of the components of $\mathbf{N}$, and the sample-function density for a Poisson process (2.51), we have that

$$p[\{\mathbf{N}_\sigma; t_0 \leqslant \sigma < t\}] = E\left\{ \exp\left[\left\langle \mathbf{1}, -\int_{t_0}^{t} \boldsymbol{\lambda}_\sigma(\mathbf{x}_\sigma)\, d\sigma + \int_{t_0}^{t} \operatorname{diag}\left[\ln \lambda_\sigma^{(k)}(\mathbf{x}_\sigma)\right] d\mathbf{N}_\sigma \right\rangle \right] \right\}, \tag{6.223}$$

where $\mathbf{1}$ is a K-dimensional vector having unity as each component. This is the multichannel extension of (6.45). From it and the multichannel differential rule, we can show that

$$p[\{\mathbf{N}_\sigma; t_0 \leqslant \sigma < t\}] = \exp\left[\left\langle \mathbf{1}, -\int_{t_0}^{t} \hat{\boldsymbol{\lambda}}_\sigma\, d\sigma + \int_{t_0}^{t} \operatorname{diag}\left(\ln \hat{\lambda}_\sigma^{(k)}\right) d\mathbf{N}_\sigma \right\rangle \right], \tag{6.224}$$

where $\hat{\boldsymbol{\lambda}}_t = E[\boldsymbol{\lambda}_t(\mathbf{x}_t) | N_\sigma; t_0 \leqslant \sigma < t]$ is the minimum mean square-error estimate of $\boldsymbol{\lambda}_t$ in terms of $\{\mathbf{N}_\sigma; t_0 \leqslant \sigma < t\}$. This is the multichannel extension of (6.71).

Equation (6.224) is derived as follows. For fixed $\{\mathbf{N}_\sigma; t_0 \leqslant \sigma < t\}$, define

$$\mathbf{H}_t(\mathbf{x}_\sigma; t_0 \leqslant \sigma < t) = -\int_{t_0}^{t} \boldsymbol{\lambda}_\sigma(\mathbf{x}_\sigma)\, d\sigma + \int_{t_0}^{t} \operatorname{diag}\left[\ln \lambda_\sigma^{(k)}(\mathbf{x}_\sigma)\right] d\mathbf{N}_\sigma. \tag{6.225}$$

Clearly,

$$d\mathbf{H}_t(\mathbf{x}_\sigma; t_0 \leqslant \sigma < t) = -\boldsymbol{\lambda}_t(\mathbf{x}_t)dt + \operatorname{diag}\left[\ln \lambda_t^{(k)}(\mathbf{x}_t)\right] d\mathbf{N}_t. \tag{6.226}$$

If

$$\varphi_t(\mathbf{x}_\sigma; t_0 \leqslant \sigma < t) \triangleq \exp\left[\langle \mathbf{1}, \mathbf{H}_t(\mathbf{x}_\sigma; t_0 \leqslant \sigma < t) \rangle\right],$$

then (6.226) and the multichannel differential rule imply that

$$d\varphi_t(\mathbf{x}_\sigma; t_0 \leqslant \sigma < t) = -\langle \mathbf{1}, \boldsymbol{\lambda}_t(\mathbf{x}_t)\varphi_t(\mathbf{x}_\sigma; t_0 \leqslant \sigma < t)\rangle dt$$
$$+ \langle [\boldsymbol{\lambda}_t(\mathbf{x}_t) - \mathbf{1}]\varphi_t(\mathbf{x}_\sigma; t_0 \leqslant \sigma < t), d\mathbf{N}_t\rangle. \quad (6.227)$$

By taking the expectation of both sides of this expression, recognizing from (6.223) that

$$E[\varphi_t(\mathbf{x}_\sigma; t_0 \leqslant \sigma < t)] = p[\{\mathbf{N}_\sigma; t_0 \leqslant \sigma < t\}], \quad (6.228)$$

and using the following relation to be established

$$E[\boldsymbol{\lambda}_t(\mathbf{x}_t)\varphi_t(\mathbf{x}_\sigma; t_0 \leqslant \sigma < t)] = \hat{\boldsymbol{\lambda}}_t E[\varphi_t(\mathbf{x}_\sigma; t_0 \leqslant \sigma < t)], \quad (6.229)$$

we obtain

$$dp[\{\mathbf{N}_\sigma; t_0 \leqslant \sigma < t\}] = -p[\{\mathbf{N}_\sigma; t_0 \leqslant \sigma < t\}]\langle \mathbf{1}, \hat{\boldsymbol{\lambda}}_t\rangle dt$$
$$+ p[\{\mathbf{N}_\sigma; t_0 \leqslant \sigma < t\}]\langle \hat{\boldsymbol{\lambda}}_t - \mathbf{1}, d\mathbf{N}_t\rangle. \quad (6.230)$$

By applying the differential rule again, we have

$$d\ln p[\{\mathbf{N}_\sigma; t_0 \leqslant \sigma < t\}] = -\langle \mathbf{1}, \hat{\boldsymbol{\lambda}}_t\rangle dt + \langle \mathbf{1}, \operatorname{diag}[\ln \hat{\lambda}_t^{(k)}] d\mathbf{N}_t\rangle. \quad (6.231)$$

Integration of this equation establishes (6.224). Equation (6.229) may be established using Bayes' rule, the sample-function density for a Poisson process, and (6.228) as follows. According to Bayes' rule, there holds

$$p_t(\mathbf{X}|\mathbf{N}_\sigma; t_0 \leqslant \sigma < t) = \frac{p(\{\mathbf{N}_\sigma; t_0 \leqslant \sigma < t\}|\mathbf{x}_t = \mathbf{X})p_t(\mathbf{X})}{E[p(\{\mathbf{N}_\sigma; t_0 \leqslant \sigma < t\}|\mathbf{x}_t)]}, \quad (6.232)$$

where $p_t(\mathbf{X})$ and $p_t(X|\mathbf{N}_\sigma; t_0 \leqslant \sigma < t)$ are the prior and posterior probability densities for $\mathbf{x}_t$, $p(\{\mathbf{N}_\sigma, t_0 \leqslant \sigma < t\}|\mathbf{x}_t = \mathbf{X})$ is the conditional sample-function density for $\mathbf{N}$ given that $\mathbf{x}_t = \mathbf{X}$, and the denominator is the required normalization factor. For the first term in the numerator, we use iterated expectations to obtain

$$p(\{\mathbf{N}_\sigma; t_0 \leqslant \sigma < t\}|\mathbf{x}_t = \mathbf{X}) = E[p(\{\mathbf{N}_\sigma; t_0 \leqslant \sigma < t\}|\mathbf{x}_\sigma; t_0 \leqslant \sigma \leqslant t)|\mathbf{x}_t = \mathbf{X}]$$
$$= E[\phi_t(\mathbf{x}_\sigma; t_0 \leqslant \sigma < t)|\mathbf{x}_t = \mathbf{X}]. \quad (6.233)$$

For the denominator of (6.232), we have from (6.228) that

$$E[p(\{\mathbf{N}_\sigma; t_0 \leqslant \sigma < t\}|\mathbf{x}_t = \mathbf{X})] = p(\{\mathbf{N}_\sigma; t_0 \leqslant \sigma < t\})$$
$$= E[\phi_t(\mathbf{x}_\sigma; t_0 \leqslant \sigma < t)]. \quad (6.234)$$

Hence, (6.232) may be rewritten using (6.233) and (6.234) as

$$p_t(\mathbf{X}|N_\sigma; t_0 \leqslant \sigma < t) E[\phi_t(\mathbf{x}_\sigma; t_0 \leqslant \sigma < t)]$$
$$= p_t(\mathbf{X}) E[\phi_t(\mathbf{x}_\sigma; t_0 \leqslant \sigma < t)|\mathbf{x}_t = \mathbf{X}]. \quad (6.235)$$

Multiplication of both sides of (6.235) by $\boldsymbol{\lambda}_t(\mathbf{X})$ and integration with respect to $\mathbf{X}$ then establishes (6.229) and, thus, the sample-function density (6.224).

Multichannel Hypothesis Testing [Extension of (6.73)]. Hypothesis testing for a multichannel doubly stochastic Poisson process is entirely analogous to that for single-channel double stochastic Poisson processes. The likelihood ratio (6.72) becomes

$$\Lambda = \frac{p^1[\{\mathbf{N}_\sigma; t_0 \leqslant \sigma < T\}]}{p^0[\{\mathbf{N}_\sigma; t_0 \leqslant \sigma < T\}]}, \quad (6.236)$$

where p^1 and p^0 are the multichannel sample-function densities under the two hypotheses. The multichannel extension of the important separation theorem of detection (6.73) is

$$\ln \Lambda = \left\langle \mathbf{1}, -\int_{t_0}^{t} (\hat{\boldsymbol{\lambda}}_\sigma^1 - \hat{\boldsymbol{\lambda}}_\sigma^0)\, d\sigma + \int_{t_0}^{t} \operatorname{diag}\left[\ln\left(\frac{\hat{\lambda}_\sigma^{1(k)}}{\hat{\lambda}_\sigma^{0(k)}} \right) \right] d\mathbf{N}_\sigma \right\rangle. \quad (6.237)$$

The interpretation and use of this representation for the likelihood ratio is the same as that for (6.73).

Multichannel Linear Filtering [Extension of (6.78)]. Let $\boldsymbol{\lambda}_t^{*l}$ be a linear estimate of $\boldsymbol{\lambda}_t$ in terms of $\{\mathbf{N}_\sigma; t_0 \leqslant \sigma < t\}$; that is,

$$\boldsymbol{\lambda}_t^{*l} = \mathbf{a}_t + \int_{t_0}^{t} \mathbf{H}(t,u)\, d\mathbf{N}_u \quad (6.238)$$

for some selection of $\mathbf{a}_t$ and $\mathbf{H}(t,u)$. Denote the linear estimate that minimizes the error covariance $E[(\boldsymbol{\lambda}_t - \boldsymbol{\lambda}_t^{*l})(\boldsymbol{\lambda}_t - \boldsymbol{\lambda}_t^{*l})']$ by $\hat{\boldsymbol{\lambda}}_t^l$; here, the minimization means that

$$E[(\boldsymbol{\lambda}_t-\boldsymbol{\lambda}_t^{*l})(\boldsymbol{\lambda}_t-\boldsymbol{\lambda}_t^{*l})']-E\big[(\boldsymbol{\lambda}_t-\hat{\boldsymbol{\lambda}}_t^l)(\boldsymbol{\lambda}_t-\hat{\boldsymbol{\lambda}}_t^l)'\big]$$

is nonnegative definite for all choices of $\boldsymbol{\lambda}_t^{*l}$. Then, the multichannel extensions of (6.78) and (6.80) are

$$\hat{\boldsymbol{\lambda}}_t^l=E(\boldsymbol{\lambda}_t)+\int_{t_0}^{t}\mathbf{H}_0(t,u)[d\mathbf{N}_u-E(\boldsymbol{\lambda}_u)\,du], \tag{6.239}$$

where the optimum impulse response matrix satisfies

$$\mathbf{H}_0(t,u)\,\mathrm{diag}\big[E(\lambda_u^{(k)})\big]+\int_{t_0}^{t}\mathbf{H}_0(t,\sigma)\mathbf{K}_\lambda(\sigma,u)\,d\sigma$$
$$=\mathbf{K}_\lambda(t,u), \qquad t_0\leqslant\sigma\leqslant t$$

and

$$E\big[(\boldsymbol{\lambda}_t-\hat{\boldsymbol{\lambda}}_t^l)(\boldsymbol{\lambda}_t-\hat{\boldsymbol{\lambda}}_t^l)'\big]=\mathbf{H}_0(t,t)\,\mathrm{diag}\big[E(\lambda_t^{(k)})\big]. \tag{6.240}$$

The derivation of these extensions is left as an exercise for the reader; see Problem 6.6.3.

In equations (6.86) and (6.87), we indicated how the Kalman–Bucy algorithm can be applied for linear filtering of single-channel doubly stochastic Poisson processes. This algorithm can also be applied for multichannel observations. Thus, suppose $\{\boldsymbol{\lambda}_t; t\geqslant t_0\}$ has a state-variable realization as

$$d\mathbf{x}_t=\mathbf{A}_t\mathbf{x}_t\,dt+\mathbf{B}_t\,d\mathbf{u}_t, \qquad t\geqslant t_0 \tag{6.241}$$
$$\boldsymbol{\lambda}_t=E(\boldsymbol{\lambda}_t)+\mathbf{C}_t\mathbf{x}_t,$$

where $\mathbf{x}$ is an n-dimensional state vector, $\mathbf{A}_t$ is a known $n\times n$ matrix, $\mathbf{B}_t$ is a known $n\times m$ matrix, $\mathbf{C}_t$ is a known $K\times n$ matrix, $\mathbf{x}_0$ is an initial random variable with zero mean and known covariance $\boldsymbol{\Sigma}_0$, and $\{\mathbf{u}_t; t\geqslant t_0\}$ is a vector of independent, zero-mean random processes with stationary, orthogonal increments and normalized so that $E(\mathbf{u}_t\mathbf{u}_\tau')=\mathbf{I}\min(t,\tau)$ where $\mathbf{I}$ is an

$m\times m$ identity matrix. Then, $\hat{\boldsymbol{\lambda}}_t^l$ is given by the following equations:

$$d\hat{\mathbf{x}}_t^l=\mathbf{A}_t\hat{\mathbf{x}}_t^l dt+\boldsymbol{\Sigma}_t\mathbf{C}_t'\operatorname{diag}^{-1}\left[E(\lambda_t^{(k)})\right]$$

$$\times\left[d\mathbf{N}_t-E(\boldsymbol{\lambda}_t)dt-\mathbf{C}_t\hat{\mathbf{x}}_t^l dt\right], \tag{6.242a}$$

$$\hat{\mathbf{x}}_{t_0}^l=\mathbf{0}$$

$$\frac{d\boldsymbol{\Sigma}_t}{dt}=\mathbf{A}_t\boldsymbol{\Sigma}_t+\boldsymbol{\Sigma}_t\mathbf{A}_t'+\mathbf{B}_t\mathbf{B}_t'$$

$$-\boldsymbol{\Sigma}_t\mathbf{C}_t\operatorname{diag}^{-1}\left[E(\lambda_t^{(k)})\right]\mathbf{C}_t'\boldsymbol{\Sigma}_t, \tag{6.242b}$$

$$\boldsymbol{\Sigma}_{t_0}=\boldsymbol{\Sigma}_0,$$

$$\hat{\boldsymbol{\lambda}}_t^l=E(\boldsymbol{\lambda}_t)+\mathbf{C}_t\hat{\mathbf{x}}_t^l. \tag{6.242c}$$

Furthermore, the error covariance matrix in estimating $\boldsymbol{\lambda}_t$ using this linear estimator is given by

$$E\left[(\boldsymbol{\lambda}_t-\hat{\boldsymbol{\lambda}}_t^l)(\boldsymbol{\lambda}_t-\hat{\boldsymbol{\lambda}}_t^l)'\right]=\mathbf{C}_t\boldsymbol{\Sigma}_t\mathbf{C}_t'. \tag{6.243}$$

Multichannel Nonlinear Filtering [Extension of (6.122)]. Let $\{\mathbf{N}_t; t\geqslant t_0\}$ be a multichannel doubly stochastic Poisson process with intensity process $\{\boldsymbol{\lambda}_t(\mathbf{x}_t);\ t\geqslant t_0\}$, where $\{\mathbf{x}_t; t\geqslant t_0\}$ is a Markovian information process characterized by the differential generator $\psi_t(\mathbf{v}|\mathbf{x}_t)$. Our main result of nonlinear filtering for single-channel doubly stochastic Poisson processes, namely (6.122), has the following extension for multichannel processes:

$$d\hat{M}_t(j\mathbf{v})=\hat{E}\left\{\exp[j\langle\mathbf{v},\mathbf{x}_t\rangle]\psi_t(\mathbf{v}|\mathbf{x}_t)\right\}dt$$

$$+\hat{E}\left\{\exp[j\langle\mathbf{v},\mathbf{x}_t\rangle]\left(\boldsymbol{\lambda}_t(\mathbf{x}_t)-\hat{\boldsymbol{\lambda}}_t\right)'\right\}\operatorname{diag}^{-1}(\hat{\lambda}_t^{(k)})\left(d\mathbf{N}_t-\hat{\boldsymbol{\lambda}}_t dt\right), \tag{6.244}$$

$$\hat{M}_{t_0}(j\mathbf{v})=E\left\{\exp[j\langle\mathbf{v},\mathbf{x}_{t_0}\rangle]\right\}$$

where $(\hat{\cdot})$ and $\hat{E}\{(\cdot)\}$ both denote the conditional expectation of $(\cdot)$ given $\{\mathbf{N}_\sigma; t_0\leqslant\sigma<t\}$. The derivation and use of this equation are entirely analogous to the derivation and use of (6.122).

Multichannel Suboptimal Filtering [Extension of (6.164) and (6.165)]. Let $\{\mathbf{N}_t; t\geqslant t_0\}$ be a multichannel doubly stochastic Poisson process with intensity process $\{\boldsymbol{\lambda}_t(\mathbf{x}_t);\ t \geqslant t_0\}$, where $\{\mathbf{x}_t; t \geqslant t_0\}$ is a Markov diffusion satisfying

$$d\mathbf{x}_t = \mathbf{a}_t(\mathbf{x}_t)dt + \mathbf{B}_t d\mathbf{w}_t, \qquad \mathbf{x}_{t_0} = \mathbf{x}_0$$

in which $\{\mathbf{w}_t; t \geqslant t_0\}$ is a vector of independent Wiener processes normalized so that $E(\mathbf{w}_t\mathbf{w}_u') = \mathbf{I}\min(t,u)$, and where $\mathbf{x}_0$ is independent of $\{\mathbf{w}_t; t \geqslant t_0\}$. The multichannel extension of the processor equation (6.164) is

$$d\mathbf{x}_t^* = \mathbf{a}_t(\mathbf{x}_t^*)dt + \tfrac{1}{2}\sum_i \sum_j \sigma_{ij}^* \frac{\partial^2 \mathbf{a}_t(\mathbf{x}_t^*)}{\partial x_i^* \partial x_j^*} dt$$

$$+ \boldsymbol{\Sigma}_t^* \left[\frac{\partial \boldsymbol{\lambda}_t(\mathbf{x}_t^*)}{\partial \mathbf{x}_t^*} \right] \operatorname{diag}^{-1}[\lambda_t^{(k)}(\mathbf{x}_t^*)][d\mathbf{N}_t - \boldsymbol{\lambda}_t(\mathbf{x}_t^*)dt],$$

$$\mathbf{x}_{t_0}^* = E(\mathbf{x}_0) \tag{6.245}$$

where $\boldsymbol{\Sigma}_t^*$ is a symmetric matrix that satisfies the multichannel variance equation, σ_{ij}^* is the i, j-element of $\boldsymbol{\Sigma}_t^*$, and $\partial\boldsymbol{\lambda}(\mathbf{x})/\partial\mathbf{x}$ denotes the Jacobian matrix having as its i, jth element $\partial\lambda_j(\mathbf{x})/\partial x_i$. The multichannel extension of the variance equation (6.165) is

$$d\boldsymbol{\Sigma}_t^* = \left[\frac{\partial \mathbf{a}_t(\mathbf{x}_t^*)}{\partial \mathbf{x}_t^*} \right]' \boldsymbol{\Sigma}_t^* dt + \boldsymbol{\Sigma}_t^* \left[\frac{\partial \mathbf{a}_t(\mathbf{x}_t^*)}{\partial \mathbf{x}_t^*} \right] dt + \mathbf{B}_t\mathbf{B}_t' dt$$

$$- \boldsymbol{\Sigma}_t^* \frac{\partial \left\{ (\partial \boldsymbol{\lambda}_t(\mathbf{x}_t^*)/\partial \mathbf{x}_t^*) \operatorname{diag}^{-1}[\lambda_t^{(k)}(\mathbf{x}_t^*)][d\mathbf{N}_t - \boldsymbol{\lambda}_t(\mathbf{x}_t^*)dt] \right\}}{\partial \mathbf{x}_t^*} \boldsymbol{\Sigma}_t^*$$

$$\boldsymbol{\Sigma}_{t_0}^* = \operatorname{cov}(\mathbf{x}_0) = E(\mathbf{x}_0\mathbf{x}_0') - E(\mathbf{x}_0)E(\mathbf{x}_0'), \tag{6.246}$$

where $\partial\mathbf{v}(\mathbf{x})/\partial\mathbf{x}$ denotes the Jacobian matrix of the vector $\mathbf{v}(\mathbf{x})$ having as its i, j-th element $\partial v_j(\mathbf{x})/\partial x_i$. The derivation of these equations closely parallels that for (6.164) and (6.165) and, therefore, will not be given here.

Multichannel Filtering Performance [Extension of (6.186)]. Let $\{\mathbf{N}_t; t \geqslant t_0\}$ be a multichannel doubly-stochastic Poisson process with intensity process $\{\boldsymbol{\lambda}_t(\mathbf{x}_t); t \geqslant t_0\}$, where $\{\mathbf{x}_t; t \geqslant t_0\}$ is a Gaussian process with zero mean and autocovariance $\mathbf{K}_{t,u} = E(\mathbf{x}_t\mathbf{x}_u')$. Let $\mathbf{x}_t^*$ denote an arbitrary causal estimate of $\mathbf{x}_t$ in terms of past multichannel data $\{\mathbf{N}_\sigma; t_0 \leqslant \sigma < t\}$. Denote the conditional mean estimate by $\hat{\mathbf{x}}_t$. Then there holds

$$E[(\mathbf{x}_t-\mathbf{x}_t^*)(\mathbf{x}_t-\mathbf{x}_t^*)'] \geqslant E[(\mathbf{x}_t-\hat{\mathbf{x}}_t)(\mathbf{x}_t-\hat{\mathbf{x}}_t)']$$

$$\geqslant \boldsymbol{\Sigma}_t \triangleq \lim_{u\uparrow t} \boldsymbol{\Gamma}_{t,u}, \tag{6.247}$$

where $\boldsymbol{\Gamma}$ satisfies the integral equation

$$\boldsymbol{\Gamma}_{t,u}+\int_{t_0}^{t}\boldsymbol{\Gamma}_{t,\tau}\mathbf{P}_\tau\mathbf{K}_{\tau,u}\,d\tau=\mathbf{K}_{t,u}, \qquad t_0\leqslant u\leqslant t \tag{6.248}$$

and where $\mathbf{P}$ is a precomputable matrix defined by

$$\mathbf{P}_t=E\left\{\left[\frac{\partial\boldsymbol{\lambda}_t(\mathbf{x}_t)}{\partial\mathbf{x}_t}\right]\operatorname{diag}^{-1}[\lambda_t^{(k)}(\mathbf{x}_t)]\left[\frac{\partial\boldsymbol{\lambda}_t(\mathbf{x}_t)}{\partial\mathbf{x}_t}\right]'\right\}, \tag{6.249}$$

where $\partial\boldsymbol{\lambda}(\mathbf{x})/\partial\mathbf{x}$ denotes the Jacobian matrix for the function $\boldsymbol{\lambda}$. This is the multichannel extension of (6.186); the only change is to incorporate the matrix $\mathbf{P}_t$ defined in (6.249) into (6.187). This same change carries over to (6.191) when $\mathbf{x}$ has a finite-dimensional state-variable realization.

6.7 SMOOTHING FOR DOUBLY STOCHASTIC POISSON PROCESSES

In Example 6.5.7 we have discussed the problem of causally estimating, or filtering, a vector-valued Markov diffusion process that modulates the intensity of an observed doubly stochastic Poisson process. Due to the causality constraint, the estimation algorithms developed do not allow the possibility of returning to update and possibly improve an estimate after additional observations are obtained. We now examine the estimation problem with the causality constraint removed. This type of estimation is termed *smoothing*, and the smoothing algorithms we develop here do allow the possibility of updating previous estimates as additional data are accumulated.

Three smoothing problems are of interest in practice. For *fixed-point smoothing*, an estimate is desired at a time t which is fixed in an observation interval $[t_0,t_1)$ and t_1 increases as additional data are accumulated. For *fixed-interval sweep smoothing*, t increases from t_0 to t_f and t_1 is fixed. For *fixed-lag smoothing*, t increases in $[t_0,t_1)$ and t_1 increases according to $t_1=t+T$, where $T\geqslant 0$ is termed the lag time. We will obtain recursive smoothing algorithms for each of these problems.

We need to introduce some terminology and notation that will be used throughout the section. Let $\{N_t; t \geqslant t_0\}$ be a doubly stochastic Poisson process with intensity process $\{\lambda_t(\mathbf{x}_t); t \geqslant t_0\}$, where $\{\mathbf{x}_t; t \geqslant t_0\}$ is an n-dimensional Markov diffusion process with a forward Kolmogorov differential operator $L(\cdot)$ defined in (6.114). Denote the counting path observed on the interval $[t_0, t_1)$ by $\mathfrak{N}_{t_0,t_1} = \{N_\sigma; t_0 \leqslant \sigma < t_1\}$. Suppose t is a time in $[t_0, t_1)$, and split the observed counting path into two subrecords $\mathfrak{N}_{t_0,t} \triangleq \{N_\sigma; t_0 \leqslant \sigma < t\}$ and $\mathfrak{N}_{t,t_1} \triangleq \{N_\sigma; t \leqslant \sigma < t_1\}$. Let $p_t(\mathbf{X}|\mathfrak{N}_{t_0,t_1})$ be the conditional probability density of $\mathbf{x}_t$ given $\mathfrak{N}_{t_0,t_1}$; this will be called the *conditional density of smoothing*. The conditional densities $p_t(\mathbf{X}|\mathfrak{N}_{t_0,t})$ and $p_t(\mathbf{X}|\mathfrak{N}_{t,t_1})$ of $\mathbf{x}_t$ given $\mathfrak{N}_{t_0,t}$ and $\mathfrak{N}_{t,t_1}$ will be called the *conditional density of filtering* and *conditional density of parameter estimation*, respectively. The conditional probability density for $\mathbf{x}_u$, $t \leqslant u < t_1$, given $\mathbf{x}_t = \mathbf{X}$ and $\mathfrak{N}_{t,u} \triangleq \{N_\sigma; t \leqslant \sigma < u\}$ will be denoted by $p_u(\boldsymbol{\xi}|\mathbf{x}_t = \mathbf{X}, \mathfrak{N}_{t,u})$ and called the *pinned conditional density of filtering*.

The following lemma gives a decomposition of the conditional density of smoothing in terms of the conditional densities of filtering and parameter estimation. The lemma is motivated by the fact that $\mathbf{x}_t$ for $t \in [t_0, t_1)$ splits the past and future of the observation $\mathfrak{N}_{t_0,t_1}$ in the sense that $\mathfrak{N}_{t_0,t}$ and $\mathfrak{N}_{t,t_1}$ are independent given $\mathbf{x}_t$. This obtains because $\mathbf{x}_t$ splits its own past and future in the sense that $\{\mathbf{x}_\sigma; t_0 \leqslant \sigma < t\}$ and $\{\mathbf{x}_\sigma; t \leqslant \sigma < t_1\}$ are independent given $\mathbf{x}_t$ by the basic properties of Markov processes and because $\{N_\sigma; \sigma \geqslant t_0\}$ has independent increments given $\{\mathbf{x}_\sigma; \sigma \geqslant t_0\}$.

Lemma 6.7.1. The conditional density of smoothing has the following decomposition:

$$p_t(\mathbf{X}|\mathfrak{N}_{t_0,t_1}) = k_t p_t(\mathbf{X}|\mathfrak{N}_{t_0,t}) p_t(\mathbf{X}|\mathfrak{N}_{t,t_1}) p_t^{-1}(\mathbf{X}), \tag{6.250}$$

where k_t is the normalization factor defined by

$$k_t^{-1} = \int_{R^n} p_t(\boldsymbol{\xi}|\mathfrak{N}_{t_0,t}) p_t(\boldsymbol{\xi}|\mathfrak{N}_{t,t_1}) p_t^{-1}(\boldsymbol{\xi})\, d\xi_1\, d\xi_2 \cdots d\xi_n$$

and where $p_t(\mathbf{X})$ is the unconditional density of $\mathbf{x}_t$.

Proof. By using Bayes' rule, we have

$$p_t(\mathbf{X}|\mathfrak{N}_{t_0,t_1}) = c_t p(\{N_\sigma; t_0 \leqslant \sigma < t_1\}|\mathbf{x}_t = \mathbf{X}) p_t(\mathbf{X}), \tag{6.251}$$

where $p(\{N_\sigma; t_0 \leqslant \sigma < t_1\}|\mathbf{x}_t = \mathbf{X})$ is the conditional sample-function density for N on $[t_0, t_1)$ given that $\mathbf{x}_t = \mathbf{X}$, and c_t is the normalization factor. We note for this conditional sample-function density that

$$
\begin{aligned}
p(\{N_\sigma; t_0 \leqslant \sigma < t_1\} | \mathbf{x}_t = \mathbf{X})& \\
&= E[p(\{N_\sigma; t_0 \leqslant \sigma < t_1\} | \mathbf{x}_t = \mathbf{X}, \mathbf{x}_\sigma; t_0 \leqslant \sigma < t_1) | \mathbf{x}_t = \mathbf{X}] \\
&= E[p(\{N_\sigma; t_0 \leqslant \sigma < t\} | \mathbf{x}_t = \mathbf{X}, \mathbf{x}_\sigma; t_0 \leqslant \sigma < t) \\
&\quad \times p(\{N_\sigma; t \leqslant \sigma < t_1\} | \mathbf{x}_t = \mathbf{X}, \mathbf{x}_\sigma; t \leqslant \sigma < t_1) | \mathbf{x}_t = \mathbf{X}] \\
&= E[p(\{N_\sigma; t_0 \leqslant \sigma < t\} | \mathbf{x}_t = \mathbf{X}, \mathbf{x}_\sigma; t_0 \leqslant \sigma < t) | \mathbf{x}_t = \mathbf{X}] \\
&\quad \times E[p(\{N_\sigma; t \leqslant \sigma < t_1\} | \mathbf{x}_t = \mathbf{X}, \mathbf{x}_\sigma; t \leqslant \sigma < t_1) | \mathbf{x}_t = \mathbf{X}].
\end{aligned}
$$

The second equality follows because N on $[t_0, t_1)$ has independent increments given $\mathbf{x}$ on $[t_0, t_1)$ and, further, because N on $[t_0, t)$ depends conditionally only on $\mathbf{x}$ on $[t_0, t)$ and, similarly, N on $[t, t_1)$ depends conditionally only on $\mathbf{x}$ on $[t, t_1)$. The third equality follows from the conditional independence of $\mathbf{x}$ on $[t_0, t)$ and on $[t, t_1)$ given $\mathbf{x}_t$. Thus, we conclude that

$$
\begin{aligned}
p(\{N_\sigma; t_0 \leqslant \sigma < t_1\} | \mathbf{x}_t = \mathbf{X})& \\
&= p(\{N_\sigma; t_0 \leqslant \sigma < t\} | \mathbf{x}_t = \mathbf{X}) p(\{N_\sigma; t \leqslant \sigma < t_1\} | \mathbf{x}_t = \mathbf{X}).
\end{aligned} \tag{6.252}
$$

By substituting (6.252) into (6.251) and using Bayes' rule, we obtain (6.250).

We need one more lemma before stating and proving the main theorems of smoothing for doubly stochastic Poisson process observations. In this lemma, we give a recursive equation for the evolution with t_1 of the conditional density of parameter estimation $p_t(\mathbf{X} | \mathcal{N}_{t,t_1})$.

Lemma 6.7.2. The conditional density of parameter estimation satisfies the differential equation

$$
\begin{aligned}
d_{t_1} p_t(\mathbf{X} | \mathcal{N}_{t,t_1}) = p_t(\mathbf{X} | \mathcal{N}_{t,t_1}) a_{t_1}(\mathbf{X}) \hat{\lambda}_{t_1 | (t,t_1)}^{-1}& \\
\times \left[dN_{t_1} - \hat{\lambda}_{t_1 | (t,t_1)} dt_1 \right]&
\end{aligned} \tag{6.253}
$$

with the initial condition $p_t(\mathbf{X})$ at $t_1 = t$. In this equation, $\hat{\lambda}_{t_1 | (t,t_1)}$ denotes the filtering estimate

$$
\hat{\lambda}_{t_1 | (t,t_1)} \triangleq E[\lambda_{t_1}(\mathbf{x}_{t_1}) | \mathcal{N}_{t,t_1}] \tag{6.254}
$$

and $a_{t_1}(\mathbf{X})$ is defined by

$$a_{t_1}(\mathbf{X})=\hat{\lambda}_{t_1|(t,t_1)}(t,\mathbf{X})-\hat{\lambda}_{t_1|(t,t_1)}, \tag{6.255}$$

where

$$\hat{\lambda}_{t_1|(t,t_1)}(t,\mathbf{X}) \triangleq E[\lambda_{t_1}(\mathbf{x}_{t_1})|\mathbf{x}_t=\mathbf{X}, \mathfrak{N}_{t,t_1}]. \tag{6.256}$$

The expectation in (6.254) implies an integration with respect to the conditional density of filtering $p_{t_1}(\mathbf{X}|\mathfrak{N}_{t,t_1})$. This density satisfies (6.142) with the initial condition $p_t(\mathbf{X})$ at $t_1=t$. The expectation in (6.256) implies an integration with respect to the pinned conditional density of filtering $p_{t_1}(\xi|\mathbf{x}_t=\mathbf{X}, \mathfrak{N}_{t,t_1})$. This density satisfies (6.142) with the initial condition $\delta(\xi-\mathbf{X})$ at $t_1=t$. We term $\hat{\lambda}_{t_1|(t,t_1)}(t,\mathbf{X})$ the pinned estimate of $\lambda_{t_1}(\mathbf{x}_{t_1})$ given the pinning constraint $\mathbf{x}_t=\mathbf{X}$. Thus, $a_{t_1}(\mathbf{X})$ in (6.253) is the difference between the pinned and unpinned estimate of the intensity at time t_1.

Proof. The proof parallels the proof of Theorem 6.5.2. Let $\hat{M}_{t|(t,t_1)}(j\mathbf{v})$ be the conditional characteristic function of $\mathbf{x}_t$ given $\mathfrak{N}_{t,t_1}$ as defined by

$$\hat{M}_{t|(t,t_1)}=E(\exp[j\langle\mathbf{v},\mathbf{x}_t\rangle]|\mathfrak{N}_{t,t_1}). \tag{6.257}$$

Corresponding to an infinitesimal time increment $[t_1,t_1+\Delta t_1)$, there is an increment $\Delta N_{t_1}=N_{t_1+\Delta t_1}-N_{t_1}$ in the observed counting path and an incremental change in the conditional characteristic function

$$\Delta_{t_1}\hat{M}_{t|(t,t_1)}(j\mathbf{v})=\hat{M}_{t|(t,t_1+\Delta t_1)}(j\mathbf{v})-\hat{M}_{t|(t,t_1)}(j\mathbf{v}) \tag{6.258}$$

where

$$\hat{M}_{t|(t,t_1+\Delta t_1)}(j\mathbf{v})=\int \exp[j\langle\mathbf{v},\mathbf{X}\rangle]\,dP_t(\mathbf{X}|\mathfrak{N}_{t,t_1+\Delta t_1}) \tag{6.259}$$

in which $P_t(\mathbf{X}|\mathfrak{N}_{t,t_1+\Delta t_1})$ is the conditional probability distribution function of $\mathbf{x}_t$ given $\mathfrak{N}_{t,t_1+\Delta t_1}$. To evaluate $\hat{M}_{t|(t,t_1+\Delta t_1)}(j\mathbf{v})$, let us define

$$\hat{M}^{\Delta}_{t|(t,t_1+\Delta t_1)}(j\mathbf{v})=\int \exp[j\langle\mathbf{v},\mathbf{X}\rangle]\,dP_t(\mathbf{X}|\mathfrak{N}_{t,t_1},\Delta N_{t_1}), \tag{6.260}$$

where $P_t(\mathbf{X}|\mathfrak{N}_{t,t_1},\Delta N_{t_1})$ is the conditional distribution function for $\mathbf{x}_t$ given $\mathfrak{N}_{t,t_1}$ and the increment ΔN_{t_1}. As in the proof of Theorem 6.5.2, we first evaluate $\hat{M}^{\Delta}$, and then get $\hat{M}$ from this evaluation because

$$\hat{M}^{\Delta}_{t|(t,t_1+\Delta t_1)} \xrightarrow{\text{a.s.}} \hat{M}_{t|(t,t_1+\Delta t_1)}$$

as $\Delta t_1 \downarrow 0$. We evaluate $\hat{M}^{\Delta}$ for fixed ΔN_{t_1}; namely, for $\Delta N_{t_1}=0$, $\Delta N_{t_1}=1$, and $\Delta N_{t_1} = m$ where $m > 1$. For $\Delta N_{t_1} = 0$, we have

$$\hat{M}^{\Delta}_{t|(t,t_1)}(j\mathbf{v}|\Delta N_{t_1}=0) = \int \exp[j\langle \mathbf{v}, \mathbf{X}\rangle]\, dP_t(\mathbf{X}|\mathcal{N}_{t,t_1}, \Delta N_{t_1}=0). \quad (6.261)$$

The finite difference

$$\delta P_t(\mathbf{X}|\mathcal{N}_{t,t_1}, \Delta N_{t_1}=0) = \Pr\left(\mathbf{x}_t \in (\mathbf{X}-\delta\mathbf{X}, \mathbf{X}]|\mathcal{N}_{t,t_1}, \Delta N_{t_1}=0\right)$$

can be evaluated as

$$\delta P_t(\mathbf{X}|\mathcal{N}_{t,t_1}, \Delta N_{t_1}=0) = \Pr\left(\Delta N_{t_1}=0|\mathbf{x}_t \in (\mathbf{X}-\delta\mathbf{X}, \mathbf{X}], \mathcal{N}_{t,t_1}\right) \times \frac{\delta P_t(\mathbf{X}|\mathcal{N}_{t,t_1})}{\Pr(\Delta N_{t_1}=0|\mathcal{N}_{t,t_1})}$$

where as $\delta\mathbf{X} \to \mathbf{0}$

$$\begin{aligned}\Pr&\left(\Delta N_{t_1}=0|\mathbf{x}_t \in (\mathbf{X}-\delta\mathbf{X}, \mathbf{X}], \mathcal{N}_{t,t_1}\right)\\ &\to E\left[\Pr(\Delta N_{t_1}=0|\mathbf{x}_t=\mathbf{X}, \mathbf{x}_{t_1}, \mathcal{N}_{t,t_1})|\mathbf{x}_t=\mathbf{X}, \mathcal{N}_{t,t_1}\right]\\ &= 1-\hat{\lambda}_{t_1|(t,t_1)}(t,\mathbf{X})\Delta t_1 + o(\Delta t_1)\end{aligned}$$

and where

$$\begin{aligned}\Pr(\Delta N_{t_1}=0|\mathcal{N}_{t,t_1}) &= E\left[\Pr(\Delta N_{t_1}=0|\mathbf{x}_{t_1}, \mathcal{N}_{t,t_1})|\mathcal{N}_{t,t_1}\right]\\ &= 1-\hat{\lambda}_{t_1|(t,t_1)}\Delta t_1 + o(\Delta t_1).\end{aligned}$$

Putting these results together, we have for $\Delta N_{t_1} = 0$ that

$$\begin{aligned}\hat{M}^{\Delta}_{t|(t,t_1)}&(j\mathbf{v}|\Delta N_{t_1}=0)\\ &= E\left\{\exp[j\langle \mathbf{v}, \mathbf{x}_t\rangle](1-a_{t_1}(\mathbf{x}_t)\Delta t_1)|\mathcal{N}_{t,t_1}\right\} + o(\Delta t_1), \quad (6.262)\end{aligned}$$

where

$$a_{t_1}(\mathbf{x}_t) \triangleq \hat{\lambda}_{t_1|(t,t_1)}(t,\mathbf{x}_t) - \hat{\lambda}_{t_1|(t,t_1)}. \quad (6.263)$$

Similarly, by the same procedure, we obtain for $\Delta N_{t_1}=1$ that

$$\hat{M}^{\Delta}_{t|(t,t_1)}(j\mathbf{v}|\Delta N_{t_1}=1)$$
$$=E\left\{\exp[j\langle\mathbf{v},\mathbf{x}_t\rangle]\left(a_{t_1}(\mathbf{x}_t)+\hat{\lambda}_{t_1|(t,t_1)}\right)|\mathcal{N}_{t,t_1}\right\}\hat{\lambda}^{-1}_{t_1|(t,t_1)}$$
$$+o(\Delta t_1) \tag{6.264}$$

and for $\Delta N_t=m$, $m>1$, that

$$\hat{M}^{\Delta}_{t|(t,t_1)}(j\mathbf{v}|\Delta N_{t_1}=m)=o(\Delta t_1). \tag{6.265}$$

Thus, for ΔN_{t_1} variable, we have

$$\hat{M}^{\Delta}_{t|(t,t_1)}(j\mathbf{v})=E\left\{\exp[j\langle\mathbf{v},\mathbf{x}_t\rangle](1-a_{t_1}(\mathbf{x}_t)\Delta t_1)|\mathcal{N}_{t,t_1}\right\}\delta_{0,\Delta N_{t_1}}$$
$$+E\left\{\exp[j\langle\mathbf{v},\mathbf{x}_t\rangle]\left(a_{t_1}(\mathbf{x}_t)+\hat{\lambda}_{t_1|(t,t_1)}\right)|\mathcal{N}_{t,t_1}\right\}\hat{\lambda}^{-1}_{t_1|(t,t_1)}\delta_{1,\Delta N_{t_1}}$$
$$+o(\Delta t_1). \tag{6.266}$$

Here, δ_{ij} is the Kronecker delta function. Because $\delta_{0,\Delta N_{t1}}$ and $\delta_{1,\Delta N_{t1}}$ approach $1-\Delta N_{t_1}$ and ΔN_{t_1}, respectively, with probability one as $\Delta t_1\downarrow 0$, we can write

$$\Delta_{t_1}\hat{M}^{\Delta}_{t|(t,t_1)}(j\mathbf{v})=\hat{M}^{\Delta}_{t|(t,t_1+\Delta t_1)}(j\mathbf{v})-\hat{M}_{t|(t,t_1)}$$
$$=E\left\{\exp[j\langle\mathbf{v},\mathbf{x}_t\rangle]a_{t_1}(\mathbf{x}_t)|\mathcal{N}_{t,t_1}\right\}\hat{\lambda}_{t_1|(t,t_1)}\left[\Delta N_{t_1}-\hat{\lambda}_{t_1|(t,t_1)}\Delta t_1\right]$$
$$+o(\Delta t_1). \tag{6.267}$$

Taking the limit as $\Delta t_1\to 0$ and then inverse Fourier transforming establishes Lemma 6.7.2.

Corollary to Lemma 6.7.2. The conditional density of parameter estimation is given by

$$p_t(\mathbf{X}|\mathcal{N}_{t,t_1})=p_t(\mathbf{X})\phi_{t,t_1}(\mathbf{X}) \tag{6.268}$$

where

$$\phi_{t,t_1}(\mathbf{X}) \triangleq \exp\left[-\int_t^{t_1} a_\sigma(\mathbf{X})\, d\sigma \right.$$
$$\left. + \int_t^{t_1} \ln\left(1 + a_\sigma(\mathbf{X})\hat{\lambda}_{\sigma|(t,\sigma)}^{-1}\right) dN_\sigma \right]. \tag{6.269}$$

Proof. Lemma 6.7.2 and the differential rule of Theorem 4.2.2 imply that

$$d_{t_1} \ln p_t(\mathbf{X}|\mathcal{N}_{t,t_1}) = -a_{t_1}(\mathbf{X})\,dt_1$$
$$+ \ln\left[1 + a_{t_1}(\mathbf{X})\hat{\lambda}_{t_1|(t,t_1)}^{-1}\right] dN_{t_1}. \tag{6.270}$$

Straightforward integration establishes the corollary to Lemma 6.7.2.

We now use Lemmas 6.7.1 and 6.7.2 to get an expression for the conditional density of smoothing in terms of conditional densities of filtering.

Theorem 6.7.1 (Smoothing in Terms of Filtering). The conditional density of smoothing for doubly stochastic Poisson process observations can be expressed as

$$p_t(\mathbf{X}|\mathcal{N}_{t_0,t_1}) = p_t(\mathbf{X}|\mathcal{N}_{t_0,t})\phi_{t,t_1}(\mathbf{X})\hat{\phi}_{t,t_1|(t_0,t)}^{-1} \tag{6.271}$$

where ϕ_{t,t_1} is defined in (6.269) and where

$$\hat{\phi}_{t,t_1|(t_0,t)} = E\left[\phi_{t,t_1}(\mathbf{x}_t)|\mathcal{N}_{t_0,t}\right]. \tag{6.272}$$

Proof. Theorem 6.7.1 follows by a straightforward substitution of (6.268) into (6.250). From the definition of k_t, we have that $k_t = \hat{\phi}_{t,t_1|(t_0,t)}^{-1}$.

Theorem 6.7.1 shows that the conditional density of smoothing can be determined from the conditional densities of filtering, $p_t(\mathbf{X}|\mathcal{N}_{t_0,t})$ and $p_\sigma(\mathbf{X}|\mathcal{N}_{t,\sigma})$, and the pinned conditional density of filtering $p_\sigma(\xi|\mathbf{x}_t = \mathbf{X}, \mathcal{N}_{t,\sigma})$. These densities satisfy (6.142) with the initial conditions $p_{t_0}(\mathbf{X})$, $p_t(\mathbf{X})$, and $\delta(\xi - \mathbf{X})$, respectively. These filtering densities, and therefore the smoothing density, can be determined recursively using the updating procedure outlined in Examples 6.5.4 and 6.5.6.

We now use the assertion of Theorem 6.7.1 to obtain differential equations that can be solved recursively for fixed-point smoothing, fixed-interval sweep smoothing, and fixed-lag smoothing.

Fixed-Point Smoothing

For fixed-point smoothing, t is a fixed time in the observation interval $[t_0, t_1)$, and the endpoint time t_1 increases as additional observations of the doubly stochastic Poisson process are accumulated. Thus, for fixed-point smoothing, it is of interest to known how the conditional density of smoothing $p_t(\mathbf{X}|\mathfrak{N}_{t_0,t_1})$ evolves with increasing t_1. This evolution is specified in the following theorem.

Theorem 6.7.2 (Fixed-Point Smoothing). The conditional density of smoothing evolves with increasing t_1, for $t_1 \geqslant t$, as

$$d_{t_1} p_t(\mathbf{X}|\mathfrak{N}_{t_0,t_1}) = p_t(\mathbf{X}|\mathfrak{N}_{t_0,t_1})\left[\hat{\lambda}_{t_1|(t,t_1)}(t,\mathbf{X}) - \hat{\lambda}_{t_1|(t_0,t_1)}\right]$$
$$\times \hat{\lambda}^{-1}_{t_1|(t_0,t_1)}\left[dN_{t_1} - \hat{\lambda}_{t_1|(t_0,t_1)}dt_1\right]. \tag{6.273}$$

The estimates of the intensity process that are required on the right side of (6.273) are all filtering estimates.

Proof. From Theorem 6.7.1, we have

$$d_{t_1} p_t(\mathbf{X}|\mathfrak{N}_{t_0,t_1}) = p_t(\mathbf{X}|\mathfrak{N}_{t_0,t})d_{t_1}\left[\phi_{t,t_1}(\mathbf{X})\hat{\phi}^{-1}_{t,t_1|(t_0,t)}\right]. \tag{6.274}$$

To evaluate the differential on the right side, we determine $d_{t_1}\phi_{t,t_1}(\mathbf{X})$ and $d_{t_1}\hat{\phi}^{-1}_{t,t_1|(t_0,t)}$ and use the differential rule of Theorem 4.2.2 on the product $\phi_{t,t_1}(\mathbf{X})\hat{\phi}^{-1}_{t,t_1|(t_0,t)}$. By using the definition of $\phi_{t,t_1}(\mathbf{X})$ in (6.269) and the differential rule of Theorem 4.2.2, we obtain

$$d_{t_1}\phi_{t,t_1}(\mathbf{X}) = \phi_{t,t_1}(\mathbf{X})a_{t_1}(\mathbf{X})\hat{\lambda}^{-1}_{t_1|(t,t_1)}\left[dN_{t_1} - \hat{\lambda}_{t|(t,t_1)}dt_1\right]. \tag{6.275}$$

Multiplication of this expression by $p_t(\mathbf{X}|\mathfrak{N}_{t_0,t})$ and integration over $\mathbf{X}$ yield

$$d\hat{\phi}_{t,t_1|(t_0,t)} = E\left[\phi_{t,t_1}(\mathbf{x}_t)a_{t_1}(\mathbf{x}_t)|\mathfrak{N}_{t_0,t}\right]\hat{\lambda}^{-1}_{t_1|(t,t_1)}\left[dN_{t_1} - \hat{\lambda}_{t|(t,t_1)}dt_1\right]. \tag{6.276}$$

It follows from (6.271) that for the expectation on the right side, we have

$$E[\phi_{t,t_1}(\mathbf{x}_t)a_{t_1}(\mathbf{x}_t)|\mathfrak{N}_{t_0,t}]$$

$$=\int_{R^n}\phi_{t,t_1}(\mathbf{X})a_{t_1}(\mathbf{X})p_t(\mathbf{X}|\mathfrak{N}_{t_0,t})\,dX_1\,dX_2\cdots dN_n$$

$$=\hat{\phi}_{t,t_1|(t_0,t)}\int_{R^n}a_{t_1}(\mathbf{X})p_t(\mathbf{X}|\mathfrak{N}_{t_0,t_1})\,dX_1\,dX_2\cdots dX_n$$

$$\triangleq\hat{\phi}_{t,t_1|(t_0,t)}\hat{a}_{t_1|(t_0,t_1)}.$$

Hence,

$$d\hat{\phi}_{t,t_1|(t_0,t)}=\hat{\phi}_{t,t_1|(t_0,t)}\hat{a}_{t_1|(t_0,t_1)}\hat{\lambda}_{t_1|}^{-1}{}_{(t,t_1)}$$

$$\times\left[dN_{t_1}-\hat{\lambda}_{t_1|(t,t_1)}dt_1\right]. \tag{6.277}$$

Straightforward application of the differential rule of Theorem 4.2.2, (6.275), and (6.277), then shows that

$$d_{t_1}\left[\phi_{t,t_1}(\mathbf{X})\hat{\phi}_{t,t_1|(t_0,t)}^{-1}\right]=\phi_{t,t_1}(\mathbf{X})\hat{\phi}_{t,t_1|(t_0,t)}^{-1}\left[a_{t_1}(\mathbf{X})-\hat{a}_{t_1|(t_0,t_1)}\right]$$

$$\times\left[\hat{\lambda}_{t_1|(t,t_1)}+\hat{a}_{t_1|(t_0,t_1)}\right]^{-1}$$

$$\times\left\{dN_{t_1}-\left[\hat{\lambda}_{t_1|(t,t_1)}+\hat{a}_{t_1|(t_0,t_1)}\right]dt_1\right\}.$$

Because

$$p_{t_1}(\xi|\mathbf{x}_t=\mathbf{X},\mathfrak{N}_{t,t_1})=p_{t_1}(\xi|\mathbf{x}_t=\mathbf{X},\mathfrak{N}_{t_0,t_1}),$$

it follows that

$$\hat{a}_{t_1|(t_0,t_1)}=\hat{\lambda}_{t_1|(t_0,t_1)}-\hat{\lambda}_{t_1|(t,t_1)}.$$

Use of this, the definition for $a_t(\mathbf{X})$, and multiplication by $p_t(\mathbf{X}|\mathfrak{N}_{t_0,t})$ then establishes Theorem 6.6.2.

The boundary condition for Theorem 6.7.2, namely $p_t(\mathbf{X}|\mathfrak{N}_{t_0,t})$ at $t_1=t$, satisfies (6.142).

Corollary to Theorem 6.7.2. The conditional density of smoothing can be expressed as

$$p_t(\mathbf{X}|\mathcal{N}_{t_0,t_1}) = p_t(\mathbf{X}|\mathcal{N}_{t_0,t})\exp[\psi_{t,t_1}(\mathbf{X})], \tag{6.278}$$

where

$$\psi_{t,t_1}(\mathbf{X}) \triangleq -\int_t^{t_1}\left[\hat{\lambda}_{\sigma|(t,\sigma)}(t,\mathbf{X}) - \hat{\lambda}_{\sigma|(t_0,\sigma)}\right]d\sigma$$
$$+\int_t^{t_1}\ln\left[\hat{\lambda}_{\sigma|(t,\sigma)}(t,\mathbf{X})\hat{\lambda}^{-1}_{\sigma|(t_0,\sigma)}\right]dN_\sigma. \tag{6.279}$$

Proof. Equation (6.278) follows easily from Theorem 6.7.2, the differential rule on $\ln p_t(\mathbf{X}|\mathcal{N}_{t_0,t_1})$, and integration.

Fixed-Interval Sweep Smoothing

For fixed-interval sweep smoothing, the observation interval $[t_0,t_1)$ is fixed and t increases from the initial observation time t_0 to the final time t_1. Thus, for fixed-interval sweep smoothing, it is of interest to know how the conditional density of smoothing $p_t(\mathbf{X}|\mathcal{N}_{t_0,t_1})$ evolves with increasing t. This evolution is specified in the following theorem.

Theorem 6.7.3 (Fixed-Interval Sweep Smoothing). The conditional density of fixed-interval sweep smoothing evolves with increasing t for $t_0 \leqslant t < t_1$ as

$$d_t p_t(\mathbf{X}|\mathcal{N}_{t_0,t_1}) = L\left[p_t(X|\mathcal{N}_{t_0,t})\right]\exp[\psi_{t,t_1}(\mathbf{X})]dt, \tag{6.280}$$

where $L[\cdot]$ is the forward Kolmogorov differential operator associated with the Markov diffusion process $\mathbf{x}$ and defined in (6.114).

Proof. To prove Theorem 6.7.3, we use (6.278) and the differential rule of Theorem 4.2.2. From (6.142), we have

$$d_t p_t(\mathbf{X}|\mathcal{N}_{t_0,t}) = L\left[p_t(\mathbf{X}|\mathcal{N}_{t_0,t})\right]dt$$
$$+p_t(\mathbf{X}|\mathcal{N}_{t_0,t})\left[\lambda_t(\mathbf{X}) - \hat{\lambda}_{t|(t_0,t)}\right]$$
$$\times\hat{\lambda}^{-1}_{t|(t_0,t)}\left[dN_t - \hat{\lambda}_{t|(t_0,t)}dt\right]. \tag{6.281}$$

From the definition of $\psi_{t,t_1}(\mathbf{X})$ in (6.279) and the differential rule of Theorem 4.2.2, we have

$$d_t \exp[\psi_{t,t_1}(\mathbf{X})] = \exp[\psi_{t,t_1}(\mathbf{X})]\{\lambda_t(\mathbf{X}) - \hat{\lambda}_{t|t_0,t}\}dt$$

$$-\exp[\psi_{t,t_1}(\mathbf{X})]\left[\lambda_t(\mathbf{X}) - \hat{\lambda}_{t|(t_0,t)}\right]\lambda_t^{-1}(\mathbf{X})dN_t. \tag{6.282}$$

Use of the differential rule to evaluate $d_t\{p_t(\mathbf{X}|\mathfrak{N}_{t_0,t})\exp[\psi_{t,t_1}(\mathbf{X})]\}$ and straightforward manipulation then establishes Theorem 6.7.3.

The boundary condition for Theorem 6.7.3, namely $p_{t_0}(\mathbf{X}|\mathfrak{N}_{t_0,t_1})$ at $t=t_0$, satisfies (6.253).

Fixed-Lag Smoothing

For fixed-lag smoothing, t increases from the initial time t_0 and the endpoint of the observation interval $[t_0,t_1)$ increases as $t_1 = t+T$, where T is termed the lag time. Thus, for fixed-lag smoothing it is of interest to know how the smoothing density $p_t(\mathbf{X}|\mathfrak{N}_{t_0,t+T})$ evolves with increasing t. This evolution is specified in the following theorem.

Theorem 6.7.4 (Fixed-Lag Smoothing). The conditional density of fixed-lag smoothing $p_t(\mathbf{X}|\mathfrak{N}_{t_0,t+T})$ evolves with increasing t according to

$$d_t p_t(\mathbf{X}|\mathfrak{N}_{t_0,t+T})$$

$$= L\left[p_t(\mathbf{X}|\mathfrak{N}_{t_0,t})\right]\exp[\psi_{t,t+T}(\mathbf{X})]dt$$

$$+p_t(\mathbf{X}|\mathfrak{N}_{t_0,t+T})\left[\hat{\lambda}_{t+T|(t,t+T)}(t,\mathbf{X}) - \hat{\lambda}_{t+T|(t_0,t+T)}\right]$$

$$\times\hat{\lambda}^{-1}_{t+T|(t_0,t+T)}\left[dN_{t+T} - \hat{\lambda}_{t+T|(t_0,t+T)}dt\right]. \tag{6.283}$$

Proof. Theorem 6.7.4 follows immediately from Theorems 6.7.2 and 6.7.3 by noting that

$$d_t p_t(\mathbf{X}|\mathfrak{N}_{t_0,t+T}) = d_t p_t(\mathbf{X}|\mathfrak{N}_{t_0,t_1}) + d_{t_1} p_t(\mathbf{X}|\mathfrak{N}_{t_0,t_1}), \tag{6.284}$$

where $t_1 = t+T$.

The boundary condition for Theorem 6.7.4, namely $p_{t_0}(\mathbf{X}|\mathcal{N}_{t_0,t_0+T})$, satisfies (6.253) for $t=t_0$ and $t_1=t_0+T$.

Summary. In this section, we have given some general representations for smoothing densities in terms of filtering densities. An important conclusion to be drawn from the representations is that the filtering problem studied in Section 6.5.2 is the essential problem to be solved for smoothing. An evolution equation for the conditional moment of smoothing is developed in Problem 6.7.1. In parallel to Section 6.5.3, this forms a possible point of departure for the development of suboptimal but implementable smoothing estimators. The representations in this section are analogs for observed point processes of those derived by J. Lo [26] for additive Gaussian observations. An integral equation approach to smoothing for normally distributed information processes is given by J. Clark [8] and D. Snyder [43].

REFERENCES

1. B.D.O. Anderson, J.B. Moore, and S.G. Loo, "Spectral Factorization of Time-Varying Covariance Functions," *IEEE Trans. Information Theory*, **IT-15**, No. 5, 550–557 (Sept. 1969).
2. H. O. Anger, "The Scintillation Camera for Radioisotope Localization," Conference on Nuclear Medicine, Heidelberg, October 1966.
3. A.B. Baggeroer, *State Variables and Communication Theory*, Research Monograph No. 61, The M.I.T. Press, Cambridge, Mass., 1970.
4. P.M. Brémaud, "A Martingale Approach to Point Processes," Memo. No. ERL-M345, Univ. of Calif., Berkeley, August 1972.
5. R.S. Bucy and P. D. Joseph, *Filtering for Stochastic Processes with Applications to Guidance*, Wiley, New York, 1968.
6. D. R. Cox, "Some Statistical Methods Connected with Series of Events," *J.Roy. Statist. Soc. B*, **17**, 129 164 (1955).
7. J. Clark and E.V. Hoversten, "Poisson Process as a Statistical Model for Photodetectors Excited by Gaussian Light," Quarterly Progress Report No. 98, Research Laboratory of Electronics, M.I.T., Cambridge, Mass., July 1970, pp. 95–101.
8. J.R. Clark, "Estimation for Poisson Processes with Applications in Optical Communication," Ph.D. Thesis, Dept. of Electrical Engineering, M.I.T., Cambridge, Mass., September 1971.
9. K. P. Dunn and I. B. Rhodes, "A Generalized Representation Theorem with Applications to Estimation and Hypothesis Testing," Proceedings of the Eleventh Annual Allerton Conference on Circuit and System Theory, Univ. of Ill., Urbana, Oct. 1973.
10. R. Esposito, "Power Scintillations Due to Wandering of the Laser Beam," *Proc. IEEE*, **55**, No. 8, 1533–1534, (Aug. 1967).
11. W. Feller, *An Introduction to Probability Theory and Its Applications: Volume II*, Wiley, New York, 1966.

12. P.M. Fishman, "Statistical Inference for Space-Time Point Processes," D.Sc. Thesis, Sever Institute of Technology, Washington University, St. Louis, Mo., June 1974.
13. P.M. Fishman and D.L. Snyder, "Estimation of a Random Space-Time Field That Influences an Observed Space-Time Point Process," Proceedings of the Eleventh Annual Allerton Conference on Circuit and System Theory, Univ. of Ill., Urbana, Oct. 1973.
14. R.H. Forrester, Jr. and D.L. Snyder, "Phase-Tracking Performance of Direct-Detection Optical Receivers," *IEEE Trans. Communications*, **COM-21**, No. 9, 1037–1039 (Sept. 1973).
15. J. W. Goodman, "Synthetic-Aperture Optics," in *Progress in Optics*, *Vol. III*, Ed. E.E. Wolf, North-Holland, Amsterdam, 1970.
16. J. Grandell, "On Stochastic Processes Generated by a Stochastic Intensity Function," *Skndinavisk Aktuarietidskrift*, 204–240 (1971).
17. J. Grandell, "Statistical Inference for Doubly Stochastic Poisson Processes," in *Stochastic Point Processes, Statistical Analysis, Theory, and Applications*, Ed. P.A.W. Lewis, Wiley, New York, 1972.
18. S. Haber, "Numerical Evaluation of Multiple Integrals," *SIAM Rev*., **12**, No. 4, 481–526 (Oct. 1970).
19. E.V. Hoversten, R.O. Harger, and S.J. Halme, "Communication Theory for the Turbulent Atmosphere," *Proc. IEEE*, **58**, No. 10, 1626–1650 (Oct. 1970).
20. A.H. Jazwinski, *Stochastic Processes and Filtering Theory*, Academic, New York, 1970.
21. S. Karp and J.R. Clark, "Photon Counting: A Problem in Classical Noise Theory," *IEEE Trans. Information Theory*, **IT-16**, No. 6, 672–680 (Nov. 1970).
22. S. Karp, E. O'Neill, and R. Gagliardi, "Communication Theory for the Free-Space Optical Channel," *Proc. IEEE*, **58**, No. 10, 1611–1626 (Oct. 1970).
23. R.S. Kennedy, "Communication Through Optical Scattering Channels: An Introduction," *Proc. IEEE*, **58**, No. 10, 1651–1665 (Oct. 1970).
24. T. Kailath, "Fredholm Resolvents, Weiner–Hopf Equations, and Riccati Differential Equations," *IEEE Trans. Information Theory*, **IT-15**, No. 6, 665–672 (Nov. 1969).
25. G. Lachs, "Theoretical Aspects of Mixtures of Thermal and Coherent Radiation," *Phys. Rev*., **138**, B1012–B1016 (1965).
26. J.T. Lo, "On Optimal Nonlinear Estimation: Continuous Observation," *Information Sci*., **6**, 19–32 (1973).
27. O. Lundberg, *On Random Processes and Their Application to Sickness and Accident Statistics*, Almqvist & Wiksells Boktryckeri A.B., Uppsala, 1964.
28. O. Macchi, "Distribution Statistique des Instants D'émission des Photoélectrons d'une Lumiére Thermique," *C.R. Acad. Sci. Paris*, *Ser. A*, **272**, 437–440 (Feb. 1971).
29. O. Macchi and B. Picinbono, "Estimation and Detection of Weak Optical Signals," *IEEE Trans. Information Theory*, **IT-18**, No. 5, 562–573 (Sept. 1972).
30. O. Macchi, "Processus Ponctuels et Coincidences," D.Sc. These, Centre D'Orsay, Univ. de Paris Sud, January 1972.
31. C.L. Mehta, "Theory of Photoelectron Counting," in *Progress in Optics*, *Vol. VIII*, Ed. E. Wolf, North-Holland, Amsterdam, 1970.
32. N.F. Moody, W. Paul, and M.L.G. Joy, "A Survey of Medical Gamma-Ray Cameras," *Proc. IEEE*, **58**, No. 2, 217–242 (Feb. 1970).
33. W.K. Pratt, *Laser Communication Systems*, Wiley, New York, 1969.
34. L. M. Robinson "Tanlock: A Phase-Lock Loop of Extended Tracking Capability," Proceedings of the National Conference on Military Electronics, 1962.

35. M. Rosenblatt, *Random Processes*, Oxford Univ. Press, New York, 1962.

36. D.A. Ross and C.C. Harris, "Measurement of Radioactivity," in *Principles of Nuclear Medicine*, Ed. H.N. Wagner, Jr., Saunders, Philadelphia, 1968.

37. M. Rudemo, "Doubly-Stochastic Poisson Processes and Process Control," *Advan. Appl. Prob.*, **4**, No. 2, 318–338 (Aug. 1972).

38. B.E.A. Saleh, "Probability Distribution of Time of Arrival of Photon Counts for a Stationary Optical Field," *IEEE Trans. Information Theory*, **IT-20**, No. 2, 262–263 (March 1974).

39. A. Segall, "A Martingale Approach to Modeling, Estimation, and Detection of Jump Processes," Information Systems Lab. Tech. Rpt. No. 7050-21, Stanford University, August 1973.

40. R.F. Serfozo, "Conditional Poisson Processes," *J. Appl. Prob.*, **9**, 288–302 (Sept. 1972).

41. D.L. Snyder and I.B. Rhodes, "Filtering and Control Performance Bounds with Implications on Asymptotic Separation," *Automatica*, **8**, 747–753 (Nov. 1972).

42. D.L. Snyder, *The State Variable Approach to Continuous Estimation with Applications to Analog Communication Theory*, M.I.T. Press, Cambridge, Mass., 1969.

43. D.L. Snyder, "Nonrecursive Smoothing for Gaussian Modulated Point Processes," Proceedings of the Ninth Annual Allerton Conference on Circuit and System Theory, Univ. of Ill., Urbana, 1971.

44. M.M. Ter-Pogossian, J.O. Eichling, D.O. Davis, and M.J. Welch, "The Measure in Vivo of Regional Cerebral Oxygen Utilization by Means of Oxyhemoglobin Labeled with Radioactive Oxygen—15," *J. Clin. Invest.*, **49**, 381–391 (1970).

45. P.J. Titterton, "Power Reduction and Fluctuations Caused by Narrow Laser Beam Motion in the Far Field," *Appl. Optics*, **12**, No. 2, 423–425 (Feb. 1973).

46. J.H. Van Schuppen, "Estimation Theory for Continuous Time Processes, A Martingal Approach," Memo. No. ERL-M405, Univ. of Calif., Berkeley, September 1973.

47. H.L. Van Trees, *Detection, Estimation, and Modulation Theory: Part I*, Wiley, New York, 1968.

48. H.L. Van Trees, *Detection, Estimation, and Modulation Theory: Part III*, Wiley, New York, 1971.

49. A.J. Viterbi, *Principles of Coherent Communication*, McGraw-Hill, New York, 1966.

50. E. Wong, *Stochastic Processes in Information and Dynamical Systems*, McGraw-Hill, New York, 1971.

51. M. Yovits and J. Jackson, "Linear Filter Optimization with Game Theory Considerations," 1955 IRE National Convention Record, Part 3, pp. 193–199, 1955.

PROBLEMS

6.2.1. Let $\{N_t; t \geqslant 0\}$ be a doubly stochastic Poisson process with intensity process x, a random variable independent of time.

a. Show that

$$\Pr(N_t = n) = (n!)^{-1}(-t)^n \frac{d^n \Pr(N_t = 0)}{dt^n}$$

b. Suppose x has the negative binomial distribution

$$\Pr(x=k)=\frac{\Gamma(k+\beta+1)}{\Gamma(k+1)\Gamma(\beta+1)}\left(\frac{\alpha}{\alpha+t}\right)^{\beta+1}\left(\frac{t}{\alpha+t}\right)^{k}$$

for $k=0,1,2,\ldots$. Show that

$$\Pr(N_t=0)=\left[\frac{1-\xi}{1-\xi\exp(-t)}\right]^{\beta+1}$$

where $\xi=\mathrm{t}(\alpha+\mathrm{t})^{-1}$. Deduce from (a) that

$$\Pr(N_t=n)=(n!)^{-1}t^n\frac{(1-\xi)^{\beta+1}}{\Gamma(\beta+1)}$$

$$\times\sum_{j=0}^{\infty}(j!)^{-1}j^n\Gamma(\beta+1+j)[\xi\exp(-t)]^j.$$

6.2.2. Let $\{N_t;\ t\geqslant t_0\}$ be a doubly stochastic Poisson process with intensity process $\{\lambda_t(\mathbf{x}_t);\ t\geqslant t_0\}$. Let

$$P_t(\Lambda)=\Pr\left(t^{-1}\int_{t_0}^{t}\lambda_\sigma(\mathbf{x}_\sigma)\,d\sigma\leqslant\Lambda\right).$$

Show that

$$\Pr(N_t=n)=\int_0^\infty(n!)^{-1}(\Lambda t)^n\exp(-\Lambda t)\,dP_t(\Lambda).$$

6.2.3. Let $\{N_t;\ t\geqslant t_0\}$ be a doubly stochastic Poisson process with intensity process $\{x\nu_t;\ t\geqslant t_0\}$ where x is a nonnegative integer-valued random variable and ν_t is a deterministic function. Suppose x is Poisson distributed with parameter μ. Show that

$$\Pr(N_t=n)=(n!)^{-1}\xi_t^n\exp[-\mu(1-e^{-\xi_t})]\}m_n(\mu e^{-\xi_t}),$$

where

$$\xi_t\triangleq\int_{t_0}^{t}\nu_\sigma\,d\sigma$$

and where $m_n(\lambda)$ denotes the nth moment about the origin of a Poisson distribution with parameter λ.

6.2.4. Let $\{N_t;\ t \geqslant t_0\}$ be a double stochastic Poisson process with intensity process $\{\lambda_t;\ t \geqslant t_0\}$. Define $\lambda_t = \int_{t_0}^{t} \lambda_\sigma \, d\sigma$.

a. Establish the following relation between the covariance functions for N and Λ:

$$\operatorname{cov}(N_t, N_u) = \operatorname{cov}(\Lambda_t, \Lambda_u) + \min[E(\Lambda_t), E(\Lambda_u)].$$

b. The index of dispersion for a random variable z is defined by $\mathrm{ID}(z) = \operatorname{var}(z)/\operatorname{mean}(z)$. Show for the processes of part (a) that

$$\mathrm{ID}(N_t) = \mathrm{ID}(\Lambda_t) + 1 \geqslant 1.$$

Conclude, therefore, that the index of dispersion for a doubly stochastic Poisson process is never smaller than that for a Poisson process with the same mean-value function.

6.2.5. Let $\{N_t;\ t \geqslant t_0\}$ be a doubly stochastic Poisson process with intensity process $\{\lambda_t;\ t \geqslant t_0\}$. Define $\Lambda_t = \int_{t_0}^{t} \lambda_\sigma \, d\sigma$. Let $M_{N_t}(s) = E[\exp(sN_t)]$ and $M_{\Lambda_t}(s) = E[\exp(s\Lambda_t)]$ be the moment generating functions for N_t and Λ_t, respectively. Similarly, let $\Gamma_{N_t}(s) = \ln M_{N_t}(s)$ and $\Gamma_{\Lambda_t}(s) = \ln M_{\Lambda_t}(s)$ be the corresponding cumulant generating functions; the cumulants are defined in Problem 2.2.3.

a. Show that

$$M_{N_t}(s) = M_{\Lambda_t}(e^s - 1) \quad \text{and} \quad \Gamma_{N_t}(s) = \Gamma_{\Lambda_t}(e^s - 1).$$

b. Use the moment generating properties of $M_{N_t}(s)$ and $M_{\Lambda_t}(s)$ to establish the following relations between the moments of N_t and Λ_t:

$$E(N_t) = E(\Lambda_t),$$

$$E(N_t^2) = E(\Lambda_t^2) + E(\Lambda_t),$$

$$E(N_t^3) = E(\Lambda_t^3) + 3E(\Lambda_t^2) + E(\Lambda_t),$$

$$E(N_t^4) = E(\Lambda_t^4) + 6E(\Lambda_t^3) + 7E(\Lambda_t^2) + E(\Lambda_t).$$

c. Use the cumulant generating properties of $\Gamma_{N_t}(s)$ and $\Gamma_{\Lambda_t}(s)$ to establish the following relations between the cumulants, $\gamma_k(N_t)$ and $\gamma_k(\Lambda_t)$, of N_t

and Λ_t:

$$\gamma_1(N_t)=\gamma_1(\Lambda_t),$$

$$\gamma_2(N_t)=\gamma_2(\Lambda_t)+\gamma_1(\Lambda_t),$$

$$\gamma_3(N_t)=\gamma_3(\Lambda_t)+3\gamma_2(\Lambda_t)+\gamma_1(\Lambda_t),$$

$$\gamma_4(N_t)=\gamma_4(\Lambda_t)+6\gamma_3(\Lambda_t)+7\gamma_2(\Lambda_t)+\gamma_1(\Lambda_t).$$

d. Use the result of (c) to show for a Poisson process with mean value function Λ_t that for all $k \geqslant 1$ there holds $\gamma_k(N_t)=\Lambda_t$.

6.2.6. Let $\{N_t;\ t\geqslant t_0\}$ be a doubly stochastic Poisson process with intensity process $\{\lambda_t;\ t\geqslant t_0\}$. Let $\{y_t;\ t\}0\}$ be a filtered doubly stochastic Poisson process,

$$y_t=\int_{t_0}^{t} h(t,u)\,dN_u.$$

Use the method of conditioning and the results of Section 4.2.1 to establish the following second-moment properties for y.

a. Show that

$$E(y_t)=\int_{t_0}^{t} h(t,u)E(\lambda_u)\,du.$$

b. Show that

$$\operatorname{var}(y_t)=\int_{t_0}^{t} h^2(t,u)E(\lambda_u)\,du$$

$$+\int_{t_0}^{t}\int_{t_0}^{t} h(t,u)K_\lambda(u,v)h(t,v)\,du\,dv,$$

where K_λ is the covariance function for λ.

c. Let

$$\operatorname{cov}(\lambda_s,y_t)=E(\lambda_s,y_t)-E(\lambda_s)E(y_t).$$

Show that

$$\operatorname{cov}(\lambda_s,y_t)=\int_{t_0}^{t} h(t,u)K_\lambda(s,u)\,du.$$

6.2.7. Suppose the covariance function $K(\sigma,u)$ for the process $\{X_t;\ t\geqslant t_0\}$ in Example 6.2.2 has the separable form

$$K(\sigma,u)=\sum_{i=1}^{m}\lambda_i\varphi_i(\sigma)\varphi_i^*(u)$$

where m is finite. In this case, K has a finite number, m, of nonzero eigenvalues. The situation when all the eigenvalues $\lambda_1,\lambda_2,\ldots,\lambda_{\mathrm{m}}$ are equal to the same constant, say σ^2, is of some importance in practice. For instance, if the background radiation $B_t=X_t-M_t$ is a white, band-limited process of bandwidth W, the constant σ^2 is the noise intensity (i.e., $\sigma^2 W$ is the total noise power) and $m\doteq 2Wt+1$ (see H. Van Trees [47, p. 193] and S. Karp and J. Clark [21].) For another example, X_t in an optical radar-ranging system can correspond to the optical field resulting from reflections off m rough, equal-strength scatterers; M_t corresponds to a specular component in the received optical radar pulse.

a. Show that the characteristic function for N_t is given by

$$M_{N_t}(ju)=\left[1-\alpha\sigma^2(e^{ju}-1)\right]^{-m}$$

$$\times\exp\left\{\alpha(e^{ju}-1)\int_{t_0}^{t}|M_\tau|^2d\tau\left[1-\alpha\sigma^2(e^{ju}-1)\right]^{-1}\right\}.$$

b. Show that when $M_t=0$, the count N_t has the negative binomial distribution

$$\Pr(N_t=n)=\frac{\Gamma(n+m)}{\Gamma(n+1)\Gamma(m)}\left[\frac{1}{1+\alpha\sigma^2}\right]^m\left[\frac{\alpha\sigma^2}{1+\alpha\sigma^2}\right]^n.$$

c. The situation $\sigma^2=0$ corresponds to the absence of background radiation or rough radar scatterers; in this situation N_t clearly has a Poisson distribution with parameter $\int_{t_0}^{t}|M_\tau|^2d\tau$. Suppose $m\gg 1$ and $\alpha\sigma^2\ll 1$ corresponding to a large number of weak scatterers or a large number of weak background-noise modes. Show that N_t is approximately Poisson distributed with parameter $\alpha[\sigma^2 m+\int_{t_0}^{t}|M_\tau|^2d\tau]$; here, $\sigma^2 m$ is the total expected energy in B_t.

6.2.8 (O. Lundberg [27]). Let $\{N_t;\ t\geqslant t_0\}$ be a homogeneous Poisson process given its intensity parameter x, where x is a nonnegative random variable. It is shown in Section 6.2.2 that the counting statistics for N are the same as those of a pure Markov birth process with birthrate $\hat{x}_t(n) = E(x|N_t = n)$. Here, you are to establish several properties of $\hat{x}_t(n)$.

a. Demonstrate that $\hat{x}_t(n)$ satisfies the following recursive equation

$$\hat{x}_t(n+1) = \hat{x}_t(n) - \frac{d[\ln \hat{x}_t(n)]}{dt}.$$

b. Show for fixed n that $\hat{x}_t(n)$ is a nonincreasing function of t. [Hint. Establish the expression

$$\hat{x}_t(n) = \frac{E[x^{n+1}\exp(-xt)]}{E[x^n \exp(-xt)]}.$$

The Schwarz inequality can be used with this to show that $\hat{x}_t(n+1) \geqslant \hat{x}_t(n)$.]

c. Show for fixed t that $\hat{x}_t(n)$ is a nondecreasing sequence in n.

d. Suppose for fixed t that $\hat{x}_t(n)$ is linear in n; that is, for some nonnegative functions, a_t and b_t, there holds $\hat{x}_t(n) = a_t + nb_t$. Show that $\hat{x}_t(n)$ must, therefore, have the form $\hat{x}_t(n) = (n + c_1)(t + c_2)^{-1}$ for some constants, c_1 and c_2. Thus, $\{N_t;\ t \geqslant 0\}$ must be a homogeneous Polya process.

6.2.9. The optical communication system shown diagrammatically in Figure P.6.2.9 is used to transmit a random variable λ to a remote user. The variable λ modulates the intensity of the transmitted light beam. The receiver consists of a photodetector (e.g., a photodiode or photomultiplier) and a counter that simply registers the total number of photoelectrons generated in any observation interval. Denote the number of photoelectrons registered by this counter in a T second interval by N_T. Make the following assumptions:

i. N_T is conditionally a Poisson random variable with parameter λT given λ; that is,

$$\Pr[N_T = n|\lambda] = (n!)^{-1}(\lambda T)^n \exp(-\lambda T)$$

for $n = 0, 1, 2, \ldots$.

ii. The probability density function for λ is

$$P_\lambda(\Lambda) = \begin{cases} \dfrac{\alpha^{\beta+1}}{\Gamma(\beta+1)} \Lambda^\beta \exp(-\alpha\Lambda), & \Lambda \geqslant 0 \\ 0, & \Lambda < 0 \end{cases}$$

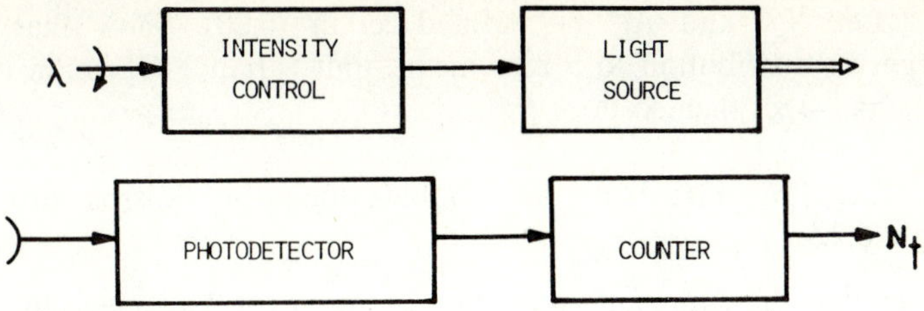

Figure P.6.2.9

where $\alpha > 0$ and $\beta < 0$ are known constants and where $\Gamma(\cdot)$ is a gamma function.

a. Determine the mean of λ.

b. Determine the variance of λ.

c. Determine the mean of N_T.

d. Determine the variance of N_T.

e. Suppose the number of registered photoelectrons in a T-second interval is observed to be n and based on this observation we want to estimate the value of λ. To do this, we first propose to use for simplicity a *linear* estimate $\hat{\lambda}_{\text{lin}}$ of the form

$$\hat{\lambda}_{\text{lin}}(n) = nc_1 + c_2$$

where c_1 and c_2 are constants. Determine these constants so that the mean square error

$$E\left[\left(\lambda - \hat{\lambda}_{\text{lin}}\right)^2\right]$$

is minimized.

f. Suppose now that the linearity restriction is removed in part (e). Determine an expression for the estimate $\hat{\lambda}(n)$ of λ in terms of n that minimizes $E[(\lambda - \hat{\lambda}(n))^2]$. Discuss your results.

6.2.10. Let $\{N_t;\ t \geqslant t_0\}$ be a doubly stochastic Poisson process with intensity process $\{xv_t;\ t \geqslant t_0\}$, where x is a nonnegative random variable with mean m_x and variance σ_x^2, and where v_t is a deterministic function. Show that if v has infinite area, then $N_t^{cn} = [N_t - m_N(t)]\sigma_N^{-1}(t)$ converges in distribution to a zero-mean, unit-variance Gaussian random variable as $t \to \infty$.

6.2.11. Let N_t^{cn} and Λ_t^{cn} be defined as in (6.30). Show that if Λ_t^{cn} converges in distribution to a zero-mean, unit-variance Gaussian random variable as $t \to \infty$, then so does N_t^{cn}.

6.2.12. Let $\{N_t;\ t \geqslant t_0\}$ be the double stochastic Poisson process of Example 6.2.5.

a. Show that the cumulant generating function for N_t is given by

$$\mu(s) = \ln E\{\exp(sN_t)\} = \left[\frac{\alpha}{1+\alpha-\exp(s)}\right]^{\beta t+1}.$$

b. By using a Chernoff bound [see (2.127)] show that

$$\Pr[N_t \leqslant \eta] \leqslant \eta^{-\eta}\left[\frac{\beta t+1+\eta}{1+\alpha}\right]^{\beta t+1+\eta}\left[\frac{\alpha}{\beta t+1}\right]^{\beta t+1}.$$

6.3.1. Let $\mathbf{x}$ be a vector of k independent, identically distributed Gaussian random variables each with mean zero and variance σ^2. Suppose $\{N_t;\ t \geqslant 0\}$ is a doubly stochastic Poisson process with intensity process $\{\lambda_t(\mathbf{x});\ t \geqslant 0\}$, where

$$\lambda_t(\mathbf{x}) = \mathbf{x}'\mathbf{x}\xi\nu_t = (x_1^2 + x_2^2 + \cdots + x_k^2)\xi\nu_t$$

where ξ is a nonnegative parameter and ν_t is a known, nonnegative function.

a. Suppose at time T, it is observed that $N_T = n$. What is the maximum likelihood estimate $\hat{\xi}_{\mathrm{ML}}(n)$ of ξ in terms of this observed count?

b. Suppose a path of N is observed on the interval $[0, T)$, and for this path, $N_T = n$ and jumps occur at times $W_1, W_2, \ldots, W_n$. What is the maximum likelihood estimate $\hat{\xi}_{\mathrm{ML}}(n;\ W_1, W_2, \ldots, W_n)$ of ξ in terms of this observed counting path?

c. Suppose now that $\xi = 1$ and that the variance σ^2 of the variables x_i is the unknown parameter. Indicate from your results in (a) and (b) what the maximum likelihood estimate of σ^2 is for both the count and count-record data.

6.3.2. Consider an orderly point process on the interval $[0, T)$. Let $\{u_i\}$

denote the *unordered* occurrence times on $[0,T)$. These unordered occurrence times and the total count N_T are equivalent to the entire counting record $\{N_\sigma;\ 0\leqslant\sigma<T\}$. Define the entropy of the counting paths by

$$H(0,T) = -E(\ln p[\{N_\sigma;\ 0\leqslant\sigma<T\}]),$$

where p is the sample-function density. Note that H can be written in terms of the statistics of the unordered occurrence times as

$$H(0,T) = -\sum_{n=0}^{\infty} \Pr(N_T=0)\ln\Pr(N_T=0)$$

$$-\sum_{n=1}^{\infty}\Pr(N_T=n)\int_0^T\cdots\int_0^T (p_{\mathbf{u}}(\mathbf{U})|N_T=n)$$

$$\times\ln(p_{\mathbf{u}}(\mathbf{U})|N_T=n)\,dU_1\,dU_2\cdots dU_n.$$

a. Show that subject to the constraint $E(N_T)=\bar{n}\triangleq\bar{\lambda}T$, the entropy $H(0,T)$ is maximized by the selection

$$p_{\mathbf{u}}(\mathbf{U}|N_T=n)\Pr(N_T=n) = \frac{1}{T^n(1+\bar{n})}\left[\frac{\bar{n}}{1+\bar{n}}\right]^n$$

for $n\geqslant 1$. Thus, conclude that the entropy is maximized when the unordered occurrence times are independently and uniformly distributed on $[0,T)$ and the counting statistics are Bose–Einstein. Hint: maximize $H(0,T)$ subject to the constraints:

(i) $\displaystyle\sum_{n=0}^{\infty}\Pr(N_T=n)=1,$

(ii) $\displaystyle\sum_{n=0}^{\infty} n\Pr(N_T=n)=\bar{n},$

(iii) $\displaystyle\int_0^T\cdots\int_0^T p_{\mathbf{u}}(\mathbf{U}|N_T=n)\,dU_1\cdots dU_n=1.$

Incorporate these constraints using Lagrange multipliers.

b. Show that

$$H_{\max}(0,T) = (1+\bar{n})\ln(1+\bar{n}) - \bar{n}\ln\left(\frac{\bar{n}}{T}\right).$$

c. Let $\{N_t;\ 0 \leqslant t < T\}$ be a doubly stochastic Poisson process with a time independent intensity process x; here, x is a nonnegative random variable. We know from Chapter 2 that for any given value of x, the unordered occurrence times of N are independently and uniformly distributed given that $N_T = n \geqslant 1$. It is natural to inquire whether there is a distribution for x for which N is the maximum entropy counting process; that is, is there a density $p_x(X)$ such that

$$\int_0^{\infty} (n!)^{-1} (XT)^n \exp(-XT) p_x(X)\,dX = \frac{1}{1+\bar{\lambda}T}\left(\frac{\bar{\lambda}T}{1+\bar{\lambda}T}\right)^n.$$

Show that this holds when x has the gamma distribution of (6.3) with $\beta = 0$ and $\alpha = \bar{\lambda}^{-1}$, that is, when x is exponentially distributed.

6.3.3. Let $\{N_t;\ t \geqslant t_0\}$ be an inhomogeneous Polya process. Use the sample-function density for this process in (6.48) to determine the probability $\Pr(N_T = n)$ that N_T equals $n \geqslant 1$.

6.3.4. Let $\{N_t;\ t \geqslant t_0\}$ be a doubly stochastic Poisson process with intensity process $\{\lambda_t(x_t);\ t \geqslant t_0\}$. Show that the sample-function density $p_t \triangleq p[\{N_\sigma;\ t_0 \leqslant \sigma < t\}]$ satisfies the stochastic differential equation

$$dp_t = -p_t\hat{\lambda}_t\,dt + p_t(\hat{\lambda}_t - 1)\,dN_t,\ p_{t_0} = 1.$$

Hint: Use the differential rule in Theorem 4.2.2 and (6.71).

6.3.5 (Interarrival-Time Distribution). Let $\{N_t;\ t \geqslant t_0\}$ be a doubly stochastic Poisson process with intensity process $\{\lambda_t(\mathbf{x}_t);\ t \geqslant t_0\}$. Define $\Lambda_t = \int_{t_0}^t \lambda_\sigma(\mathbf{x}_\sigma)\,d\sigma$, and denote the interarrival-time sequence by $\{t_n\}$.

a. The forward-occurrence density is given in (2.47) for a Poisson process. Show for $\{N_t;\ t \geqslant t_0\}$ that

$$p_{t_n|t_{n-1},\ldots,t_1}(T|T_{n-1},\ldots,T_1)$$

$$= \frac{-\partial E\left[\exp(-\Lambda_{t_0+T_1+\cdots+T_{n-1}+T} + \Lambda_{t_0+T_1+\cdots+T_{n-1}})\right]}{\partial T}.$$

b. Conclude from (a) that if $\{\Lambda_t;\ t \geqslant t_0\}$ is a stationary process then

$$p_{t_n}(T) = \frac{-\partial M_\Lambda(1;\ T)}{\partial T},$$

where $p_{t_n}(T)$ is the unconditional probability density function for the nth interarrival and where

$$M_\Lambda(s;\ T) = E[\exp(s\Lambda_T)]$$

is the moment-generating function for Λ_T. This relation has been noted by B. Saleh [38]. It is useful because $M_\Lambda(s;\ t)$ can be evaluated in many applications of interest.

6.4.1. Let $\{N_t;\ t \geqslant 0\}$ be a doubly stochastic Poisson process with an intensity process $\{\lambda_t;\ t \geqslant 0\}$ constructed as follows. At $t=0$ a random variable x_0 is selected and $\lambda_t = x_0$ for $0 \leqslant t < 1$; at $t=1$ a random variable x_1 is selected and $\lambda_t = x_1$ for $1 \leqslant t < 2$; and so on. This procedure is continued indefinitely with a new random variable being selected each second and λ_t being equal to it for the subsequent one-second interval. Assume the random variables are selected independently and all are identically distributed as an exponential random variable x; the probability density for x is $p_x(X) = \alpha \exp(-\alpha X)$ for $X \geqslant 0$. A typical sample function for the intensity process is shown in Figure P.6.4.1.

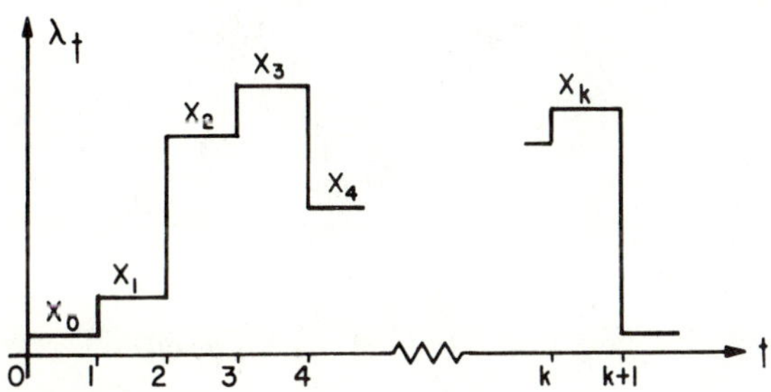

Figure P.6.4.1

a. Suppose $k \leqslant T < k+1$ where k is an integer, $k \geqslant 0$. Determine the mean and variance for N_T, the number of points on $[0, T)$.

b. Determine the sample-function density $p[\{N_\sigma;\ 0 \leqslant \sigma < T\}]$ for paths of N on the interval $[0, T)$ for $k \leqslant T < k+1$.

c. Consider the binary hypothesis testing problem in which $\lambda_t^{(1)}$ is the intensity process described above and $\lambda_t^{(0)}$ is the constant α^{-1}. What is the likelihood ratio for deciding which of these two intensities govern an observed counting path on $[0, T)$?

6.4.2. Suppose in the model of Problem 6.3.1 that ξ has one of two possible values, ξ_0 or ξ_1. A path of N is observed on the interval $[0, T)$, and for this path $N_T = n$ and the jumps occur at times $W_1, W_2, \ldots,$ and W_n. What is the test for deciding between ξ_0 and ξ_1 with minimum probability of error in terms of this observed path? Assume the prior probabilities of ξ_0 and ξ_1 are equal.

6.5.1. In the linear filtering problem discussed in Section 6.5.1, it is assumed that individual occurrence times can be observed. This is not feasible in certain applications. In this problem, we examine the linear estimation problem when counters are used to collect the data.

Let $\{N_t;\ t \geqslant t_0\}$ be a doubly stochastic Poisson process with intensity process $\{\lambda_t;\ t \geqslant t_0\}$. Assume that the mean $E(\lambda_t)$ and covariance function $K_\lambda(t,u)$ for the intensity process are known. An observation interval $[t_0, T)$ is partitioned into m subintervals according to the times $t_0 < t_1 < t_2 < \cdots < t_m = T$ and the number of points occurring in each subinterval is observed. Denote these observables by $N_1, N_2, \ldots, N_m$ where

$$N_k = N_{t_k} - N_{t_{k-1}}, \qquad k = 1, 2, \ldots, m.$$

Define ν_k by

$$\nu_k = (t_k - t_{k-1})^{-1} \int_{t_{k-1}}^{t_k} \lambda_\sigma \, d\sigma$$

a. Show that

$$\operatorname{cov}(\nu_k, \nu_l) = (t_k - t_{k-1})^{-1} (t_l - t_{l-1})^{-1}$$

$$\times \int_{t_{k-1}}^{t_k} \int_{t_{l-1}}^{t_l} K_\lambda(u, v) \, du \, dv$$

b. Show that

$$E(N_k) = (t_k - t_{k-1}) E(\nu_k)$$

and

$$\operatorname{cov}(N_k, N_l) = (t_k - t_{k-1})(t_l - t_{l-1}) \operatorname{cov}(\nu_k, \nu_l)$$

$$+ (t_k - t_{k-1}) E(\nu_k) \delta_{kl}$$

where δ_{kl} is the Kronecker delta function.

c. Define the m-vectors $\mathbf{N}$ and $\boldsymbol{\nu}$ by

$$\mathbf{N}' = [N_1 N_2 \ldots N_m]' \quad \text{and} \quad \boldsymbol{\nu}' = [\nu_1 \nu_2 \ldots \nu_m]'$$

and the $m\times m$ diagonal matrix of subinterval durations $\mathbf{T}$ by

$$\mathbf{T}=\operatorname{diag}(t_k-t_{k-1})=\begin{bmatrix} t_1-t_0 & & & 0 \\ & t_2-t_1 & & \\ 0 & & \ddots & \\ & & & t_m-t_{m-1} \end{bmatrix}.$$

Furthermore, let $\mathbf{K}_N$ and $\mathbf{K}_\nu$ be the covariance matrices for $\mathbf{N}$ and $\boldsymbol{\nu}$; the (k,l)-elements of $\mathbf{K}_N$ and $\mathbf{K}_\nu$ are $\operatorname{cov}(N_k,N_l)$ and $\operatorname{cov}(\nu_k,\nu_l)$, respectively. Notice from part (b) that $E(\mathbf{N})=\mathbf{T}E(\boldsymbol{\nu})$ and

$$\mathbf{K}_N=\mathbf{T}\mathbf{K}_\nu\mathbf{T}+\mathbf{T}\operatorname{diag}(E(\nu_k)).$$

Suppose we observe $\mathbf{N}$ and want to estimate $\boldsymbol{\nu}$. Denote the estimate by $\boldsymbol{\nu}^*=\boldsymbol{\nu}^*(\mathbf{N})$, and suppose it is constrained to be a linear function of $\mathbf{N}$. Thus, $\boldsymbol{\nu}^*$ can be written in the form

$$\boldsymbol{\nu}^*=\mathbf{a}+\mathbf{H}\{\mathbf{N}-\mathbf{T}E(\boldsymbol{\nu})\}$$

for some selection of $\mathbf{a}$ and $\mathbf{H}$, where $\mathbf{a}$ is an m vector and $\mathbf{H}$ is an $m\times m$ matrix. Show that the selection of $\mathbf{a}$ and $\mathbf{H}$ that minimizes the mean-square error $E[|\boldsymbol{\nu}-\boldsymbol{\nu}^*|^2]$ is $\mathbf{a}=E(\boldsymbol{\nu})$ and $\mathbf{H}=\boldsymbol{\Sigma}\mathbf{T}\mathbf{K}_N^{-1}$, where

$$\boldsymbol{\Sigma}^{-1}=\mathbf{T}\mathbf{K}_N^{-1}\mathbf{T}+\mathbf{K}_\nu^{-1}.$$

d. Show that the minimum mean-square error obtained in part (c) is

$$\min_{\boldsymbol{\nu}^*} E[|\boldsymbol{\nu}-\boldsymbol{\nu}^*|^2]=\operatorname{trace}(\boldsymbol{\Sigma}).$$

6.5.2. Let $\{N_t;\ t\geqslant t_0\}$ be a doubly stochastic Poisson process with intensity process $\{\lambda_t;\ t\geqslant t_0\}$. Suppose a path of N is observed on the interval $[t_0,T)$ and based on these observations it is desired to estimate some function of the intensity λ_t at time t. Denote this desired estimate by d_t; d_t could possibly equal λ_t, the integral of λ up to time t, the derivative of λ at time t, or some nonlinear function of λ_t. We will need to know the cross-covariance function $K_{d\lambda}(t,u)=E(d_t\lambda_u)$ and the mean value function $E(d_t)$. The time t may or may not equal T. The situation $t=T$ is the filtering problem of Section 6.5. The *prediction problem* occurs for $t>T$

and the *smoothing problem* for $t<T$. If $t=T-\tau$ and T increases as a real-time parameter, the smoothing problem is termed *fixed lag* with lag time τ. It t is fixed and T increases, the smoothing problem is termed *fixed point*. An if T is fixed and t increases in $[t_0, T)$, the smoothing problem is termed a *fixed interval sweep*. Suppose the estimator of d_t is constrained to be linear; that is, the estimate d_t^{*l} of d_t must have the form

$$d_t^{*l} = a_t + \int_{t_0}^{T} h(t,u)\, dN_u$$

for some deterministic function a_t and (possible noncausal) impulse response $h(t,u)$. Denote by $\hat{d}_t^l$ the linear estimate d_t^{*l} that results when a_t and $h(t,u)$ are selected to minimize the mean-square error $E[(d_t - d_t^{*l})^2]$.

a. Show that

$$\hat{d}_t^l = E(d_t) + \int_{t_0}^{T} h_0(t,u)[dN_u - E(\lambda_u)\,du],$$

where the optimum impulse response $h_0(t,u)$ is the solution to the integral equation

$$h_0(t,u)E(\lambda_u) + \int_{t_0}^{T} h_0(t,\sigma)K_\lambda(\sigma,u)\,d\sigma = K_{d\lambda}(t,u), \qquad t_0 \leqslant u < t$$

where $E(\lambda_u)$ and $K_\lambda(\sigma,u)$ are the mean on covariance functions for the intensity process.

b. Show that the minimum mean-square error is given by

$$E\left[\left(d_t - \hat{d}_t^l\right)^2\right] = K_d(t,t) - \int_{t_0}^{T} h_0(t,\sigma)K_{d\lambda}(t,\sigma)\,d\sigma.$$

6.5.3. Let $\{N_t;\ t \geqslant t_0\}$ be a doubly stochastic Poisson process with intensity process $\{\lambda_t(\mathbf{x}_t);\ t \geqslant t_0\}$. Suppose $\mathbf{h}_t^*$ is an arbitrary filtering estimate of $\mathbf{h}_t(\mathbf{x}_t)$ in terms of an observed record of N on the interval $[t_0, t)$ and that $\hat{\mathbf{h}}_t = E[\mathbf{h}_t(\mathbf{x}_t) | N_\sigma;\ t_0 \leqslant \sigma < t]$. Show for all choices of $\mathbf{h}_t^*$ and for any nonnegative definite matrix $\mathbf{M}_t$ that

$$E[(\mathbf{h}_t(\mathbf{x}_t) - \mathbf{h}_t^*)'\mathbf{M}_t(\mathbf{h}_t(\mathbf{x}_t) - \mathbf{h}_t^*)] \geqslant E\left[\left(\mathbf{h}_t(\mathbf{x}_t) - \hat{\mathbf{h}}_t\right)'\mathbf{M}_t\left(\mathbf{h}_t(\mathbf{x}_t) - \hat{\mathbf{h}}_t\right)\right].$$

Conclude, therefore, that the minimum mean square-error estimate of $\mathbf{h}_t(\mathbf{x}_t)$ also minimizes the weighted mean-square error. Also, use this result to argue that the minimum mean square-error estimate of a linear combination of the components of $\mathbf{h}_t(\hat{\mathbf{x}}_t)$ equals the corresponding linear combination of the components of $\hat{\mathbf{h}}_t$.

6.5.4. Let $\{N_t;\ t \geq t_0\}$ be a doubly stochastic Poisson process with intensity process $\{\lambda_t(x_t);\ t \geq t_0\}$. Suppose x_t^* is an arbitrary filtering estimate of x_t in terms of an observed record of N on the interval $[t_0, t)$ and that $\hat{x}_t^{(\text{abs})}$ is the value of X for which the posterior distribution $P_{x_t}(X|N_\sigma;\ t_0 \leq \sigma < t)$ for x_t is equal to 0.5. Thus, $\hat{x}_t^{(\text{abs})}$ is the conditional median of x_t given $\{N_\sigma;\ t_0 < \sigma < t\}$. Show that

$$E[|x_t - x_t^*|] \geq E[|x_t - \hat{x}_t^{(\text{abs})}|].$$

6.5.5. Consider the finite difference version of (6.111),

$$\Delta\mathbf{x}_t = \mathbf{a}_t(\mathbf{x}_t)\Delta t + \mathbf{B}_t(\mathbf{x}_t)\Delta\mathbf{w}_t$$

where $\Delta\mathbf{x}_t = \mathbf{x}_{t+\Delta t} - \mathbf{x}_t$, $\Delta\mathbf{w}_t = \mathbf{w}_{t+\Delta t} - \mathbf{w}_t$, and $\{\mathbf{w}_t;\ t \geq t_0\}$ is a vector of independent Wiener processes such that $E(\mathbf{w}_t\mathbf{w}_u') = \mathbf{I}\min(t,u)$. In this case,

$$\begin{aligned} &E\{(\exp[j\langle\mathbf{v}, \Delta\mathbf{x}_t\rangle] - 1)|\mathbf{x}_t\} \\ &\qquad = \exp[j\langle\mathbf{v}, \mathbf{a}_t(\mathbf{x}_t)\rangle\Delta t]E\{\exp[j\langle\mathbf{v}, \mathbf{B}_t(\mathbf{x}_t)\Delta\mathbf{w}_t\rangle]|\mathbf{x}_t\} \quad 1. \end{aligned}$$

Use the fact that $\Delta\mathbf{w}_t$ is normally distributed with zero mean and covariance $\mathbf{I}\Delta t$ to deduce that

$$E\{\exp[j\langle\mathbf{v}, \mathbf{B}_t(\mathbf{x}_t)\Delta\mathbf{w}_t\rangle|\mathbf{x}_t]\} = \exp[-\tfrac{1}{2}\langle\mathbf{v}, \mathbf{B}_t(\mathbf{x}_t)\mathbf{B}_t'(\mathbf{x}_t)\mathbf{v}\rangle\Delta t].$$

Conclude, therefore, that

$$\begin{aligned} \psi_t(\mathbf{v}|\mathbf{x}_t) &\triangleq \lim_{\Delta t \downarrow 0} (\Delta t)^{-1} E\{(\exp[j\langle\mathbf{v}, \Delta\mathbf{x}_t\rangle] - 1)|\mathbf{x}_t\} \\ &= j\langle\mathbf{v}, \mathbf{a}_t(\mathbf{x}_t)\rangle - \tfrac{1}{2}\langle\mathbf{v}, \mathbf{B}_t(\mathbf{x}_t)\mathbf{B}_t'(\mathbf{x}_t)\mathbf{v}\rangle. \end{aligned}$$

6.5.6. Derive equation (6.119).

6.5.7 (The Separation Theorem of Detection). The separation theorem of detection, Equation (6.73), was derived in Section 6.4 using expressions for the sample-function density. This important theorem is derived here using the representation in (6.122) for the conditional characteristic function $\hat{M}_t(j\mathbf{v})$ and the differential rule of Theorem 4.2.2.

Let $\{N_t;\ t \geqslant t_0\}$ be a doubly stochastic Poisson process with either a known, positive, constant intensity of unit amplitude or an intensity process $\{\lambda_t(\mathbf{x}_t);\ t \geqslant t_0\}$ in which $\{\mathbf{x}_t;\ t \geqslant t_0\}$ is a Markovian information process. Thus, the intensity process for N can be written as $1+\alpha\lambda_t(\mathbf{x}_t)$, where α is a discrete random variable that is either one or zero. Let the prior probabilities of α be $P_1 = \Pr(\alpha=1)$ and $P_0 = 1-P_1 = \Pr(\alpha=0)$, and assume α and $\{\mathbf{x}_t;\ t \geqslant t_0\}$ are independent. The hypothesis testing problem is to decide whether α is zero or one given a record $\{N_\sigma;\ t_0 \leqslant \sigma < t\}$. Let $l_t = \ln[p_t/(1-p_t)]$, where

$$p_t \triangleq \Pr(\alpha=1|N_\sigma;\ t_0 \leqslant \sigma < t).$$

The log-likelihood ratio l_t is a sufficient statistic for making the decision at time t. We want to establish that

$$l_t = \ln\left(\frac{P_1}{P_0}\right) - \int_{t_0}^{t}(\hat{\lambda}_\sigma - 1)\,d\sigma + \int_{t_0}^{t}\ln\hat{\lambda}_\sigma\,dN_\sigma,$$

where $\hat{\lambda}_\sigma$ is the pseudo estimate defined by

$$\hat{\lambda}_t = E[\lambda_t(\mathbf{x}_t)|\alpha=1,\ N_\sigma;\ t_0 \leqslant \sigma < t].$$

a. As $\{\mathbf{x}_t;\ t \geqslant t_0\}$ is a Markov process, so is the augmented process $\{(\alpha,\mathbf{x}_t);\ t \geqslant t_0\}$. Use this and (6.122) to show that the conditional characteristic function

$$\hat{M}_t(ju,j\mathbf{v}) \triangleq E\{\exp(ju\alpha)\exp[j\langle\mathbf{v},\mathbf{x}_t\rangle]|N_\sigma;\ t_0 \leqslant \sigma < t\}$$

satisfies

$$d\hat{M}_t(ju,j\mathbf{v}) = \hat{E}\{\exp(ju\alpha)\exp[j\langle\mathbf{v},\mathbf{x}_t\rangle]\psi_t(\mathbf{v}|\mathbf{x}_t)\}\,dt$$

$$+\hat{E}\{\exp(ju\alpha)\exp[j\langle\mathbf{v},\mathbf{x}_t\rangle](1+\alpha\lambda_t(\mathbf{x}_t)-e_t)\}e_t^{-1}(dN_t - e_t\,dt),$$

$$\hat{M}_{t_0}(ju,j\mathbf{v}) = (P_0 + P_1\exp(ju))E\{\exp[j\langle\mathbf{v},\mathbf{x}_{t_0}\rangle]\},$$

where e_t is the estimate

$$e_t \triangleq E[1+\alpha\lambda_t(\mathbf{x}_t)|N_\sigma;t_0 \leqslant \sigma < t].$$

Hint: You will need to show that the characteristic form for the differential generator for $(\alpha,\mathbf{x}_t)$ is the same as that for $\mathbf{x}_t$ alone.

b. Use the fact that α is zero or one to show that

$$e_t = 1 + p_t\hat{\lambda}_t,$$

where p_t is the posterior probability that $\alpha = 1$ and where $\hat{\lambda}_t$ is the pseudo estimate defined above.

c. By using the results of (a) and (b) show that

$$dp_t = -p_t(1-p_t)\hat{\lambda}_t dt + p_t(1-p_t)\hat{\lambda}_t\left(1+p_t\hat{\lambda}_t\right)^{-1} dN_t,$$

$$p_{t_0} = P_1.$$

Hint: Set $\mathbf{v} = \mathbf{0}$ in part (a) and then inverse Fourier transform.

d. Use the differential rule on $l_t = \ln[p_t/(1-p_t)]$ to deduce that

$$dl_t = -\left(\hat{\lambda}_t - 1\right) + \ln\hat{\lambda}_t dN_t,$$

$$l_{t_0} = \ln\left(\frac{P_1}{P_0}\right).$$

6.5.8 (Continuation). Suppose that the intensity of the doubly-stochastic Poisson process of Problem 6.5.7 can now be written as $\alpha\lambda_t^1(\mathbf{x}_t) + (1-\alpha) \times\lambda_t^0(\mathbf{x}_t)$, where α is either zero or one as before. Define

$$l_t = \ln\left(\frac{\Pr(\alpha = 1|N_\sigma;\ t_0 \leqslant \sigma < t)}{\Pr(\alpha = 0|N_\sigma; t_0 \leqslant \sigma < t)}\right).$$

Use the results of Problem 6.5.7 to establish the separation theorem of detection,

$$l_t = -\int_{t_0}^{t}\left(\hat{\lambda}_\sigma^{(1)} - \hat{\lambda}\sigma^{(0)}\right) d\sigma + \int_{t_0}^{t} \ln\left(\frac{\hat{\lambda}_\sigma^{(1)}}{\hat{\lambda}_\sigma^{(0)}}\right) dN_\sigma,$$

$$l_{t_0} = \ln\left(\frac{P_1}{P_0}\right),$$

where $\hat{\lambda}_t^{(i)}$ are the pseudo estimates defined by

$$\hat{\lambda}_t^{(i)} = E\left(\lambda_t^{(i)}(\mathbf{x}_t)|\alpha = i, N_\sigma;\ t_0 \leqslant \sigma < t\right)$$

for $i = 0$ and $i = 1$. Hint: introduce a dummy hypothesis that the intensity is simply unity and express l_t as a sum of two terms involving the posterior probability of this hypothesis.

6.5.9 (Filtering in the Presence of Intensity Uncertainty). Let $\{N_t; t \geq t_0\}$ be a doubly stochastic Poisson process with an intensity process $\{\lambda_0 + \alpha\lambda_t(\mathbf{x}_t); t \geq t_0\}$ where λ_0 is a known constant, $\{\mathbf{x}_t; t \geq t_0\}$ is a Markovian information process, and α is a discrete random variable independent of $\{\mathbf{x}_t; t \geq t_0\}$ and of value either $\alpha = 1$ of $\alpha = 0$ with prior probabilities P_1 and $P_0 = 1 - P_1$, respectively. Thus, depending on the value of α, the intensity may or may not contain the information process. The problem is to describe the posterior statistics of $\mathbf{x}_t$ given $\{N_\sigma; t_0 \leq \sigma < t\}$ in the presence of this intensity uncertainty.

a. Show that

$$\hat{M}_t(j\mathbf{v}) = \left[M_t(j\mathbf{v}) + \Lambda_t \hat{M}_t^{(1)}(j\mathbf{v}) \right](1 + \Lambda_t)^{-1},$$

where

$$M_t(j\mathbf{v}) = E\{\exp[j\langle \mathbf{v}, \mathbf{x}_t \rangle]\},$$

$$\hat{M}_t(j\mathbf{v}) = E\{\exp[j\langle \mathbf{v}, \mathbf{x}_t \rangle] | N_\sigma; t_0 \leq \sigma < t\},$$

and

$$\hat{M}_t^{(1)}(j\mathbf{v}) = E\{\exp[j\langle \mathbf{v}, \mathbf{x}_t \rangle] | \alpha = 1, N_\sigma; t_0 \leq \sigma < t\},$$

and where Λ_t is the likelihood ratio

$$\Lambda_t = \frac{\Pr(\alpha = 1 | N_\sigma; t_0 \leq \sigma < t)}{\Pr(\alpha = 0 | N_\sigma; t_0 \leq \sigma < t)}.$$

b. Suppose the information process is simply a collection of time independent random variables $\mathbf{x}$ with prior probability density $p_{\mathbf{x}}(\mathbf{X})$. Show that the posterior density is given by

$$p_t(\mathbf{X} | N_\sigma; t_0 \leq \sigma < t)$$

$$= p_{\mathbf{x}}(\mathbf{X}) \left\{ \frac{1 + \exp\left(-\int_{t_0}^{t} \lambda_\sigma(\mathbf{X})\, d\sigma + \int_{t_0}^{t} \ln\left(1 + \lambda_0^{-1}\lambda_\sigma(\mathbf{X})\right) dN_\sigma \right)}{1 + E\left[\exp\left(-\int_{t_0}^{t} \lambda_\sigma(\mathbf{x})\, d\sigma + \int_{t_0}^{t} \ln\left(1 + \lambda_0^{-1}\lambda_\sigma(\mathbf{x})\right) dN_\sigma \right) \right]} \right\}.$$

The expectation appearing in the denominator is with respect to the prior statistics of $\mathbf{x}$.

6.5.10. Define

$$f_t(\mathbf{X})=\frac{\exp[H_t(\mathbf{X})]p_{\mathbf{x}}(\mathbf{X})}{E\{\exp[H_t(\mathbf{x})]\}},$$

where

$$H_t(\mathbf{x})=-\int_{t_0}^{t}\lambda_\sigma(\mathbf{x})\,d\sigma+\int_{t_0}^{t}\ln\lambda_\sigma(\mathbf{x})\,dN_\sigma$$

and the expectation in the denominator is with respect to the probability density $p_{\mathbf{x}}(\mathbf{X})$ of $\mathbf{x}$. According to (6.138), $f_t(\mathbf{X})$ is the posterior density of $\mathbf{x}$ given observations $\{N_\sigma;\ t_0\leqslant\sigma<t\}$ of a doubly stochastic Poisson process with intensity process $\{\lambda_t(\mathbf{x});\ t\geqslant t_0\}$. This can be verified directly by using the differential rule of Theorem 4.2.2 to show that $f_t(\mathbf{X})$ satisfies (6.133). Let $P_t(\mathbf{X})$ and Q_t be the numerator and denominator of $f_t(\mathbf{X})$, respectively.

a. Show that

$$dP_t(\mathbf{X})=-\lambda_t(\mathbf{X})P_t(\mathbf{X})dt+(\lambda_t(\mathbf{X})-1)P_t(\mathbf{X})dN_t,$$

$$P_{t_0}(\mathbf{X})=p_{\mathbf{x}}(\mathbf{X}).$$

b. Show that

$$dQ_t=-\hat{\lambda}_tQ_t\,dt+(\hat{\lambda}_t-1)Q_t\,dN_t,$$

$$Q_{t_0}=1,$$

where $\hat{\lambda}_t=E[\lambda_t(\mathbf{x})|N_\sigma;\ t_0\leqslant\sigma<t]$.

c. Use the results of (a) and (b) to show that

$$df_t(\mathbf{X})=f_t(\mathbf{X})\left[\lambda_t(\mathbf{X})-\hat{\lambda}_t\right]\hat{\lambda}_t^{-1}\left(dN_t-\hat{\lambda}_t\,dt\right),$$

$$f_{t_0}(\mathbf{X})=p_{\mathbf{x}}(\mathbf{X}).$$

6.5.11. Let $\{N_t;\ t\geqslant t_0\}$ be an inhomogeneous Poisson process having an intensity $\{\nu_t;\ t\geqslant t_0\}$ that is randomly scaled by a nonnegative random variable x. The result is a doubly stochastic Poisson process with intensity process $\{x\nu_t;\ t\geqslant t_0\}$. Suppose x is gamma distributed with parameters α

and β,

$$p_x(X)=\frac{\alpha^{\beta+1}}{\Gamma(\beta+1)}X^{\beta}\exp(-\alpha X).$$

a. Show that x is also conditionally gamma distributed given $\{N_\sigma;\ t_0\leqslant\sigma<t\}$ and that in the conditional distribution, α and β become $\alpha+\int_{t_0}^{t}\nu_\sigma\,d\sigma$ and $\beta+N_t$, respectively.

b. Evaluate the estimate $\hat{x}_t=E(x|N_\sigma;\ t_0\leqslant\sigma<t)$.

c. Show that $\{N_t;\ t\geqslant t_0\}$ is a Markov birth process and determine the birth rate $\mu_t(n)$, where

$$\mu_t(n)=\lim_{\Delta t\downarrow 0}(\Delta t)^{-1}\quad \Pr(N_{t,t+\Delta t}=1|N_t=n).$$

Hint: Use Theorems 5.3.1 and 6.2.1.

6.5.12 (Continuation). Let the model be that of Problem 6.5.11 with $\hat{x}_t$ as determined in part (b).

a. Evaluate the conditional mean-square error

$$E\left[(x-\hat{x}_t)^2|N_\sigma;\ t_0\leqslant\sigma<t\right].$$

b. Evaluate the minimum mean-square error

$$E\left[(x-\hat{x}_t)^2\right].$$

c. Evaluate the conditional bias $B(x)$ of the minimum mean square-error estimate of x, where

$$B(x)=E(\hat{x}_t-x|x).$$

d. Evaluate the bias $B=E(B(x))$ of the minimum mean square-error estimate of x.

6.5.13. Let $\{N_t;\ t\geqslant t_0\}$ be a doubly stochastic Poisson process with intensity process $\{\nu_t\exp(-xt);\ t\geqslant t_0\}$ where x is a nonnegative random variable with prior probability density $p_x(X)$. Show that the minimum mean square-error estimate $\hat{x}_t$ of x given $\{N_\sigma;\ t_0 \leqslant \sigma < t\}$ is given by

$$\hat{x}_t = f_t\left(\int_{t_0}^{t} \sigma\, dN_\sigma\right)$$

where $f_t(z)$ is the precomputable function

$$\left[\exp\left(-zx - \int_{t_0}^{t} \nu_\sigma \exp(-x\sigma)\, d\sigma\right)\right]$$

in which $E(\cdot)$ indicates an integration with respect to the prior density of x. Conclude, therefore, that the sum of the occurrence times is a sufficient statistic for estimating x.

6.5.14. Let $\{N_t;\ t \geqslant 0\}$ be a homogeneous, doubly stochastic Poisson process with a time independent intensity $\lambda(\mathbf{x}) = \Sigma_{i=1}^{k} x_i^2$, where $x_1, x_2, \ldots,$ and x_k are independent, identically distributed Gaussian random variables with zero mean and variance σ^2. Suppose at time T that N_T is observed to have the value n.

a. Determine the minimum mean square-error estimate $\hat{\lambda}_{\text{MMSE}}(n)$ of the intensity given $N_T = n$; that is, evaluate the conditional mean of λ given $N_T = n$.

b. Determine the maximum *a posteriori* probability estimate $\hat{\lambda}_{\text{MAP}}(n)$ of the intensity given $N_t = n$; this estimate is the conditional mode of λ given $N_T = n$.

c. Show that the maximum likelihood estimate $\hat{k}_{\text{ML}}(n)$ of the number of independent components in λ given $N_T = n$ is the value of k that maximizes

$$\Gamma\left(n + \left(\frac{k}{2}\right)\right)\Gamma^{-1}\left(\frac{k}{2}\right)[1 + 2\sigma^2 T]^{-(k/2)},$$

where $\Gamma(\cdot)$ denotes the gamma function.

d. Determine the minimum mean square-error estimate $\hat{\lambda}_{\text{MMSE}}(N_\sigma;\ 0 \leqslant \sigma < T)$ of the intensity in terms of an entire counting path of N observed on an interval $[0, T)$.

6.5.15 (On–Off Modulated Light). Suppose the random telegraph wave of Examples 6.5.3 and 6.5.9 is modified so that the transition rates from state ζ_0 (off) to state ζ_1 (on) and from state ζ_1 to state ζ_0 are not equal. In particular, let the rate from ζ_0 to ζ_1 be ν_{01} and from ζ_1 to ζ_0 be ν_{10}. The model is unchanged otherwise.

a. Specify a differential equation for the conditional occupancy probability $\hat{\pi}_1(t)$ of the on state ζ_1 between observed photoelectron emission times.

b. Solve the differential equation of part (a).

c. Indicate how $\hat{\pi}_1(t)$ should be updated at the instant a photoelectron emission occurs.

6.5.16. Let $\{N_t;\ t \geqslant t_0\}$ be a doubly stochastic Poisson process with intensity process $\{\lambda_t(x_t);\ t \geqslant t_0\}$, where $\{x_t;\ t \geqslant t_0\}$ is a continuous time Markov jump process. Let $\{\zeta_i\}$ denote the states of x_t and $\pi_i(t) = \Pr(x_t = \zeta_i)$ the state occupancy probabilities at time t.

a. Show that the conditional state occupancy probabilities given an observed counting path $\{N_\sigma;\ t_0 \leqslant \sigma < t\}$ have the representation

$$\hat{\pi}_i(t) = \frac{E\{\exp[H_t(x_\sigma;\ t_0 \leqslant \sigma \leqslant t)] | x_t = \zeta_i\}\pi_i(t)}{E\{\exp[H_t(x_\sigma;\ t_0 \leqslant \sigma \leqslant t)]\}},$$

where

$$H_t(x_\sigma;\ t_0 \leqslant \sigma \leqslant t) = -\int_{t_0}^{t} \lambda_\sigma(x_\sigma)\,d\sigma + \int_{t_0}^{t} \ln \lambda_\sigma(x_\sigma)\,dN_\sigma.$$

b. Use the result of (a) and the differential rule of Theorem 4.2.2 to obtain an alternative derivation of the equation (6.155) for the time evolution of $\hat{\pi}_i(t)$.

6.5.17. Let $\{N_t;\ t \geqslant t_0\}$ be a doubly stochastic Poisson process with intensity process $\{\lambda_t(y_t);\ t \geqslant t_0\}$, where $\lambda_t(y_t) = \exp(y_t)$ and $\{y_t;\ t \geqslant t_0\}$ is a Gaussian random process formally satisfying

$$\frac{d^2 y_t}{dt^2} + 2\alpha \frac{dy_t}{dt} + y_t = u_t,$$

where $\{u_t;\ t \geqslant t_0\}$ is a white Gaussian process with unit intensity. Thus, λ is a lognormal random process.

a. Show that y can be represented in state-variable form as

$$d\mathbf{x}_t = \mathbf{A}\mathbf{x}_t\,dt + \mathbf{b}\,dw_t, \qquad y_t = \mathbf{c}'\mathbf{x}_t$$

where $\mathbf{x}_t$ is a two-dimensional vector, $\{w_t;\ t \geqslant t_0\}$ is a Wiener process with unit parameter, and $\mathbf{A}$, $\mathbf{b}$, and $\mathbf{c}$ are given by

$$\mathbf{A}=\begin{bmatrix} 1 & 0 \\ -1 & -2\alpha \end{bmatrix}, \quad \mathbf{b}=\begin{bmatrix} 0 \\ 1 \end{bmatrix}, \quad \mathbf{c}=\begin{bmatrix} 1 \\ 0 \end{bmatrix}.$$

b. Specialize the processor and variance equations, (6.164) and (6.165), to determine the suboptimal estimator for y_t.

c. Suppose now that the damping constant α is a nonnegative random variable to be estimated from the observed point process. Let $x_{0t}=\alpha$; then $dx_{0t}=0$. Adjoin x_{0t} to the two-dimensional state vector of (a) and specialize the processor and variance equations, (6.164) and (6.165), to determine the suboptimal estimator for α and y_t.

6.5.18 (Subcarrier Amplitude Modulation). The model for subcarrier amplitude modulation is given in Example 6.5.2. The linearly constrained minimum mean square-error estimator for the state $\mathbf{x}_t$ is specified by (6.95) and (6.96). Suppose now that the linearity constraint on the estimator is removed. Specialize the processor and variance equations, (6.164) and (6.165), for this amplitude estimation model to get a specification for the suboptimal minimum mean square-error estimator. Compare the linear and nonlinear estimators; in particular, indicate where the nonlinearities in the suboptimal estimator enter.

6.5.19. The suboptimal estimator defined by the processor and variance equations, (6.164) and (6.165), does not necessarily become optimal when the functions $\mathbf{a}_t(\mathbf{x}_t)$ and $\lambda_t(\mathbf{x}_t)$ are linear in $\mathbf{x}_t$. The linearly constrained estimators of Section 6.5.1 can, in fact, have a superior mean-square error performance in this case and perhaps others as well. These are points following from the results of this problem that lend further emphasis to our caution that (6.164) and (6.165) should be tested for their utility before being used in particular applications.

Let $\{N_t;\ t\geqslant 0\}$ be a doubly stochastic Poisson process with intensity process $\{x\nu_t;\ t\geqslant 0\}$, where ν_t is a known nonnegative function of time and where x is gamma distributed. The probability density function for x is

$$p_x(X)=\frac{\alpha^{\beta+1}}{\Gamma(\beta+1)}X^{\beta}\exp(-\alpha X).$$

In Problem 6.5.11(b), it is shown that the optimal minimum mean square-error estimate $\hat{x}_t$ or x_t given $\{N_\sigma;\ 0 \leqslant \sigma < t\}$ is given by

$$\hat{x}_t = \frac{N_t + \beta + 1}{\alpha + \int_0^t v_\sigma\, d\sigma}.$$

Since $\hat{x}_t$ is linear in N_t it follows that $\hat{x}_t = \hat{x}_t^l$, where $\hat{x}_t^l$ is the linearly constrained estimate of Section 6.5.1. It is further shown in Problem 6.5.12(a) that the conditional error covariance $\hat{\Sigma}_t$ associated with $\hat{x}_t$ is given by

$$\hat{\Sigma}_t = \frac{N_t + \beta + 1}{\left(\alpha + \int_0^t v_\sigma\, d\sigma\right)^2}.$$

a. By using the differential rule of Theorem 4.2.2 and the results cited above from Problems 6.5.11(b) and 6.5.12(a), show that $\hat{x}_t$ and $\hat{\Sigma}_t$ satisfy

$$d\hat{x}_t = \hat{\Sigma}_t \hat{x}_t^{-1}[dN_t - \hat{x}_t v_t\, dt], \qquad \hat{x}_0 = \frac{\beta + 1}{\alpha}$$

$$d\hat{\Sigma}_t = -2\hat{\Sigma}_t^2 \hat{x}_t^{-1} dt + \hat{\Sigma}_t^2 \hat{x}_t^{-2} dN_t, \qquad \hat{\Sigma}_0 = \frac{\beta + 1}{\alpha^2}.$$

b. Show that the suboptimal estimator of (6.164) and (6.165) satisfies

$$dx_t^* = \Sigma_t^* x_t^{*-1}[dN_t - x_t^* v_t\, dt], \qquad x_0^* = \frac{\beta + 1}{\alpha}$$

$$d\Sigma_t^* = -\Sigma_t^{*2} x_t^{*-2} dN_t, \qquad \Sigma_0^* = \frac{\beta + 1}{\alpha^2}.$$

Conclude, therefore, that the optimal and suboptimal estimators are not identical and that a linear estimator can have a performance that is superior to that of the suboptimal estimator for this problem.

Problem 6.5.20 (Performance for Phase-Unsynchronized Subcarrier Angle-Modulation). Suppose the intensity process of a doubly stochastic Poisson process is given by

$$\lambda_t(\theta, \mathbf{x}_t) = G[1 + m\cos(\omega t + \theta + \mathbf{c}_t'\mathbf{x}_t)],$$

where θ is a Gaussian random variable with zero mean and variance σ_θ^2

and $\{\mathbf{x}_t;\ t \geqslant t_0\}$ is a Gaussian process satisfying

$$d\mathbf{x}_t = \mathbf{A}_t \mathbf{x}_t\, dt + \mathbf{B}_t\, dw_t, \qquad \mathbf{x}_{t_0} = \mathbf{x}_0$$

where $\{\mathbf{w}_t;\ t \geqslant t_0\}$ is a normalized, vector Wiener process and $\mathbf{x}_0$ is a zero mean Gaussian random variable with covariance $\boldsymbol{\Lambda}_0$. This intensity can arise in a direct-detection optical communication system employing subcarrier angle modulation when the subcarrier is phase unsynchronized; here, θ represents the lack of synchronization, and the situation $\sigma_\theta^2 \to \infty$ corresponds to θ being uniformly distributed on $[-\pi, \pi]$ when measured modulo 2π.

a. Let $\tilde{\mathbf{x}}_t' = [\theta, \mathbf{x}_t']$ and adopt the notation $\lambda_t(\theta, \mathbf{x}_t) = \lambda_t(\tilde{\mathbf{x}}_t)$. Show that

$$\lambda_t^{-1}(\tilde{\mathbf{x}}_t)\left[\frac{\partial \lambda_t(\tilde{\mathbf{x}}_t)}{\partial \tilde{\mathbf{x}}_t}\right]\left[\frac{\partial \lambda_t(\tilde{\mathbf{x}}_t)}{\partial \tilde{\mathbf{x}}_t}\right]' = Gm^2 f(\omega t + \theta + \mathbf{c}_t'\mathbf{x}_t)[1 \quad \mathbf{c}_t']'[1 \quad \mathbf{c}_t'],$$

where f is defined in (6.178).

b. By using the expansion for f in (6.207), show that the matrix $\mathbf{P}_t$ defined by

$$\mathbf{P}_t = E\left\{\lambda_t^{-1}(\tilde{\mathbf{x}}_t)\left[\frac{\partial \lambda_t(\tilde{\mathbf{x}}_t)}{\partial \tilde{\mathbf{x}}_t}\right]\left[\frac{\partial \lambda_t(\tilde{\mathbf{x}})}{\partial \tilde{\mathbf{x}}_t}\right]'\right\}$$

has the evaluation

$$\mathbf{P}_t = \Bigg\{G\left[1 - (1 - m^2)^{1/2}\right]$$

$$+ \sum_{n=1}^{\infty} a_n G \exp\left[-\tfrac{1}{2} n^2(\sigma_\theta^2 + \mathbf{c}_t'\boldsymbol{\Lambda}_t \mathbf{c}_t)\right]\cos(n\omega t)\Bigg\}[1 \quad \mathbf{c}_t']'[1 \quad \mathbf{c}_t'],$$

where $\boldsymbol{\Lambda}_t$ is the covariance of $\mathbf{x}_t$. Consequently, conclude that as $\sigma_\theta^2 \to \infty$, $\mathbf{P}_t$ becomes

$$\mathbf{P}_t = \left\{G\left[1 - (1 - m^2)^{1/2}\right]\right\}[1 \quad \mathbf{c}_t']'[1 \quad \mathbf{c}_t'].$$

c. Show that for any causal estimate y_t^* of $y_t \triangleq \mathbf{c}_t'\mathbf{x}_t$ in the presence of total phase incoherence (i.e., $\sigma_\theta^2 \to \infty$), there holds

$$E\left[(y_t - y_t^*)^2\right] \geqslant [0 \quad \mathbf{c}_t']\tilde{\mathbf{\Sigma}}_t[0 \quad \mathbf{c}_t']'$$

where $\tilde{\mathbf{\Sigma}}_t$ satisfies

$$\frac{d\tilde{\mathbf{\Sigma}}_t}{dt} = \tilde{\mathbf{A}}_t\tilde{\mathbf{\Sigma}}_t + \tilde{\mathbf{\Sigma}}_t\tilde{\mathbf{A}}_t' + \tilde{\mathbf{B}}_t\tilde{\mathbf{B}}_t'$$

$$- G\left[1 - (1 - m^2)^{1/2}\right]\tilde{\mathbf{\Sigma}}_t[1 \quad \mathbf{c}_t']'[1 \quad \mathbf{c}_t']\tilde{\mathbf{\Sigma}}_t,$$

where

$$\tilde{\mathbf{A}}_t \triangleq \left[\begin{array}{c|c} 0 & \mathbf{0}' \\ \hline \mathbf{0} & \mathbf{A}_t \end{array}\right], \qquad \tilde{\mathbf{B}}_t = \left[\begin{array}{c} \mathbf{0}' \\ \hline \mathbf{B}_t \end{array}\right]$$

and the intial condition is

$$\tilde{\mathbf{\Sigma}}_{t_0} = \left[\begin{array}{c|c} 0 & \mathbf{0}' \\ \hline \mathbf{0} & \Lambda_0 \end{array}\right].$$

6.5.21 (Automatic Gain-Control for Lognormal Fading). The problem of estimating a fading process is equivalent to the problem of designing an automatic gain control in the receiver. The problem of binary detection in the presence of lognormal fading is examined for the optical radar context in Example 6.5.13. In this problem, we examine the estimation performance associated with the automatic gain control used to compensate for the fading process y_t.

Suppose the intensity process is given by

$$\lambda_t(\mathbf{x}_t) = \alpha |S_t|^2 \exp(y_t)$$

where $\mathbf{x}_t$, α, S_t, and y_t are defined in Example 6.5.13. Here, we assume that the dark current count rate λ_0 is sufficiently small that it can be neglected, and we set $\lambda_0 = 0$. Show that for any causal estimate y_t^* of y_t there holds

$$E\left[(y_t - y_t^*)^2\right] \geqslant \mathbf{c}_t'\mathbf{\Sigma}_t\mathbf{c}_t,$$

where Σ_t satisfies

$$\frac{d\Sigma_t}{dt}=\mathbf{A}_t\Sigma_t+\Sigma_t\mathbf{A}_t'+\mathbf{B}_t\mathbf{B}_t'$$

$$-\alpha|S_t|^2\exp\left(\tfrac{1}{2}\sigma_t^2\right)\Sigma_t\mathbf{c}_t\mathbf{c}_t'\Sigma_t,\qquad \Sigma_{t_0}=\Sigma_0$$

where σ_t^2 is the variance of $\mathbf{y}_t$. It can be shown (see H. Van Trees [47, p. 533]) that $\sigma_t^2=\mathbf{c}_t'\Lambda_t\mathbf{c}_t$, where

$$\frac{d\Lambda_t}{dt}=\mathbf{A}_t\Lambda_t+\Lambda_t\mathbf{A}_t'+\mathbf{B}_t\mathbf{B}_t',\qquad \Lambda_{t_0}=\Sigma_0.$$

6.6.1. Parallel the proof of Theorem 4.2.2 to establish the multichannel differential rule (6.222).

6.6.2. Let $\{\mathbf{N}_t; t\geqslant t_0\}$ be a multichannel doubly stochastic Poisson process with intensity processs $\{\boldsymbol{\lambda}_t; t\geqslant t_0\}$. Let $\{\mathbf{y}_t; t\geqslant t_0\}$ be a filtered multichannel doubly stochastic Poisson process,

$$\mathbf{y}_t=\int_{t_0}^{t}\mathbf{H}(t,u)\,d\mathbf{N}_u.$$

Establish the following extension of the results in Problem 6.2.6.

a. Show that

$$E(\mathbf{y}_t)=\int_{t_0}^{t}\mathbf{H}(t,u)E(\boldsymbol{\lambda}_u)\,du.$$

b. Let $\mathbf{K}_y(t,t)=E(\mathbf{y}_t\mathbf{y}_t')-E(\mathbf{y}_t)E(\mathbf{y}_t')$ be the covariance of $\mathbf{y}$. Show that

$$\mathbf{K}_y(t,t)=\int_{t_0}^{t}\mathbf{H}(t,u)\operatorname{diag}\left[E(\lambda_u^{(k)})\right]\mathbf{H}'(t,u)\,du$$

$$+\int_{t_0}^{t}\int_{t_0}^{t}\mathbf{H}(t,u)\mathbf{K}_\lambda(u,v)\mathbf{H}'(t,v)\,du\,dv,$$

where $\mathbf{K}_\lambda$ is the covariance of $\boldsymbol{\lambda}$.

c. Suppose $\mathbf{H}$ is a square matrix so that the dimension of $\mathbf{y}$ and $\mathbf{N}$ are the same. Show that

$$\operatorname{cov}(\mathbf{y}_t, \boldsymbol{\lambda}_s) \triangleq E(\mathbf{y}_t \boldsymbol{\lambda}_s') - E(\mathbf{y}_t)E(\boldsymbol{\lambda}_s')$$

has the evaluation

$$\operatorname{cov}(\mathbf{y}_t, \boldsymbol{\lambda}_s) = \int_{t_0}^{t} \mathbf{H}(t,u)\mathbf{K}_\lambda(u,s)\,du.$$

6.6.3. Derive the equations, (6.239) and (6.240), of multichannel linear filtering by using the results of Problem 6.6.2 and paralleling the proof of Theorem 6.5.1.

6.6.4. Let $\{\mathbf{N}_t; t \geqslant t_0\}$ be a multichannel doubly stochastic Poisson process with intensity process $\{\boldsymbol{\lambda}_t(\mathbf{x}); t \geqslant t_0\}$, where $\mathbf{x}$ is a random variable with probability density function $p_{\mathbf{x}}(\mathbf{X})$. Let $p_t(\mathbf{X}|\mathbf{N}_\sigma; t_0 \leqslant \sigma < t)$ be the posterior probability density of $\mathbf{x}$ given $\{\mathbf{N}_\sigma; t_0 \leqslant \sigma < t\}$.

a. By using the multichannel differential rule and (6.244), show as an extension of (6.136) that

$$p_t(\mathbf{X}|\mathbf{N}_\sigma; t_0 \leqslant \sigma < t)$$

$$= p_{\mathbf{x}}(\mathbf{X}) \exp\left[\left\langle \mathbf{1}, -\int_{t_0}^{t} \left(\boldsymbol{\lambda}_\sigma(\mathbf{X}) - \hat{\boldsymbol{\lambda}}_\sigma\right) d\sigma \right.\right.$$

$$\left.\left. + \int_{t_0}^{t} \operatorname{diag}\left[\ln\left(\lambda_\sigma^{(k)}(\mathbf{X})/\hat{\lambda}_\sigma^{(k)}\right)\right] d\mathbf{N}_\sigma \right\rangle\right].$$

b. What is the multichannel extension of (6.138)?

6.7.1. Let $\{N_t; t \geqslant t_0\}$ be a doubly stochastic Poisson process with intensity process $\{\lambda_t(\mathbf{x}_t); t \geqslant t_0\}$, where $\{\mathbf{x}_t; t \geqslant t_0\}$ is a Markovian information process. Denote by $\hat{\mathbf{x}}_{t/u} = E[\mathbf{x}_t | N_\sigma; t_0 \leqslant \sigma < u]$ the minimum mean square-error estimate fo $\mathbf{x}_t$ in terms of $\{N_\sigma; t_0 \leqslant \sigma < u\}$, where $u \geqslant t$. Use Theorem 6.6.2, to show that

$$\hat{\mathbf{x}}_{t/u} = \hat{\mathbf{x}}_{t/t} + \int_{t}^{u} E\left\{\mathbf{x}_t\left[\lambda_\sigma(\mathbf{x}) - \hat{\lambda}_{\sigma|(t_0,\sigma)}\right] | \mathcal{N}_{t_0,\sigma}\right\} \hat{\lambda}_{\sigma|}^{-1}{}_{(t_0,\sigma)}\, dI_\sigma,$$

where

$$I_t = N_t - \int_{t_0}^{t} \hat{\lambda}_{\sigma|(t_0,\sigma)}\, d\sigma.$$

MARKED POINT PROCESSES

CHAPTER SEVEN

7.1 INTRODUCTION

A hierarchy of useful point process models has been developed in Chapters 3, 5, and 6 by making a series of modifications of the Poisson process of Chapter 2. These modifications, which previously have been treated separately, are all incorporated simultaneously in the model of this final chapter. The following remarks indicate the relationship between the new model and those of earlier chapters:

a. In Chapter 3, we endowed each point of a Poisson process with an auxiliary variable called a mark. For the compound Poisson process, which is the primary focus of Chapter 3, the marks are identically distributed, mutually independent, and independent of the points of the Poisson process they label. The point-process model we now study may also have a mark associated with each point, but there is no requirement that the marks be identically distributed and mutually independent. Nor is there a requirement that they be independent of the point process they label. Thus, in general, the distribution of the nth mark may depend upon the mark and occurrence time of each of the $n-1$ previous marked points.
b. The self-exciting point processes of Chapter 5 are obtained by randomizing the intensity of the Poisson process. This randomization is effected by allowing the intensity to become a function of past occurrence times. Similarly the point process in the model of this chapter may be self-exciting but with the additional generalization that the intensity can be a function of past marks as well as past occurrence times.
c. The doubly stochastic Poisson processes of Chapter 6 are also obtained by randomizing the intensity of the Poisson process, but in this case, the randomization is effected by allowing the intensity to become a function of an "outside" information process rather than the point process itself. So may the point process of this chapter be doubly stochastic. More generally, however, we allow for the possibility that the information process influences the marks as well as the rate at which points occur.

It is evident from the above remarks that the point processes of interest in this final chapter may be termed doubly stochastic, self-exciting, marked point processes. For brevity, we call them simply *marked point processes*.

Our discussion of marked point processes is divided into two main sections corresponding to whether the mark space is denumerable or nondenumerable. We have kept our objectives in both sections modest in order to limit the length of the chapter. Thus, our principal goal is to develop certain limited aspects of the theory that are required for decision and estimation problems in which marked point processes model observed data. In this regard, we obtain representations for the sample-function density of a marked point process and for the posterior statistics of an information process that influences an observed marked point process. The form and use of these quantities are entirely analogous to those for their counterparts appearing in Chapters 2, 3, 5, and 6. Consequently, all the techniques we have previously developed for solving decision and estimation problems can be applied immediately for the more general point processes of this chapter.

The approach to estimation and decision that is developed in Sections 7.2 and 7.3 was initiated by D. Snyder [17], I. Rubin [16], P. Fishman [4], and P. Fishman and D. Snyder [5]. Denumerable mark spaces were assumed by D. Snyder and nondenumerable spaces by I. Rubin and P. Fishman. P. Fishman gives a rigorous, measure-theoretic treatment of the problem and develops several results more general than those given here.

The notation for a marked point process is indicated in Figure 3.1. Each occurrence time w_i in the point process has associated with it a mark $\mathbf{u}_i$ having random values in a mark space $\mathfrak{U}$. In our development, this mark space will be either a denumerable collection $\{\mathbf{U}_1, \mathbf{U}_2 \ldots\}$ or a region of a finite-dimensional Euclidean space. More abstract mark spaces may be accommodated, but they will not be required here. Let T be a time interval in $[t_0, \infty)$ and A a subset of $\mathfrak{U}$. The number of points occurring in T and having marks in A will be denoted by $N(T \times A)$. Thus, for example, $N([s,t) \times \{\mathbf{U}\})$ is the total number of points in $[s,t)$ with mark equal to $\mathbf{U}$, where $\mathbf{U}$ is an element of $\mathfrak{U}$. Similarly, $N([t_0, t) \times \mathfrak{U})$ is the total number of points in $[t_0, t)$ regardless of their mark. We will also use N_t for this latter quantity; thus, we define $N_t = N([t_0, t) \times \mathfrak{U})$ and $N_{s,t} = N_t - N_s$ for $t_0 \leqslant s < t$.

As in Chapter 3, it is convenient to introduce a *mark-accumulator process*, which we will denote by $\{\mathbf{z}_t; t \geqslant t_0\}$. This is defined for each $t \geqslant t_0$ by

$$\mathbf{z}_t = \sum_{i=0}^{N_t} \mathbf{u}_i, \tag{7.1}$$

where we set $\mathbf{u}_0 = \mathbf{0}$. The mark accumulator process takes values in a state space $\mathfrak{Z}$ comprised of all vectors $\mathbf{Z}$ of the form $\mathbf{Z} = \sum_{i=0}^{n} \mathbf{U}_i$ for $n = 0, 1, 2, \ldots$, where $\mathbf{U}_0 = \mathbf{0}$ and $\mathbf{U}_i \in \mathfrak{U}$ for $i = 1, 2, \ldots, n$. Thus, for example, if $\mathfrak{U} = \{1\}$, in which case all marks are unity, then $\mathfrak{Z} = \{0, 1, 2, \ldots\}$, the set of nonnegative integers. When viewed in the multidimensional space $[t_0, \infty) \times \mathfrak{Z}$, the accumulator process has piecewise constant paths with jump discontinuities occurring at random times in $[t_0, \infty)$ and having random amplitudes in $\mathfrak{U}$; a representative sample path for an accumulator process with a one-dimensional state space may be seen in Figure 3.2. It is evident that a marked point process and its accumulator process are equivalent whenever $\mathbf{0}$ is not an element of $\mathfrak{U}$.

7.2 MARKED POINT PROCESSES HAVING A DENUMERABLE MARK SPACE

Throughout this section, $\mathfrak{U}$ will be a denumerable collection $\{\mathbf{U}_1, \mathbf{U}_2, \ldots\}$. We assume that:

a. $\{N_t; t \geq t_0\}$ is conditionally orderly;
b. the limit of the function $a(\Delta t, N_t, \mathbf{U}_i)$ defined by

$$a(\Delta t, N_t, \mathbf{U}_i) = \begin{cases} (\Delta t)^{-1} Pr[N([t, t+\Delta t) \times \{\mathbf{U}_i\}) = 1 | N_t], & N_t = 0 \\ (\Delta t)^{-1} Pr\left[N_t([t, t+\Delta t) \times \{\mathbf{U}_i\}) = 1 | N_t; w_1, \ldots, w_{N_t}, \mathbf{u}_1, \ldots, \mathbf{u}_{N_t}\right], \\ \qquad N_t \geq 1 \end{cases} \tag{7.2}$$

exists as Δt tends to zero for each i and almost every realization of the marked point process;
c. $Pr[N_{t_0} = 0] = 1$.

Qualitatively, assumption (a) means that the conditional probability of two or more points occurring in $[t, t+\Delta t)$ given the marked point process on $[t_0, t)$ can be made an arbitrarily small fraction of there being only one point by selecting Δt sufficiently small. We denote the limit in assumption (b) by $\mu_{t,\mathbf{U}_i}(0)$ and $\mu_{t,\mathbf{U}_i}(N_t; w_1, \ldots, w_{N_t}, \mathbf{u}_1, \ldots, \mathbf{u}_{N_t})$ for $N_t = 0$ and $N_t \geq 1$, respectively; thus,

$$\mu_{t,\mathbf{U}_i}(0) = \lim_{\Delta t \downarrow 0} (\Delta t)^{-1} Pr[N([t, t+\Delta t) \times \{\mathbf{U}_i\}) = 1 | N_t], \tag{7.3a}$$

for $N_t = 0$, and

$$\mu_{t,\mathbf{U}_i}(N_t; w_1, \ldots, w_{N_t}, \mathbf{u}_1, \ldots, \mathbf{u}_{N_t})$$

$$= \lim_{\Delta t \downarrow 0} (\Delta t)^{-1} Pr\left[N([t, t+\Delta t) \times \{\mathbf{U}_i\}) = 1 | N_t; w_1, \ldots \right.$$

$$\left. \ldots, w_{N_t}, \mathbf{u}_1, \ldots, \mathbf{u}_{N_t}\right], \tag{7.3b}$$

for $N_t \geqslant 1$. These limiting functions generalize (5.2) to account for marks, and they characterize the infinitesimal behavior of the marked point process in terms of its previous global behavior. It is convenient to introduce a random process $\{\lambda_{t,\mathbf{U}_i}; t \geqslant t_0\}$ defined by

$$\lambda_{t,\mathbf{U}_i} = \begin{cases} \mu_{t,\mathbf{U}_i}(0), & t_0 \leqslant t \leqslant w_1 \\ \mu_{t,\mathbf{U}_i}(N_t; w_1, \ldots, w_{N_t}, \mathbf{u}_1, \ldots, \mathbf{u}_{N_t}), & w_{N_t} < t \leqslant w_{N_t+1} \end{cases}. \tag{7.4}$$

We term this process the *intensity process for the* ith *mark*. It is a left-continuous function of t. The most immediate interpretation of this intensity process is that the conditional probability of there being a point in $[t, t+\Delta t)$ with mark $\mathbf{U}_i$ given the number of points in $[t_0, t)$, their occurrence times, and their marks is approximated to order Δt by $\lambda_{t,\mathbf{U}_i}\Delta t$.

Further interpretation for the intensity process $\{\lambda_{t,\mathbf{U}_i}; t \geqslant t_0\}$ for the ith mark is obtained by examining separately the evolution of the occurrence times and marks. Consider, first, the evolution of the occurrence times. We have

$$\begin{aligned} Pr[N_{t,t+\Delta t} = 1 | N_t = 0] &= Pr[N([t, t+\Delta t) \times \mathfrak{U}) = 1 | N_t = 0] \\ &= \sum_i Pr[N([t, t+\Delta t) \times \{\mathbf{U}_i\}) = 1 | N_t = 0] \\ &= \mu_t(0)\Delta t + o(\Delta t), \end{aligned} \tag{7.5}$$

where we define

$$\mu_t(0) = \sum_i \mu_{t,\mathbf{U}_i}(0),$$

and where $o(\Delta t)/\Delta t$ tends to zero as Δt tends to zero. Similarly, for $N_t \geqslant 1$, we have

$$\begin{aligned} Pr[N_{t,t+\Delta t} = 1 | N_t; w_1, \ldots, w_{N_t}, \mathbf{u}_1, \ldots, \mathbf{u}_{N_t}] \\ = \mu_t(N_t; w_1, \ldots, w_{N_t}, \mathbf{u}_1, \ldots, \mathbf{u}_{N_t})\Delta t + o(\Delta t), \end{aligned} \tag{7.6}$$

where we define

$$\mu_t(N_t; w_1, \ldots, w_{N_t}, \mathbf{u}_1, \ldots, \mathbf{u}_{N_t}) = \sum_i \mu_{t,\mathbf{U}_i}(N_t; w_1, \ldots, w_{N_t}, \mathbf{u}_1, \ldots, \mathbf{u}_{N_t}).$$

Thus, we conclude that the conditional probability of there being a point in $[t,t+\Delta t)$ regardless of its mark, given the number of points in $[t_0,t)$, their occurrence times, and their marks, is approximated to order Δt by $\lambda_t\Delta t$, where

$$\lambda_t = \sum_i \lambda_{t,\mathbf{U}_i}$$

$$= \begin{cases} \sum_i \mu_{t,\mathbf{U}_i}(0), & t_0 \leqslant t \leqslant w_1 \\ \sum_i \mu_{t,\mathbf{U}_i}(N_t; w_1,\ldots,w_{N_t},\mathbf{u}_1,\ldots,\mathbf{u}_{N_t}), & w_{N_t} < t \leqslant w_{N_t+1} \end{cases}. \tag{7.7}$$

By virtue of the conditional orderliness assumption (a), the conditional probability of there being no point in $[t,t+\Delta t)$ is approximated by $1-\lambda_t\Delta t$. The intensity process $\{\lambda_t; t\geqslant t_0\}$ generalizes (5.3) to account for a possible dependency of the occurrence times on the marks. We assume that

$$E(\lambda_t)<\infty \tag{7.8}$$

for $t\geqslant t_0$.

By paralleling the derivation of (5.5) and (5.6), we see for $N_s=0$ that

$$Pr[N_{s,t}=0|N_s=0]=\exp\left[-\int_s^t \mu_\sigma(0)\,d\sigma\right] \tag{7.9}$$

and for $N_s\geqslant 1$ that

$$Pr[N_{s,t}=0|N_s; w_1,\ldots,w_{N_s},\mathbf{u}_1,\ldots,\mathbf{u}_{N_s}]$$

$$=\exp\left[-\int_s^t \mu_\sigma(N_s; w_1,\ldots,w_{N_s},\mathbf{u}_1,\ldots,\mathbf{u}_{N_s})\,d\sigma\right], \tag{7.10}$$

where $t_0\leqslant s<t$. Thus, the functions $\mu_t(0)$ and $\mu_t(N_t; w_1,\ldots,w_{N_t},\mathbf{u}_1,\ldots,\mathbf{u}_{N_t})$, which are defined for infinitesimal intervals in (7.5) and (7.6), specify and are specified by conditional probabilities of there being no point in the global interval $[s,t)$ in (7.9) and (7.10).

Consider, next, the evolution of the marks. Suppose a point occurs in $[t,t+\Delta t)$ and that we are given the marked point process on $[t_0,t)$. Then, the conditional probability that the mark on the point in $[t,t+\Delta t)$ is $\mathbf{U}_i$ may be approximated to order Δt by

$$\lambda_{t,\mathbf{U}_i}\lambda_t^{-1}. \tag{7.11}$$

In order to verify this assertion for $N_t=0$, we note by Bayes' rule that

$$Pr[N([t,t+\Delta t)\times\{\mathbf{U}_i\})=1|N_{t,t+\Delta t}=1,N_t=0]$$

$$=\frac{Pr[N([t,t+\Delta t)\times\{\mathbf{U}_i\})=1|N_t=0]}{Pr[N_{t,t+\Delta t}=1|N_t=0]}$$

$$=\frac{\lambda_{t,\mathbf{U}_i}\Delta t+o(\Delta t)}{\lambda_t\Delta t+o(\Delta t)}=\lambda_{t,\mathbf{U}_i}\lambda_t^{-1}+o(\Delta t). \quad (7.12)$$

The argument is identical when $N_t\geqslant 1$.

We conclude from these several remarks that the intensity processes $\{\lambda_{t,\mathbf{U}_i};t\geqslant t_0\}$ have the following interpretations:

a. $\lambda_{t,\mathbf{U}_i}\Delta t$ is approximately the conditional probability a point having mark $\mathbf{U}_i$ occurs in $[t,t+\Delta t)$ given $N_t,w_1,\ldots,w_{N_t},\mathbf{u}_1,\ldots,$ and $\mathbf{u}_{N_t}$;
b. $[\Sigma_i\lambda_{t,\mathbf{U}_i}]\Delta t$ is approximately the conditional probability a point having any mark occurs in $[t,t+\Delta t)$ given $N_t,w_1,\ldots,w_{N_t},\mathbf{u}_1,\ldots,$ and $\mathbf{u}_{N_t}$;
c. $\lambda_{t,\mathbf{U}_i}[\Sigma_j\lambda_{t,\mathbf{U}_j}]^{-1}$ is approximately the conditional probability the mark on a point in $[t,t+\Delta t)$ is $\mathbf{U}_i$ given that a point does occur in $[t,t+\Delta t)$ and given $N_t,w_1,\ldots,w_{N_t},\mathbf{u}_1,\ldots,$ and $\mathbf{u}_{N_t}$.

The marked point process we have defined generalizes the compound Poisson process of Chapter 3 in that (1) the intensity governing the occurrence times may depend on previous occurrence times and marks, and (2) the distribution of a mark may depend on previous occurrence times and marks. The intensity processes $\{\lambda_{t,\mathbf{U}_i};\ t\geqslant t_0\}$ are the basic quantities that characterize a marked point process in our development. We assume throughout the discussion that these are known functions. In applications, $\lambda_{t,\mathbf{U}_i}$ may be postulated under favorable circumstances when there is an understanding of the physical mechanisms giving rise to the marked point process; otherwise, $\lambda_{t,\mathbf{U}_i}$ must be estimated from measurements. Later in the section, we will investigate procedures for estimating $\lambda_{t,\mathbf{U}_i}$ when it is parametrized by unknown variables or a random process.

Some examples of the type of processes included as special cases in this *self-exciting, marked point process* model are given in the following.

Example 7.2.1 (Generalized Compound Poisson Process). Points occur in a compound Poisson process as a Poisson process and marks are assigned independently to each point. If ν_t is the intensity of the Poisson process and if $P(\mathbf{U}_i)$ is the probability mark $\mathbf{U}_i$ is assigned to a point, then the function $\lambda_{t,\mathbf{U}_i}$ has the form

$$\lambda_{t,\mathbf{U}_i} = P(\mathbf{U}_i)\nu_t.$$

There are various straightforward generalizations of this situation that are of interest in applications. For instance, if marks are assigned independently to the points of a general self-exciting point process of the type studied in Chapter 5, then $\lambda_{t,\mathbf{U}_i}$ has the form

$$\lambda_{t,\mathbf{U}_i} = \begin{cases} P(\mathbf{U}_i)\mu_t(0), & t_0 \leqslant t \leqslant w_1 \\ P(\mathbf{U}_i)\mu_t(N_t; w_1,\ldots,w_{N_t}), & w_{N_t} < t \leqslant w_{N_t+1} \end{cases}$$

where the functions μ are those defined in (5.2). Special cases of this situation include all the limited memory, self-exciting point processes of Section 5.3. Of these, the renewal process is a notable example. Marks do not have to be assigned independently. For example, they can be assigned according to a Markov sequence so that the mark on a particular point given all previous marks depends only on the mark of the preceding point. If $P(\mathbf{U}_i|\mathbf{U}_j)$ is the probability mark $\mathbf{U}_i$ is assigned given that the preceding mark is $\mathbf{U}_j$, and if the point process which the marks label is a general self-exciting point process, then $\lambda_{t,\mathbf{U}_i}$ has the form

$$\lambda_{t,\mathbf{U}_i} = \begin{cases} P(\mathbf{U}_i)\mu_t(0), & t_0 \leqslant t \leqslant w_1 \\ P(\mathbf{U}_i|\mathbf{U}_j)\mu_t(N_t; w_1,\ldots,w_{N_t}), & w_{N_t} < t \leqslant w_{N_t+1}. \end{cases}$$

For these examples, the influence of the marks and of the occurrence times is reflected by separate factors in $\lambda_{t,\mathbf{U}_i}$. In our general model, such separation does not need to occur because the marks assigned to points may depend on $N_t, w_1,\ldots,$ and w_{N_t} and, further, the intensity of the point process may depend on $\mathbf{u}_1,\ldots,$ and $\mathbf{u}_{N_t}$.

Example 7.2.2 (Self-Exciting Counting Process). The accumulator process $\{z_t; t \geqslant t_0\}$ is a self-exciting counting process of the type studied in Chapter 5 when $\mathfrak{U} = \{1\}$.

Example 7.2.3 (Self-Exciting Birth-Death Process). The accumulator process $\{z_t; t \geqslant t_0\}$ is a self-exciting birth-death process when $\mathfrak{U} = \{-1, 1\}$. In this case, z increases or decreases by unity corresponding to a birth and death, respectively, at each point occurrence. The birth and death rates are $\lambda_{t,1}$ and $\lambda_{t,-1}$, respectively. In general, these rates may depend on each of the quantities $N_t, w_1,\ldots,w_{N_t}, u_1,\ldots,$ and u_{N_t} independently. A Markov birth-death process is a widely used model and results when this dependency is only on the sum $z_t = u_0 + u_1 + \cdots + u_{N_t}$, where $u_0 = 0$. A detailed treatment of Markov birth-death processes is given by A. Bharucha-Reid [2, p. 86].

Example 7.2.4 (Denumerable-State Markov Jump Processes). Suppose $\{\mathbf{x}_t; t \geqslant t_0, \mathbf{x}_{t_0} = \mathbf{X}_0\}$ is a Markov jump process of the type described in Section 6.5.2, but assume that the states may be vector valued. The instantaneous rate $a_{\zeta_i,\zeta_k}(t)$ of transitions from state ζ_i to state ζ_k, which is defined from (6.115) by

$$a_{\zeta_i,\zeta_k}(t) = \begin{cases} \lim\limits_{\Delta t \downarrow 0} (\Delta t)^{-1} Pr[\mathbf{x}_{t+\Delta t} = \zeta_k | \mathbf{x}_t = \zeta_i], & i \neq k \\ -\sum\limits_{j(j \neq i)} a_{\zeta_i,\zeta_j}(t), & i = k \end{cases} \tag{7.13}$$

characterizes $\{\mathbf{x}_t; t \geqslant t_0, \mathbf{x}_{t_0} = \mathbf{X}_0\}$ statistically given that the initial state is $\mathbf{X}_0$. This process may be imbedded in a marked point process as follows. Let the mark space $\mathfrak{U}$ be the set of all possible state transitions of $\mathbf{x}$; thus, $\mathfrak{U}$ is the set of all nonzero vectors $\mathbf{U}$ such that $a_{\zeta_i,\zeta_i+\mathbf{U}}(t) \neq 0$ for some $t \geqslant t_0$ and ζ_i and $\zeta_i + \mathbf{U}$ are possible states of $\mathbf{x}$. Now construct a marked point process with occurrence times at the transitions of $\mathbf{x}$ and marks equal to the corresponding transitions in $\mathbf{x}$; thus, if $\mathbf{x}$ makes a transition at time t from $\mathbf{x}_t$ to $\mathbf{x}_{t+}$, then the marked point process will have a point at time t with mark $\mathbf{x}_{t+} - \mathbf{x}_t$. It is evident that the mark intensity processes $\lambda_{t,\mathbf{U}_i}$ are given in terms of the transition rates $a_{\zeta_i,\zeta_j}(t)$ according to

$$\lambda_{t,\mathbf{U}_i} = a_{\mathbf{z}_t + \mathbf{X}_0, \mathbf{z}_t + \mathbf{X}_0 + \mathbf{U}_i}(t), \tag{7.14}$$

where $\mathbf{z}_t$ is the sum of the marks for all points in $[t_0, t)$ and where $\mathbf{z}_t + \mathbf{X}_0 = \mathbf{x}_t$. The converse will also hold in that a marked point process with mark intensities in the form of (7.14) will have a mark-accumulator process that is a Markov jump process. For detailed properties and applications of Markov jump processes, we refer to A. Bharucha-Reid [2, Chapters 2, 4–9] and E. Parzen [11, Chapter 7].

Example 7.2.5 (Semi-Markov Jump Process). A semi-Markov jump process is a generalized form of renewal counting process and is discussed in the basic papers of R. Pyke [13, 14]. S. Ross [15, Chapters 5, 7] gives applications and further references for these processes. A semi-Markov process is constructed as follows. Let $\{\mathbf{x}_t; t \geqslant t_0, \mathbf{x}_{t_0} = \mathbf{X}_0\}$ be a random process with a denumerable state space $\{\zeta_1, \zeta_2, \ldots\}$. Denote by $P(\zeta_j | \zeta_i)$ the probability that state ζ_j succeeds state ζ_i and by $P_{\zeta_i,\zeta_j}(t)$ the conditional probability that a transition from state ζ_i to state ζ_j takes place within an amount of time t given that the process has just entered state ζ_i and will next enter ζ_j. The process may be constructed in two stages. When state ζ_i is entered, the next state is chosen according to the state transition probabilities $P(\zeta_j|\zeta_i)$. Then, given that the next state is ζ_j, the time to the transition is chosen according to the interval distribution function $P_{\zeta_i,\zeta_j}(t)$.

A semi-Markov process may be imbedded in a marked point process by following the procedure outlined in Example 7.2.4. Let the mark space $\mathfrak{U}$ be the set of all possible state transitions of $\mathbf{x}$; thus, $\mathfrak{U}$ is the set of all nonzero vectors $\mathbf{U}$ such that $P(\zeta_i+\mathbf{U}|\zeta_i)$ is nonzero for some state ζ_i of $\mathbf{x}$. Now construct a marked point process with occurrence times at the transitions of $\mathbf{x}$ and with marks equal to the corresponding transition in $\mathbf{x}$. It can be shown that the intensity process for the ith mark is given by

$$\lambda_{t,\mathbf{U}_i}=\begin{cases} P(\mathbf{X}_0+\mathbf{U}_i|\mathbf{X}_0)h_{\mathbf{X}_0,\mathbf{X}_0+\mathbf{U}_i}(t-t_0), & t_0\leqslant t\leqslant w_1\\ P(\mathbf{u}_{N_t}+\mathbf{U}_i|\mathbf{u}_{N_t})\,h_{\mathbf{u}_{N_t},\mathbf{u}_{N_t}+\mathbf{U}_i}(t-w_{N_t}), & w_{N_t}<t\leqslant w_{N_t+1}\end{cases}$$

where $\mathbf{U}_i\in\mathfrak{U}$ is a possible transition in $\mathbf{x}$ and where $h_{\zeta_i,\zeta_j}(t)$ is the generalized hazard function defined by

$$h_{\zeta_i,\zeta_j}(t)=p_{\zeta_i,\zeta_j}(t)\left[1-P_{\zeta_i,\zeta_j}(t)\right]^{-1},$$

in which $p_{\zeta_i,\zeta_j}(t)=dP_{\zeta_i,\zeta_j}(t)/dt$ is the probability density of the interarrival time for a transition in $\mathbf{x}$ from state ζ_i to ζ_j. It is evident that if $\mathbf{x}$ has only unit positive transitions, then the mark accumulator process $\{z_t;t\geqslant t_0\}$ is a renewal counting process with intensity given by (5.84). For detailed properties and applications of semi-Markov jump processes, we refer to R. Pyke [13, 14] and S. Ross [15, Chapters 5, 7].

Example 7.2.6 (Mutually Exciting Point Processes). Point processes that interact are called *mutually exciting*. The interaction is in the sense that the number and occurrence times of points in all the processes together can influence the evolution of any (or, possible, all) of the component point processes. Mutually exciting point processes are a natural, multichannel extension of the self-exciting point processes of Chapter 5. The multichannel doubly stochastic Poisson process of Section 6.6 is an example of a collection of mutually exciting point processes found useful in applications. Additional applications in electrophysiology are given by D. Perkel, G. Gerstein, and G. Moore [12], M. Ten-Hoppen and H. Reuver [18], and by A. Lawrance [9, 10]. Mutually exciting point processes are special marked point processes in which a mark on a point identifies the component process with which it is to be associated.

The above six examples show that the marked point-process model encompasses a wide variety of point and jump processes that have proven useful in practice. We will further extend the model below by allowing the mark intensity processes $\lambda_{t,\mathbf{U}_i}$ to become functions of an information process as well as past marks and occurrence times.

Sample-Function Density: Denumerable Mark Space

The *sample-function density* has been a recurring and fundamental quantity in previous chapters. We have seen in Chapters 2, 3, 5, and 6 the role of the sample-function density in estimation and decision problems for the Poisson process and its generalizations, the compound Poisson process, the self-exciting point process and the doubly stochastic Poission process. The sample-function density for the more general marked point processes of this chapter is equally important. Its role in the estimation of unknown parameters that influence the mark intensities $\lambda_{t,\mathbf{U}_i}$, and hence the evolution of the occurrence times and marks, will be the same as in earlier chapters. In particular, the maximum likelihood estimate of the unknown parameters in terms of an observed realization of the marked point process is the value of the parameters maximizing the sample-function density evaluated at the observed realization. Furthermore, the likelihood ratio, which from Chapter 2 is a sufficient statistic for deciding between two collections of mark intensities $\{\lambda_{t,\mathbf{U}_i}^{(0)}\}$ and $\{\lambda_{t,\mathbf{U}_i}^{(1)}\}$ that may govern a marked point process, is also determined by the sample-function density. With this as motivation, we now evaluate the sample-function density for a marked point process. We will see that it bears a striking resemblance to the sample-function density in (6.35) and (3.36) for a compound Poisson process.

Consider a marked point process for $t \geqslant t_0$ with a denumerable mark space $\mathfrak{U} = \{\mathbf{U}_1, \mathbf{U}_2, \ldots\}$. Denote the mark-accumulator process defined in (7.1) by $\{\mathbf{z}_t; t \geqslant t_0\}$. In parallel with (3.32), the quantity $p[\{\mathbf{z}_\sigma; t_0 \leqslant \sigma < t\}]$ defined below is the sample-function density for the marked point process on the interval $[t_0, t)$:

$$p[\{\mathbf{z}_\sigma; t_0 \leqslant \sigma < t\}] = \begin{cases} Pr(N_t = 0), & N_t = 0 \\ p_{\mathbf{w}}(\mathbf{W}, \mathbf{u}_1 = \xi_1, \ldots, \mathbf{u}_n = \xi_n, N_t = n), & N_t = n \geqslant 1 \end{cases} \tag{7.15}$$

where $\xi_i \in \mathbf{U}$ for each i, $1 \leqslant i \leqslant n$, and where

$$p_{\mathbf{w}}(\mathbf{W}, \mathbf{u}_1 = \xi_1, \ldots, \mathbf{u}_n = \xi_n, N_t = n)$$

$$= \lim_{\substack{\max(\Delta W_i) \downarrow 0 \\ 1 \leqslant i \leqslant n}} \left(\prod_{i=1}^{n} \Delta W_i \right)^{-1} Pr[E_n, N_t = n],$$

in which E_n denotes the event

$$\{N([t_0, W_1)\times \mathfrak{U})=0, N([W_1, W_1+\Delta W_1)\times\{\xi_1\})=1\}$$

for $n=1$ and the event

$$\{N([t_0, W_1)\times \mathfrak{U})=0, N([W_1, W_1+\Delta W_1)\times\{\xi_1\})=1,\ldots$$

$$\ldots, N([W_{n-1}+\Delta W_{n-1}, W_n)\times \mathfrak{U})=0, N([W_n, W_n+\Delta W_n)\times\{\xi_n\})=1\}$$

for $n\geqslant 2$. It is evident from (7.15) that $p[\{\mathbf{z}_\sigma; t_0\leqslant\sigma<t\}]\Delta W_1\Delta W_2\cdots\Delta W_n$ may be interpreted roughly as the probability of obtaining a particular realization of the marked point process on the interval $[t_0, t)$ with $N_t=n$ points located at times $w_1=W_1,\ldots,w_n=W_n$ and having corresponding marks $\mathbf{u}_1=\xi_1,\ldots,\mathbf{u}_n=\xi_n$.

The factors in (7.15) that make up the sample-function density can all be expressed in terms of the functions $\mu_{t,\mathbf{U}_i}$ defined in (7.3) as follows. We first note from (7.9) that for $N_t=0$,

$$p[\{\mathbf{z}_\sigma; t_0\leqslant\sigma<t\}]=\exp\left[-\int_{t_0}^{t}\mu_\sigma(0)\,d\sigma\right], \tag{7.16}$$

where $\mu_t(0)=\Sigma_i\mu_{t,\mathbf{U}_i}(0)$ for $t\geqslant t_0$. For $N_t=1$, we have

$$Pr(E_1, N_t=1)=Pr(N_{W_1}=0)Pr\left[N([W_1, W_1+\Delta W_1)\times\{\xi_1\})=1|N_{W_1}=0\right]$$

$$\times Pr[N_{W_1+\Delta W_1, t}=0|E_1].$$

Hence, by using (7.3) and (7.10), we deduce for $N_t=1$ that

$$p[\{\mathbf{z}_\sigma; t_0\leqslant\sigma<t\}]$$

$$=\mu_{W_1,\xi_1}(0)\exp\left[-\int_{t_0}^{W_1}\mu_\sigma(0)\,d\sigma-\int_{W_1}^{t}\mu_\sigma(1; W_1,\xi_1)\,d\sigma\right], \tag{7.17}$$

where $\mu_t(1; W_1,\xi_1)=\Sigma_i\mu_{t,\mathbf{U}_i}(1; W_1,\xi_1)$ for $t\geqslant t_0$. By this same procedure, we deduce for $N_t=n\geqslant 2$ that

$$Pr(E_n, N_t=n)=Pr(N_{W_1}=0)Pr\left[N([W_1, W_1+\Delta W_1)\times\{\xi_1\})=1|N_{W_1}=0\right]$$

$$\times\left\{\prod_{i=2}^{n}Pr\left[N([W_i, W_i+\Delta W_i)\times\{\xi_i\})=1|E_{i-1}, N_{W_i}=i-1\right]\right.$$

$$\left.\times Pr(N_{W_{i-1}+\Delta W_{i-1}, W_i}=0|E_{i-1})\right\}Pr(N_{W_n+\Delta W_n, t}=0|E_n)$$

and, hence, for $N_t = n \geqslant 2$ that

$$p[\{\mathbf{z}_\sigma; t_0 \leqslant \sigma < t\}]$$

$$= \mu_{W_1,\xi_1}(0)\left\{ \prod_{i=2}^{n} \mu_{W_i,\xi_i}(i-1; W_1,\dots,W_{i-1},\xi_1,\dots,\xi_{i-1}) \right\}$$

$$\times \exp\left[-\int_{t_0}^{W_1} \mu_\sigma(0)\,d\sigma - \sum_{i=2}^{n} \int_{W_{i-1}}^{W_i} \mu_\sigma(i-1; W_1,\dots,W_{i-1},\xi_1,\dots,\xi_{i-1})\,d\sigma \right.$$

$$\left. - \int_{W_n}^{t} \mu_\sigma(n; W_1,\dots,W_n,\xi_1,\dots,\xi_n)\,d\sigma \right]. \tag{7.18}$$

Equations (7.16)–(7.18) are seen to generalize (5.26) to account for marks. These equations can be placed in a more compact and suggestive form by incorporating into them the mark intensity processes $\lambda_{t,\mathbf{U}_i}$ and λ_t defined in (7.4) and (7.7), respectively, and the counting integrals of Chapter 4. For this purpose, we note that

$$\sum_i \int_{t_0}^{t} \lambda_{\sigma,\mathbf{U}_i}\,d\sigma$$

$$= \begin{cases} \int_{t_0}^{t} \mu_\sigma(0)\,d\sigma, & N_t = 0 \\ \int_{t_0}^{w_1} \mu_\sigma(0)\,d\sigma + \int_{w_1}^{t} \mu_\sigma(1; w_1, \mathbf{u}_1)\,d\sigma, & N_t = 1 \\ \int_{t_0}^{w_1} \mu_\sigma(0)\,d\sigma + \sum_{i=2}^{N_t} \int_{w_{i-1}}^{w_i} \mu_\sigma(i-1; w_1,\dots,w_{i-1},\mathbf{u}_1,\dots,\mathbf{u}_{i-1})\,d\sigma & \\ \quad + \int_{w_{N_t}}^{t} \mu_\sigma(N_t; w_1,\dots,w_{N_t},\mathbf{u}_1,\dots,\mathbf{u}_{N_t})\,d\sigma, & N_t \geqslant 2. \end{cases} \tag{7.19}$$

Similarly, for the counting integrals defined in Section 4.2.1, we have that

$$\sum_k \int_{t_0}^{t} \ln(\lambda_{\sigma,\mathbf{U}_k}) N(d\sigma \times \{\mathbf{U}_k\})$$

$$= \begin{cases} 0, & N_t = 0 \\ \ln \mu_{w_1,\mathbf{u}_1}(0), & N_t = 1 \\ \ln \mu_{w_1,\mathbf{u}_1}(0) + \sum_{i=2}^{N_t} \ln \mu_{w_i,\mathbf{u}_i}(i-1; w_1,\dots,w_{i-1},\mathbf{u}_1,\dots,\mathbf{u}_{i-1}), & N_t \geqslant 2. \end{cases} \tag{7.20}$$

By combining (7.16)–(7.20), we conclude that the sample-function density for a marked point process on the interval $[t_0,t)$ can be expressed in terms of the mark counting processes $\{N([t_0,t)\times\{\mathbf{U}_i\}); t\geqslant t_0, \mathbf{U}_i\in\mathfrak{U}\}$ and the mark intensity processes $\{\lambda_{t,\mathbf{U}_i}; t\geqslant t_0, \mathbf{U}_i\in\mathfrak{U}\}$ as

$$p[\{\mathbf{z}_\sigma; t_0\leqslant\sigma<t\}]$$

$$=\prod_i \exp\left[-\int_{t_0}^t \lambda_{\sigma,\mathbf{U}_i}\,d\sigma+\int_{t_0}^t \ln(\lambda_{\sigma,\mathbf{U}_i})N(d\sigma\times\{\mathbf{U}_i\})\right]. \tag{7.21}$$

From Example 7.2.1, the functions $\lambda_{t,\mathbf{U}_i}$ have the factored form $\lambda_{t,\mathbf{U}_i} = P(\mathbf{U}_i)\nu_t$ for a compound Poisson process. In this case, (7.21) reduces to (3.36). It is interesting to note that the factoring of the sample-function density in (3.36) carries over to the more general marked point processes of this chapter even though the mark counting processes $\{N([t_0,t)\times\{\mathbf{U}_i\}); t \geqslant t_0, \mathbf{U}_i\in\mathfrak{U}\}$ are not mutually independent; they are, in fact, mutually exciting in the sense of Example 7.2.6.

The sample-function density in (7.21) can be written alternatively as

$$p[\{\mathbf{z}_\sigma; t_0\leqslant\sigma<t\}]=\exp\left[-\int_{t_0}^t \lambda_\sigma\,d\sigma+\int_{t_0}^t \ln\lambda_\sigma\,dN_\sigma\right.$$

$$\left.+\sum_i\int_{t_0}^t \ln\left(\lambda_{\sigma,\mathbf{U}_i}\lambda_\sigma^{-1}\right)N(d\sigma\times\{\mathbf{U}_i\})\right]. \tag{7.22}$$

This form generalizes (3.35) to which it reduces for a compound Poisson process. It is evident from (7.22) that the sample-function density for a marked point process splits into two factors,

$$\exp\left[-\int_{t_0}^t \lambda_\sigma\,d\sigma+\int_{t_0}^t \ln\lambda_\sigma\,dN_\sigma\right]$$

and

$$\sum_i\int_{t_0}^t \ln\left(\lambda_{\sigma,\mathbf{U}_i}\lambda_\sigma^{-1}\right)N(d\sigma\times\{\mathbf{U}_i\}).$$

The first factor is contributed by the number of points in $[t_0,t)$ and their times of occurrence. This factor has the form of the sample-function density for a self-exciting point process, as given in (5.29). The second factor is contributed by the marks on the points in $[t_0,t)$.

The use of (7.21) in estimation and decision problems involving observed marked point processes is precisely the same as the use of the sample-function densities in Chapters 2, 3, 5, and 6. Here, we assume that observations of the marked point process on the interval $[t_0,t)$ include a measurement of the total number of points in the interval, their times of occurrence, and their marks. The use of the sample-function density (7.21) is outlined in the following examples.

Example 7.2.7 (Parameter Estimation for Marked Point Processes with a Denumerable Mark Space). The problem of estimating parameters arises frequently in applications where a marked point process is used as a model for observed data. This problem is discussed in Example 3.3.2 for the compound Poisson process, and the situation is entirely analogous here. We consider as observed data a realization $\{N_t; w_1,\ldots,w_{N_t},\mathbf{u}_1,\ldots,\mathbf{u}_{N_t}\}$ of a marked point process on the interval $[t_0,t)$. The intensity processes $\lambda_{t,\mathbf{U}_i}(\boldsymbol{\theta})$ are presumed to be a function of a collection of unknown, nonrandom parameters, which are denoted by the parameter vector $\boldsymbol{\theta}$. As in Example 3.3.2, the maximum likelihood estimate $\hat{\boldsymbol{\theta}}_{\mathrm{ML}}$ of $\boldsymbol{\theta}$ in terms of the observations $N_t=n$, $w_1=W_1,\ldots,$ $w_n=W_n$, $\mathbf{u}_1=\boldsymbol{\xi}_1,\ldots,$ and $\mathbf{u}_n=\boldsymbol{\xi}_n$ is the value of $\boldsymbol{\theta}$ that maximizes the probability of having observed these data. We conclude that $\hat{\boldsymbol{\theta}}_{\mathrm{ML}}$ maximizes the sample-function density (7.21), or equivalently its logarithm, evaluated for the given data. Thus, from (7.21), $\hat{\boldsymbol{\theta}}_{\mathrm{ML}}$ maximizes

$$l(\boldsymbol{\theta})=\sum_i\left[-\int_{t_0}^{t}\lambda_{\sigma,\mathbf{U}_i}(\boldsymbol{\theta})\,d\sigma+\int_{t_0}^{t}\ln\left[\lambda_{\sigma,\mathbf{U}_i}(\boldsymbol{\theta})\right]N(d\sigma\times\{\mathbf{U}_i\})\right] \quad (7.23)$$

as a function of $\boldsymbol{\theta}$, where the right side is to be evaluated for the observed realization of the marked point process. This maximization is subject to any constraints on the possible values of $\boldsymbol{\theta}$. A particular application of this general principle is given in the next example.

Example 7.2.8 (Estimating the Transition Rates of a Markov Jump Process). With reference to Example 7.2.4, let $\{\mathbf{x}_t; t\geqslant t_0, \mathbf{x}_{t_0}=\mathbf{X}_0\}$ be a Markov jump process with initial state $\mathbf{X}_0$ and homogeneous transition rates a_{ζ_i,ζ_j}, $i\neq j$, where $\zeta_1,\zeta_2,\ldots$ are the possible states of $\mathbf{x}_t$ for any $t\geqslant t_0$. Suppose $\mathbf{x}$ is observed on the interval $[t_0,T)$, and based on these data, it is desired to estimate the rates a_{ζ_i,ζ_j}. The maximum likelihood estimates of these rates can be determined by using (7.23). For this purpose, we incorporate (7.14) into (7.23) and obtain

$$\begin{aligned} l=-\sum_i\big[&(W_1-t_0)a_{\mathbf{X}_0,\mathbf{X}_0+\mathbf{U}_i}+(W_2-W_1)a_{\mathbf{X}_0+\boldsymbol{\xi}_1,\mathbf{X}_0+\boldsymbol{\xi}_1+\mathbf{U}_i}+\cdots\\ &+(T-W_n)a_{\mathbf{X}_0+\boldsymbol{\xi}_1+\cdots+\boldsymbol{\xi}_n,\mathbf{X}_0+\boldsymbol{\xi}_1+\cdots+\boldsymbol{\xi}_n+\mathbf{U}_i}\big]\\ &+\big[\ln a_{\mathbf{X}_0,\mathbf{X}_0+\boldsymbol{\xi}_1}+\ln a_{\mathbf{X}_0+\boldsymbol{\xi}_1,\mathbf{X}_0+\boldsymbol{\xi}_1+\boldsymbol{\xi}_2}+\cdots\\ &+\ln a_{\mathbf{X}_0+\boldsymbol{\xi}_1+\cdots+\boldsymbol{\xi}_{n-1},\mathbf{X}_0+\boldsymbol{\xi}_1+\cdots+\boldsymbol{\xi}_{n-1}+\boldsymbol{\xi}_n}\big], \end{aligned}$$

where n is the number of transitions observed in $\mathbf{x}$ during $[t_0, T)$, W_1, $W_2, \ldots$, and W_n are the observed times of these transitions, and $\xi_1, \xi_2, \ldots$, and ξ_n are the observed transitions. Let us define

$$1_{\mathbf{U}}(\xi) = \begin{cases} 1, & \xi = \mathbf{U} \\ 0, & \xi \neq \mathbf{U}. \end{cases}$$

Then, the likelihood function can be rewritten as

$$l = -\sum_i \sum_{j(\neq i)} \left[(W_1 - t_0) 1_{\mathbf{x}_0}(\zeta_i) + (W_2 - W_1) u_{\mathbf{x}_0+\xi_1}(\zeta_i) + \cdots \right.$$
$$\left. + (T - W_n) 1_{\mathbf{x}_0+\xi_1+\cdots+\xi_n}(\xi_i) \right] a_{\zeta_i,\zeta_j}$$
$$+ \sum_i \sum_{j(\neq i)} \left[1_{\mathbf{x}_0}(\zeta_i) 1_{\mathbf{x}_0+\xi_1}(\zeta_j) + 1_{\mathbf{x}_0+\xi_1}(\zeta_i) 1_{\mathbf{x}_0+\xi_1+\xi_2}(\zeta_j) + \cdots \right.$$
$$\left. + 1_{\mathbf{x}_0+\xi_1+\cdots+\xi_{n-1}}(\zeta_i) 1_{\mathbf{x}_0+\xi_1+\cdots+\xi_n}(\zeta_j) \right] \ln(a_{\zeta_i,\zeta_j}).$$

Hence, for $i \neq k$,

$$\frac{\partial l}{\partial a_{\zeta_i,\zeta_j}} = -T(\zeta_i) + \frac{N(\zeta_i \to \zeta_j)}{a_{\zeta_i,\zeta_j}},$$

where we define

$$T(\zeta_i) = (W_1 - t_0) 1_{\mathbf{x}_0}(\zeta_i) + (W_2 - W_1) 1_{\mathbf{x}_0+\xi_1}(\zeta_i) + \cdots$$
$$+ (T - W_n) 1_{\mathbf{x}_0+\xi_1+\cdots+\xi_n}(\zeta_i)$$

and

$$N(\zeta_i \to \zeta_j) = 1_{\mathbf{x}_0}(\zeta_i) 1_{\mathbf{x}_0+\xi_1}(\zeta_j) + 1_{\mathbf{x}_0+\xi_1}(\zeta_i) 1_{\mathbf{x}_0+\xi_1+\xi_2}(\zeta_j) + \cdots$$
$$+ 1_{\mathbf{x}_0+\xi_1+\cdots+\xi_{n-1}}(\zeta_i) 1_{\mathbf{x}_0+\xi_1+\cdots+\xi_n}(\zeta_j).$$

The quantity $T(\zeta_i)$ is the total amount of time during $[t_0, T)$ that $\mathbf{x}$ is observed in state ζ_i, and $N(\zeta_i \to \zeta_j)$ is the observed number of transitions in $\mathbf{x}$ from state ζ_i to state ζ_j. We conclude that the maximum likelihood estimate $\hat{a}_{\zeta_i,\zeta_j}$ of a_{ζ_i,ζ_j} is given by

$$\hat{a}_{\zeta_i,\zeta_j} = \frac{N(\zeta_i \to \zeta_j)}{T(\zeta_i)}$$

provided $T(\zeta_i) \neq 0$. We set $\hat{a}_{\zeta_i,\zeta_j} = 0$ if $T(\zeta_i) = 0$.

Doubly Stochastic Marked Point Processes: Denumerable Mark Space

We now extend the above model to include an information process. This extension is the analog for marked point processes of the extension in Chapter 6 of the Poisson process to the doubly stochastic Poisson process. Thus, we will now allow the evolution of the marks and the occurrence times to be influenced by an "outside" random process as well as by self-excitation from previous marks and occurrence times. The issues of interest for this extended class of marked point processes parallel those in Chapter 6 for the doubly stochastic Poisson process. In particular, we will derive representations for the sample-function density of the doubly stochastic, marked point process and for the posterior statistics of the information process given an observation of the marked point process it influences. These two quantities will then be used in decision and estimation problems in which a marked point process models observed data.

Following the notation of Chapter 6, we denote the information process by $\{\mathbf{x}_t; t \geqslant t_0\}$. This is a left-continuous, vector-valued random process which, for much of our discussion, can have essentially arbitrary statistics. We now allow the probability that a point having mark $\mathbf{U}_i$ occurs in the infinitesimal interval $[t, t+\Delta t)$ to depend on $\mathbf{x}_t$ and the marked point process on $[t_0, t)$. In particular, in place of (7.4), we now assume that

$$\lambda_{t,\mathbf{U}_i}(\mathbf{x}_t) = \begin{cases} \mu_{t,\mathbf{U}_i}(0, \mathbf{x}_t), & t_0 \leqslant t \leqslant w_1 \\ \mu_{t,\mathbf{U}_i}(N_t; w_1, \ldots, w_{N_t}, \mathbf{u}_1, \ldots, \mathbf{u}_{N_t}, \mathbf{x}_t), & w_{N_t} < t \leqslant w_{N_t+1} \end{cases} \tag{7.24}$$

where

$$\mu_{t,\mathbf{U}_i}(0, \mathbf{x}_t) = \lim_{\Delta t \downarrow 0} (\Delta t)^{-1} Pr[N([t, t+\Delta t) \times \{\mathbf{U}_i\}) = 1 | N_t, \mathbf{x}_\sigma; \sigma \geqslant t_0], \qquad N_t = 0 \tag{7.25a}$$

and where

$$\begin{aligned} &\mu_{t,\mathbf{U}_i}(N_t; w_1, \ldots, w_{N_t}, \mathbf{u}_1, \ldots, \mathbf{u}_{N_t}, \mathbf{x}_t) \\ &\quad = \lim_{\Delta t \downarrow 0} (\Delta t)^{-1} Pr\Big[N([t, t+\Delta t) \times \{\mathbf{U}_i\}) = 1 | N_t; w_1, \ldots \\ &\qquad \ldots, w_{N_t}, \mathbf{u}_1, \ldots, \mathbf{u}_{N_t}, \mathbf{x}_\sigma; \sigma \geqslant t_0\Big], \qquad N_t \geqslant 1. \end{aligned} \tag{7.25b}$$

The limits in (7.25) are assumed to exist with probability one.

The quantity $\lambda_{t,\mathbf{U}_i}(\mathbf{x}_t)\Delta \mathrm{t}$ is approximately the conditional probability a point having mark $\mathbf{U}_i$ occurs in $[t, t+\Delta t)$ given the information process for all time and given the number of points in $[t_0, t)$, their times of occurrence, and their marks.

Example 7.2.9 (Optical Position Sensing, continued). Consider Example 3.3.3 again, but suppose now that the beam of light incident on the four-quadrant photodetector wanders about in a random walk due to temporal fluctuations in the propagation medium between the transmitter and the receiver. Thus, as indicated in Figure 7.1, the location parameters α and β of the center of the light beam will now be random processes $\{\alpha_t, \beta_t; t \geq t_0\}$, and we have

$$\lambda_{t,Q_i}(\alpha_t, \beta_t) = \int_{\mathcal{Q}_i}\int I(a,b|\alpha_t,\beta_t)\,da\,db, \tag{7.26}$$

where Q_i is a mark indicating the ith quadrant, $\mathcal{Q}_i$ is the region of the ith quadrant, and from (3.39)

$$I(a,b|\alpha_t,\beta_t) = I_0 \exp\left\{-\frac{1}{2\rho^2}\left[(a-\alpha_t)^2 + (b-\beta_t)^2\right]\right. . \tag{7.27}$$

Here, the probability that a photoevent occurs in the ith quadrant in the interval $[t, t+\Delta t)$ is approximated to order Δt by $\lambda_{t,Q_i}(\alpha_t,\beta_t)\Delta t$. It is seen in this example that $\lambda_{t,Q_i}(\alpha_t,\beta_t)$ does not depend upon past occurrence times

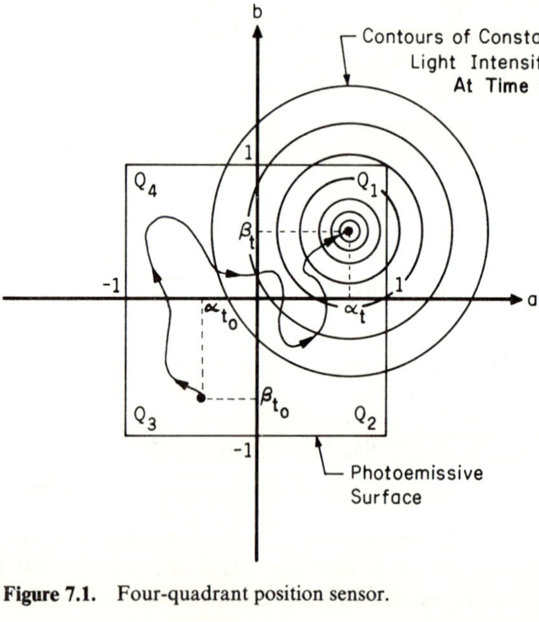

Figure 7.1. Four-quadrant position sensor.

and marks. Such dependence would occur if the output of the position sensor were used to generate feedback signals to control the position of the photoemissive surface. For instance, suppose

$$\alpha_t^* = \alpha_t^*(N_t; w_1, \ldots, w_{N_t}, q_1, \ldots, q_{N_t})$$

and

$$\beta_t^* = \beta_t^*(N_t; w_1, \ldots, w_{N_t}, q_1, \ldots, q_{N_t})$$

are estimates of α_t and β_t derived in some fashion from the number N_t of photoevents in $[t_0, t)$, their occurrence times $w_1, \ldots, w_{N_t}$, and their quadrants $q_1, \ldots, q_{N_t}$. If the photoemissive surface in Figure 7.1 is centered by a feedback mechanism to the point $(a = \alpha_t^*, b = \beta_t^*)$ in an effort to maintain alignment between the transmitter and receiver, then (7.26) becomes

$$\lambda_{t,Q_i}(\alpha_t, \beta_t) = \int_{\mathcal{Q}_i^*}\int I(a, b | \alpha_t, \beta_t)\, da\, db, \tag{7.28}$$

where

$$\mathcal{Q}_1^* = \{a, b | 0 \leqslant a - \alpha_t^* \leqslant 1, 0 \leqslant b - \beta_t^* \leqslant 1\},$$

$$\mathcal{Q}_2^* = \{a, b | 0 \leqslant a - \alpha_t^* \leqslant 1, -1 \leqslant b - \beta_t^* \leqslant 0\},$$

$$\mathcal{Q}_3^* = \{a, b | -1 \leqslant a - \alpha_t^* \leqslant 0, -1 \leqslant b - \beta_t^* \leqslant 0\},$$

$$\mathcal{Q}_4^* = \{a, b | -1 \leqslant a - \alpha_t^* \leqslant 0, 0 \leqslant b - \beta_t^* \leqslant 1\}.$$

In this case, $\lambda_{t,Q_i}(\alpha_t, \beta_t)$ not only depends on the position processes, α_t and β_t, but also on the photoevents observed during $[t_0, t)$. General principles for forming the estimates α_t^* and β_t^* are developed below.

From (7.21), the *conditional* sample function density of a doubly stochastic, marked point process is given by

$$\begin{aligned} p[\{\mathbf{z}_\sigma; t_0 \leqslant \sigma < t\} | \mathbf{x}_\tau; \tau \geqslant t_0] &= \prod_i \exp\bigg[-\int_{t_0}^t \lambda_{\sigma, \mathbf{U}_i}(\mathbf{x}_\sigma)\, d\sigma \\ &\quad + \int_{t_0}^t \ln[\lambda_{\sigma,\mathbf{U}_i}(\mathbf{x}_\sigma)] N(d\sigma \times \{\mathbf{U}_i\})\bigg]. \end{aligned} \tag{7.29}$$

The *unconditional* sample function density is not as easily specified. At least in principle, it can be determined by evaluating the expectation

$$p[\{\mathbf{z}_\sigma; t_0 \leqslant \sigma < t\}] = E(p[\{\mathbf{z}_\sigma; t_0 \leqslant \sigma < t\} | \mathbf{x}_\tau; \tau \geqslant t_0]) \tag{7.30}$$

of (7.29) with respect to the distribution of the random variables $\{\mathbf{x}_\tau; \tau \geq t_0\}$. However, this method, termed the *method of conditioning* in Section 6.3, is usually beset with analytical complications so that $p[\{\mathbf{z}_\sigma; t_0 \leq \sigma < t\}]$ can only be determined in especially simple situations. We now show that under the condition

$$E\left[\int_t^{t+\tau} \lambda_{\sigma,\mathbf{U}_i}(\mathbf{x}_\sigma)\, d\sigma\right] < \infty, \tag{7.31}$$

for $t \geq t_0$ and $\mathbf{U}_i \in \mathfrak{U}$, the sample function density $p[\{\mathbf{z}_\sigma; t_0 \leq \sigma < t\}]$ can be represented as

$$p[\{\mathbf{z}_\sigma; t_0 \leq \sigma < t\}]$$

$$= \prod_i \exp\left[-\int_{t_0}^{t} \hat{\lambda}_{\sigma,\mathbf{U}_i}\, d\sigma + \int_{t_0}^{t} \ln \hat{\lambda}_{\sigma,\mathbf{U}_i} N(d\sigma \times \{\mathbf{U}_i\})\right], \tag{7.32}$$

where

$$\hat{\lambda}_{t,\mathbf{U}_i} = E\left[\lambda_{t,\mathbf{U}_i}(\mathbf{x}_t) | N_t; w_1, \ldots, w_{N_t}, \mathbf{u}_1, \ldots, \mathbf{u}_{N_t}\right]$$

is the minimum mean square-error estimate of $\lambda_{t,\mathbf{U}_i}(\mathbf{x}_t)$ in terms of observations of the marked point process during $[t_0, t)$. In a certain sense, (7.32) shows how to interchange the expectation and exponential function when (7.29) is introduced into (7.30). Equation (7.32) will not be an explicit expression for $p[\{\mathbf{z}_\sigma; t_0 \leq \sigma < t\}]$ unless the conditional expectation $\hat{\lambda}_{t,\mathbf{U}_i}$ can be evaluated, a task that is no less difficult than evaluating the expectation in (7.30). Even so, the representation (7.32) is quite useful conceptually and in application as well. In parallel with the philosophy in Section 6.4 for the doubly stochastic Poisson process, it leads immediately to a *separation theorem of detection* for observed, doubly stochastic, marked point processes. General principles for forming the estimates $\hat{\lambda}_{t,\mathbf{U}_i}$ are developed below when we study the posterior statistics of the information process given observations of the marked point process it influences.

Our approach for establishing (7.32) is to show that the function defined in (7.4) equals $\hat{\lambda}_{t,\mathbf{U}_i}$ for a doubly stochastic marked point process. This fact with (7.21) implies (7.32). Thus, we need to evaluate the functions $\mu_{t,\mathbf{U}_i}(0)$ and $\mu_{t,\mathbf{U}_i}(N_t; w_1, \ldots, \mathbf{w}_{N_t}, \mathbf{u}_1, \ldots, \mathbf{u}_{N_t})$ defined in (7.3). First, consider $N_t \geq 1$. Let $\{\rho_n\}$ be a sequence of positive numbers converging to zero as n tends to infinity, and define

$$f_n = \rho_n^{-1} Pr\Big[N([t, t+\rho_n) \times \{\mathbf{U}_i\}) = 1 | N_t; w_1, \ldots, w_{N_t}, \mathbf{u}_1, \ldots$$

$$\ldots, \mathbf{u}_{N_t}, \mathbf{x}_\tau; \tau \geq t_0\Big]. \tag{7.33}$$

Then, $E(f_n) < \infty$ and, according to (7.25b), the sequence $\{f_n\}$ converges to $\mu_{t,\mathbf{U}_i}(N_t; w_1, \dots, w_{N_t}, \mathbf{u}_1, \dots, \mathbf{u}_{N_t}, \mathbf{x}_t)$ with probability one as n tends to infinity. Also define

$$g_n = \rho_n^{-1} \int_t^{t+\rho_n} \mu_{\sigma,\mathbf{U}_i}(N_t; w_1, \dots, w_{N_t}, \mathbf{u}_1, \dots, \mathbf{u}_{N_t}, \mathbf{x}_t)\, d\sigma. \tag{7.34}$$

Then, $E(g_n) < \infty$ by assumption (7.31). Clearly the sequence $\{g_n\}$ converges to $\mu_{t,\mathbf{U}_i}(N_t; w_1, \dots, w_{N_t}, \mathbf{u}_1, \dots, \mathbf{u}_{N_t}, \mathbf{x}_t)$ with probability one as n tends to infinity. Also,

$$\lim_{n\to\infty} E(g_n) = E\left[\mu_{t,\mathbf{U}_i}(N_t; w_1, \dots, w_{N_t}, \mathbf{u}_1, \dots, \mathbf{u}_{N_t}, \mathbf{x}_t)\right].$$

Finally, we note by using the inequality $x \geqslant 1 - \exp(-x)$ and (7.10) that

$$g_n \geqslant \rho_n^{-1}\Big\{1 - Pr\big[N([t, t+\rho_n) \times \{\mathbf{U}_i\}) = 0 | N_t; w_1, \dots, w_{N_t}, \mathbf{u}_1, \dots$$
$$\dots, \mathbf{u}_{N_t}, \mathbf{x}_\sigma; \sigma \geqslant t_0\Big\} \geqslant f_n. \tag{7.35}$$

Consequently, the sequences $\{f_n\}$ and $\{g_n\}$ satisfy the conditions of the generalized bounded convergence theorem, Theorem 1.4.1, and we conclude, therefore, that

$$\begin{aligned}
&\mu_{t,\mathbf{U}_i}(N_t; w_1, \dots, w_{N_t}, \mathbf{u}_1, \dots, \mathbf{u}_{N_t}) \\
&\triangleq \lim_{n\to\infty} \rho_n^{-1} Pr\left[N([t, t+\rho_n) \times \{\mathbf{U}_i\}) = 1 | N_t; w_1, \dots, w_{N_t}, \mathbf{u}_1, \dots, \mathbf{u}_{N_t}\right] \\
&= \lim_{n\to\infty} \rho_n^{-1} E\Big\{Pr\big[N([t, t+\rho_n) \times \{\mathbf{U}_i\}) = 1 | N_t; w_1, \dots, w_{N_t}, \mathbf{u}_1, \dots \\
&\qquad \dots, \mathbf{u}_{N_t}, \mathbf{x}_\sigma; \sigma \geqslant t_0\big] | N_t; w_1, \dots, w_{N_t}, \mathbf{u}_1, \dots, \mathbf{u}_{N_t}\Big\} \\
&= \lim_{n\to\infty} E[f_n | N_t; w_1, \dots, w_{N_t}, \mathbf{u}_1, \dots, \mathbf{u}_{N_t}] \\
&= E\left[\lim_{n\to\infty} f_n | N_t; w_1, \dots, w_{N_t}, \mathbf{u}_1, \dots, \mathbf{u}_{N_t}\right] \\
&= E\left[\mu_{t,\mathbf{U}_i}(N_t; w_1, \dots, w_{N_t}, \mathbf{u}_1, \dots, \mathbf{u}_{N_t}, \mathbf{x}_t) | N_t; w_1, \dots, w_{N_t}, \mathbf{u}_1, \dots, \mathbf{u}_{N_t}\right] \\
&= \hat{\lambda}_{t,\mathbf{U}_i}.
\end{aligned} \tag{7.36}$$

Thus, the function defined in (7.4) equals $\hat{\lambda}_{t,\mathbf{U}_i}$ for $N_t \geqslant 1$. The derivation for $N_t = 0$ is analogous when f_n is defined to be

$$f_n = \rho_n^{-1} Pr\left[N([t,t+\rho_n)\times\{\mathbf{U}_i\}) = 1 \mid N_t = 0,\ \mathbf{x}_\tau; \tau \geqslant t_0\right].$$

The function defined in (7.4) therefore equals the minimum mean square-error estimate $\hat{\lambda}_{t,\mathbf{U}_i}$ of the randomized mark-intensity $\lambda_{t,\mathbf{U}_i}(\mathbf{x}_t)$. That (7.32) holds is then an immediate consequence of the expression for the sample-function density in (7.21).

Example 7.2.10 (Hypothesis Testing for Marked Point Processes). Suppose a marked point process observed on the interval $[t_0, T)$ has mark intensity $\lambda_{t,\mathbf{U}_i}^{(0)}(\mathbf{x}_t)$ or $\lambda_{t,\mathbf{U}_i}^{(1)}(\mathbf{x}_t)$ on hypotheses H_0 and H_1, respectively, where $\{\mathbf{x}_t; t \geqslant t_0\}$ is an underlying random process that influences the observed marks and occurrence times. Thus, given all the quantities $N_t, w_1, \ldots, w_{N_t}, \mathbf{u}_1, \ldots, \mathbf{u}_{N_t}, \{\mathbf{x}_\tau; \tau \geqslant t_0\}$, the probability a point occurs in $[t, t+\Delta t)$ is approximated to order Δt by $\Sigma_i \lambda_{t,\mathbf{U}_i}^{(0)}(\mathbf{x}_t)\Delta t$ on H_0 and by $\Sigma_i \lambda_{t,\mathbf{U}_i}^{(1)}(\mathbf{x}_t)\Delta t$ on H_1. If, in addition, it is given that a point does occur in $[t, t+\Delta t)$, the probability its mark is $\mathbf{U}_i$ is approximated by

$$\lambda_{t,\mathbf{U}_i}^{(0)}(\mathbf{x}_t)\left[\sum_j \lambda_{t,\mathbf{U}_j}^{(0)}(\mathbf{x}_t)\right]^{-1}$$

on H_0 and by

$$\lambda_{t,\mathbf{U}_i}^{(1)}(\mathbf{x}_t)\left[\sum_j \lambda_{t,\mathbf{U}_j}^{(1)}(\mathbf{x}_t)\right]^{-1}$$

on H_1. On the other hand, if we are given only $N_t, w_1, \ldots, w_{N_t}, \mathbf{u}_1, \ldots, \mathbf{u}_{N_t}$, which would be the data obtained from an observation of the marked point process, then the probability a point occurs in $[t, t+\Delta t)$ is approximated by $\Sigma_i \hat{\lambda}_{t,\mathbf{U}_i}^{(0)}\Delta t$ on H_0 and by $\Sigma_i \hat{\lambda}_{t,\mathbf{U}_i}^{(1)}\Delta t$ on H_1 where

$$\hat{\lambda}_{t,\mathbf{U}_i}^{(j)} = E\left[\lambda_{t,\mathbf{U}_i}^{(j)}(\mathbf{x}_t) \mid N_t; w_1, \ldots, w_{N_t}, \mathbf{u}_1, \ldots, \mathbf{u}_{N_t}; H_j\right]$$

for $j = 0, 1$. If, in addition, a point is known to occur in $[t, t+\Delta t)$, then the probability its mark is $\mathbf{U}_i$ is approximated by $\hat{\lambda}_{t,\mathbf{U}_i}^{(j)}[\Sigma_k \hat{\lambda}_{t,\mathbf{U}_k}^{(j)}]^{-1}$ on H_j, $j = 0, 1$.

From Section 2.5 of Chapter 2, we know that the strategy which minimizes the average cost of deciding between H_0 and H_1 is to evaluate the likelihood ratio

$$\Lambda[\{\mathbf{z}_\sigma; t_0 \leqslant \sigma < T\}] = \frac{p[\{\mathbf{z}_\sigma; t_0 \leqslant \sigma < T\} \mid H_1] Pr(H_1)}{p[\{\mathbf{z}_\sigma; t_0 \leqslant \sigma < T\} \mid H_0] Pr(H_0)}$$

for the observed realization of the marked point process and compare the result to a predetermined constant that depends only on the costs and not the observations. By using (7.32), we see that the test statistic is given by

$$\ln \Lambda[\{\mathbf{z}_\sigma; t_0 \leqslant \sigma < T\}]$$

$$= \sum_i \left\{ -\int_{t_0}^{T} \left[\hat{\lambda}^{(1)}_{\sigma,\mathbf{U}_i} - \hat{\lambda}^{(0)}_{\sigma,\mathbf{U}_i}\right] d\sigma \right.$$

$$\left. + \int_{t_0}^{T} \ln\left[\frac{\hat{\lambda}^{(1)}_{\sigma,\mathbf{U}_i}}{\hat{\lambda}^{(0)}_{\sigma,\mathbf{U}_i}}\right] N(d\sigma \times \{\mathbf{U}_i\}) \right\}. \tag{7.37}$$

This expression generalizes (3.44) for general marked point processes. In parallel with the philosophy of Section 6.4 of Chapter 6, it may be interpreted as a *separation theorem of detection* for observed, marked point processes.

Filtering for Marked Point Processes: Denumerable Mark Space

Estimation problems for observed marked point processes are of basic theoretical and practical importance. As we have just seen, minimum mean square-error estimates of the randomized mark intensities $\lambda_{t,\mathbf{U}_i}(\mathbf{x}_t)$ appear as a natural component in the sample function density of a marked point process. The separation theorem of detection brings out the fundamental role of these estimates in hypothesis testing. The estimation problem arises in practice whenever it is desired to determine the information process that modulates the mark intensities.

We now develop general principles for estimating the information process, or specified functions of the information process, in terms of observations of the marked point process it influences. We emphasize the *filtering problem*, which is the problem of estimating $\mathbf{x}_t$ given the marked point process on $[t_0, t)$. The development generalizes Section 6.5, where we studied the problem for doubly stochastic Poisson processes. The present model includes as special cases doubly stochastic versions of all the following processes: Poisson processes, compound Poisson processes, self-exciting and mutually exciting point processes, renewal point processes, birth-death processes, Markov jump processes, and semi-Markov jump processes. We will assume that $\{\mathbf{x}_t; t \geqslant t_0\}$ is a Markov process; this is not overly restrictive since Markov processes include a wide variety of useful special cases (see Section 6.5.2). As in Section 6.5.2, our objective is to specify the conditional characteristic function of $\mathbf{x}_t$ given $N_t, w_1, \ldots, w_{N_t}$, $\mathbf{u}_1, \ldots, \mathbf{u}_{N_t}$. This function completely characterizes $\mathbf{x}_t$ statistically in terms of the marked point process on $[t_0, t)$. Its use in estimating $\mathbf{x}_t$ under a variety of performance criteria will be identical to that in Section 6.5.2 for the doubly-stochastic Poisson process. We emphasize the minimum mean-square error criterion because of the basic role it seems to play in the theory.

Denote the conditional characteristic function of $\mathbf{x}_t$ given $N_t, w_1, \ldots, w_{N_t}, \mathbf{u}_1, \ldots, \mathbf{u}_{N_t}$ by $\hat{M}_t(jv)$, where

$$\hat{M}_t(j\mathbf{v}) = E\{\exp[j\langle \mathbf{v}, \mathbf{x}_t\rangle] | N_t; w_1, \ldots, w_{N_t}, \mathbf{u}_1, \ldots, \mathbf{u}_{N_t}\}. \tag{7.38}$$

Recall that $\langle \mathbf{v}_1, \mathbf{v}_2\rangle$ denotes the inner or dot product of the vectors $\mathbf{v}_1$ and $\mathbf{v}_2$. The main result characterizing $\hat{M}_t(j\mathbf{v})$ is contained in the following theorem, which generalizes Theorem 6.5.2.

Theorem 7.2.1 (Evolution of Posterior Statistics). Consider a marked point process with occurrence times in $[t_0, \infty)$ and marks in $\mathfrak{U} = \{\mathbf{U}_1, \mathbf{U}_2, \ldots\}$. Suppose the mark intensities $\lambda_{t,\mathbf{U}_i}(\mathbf{x}_t)$ are positive functions of an information process $\{\mathbf{x}_t; t \geq t_0\}$ and that $E[\Sigma_i \lambda_{t,\mathbf{U}_i}(\mathbf{x}_t)] < \infty$. Here, $\{\mathbf{x}_t; t \geq t_0\}$ is a Markov process satisfying the assumptions of Section 6.5.2 and is completely characterized by the differential generator $\psi_t(\mathbf{v}|\mathbf{x}_t)$ defined in (6.109). Then the conditional characteristic function $\hat{M}_t(j\mathbf{v})$ of $\mathbf{x}_t$ satisfies the following stochastic differential-equation:

$$d\hat{M}_t(j\mathbf{v}) = \hat{E}\{\exp[j\langle \mathbf{v}, \mathbf{x}_t\rangle]\psi_t(\mathbf{v}|\mathbf{x}_t)\}\, dt$$

$$+ \sum_i \hat{E}\left\{\exp[j\langle \mathbf{v}, \mathbf{x}_t\rangle]\left(\lambda_{t,\mathbf{U}_i}(\mathbf{x}_t) - \hat{\lambda}_{t,\mathbf{U}_i}\right)\right\}\hat{\lambda}_{t,\mathbf{U}_i}^{-1}\left[N(dt \times \{\mathbf{U}_i\}) - \hat{\lambda}_{t,\mathbf{U}_i}\, dt\right],$$

$$M_{t_0}(j\mathbf{v}) = E\{\exp[j\langle \mathbf{v}, \mathbf{x}_{t_0}\rangle], \tag{7.39}$$

where $(\hat{\cdot})$ and $\hat{E}\{(\cdot)\}$ denote the conditional expectation of $(\cdot)$ given $N_t, w_1, \ldots, w_{N_t}, \mathbf{u}_1, \ldots, \mathbf{u}_{N_t}$.

Proof. The proof of Theorem 7.2.1 closely parallels the proof of Theorem 6.5.2, which gives a similar result for the doubly-stochastic Poisson process. In the proof, we use the notation $\mathfrak{B}_t$ for the realization of the marked point process on $[t_0, t)$; thus $\mathfrak{B}_t$ denotes the random variables $N_t, w_1, \ldots, w_{N_t}, \mathbf{u}_1, \ldots, \mathbf{u}_{N_t}$. Corresponding to an infinitesimal time interval $[t, t+\Delta t)$ there is an increment $\Delta\hat{M}_t(j\mathbf{v}) = \hat{M}_{t+\Delta t}(j\mathbf{v}) - \hat{M}_t(j\mathbf{v})$ in the conditional characteristic function for the information process, where

$$\hat{M}_{t+\Delta t}(j\mathbf{v}) = E\{\exp[j\langle \mathbf{v}, \mathbf{x}_t\rangle]\exp[j\langle \mathbf{v}, \Delta\mathbf{x}_t\rangle] \,|\, \mathcal{B}_{t+\Delta t}\}$$

$$= \int \exp[j\langle \mathbf{v}, \mathbf{X}\rangle]\exp[j\langle \mathbf{v}, \mathbf{Y}\rangle]\, dP(\mathbf{X}, \mathbf{Y} \,|\, \mathcal{B}_{t+\Delta t}). \tag{7.40}$$

Here, $\Delta\mathbf{x}_t = \mathbf{x}_{t+\Delta t} - \mathbf{x}_t$ is the increment in the information process and $P(\mathbf{X}, \mathbf{Y} \,|\, \mathcal{B}_{t+\Delta t})$ is the joint conditional probability distribution function for $\mathbf{x}_t$ and $\Delta\mathbf{x}_t$ given $\mathcal{B}_{t+\Delta t}$, the realization of the marked point process on $[t_0, t+\Delta t)$.

In order to evaluate $\hat{M}^{\Delta}_{t+\Delta t}(j\mathbf{v})$, let us define

$$\hat{M}^{\Delta}_{t+\Delta t}(j\mathbf{v}) = \int \exp[j\langle \mathbf{v}, \mathbf{X}\rangle]\exp[j\langle \mathbf{v}, \mathbf{Y}\rangle]\, dP(\mathbf{X}, \mathbf{Y} \,|\, \mathcal{B}_t, (\Delta N_t, \Delta\mathbf{z}_t)), \tag{7.41}$$

in which $P(\mathbf{X}, \mathbf{Y} \,|\, \mathcal{B}_t, (\Delta N_t, \Delta\mathbf{z}_t))$ is the joint, conditional probability distribution function for $\mathbf{x}_t$ and $\Delta\mathbf{x}_t$ given $\mathcal{B}_t$ and $(\Delta N_t, \Delta\mathbf{z}_t)$, where $\Delta N_t = N_{t+\Delta t} - N_t$ is the number of points in $[t, t+\Delta t)$ regardless of their mark and $\Delta\mathbf{z}_t = \mathbf{z}_{t+\Delta t} - \mathbf{z}_t$ is the increment in the mark-accumulator process on $[t, t+\Delta t)$. We first evaluate $\hat{M}^{\Delta}_{t+\Delta t}$ and then get $\hat{M}_{t+\Delta t}$ from this evaluation by using the fact that $\hat{M}^{\Delta}_{t+\Delta t}$ approaches $\hat{M}_{t+\Delta t}$ with probability one as Δt tends to zero; this holds because $(\mathcal{B}_t, (\Delta N_t, \Delta\mathbf{z}_t))$ converges to $\mathcal{B}_{t+\Delta t}$ as Δt tends to zero, which implies by the Martingale convergence theorem (see W. Feller [3, p. 236]) that $P(\mathbf{X}, \mathbf{Y} \,|\, \mathcal{B}_t, (\Delta N_t, \Delta\mathbf{z}_t))$ approaches $P(\mathbf{X}, \mathbf{Y} \,|\, \mathcal{B}_{t+\Delta t})$ with probability one as $\Delta t \downarrow 0$.

We evaluate $\hat{M}^{\Delta}_{t+\Delta t}$ for fixed $(\Delta N_t, \Delta\mathbf{z}_t)$, namely for $(\Delta N_t, \Delta\mathbf{z}_t) = (0, \mathbf{0})$, $(\Delta N_t, \Delta\mathbf{z}_t) = (1, \mathbf{U}_i)$, and for $(\Delta N_t, \Delta\mathbf{z}_t) = (m, \Sigma_{j=1}^{m} \mathbf{U}_{i_j})$, where $m \geq 2$. For $(\Delta N_t, \Delta\mathbf{z}_t) = (0, \mathbf{0})$, we have

$$\hat{M}^{\Delta}_{t+\Delta t}(j\mathbf{v}) = \int \exp[j\langle \mathbf{v}, \mathbf{X}\rangle]\exp[j\langle \mathbf{v}, \mathbf{Y}\rangle]\, dP(\mathbf{X}, \mathbf{Y} \,|\, \mathcal{B}_t, \Delta N_t = 0).$$

The finite difference

$$\delta P(\mathbf{X}, \mathbf{Y} \,|\, \mathcal{B}_t, \Delta N_t = 0) \triangleq Pr(\mathbf{x}_t \in (\mathbf{X} - \delta\mathbf{X}, \mathbf{X}], \Delta\mathbf{x}_t \in (\mathbf{Y} - \delta\mathbf{Y}, \mathbf{Y}] \,|\, \mathcal{B}_t, \Delta N_t = 0)$$

can be evaluated using Bayes' rule as

$$\delta P(\mathbf{X}, \mathbf{Y} \,|\, \mathcal{B}_t, \Delta N_t = 0) = Pr(\Delta N_t = 0 \,|\, \mathbf{x}_t \in (\mathbf{X} - \delta\mathbf{X}, \mathbf{X}], \Delta\mathbf{x}_t \in (\mathbf{Y} - \delta\mathbf{Y}, \mathbf{Y}], \mathcal{B}_t)$$

$$\times \frac{\delta P(\mathbf{X}, \mathbf{Y} \,|\, \mathcal{B}_t)}{Pr(\Delta N_t = 0 \,|\, \mathcal{B}_t)},$$

where as $\delta\mathbf{X} \to \mathbf{0}$ and $\delta\mathbf{Y} \to \mathbf{0}$

$$Pr(\Delta N_t = 0 \,|\, \mathbf{x}_t \in (\mathbf{X} - \delta\mathbf{X}, \mathbf{X}], \Delta\mathbf{x}_t \in (\mathbf{Y} - \delta\mathbf{Y}, \mathbf{Y}], \mathcal{B}_t)$$

$$= 1 - \sum_i \lambda_{t, \mathbf{U}_i}(\mathbf{X})\Delta t + o(\Delta t). \tag{7.42}$$

and where

$$Pr(\Delta N_t=0|\mathcal{B}_t) = E\{Pr(\Delta N_t=0|\mathbf{x}_t,\mathcal{B}_t)|\mathcal{B}_t\}$$
$$=1-\sum_i \hat{\lambda}_{t,\mathbf{U}_i}\Delta t+o(\Delta t), \tag{7.43}$$

where $o(\Delta t)/\Delta t \to 0$ as $\Delta t \downarrow 0$. By putting these results together, we obtain for $\Delta N_t=0$ that

$$\hat{M}^{\Delta}_{t+\Delta t}(j\mathbf{v})$$
$$=E\Bigg\{\exp[j\langle\mathbf{v},\mathbf{x}_t\rangle]\exp[j\langle\mathbf{v},\Delta\mathbf{x}_t\rangle]$$
$$\times\Big(1-\sum_i \lambda_{t,\mathbf{U}_i}(\mathbf{x}_t)\Delta t\Big)|\mathcal{B}_t\Bigg\}\Big(1-\sum_i \hat{\lambda}_{t,\mathbf{U}_i}\Delta t\Big)^{-1}+o(\Delta t)$$
$$=E\Bigg\{\exp[j\langle\mathbf{v},\mathbf{x}_t\rangle]\exp[j\langle\mathbf{v},\Delta\mathbf{x}_t\rangle]$$
$$\times\Big(1-\sum_i\Big[\lambda_{t,\mathbf{U}_i}(\mathbf{x}_t)-\hat{\lambda}_{t,\mathbf{U}_i}\Big]\Delta t\Big)|\mathcal{B}_t\Bigg\}+o(\Delta t). \tag{7.44}$$

Similarly, by using this same procedure, we obtain for $(\Delta N_t,\Delta\mathbf{z}_t)=(1,\mathbf{U}_i)$ that

$$\hat{M}^{\Delta}_{t+\Delta t}(j\mathbf{v})$$
$$=E\{\exp[j\langle\mathbf{v},\mathbf{x}_t\rangle]\exp[j\langle\mathbf{v},\Delta\mathbf{x}_t\rangle]\lambda_{t,\mathbf{U}_i}(\mathbf{x}_t)|\mathcal{B}_t\}\hat{\lambda}^{-1}_{t,\mathbf{U}_i}+o(\Delta t) \tag{7.45}$$

and for $(\Delta N_t,\Delta\mathbf{z}_t)=(m,\Sigma_{j=1}^{m}\mathbf{U}_{i_j})$, $m\geqslant 2$, that

$$\hat{M}^{\Delta}_{t+\Delta t}(j\mathbf{v})=o(\Delta t). \tag{7.46}$$

It then follows that for $(\Delta N_t,\Delta\mathbf{z}_t)$ variable, we have

$$\hat{M}^{\Delta}_{t+\Delta t}(j\mathbf{v})$$
$$=E\Bigg\{\exp[j\langle\mathbf{v},\mathbf{x}_t\rangle]\exp[j\langle\mathbf{v},\Delta\mathbf{x}_t\rangle]$$
$$\times\Big(1-\sum_i\Big[\lambda_{t,\mathbf{U}_i}(\mathbf{x}_t)-\hat{\lambda}_{t,\mathbf{U}_i}\Big]\Delta t\Big)|\mathcal{B}_t\Bigg\}\delta_{0,\Delta N_t}$$
$$+E\{\exp[j\langle\mathbf{v},\mathbf{x}_t\rangle]\exp[j\langle\mathbf{v},\Delta\mathbf{x}_t\rangle]\lambda_{t,\Delta\mathbf{z}_t}(\mathbf{x}_t)|\mathcal{B}_t\}$$
$$\times\hat{\lambda}^{-1}_{t,\Delta\mathbf{z}_t}\delta_{1,\Delta N_t}+o(\Delta t). \tag{7.47}$$

Here, δ_{ij} is the Kronecker delta function. Because $\delta_{0,\Delta N_t}$ approaches $1-\Delta N_t$ and $\delta_{1,\Delta N_t}$ approaches $\Delta N_t = \Sigma_i N([t,t+\Delta t)\times\{\mathbf{U}_i\})$ with probability one as Δt approaches zero, we can rewrite (7.47) as

$$\hat{M}^{\Delta}_{t+\Delta t}(j\mathbf{v}) = E\{\exp[j\langle\mathbf{v},\mathbf{x}_t\rangle]\exp[j\langle\mathbf{v},\Delta\mathbf{x}_t\rangle]|\mathcal{B}_t\}$$
$$+\sum_i E\Big\{\exp[j\langle\mathbf{v},\mathbf{x}_t\rangle]\exp[j\langle\mathbf{v},\Delta\mathbf{x}_t\rangle]\big(\lambda_{t,\mathbf{U}_i}(\mathbf{x}_t)-\hat{\lambda}_{t,\mathbf{U}_i}\big)|\mathcal{B}_t\Big\}$$
$$\times\hat{\lambda}^{-1}_{t,\mathbf{U}_i}\Big[N([t,t+\Delta t)\times\{\mathbf{U}_i\})-\hat{\lambda}_{t,\mathbf{U}_i}\Delta t\Big]+o(\Delta t). \quad (7.48)$$

By subtracting the conditional characteristic function for $\mathbf{x}_t$ given $\mathcal{B}_t$, we therefore have

$$\hat{M}^{\Delta}_{t+\Delta t}(j\mathbf{v})-\hat{M}_t(j\mathbf{v})$$
$$= E\Big\{\exp[j\langle\mathbf{v},\mathbf{x}_t\rangle](\Delta t)^{-1}E(\exp[j\langle\mathbf{v},\Delta\mathbf{x}_t\rangle]-1|\mathbf{x}_t,\mathcal{B}_t)|\mathcal{B}_t\Big\}\Delta t$$
$$+\sum_i E\Big\{\exp[j\langle\mathbf{v},\mathbf{x}_t\rangle]\exp[j\langle\mathbf{v},\Delta\mathbf{x}_t\rangle]\big(\lambda_{t,\mathbf{U}_i}(\mathbf{x}_t)-\hat{\lambda}_{t,\mathbf{U}_i}\big)|\mathcal{B}_t\Big\}$$
$$\times\hat{\lambda}^{-1}_{t,\mathbf{U}_i}\Big[N([t,t+\Delta t)\times\{\mathbf{U}_i\})-\hat{\lambda}_{t,\mathbf{U}_i}\Delta t\Big]+o(\Delta t). \quad (7.49)$$

However,

$$E(\exp[j\langle\mathbf{v},\Delta\mathbf{x}_t\rangle]-1|\mathbf{x}_t,\mathcal{B}_t)=E\{\exp[j\langle\mathbf{v},\Delta\mathbf{x}_t\rangle]-1|\mathbf{x}_t\}$$

because $\{\mathbf{x}_t; t\geqslant t_0\}$ is a Markov process. Thus, we can now establish Theorem 7.2.1 as follows:

$$d\hat{M}_t(j\mathbf{v}) = \lim_{\Delta t\downarrow 0}\Big[\hat{M}_{t+\Delta t}(j\mathbf{v})-\hat{M}_t(j\mathbf{v})\Big]$$
$$= \lim_{\Delta t\downarrow 0}\Big[\hat{M}^{\Delta}_{t+\Delta t}(j\mathbf{v})-\hat{M}_t(j\mathbf{v})\Big]$$
$$= E\{\exp[j\langle\mathbf{v},\mathbf{x}_t\rangle]\psi_t(\mathbf{v}|\mathbf{x}_t)|\mathcal{B}_t\}\,dt$$
$$+\sum_i E\Big\{\exp[j\langle\mathbf{v},\mathbf{x}_t\rangle]\big(\lambda_{t,\mathbf{U}_i}(\mathbf{x}_t)-\hat{\lambda}_{t,\mathbf{U}_i}\big)|\mathcal{B}_t\Big\}\hat{\lambda}^{-1}_{t,\mathbf{U}_i}$$
$$\times\Big[N(dt\times\{\mathbf{U}_i\})-\hat{\lambda}_{t,\mathbf{U}_i}dt\Big], \quad (7.50)$$

where the interchange of limit and expectation required for the first term in the last equality is justified by assumption (6.108) and the bounded convergence theorem, Theorem 1.4.1. This ends the proof of Theorem 7.2.1.

Equation (7.39) is our basic result for solving the nonlinear filtering problem for observed, doubly stochastic, marked point processes. It generalizes (6.122) to include marks and self-excitation. The procedure for implementing an estimate of some prescribed function $\mathbf{h}_t(\mathbf{x}_t)$ of $\mathbf{x}_t$ is clearly very complicated in general. First, it is necessary to solve (7.39) for $\hat{M}_t(j\mathbf{v})$ for the given observations of the marked point process on $[t_0, t)$. Then this characteristic function must be inverse Fourier transformed to determine the posterior probability density or distribution function for $\mathbf{x}_t$. Finally, the desired estimate must be determined by evaluating the expectation

$$\hat{\mathbf{h}}_t = E\left[\mathbf{h}_t(\mathbf{x}_t) | N_t; w_1, \ldots, w_{N_t}, \mathbf{u}_1, \ldots, \mathbf{u}_{N_t}\right].$$

These steps are generally intractable analytically so that numerical techniques will be required. By virtue of the similarity of (6.122) and (7.39), it should be evident that (7.39) is well suited for the development of numerical implementations of both optimal and suboptimal estimators. This is explored briefly in the following examples.

Example 7.2.11 (Estimation of a Markov Jump Process That Influences an Observed Marked Point Process). Suppose, as in Example 6.5.8, that the information process $\{x_t; t \geq t_0\}$ is a Markov jump process with states $\{\zeta_1, \zeta_2, \ldots\}$. Then, the function $\psi_t(v|x_t)$ required in (7.39) is given in (6.119). Substitution of this function into (7.39) shows that the conditional state-occupancy probabilities

$$\hat{\pi}_j(t) = Pr\left(x_t = \zeta_j | N_t; w_1, \ldots, w_{N_t}, \mathbf{u}_1, \ldots, \mathbf{u}_{N_t}\right)$$

of x_t given the marked point process on $[t_0, t)$ satisfy

$$\begin{aligned} d\hat{\pi}_j(t) &= \sum_k a_{kj}(t)\hat{\pi}_k(t)\,dt \\ &\quad + \hat{\pi}_j(t)\sum_i \left(\lambda_{t,\mathbf{U}_i}(\zeta_j) - \hat{\lambda}_{t,\mathbf{U}_i}\right)\hat{\lambda}_{t,\mathbf{U}_i}^{-1}\left[N(dt \times \{\mathbf{U}_i\}) - \hat{\lambda}_{t,\mathbf{U}_i}\,dt\right], \end{aligned}$$

$$\hat{\pi}_j(t_0) = Pr\left[x_{t_0} = \zeta_j\right], \tag{7.51}$$

where

$$\hat{\lambda}_{t,\mathbf{U}_i} = \sum_j \lambda_{t,\mathbf{U}_i}(\zeta_j)\hat{\pi}_j(t). \tag{7.52}$$

This equation for the evolution of $\hat{\pi}_j(t)$ generalizes (6.155) to accomodate both marks and self-excitation in the observed point process. Its use in defining an updating algorithm by which the conditional state-occupancy probabilities and estimates of the mark intensity processes can be generated numerically is precisely the same as that outlined in Example 6.5.8. A Yashin [19] examines a special case of the present model in which the observed marked point process corresponds to a Markov jump process.

Example 7.2.12 (Extimation of a Markov Diffusion Process That Influences an Observed Marked Point Process). Suppose the information process $\{\mathbf{x}_t; t \geqslant t_0\}$ is a Markov diffusion process satisfying (6.111). The function $\psi_t(\mathbf{v}|\mathbf{x}_t)$ required in (7.39) is given in (6.112). Substitution of this function into (7.39) and inverse Fourier transformation shows that the posterior probability density function $p_t(\mathbf{X}|\mathfrak{B}_t)$ of $\mathbf{x}_t$ given $\mathfrak{B}_t = \{N_t; w_1,\ldots,w_{N_t},\mathbf{u}_1,\ldots,\mathbf{u}_{N_t})$ satisfies

$$dp_t(\mathbf{x}|\mathfrak{B}_t) = L[p_t(\mathbf{X}|\mathfrak{B}_t)]\,dt$$

$$+p_t(\mathbf{X}|\mathfrak{B}_t)\sum_i \left(\lambda_{t,\mathbf{U}_i}(\mathbf{X}) - \hat{\lambda}_{t,\mathbf{U}_i}\right)\hat{\lambda}_{t,\mathbf{U}_i}^{-1}\left[N(dt\times\{\mathbf{U}_i\}) - \hat{\lambda}_{t,\mathbf{U}_i}dt\right], \tag{7.53}$$

where $L(\cdot)$ is the forward Kolmogorov differential operator defined in (6.14). This equation generalizes (6.142) to accomodate both marks and self-excitation in the observed point process. It is difficult to solve analytically but can lead to updating algorithms that are feasible to implement in situations of limited complexity. Moreover, both (7.53) and (7.39) provide possible points of departure in the development of suboptimal estimators that are useful in practice.

As an example illustrating this latter idea, suppose as in Section 6.5.3 that the information process is a Markov diffusion satisfying the stochastic differential equation

$$d\mathbf{x}_t = \mathbf{a}_t(\mathbf{x}_t)\,dt + \mathbf{B}_t\,d\mathbf{w}_t,$$

$$\mathbf{x}_{t_0} = \mathbf{x}_0, \tag{7.54}$$

in which $\{\mathbf{w}_t; t \geqslant t_0\}$ is a vector of independent Wiener processes such that $E(\mathbf{w}_t\mathbf{w}_u') = \mathbf{I}\min(t,u)$, and where the initial condition $\mathbf{x}_0$ is a random variable independent of $\{\mathbf{w}_t; t \geqslant t_0\}$. Denote the conditional mean and covariance of $\mathbf{x}_t$ given $\mathfrak{B}_t = \{N_t; w_1,\ldots,w_{N_t},\mathbf{u}_1,\ldots,\mathbf{u}_{N_t}\}$ by $\hat{\mathbf{x}}_t$ and $\hat{\boldsymbol{\Sigma}}_t$, respectively; thus, $\hat{\mathbf{x}}_t = E[\mathbf{x}_t|\mathfrak{B}_t]$ and $\hat{\boldsymbol{\Sigma}}_t = E[\mathbf{x}_t\mathbf{x}_t'|\mathfrak{B}_t] - \hat{\mathbf{x}}_t\hat{\mathbf{x}}_t'$. Then, by using

(7.39) and paralleling the development of (6.153) and (6.154), we obtain

$$d\hat{\mathbf{x}}_t = \hat{\mathbf{a}}_t\, dt + \sum_i \hat{E}\left\{(\mathbf{x}_t - \hat{\mathbf{x}}_t)\left(\lambda_{t,\mathbf{U}_i}(\mathbf{x}_t) - \hat{\lambda}_{t,\mathbf{U}_i}\right)\right\}$$

$$\times \hat{\lambda}_{t,\mathbf{U}_i}^{-1}\left[N(dt \times \{\mathbf{U}_i\}) - \hat{\lambda}_{t,\mathbf{U}_i}\, dt\right], \qquad \hat{\mathbf{x}}_{t_0} = E(\mathbf{x}_0) \quad (7.55)$$

and

$$d\hat{\boldsymbol{\Sigma}}_t = \hat{E}\left\{[\mathbf{a}_t(\mathbf{x}_t) - \hat{\mathbf{a}}_t](\mathbf{x}_t - \hat{\mathbf{x}}_t)' + (\mathbf{x}_t - \hat{\mathbf{x}}_t)[\mathbf{a}_t(\mathbf{x}_t) - \hat{\mathbf{a}}_t]' + \mathbf{B}_t\mathbf{B}_t'\right\} dt$$

$$+ \sum_i E\left\{(\mathbf{x}_t - \hat{\mathbf{x}}_t)(\mathbf{x}_t - \hat{\mathbf{x}}_t)'\left(\lambda_{t,\mathbf{U}_i}(\mathbf{x}_t) - \hat{\lambda}_{t,\mathbf{U}_i}\right)\right\}$$

$$\times \hat{\lambda}_{t,\mathbf{U}_i}^{-1}\left[N(dt \times \{\mathbf{U}_i\}) - \hat{\lambda}_{t,\mathbf{U}_i}\, dt\right]$$

$$- \sum_i \hat{E}\left\{(\mathbf{x}_t - \hat{\mathbf{x}}_t)\left(\lambda_{t,\mathbf{U}_i}(\mathbf{x}_t) - \hat{\lambda}_{t,\mathbf{U}_i}\right)\right\} \hat{E}\left\{(\mathbf{x}_t - \hat{\mathbf{x}}_t)'\left(\lambda_{t,\mathbf{U}_i}(\mathbf{x}_t) - \hat{\lambda}_{t,\mathbf{U}_i}\right)\right\}$$

$$\times \hat{\lambda}_{t,\mathbf{U}_i}^{-2} N(dt \times \{\mathbf{U}_i\}), \quad (7.56)$$

$$\hat{\boldsymbol{\Sigma}}_{t_0} = E(\mathbf{x}_0\mathbf{x}_0') - \quad E(\mathbf{x}_0)E(\mathbf{x}_0').$$

Now, by duplicating the approximations made in Section 6.5.3 to derive the suboptimal estimator defined by (6.164) and (6.165), we obtain from (7.55) and (7.56) that

$$d\mathbf{x}_t^* = \mathbf{a}_t(\mathbf{x}_t^*)\, dt + \frac{1}{2}\sum_j \sum_k \sigma_{jk}^* \frac{\partial^2 \mathbf{a}_t(\mathbf{x}_t^*)}{\partial x_j^* \partial x_k^*}\, dt$$

$$+ \boldsymbol{\Sigma}_t^* \left\{ \sum_i \left[\frac{\partial \lambda_{t,\mathbf{U}_i}(\mathbf{x}_t^*)}{\partial \mathbf{x}_t^*} \right] \lambda_{t,\mathbf{U}_i}^{-1}(\mathbf{x}_t^*) \right.$$

$$\left. \times \left[N(dt \times \{\mathbf{U}_i\}) - \lambda_{t,\mathbf{U}_i}(\mathbf{x}_t^*)\, dt \right] \right\}, \quad (7.57)$$

$$\mathbf{x}_{t_0}^* = E(\mathbf{x}_0)$$

where $\boldsymbol{\Sigma}_t^*$ is a symmetric matrix that satisfies the equation below, σ_{ij}^* is the i,j-element of $\boldsymbol{\Sigma}_t^*$, and $\partial\lambda(\mathbf{x})/\partial\mathbf{x}$ denotes the gradiant vector having as its

ith element $\partial\lambda(\mathbf{x})/\partial x_i$. The equation for $\boldsymbol{\Sigma}_t^*$ is

$$d\boldsymbol{\Sigma}_t^* = \left[\frac{\partial \mathbf{a}_t(\mathbf{x}_t^*)}{\partial \mathbf{x}_t^*}\right]'\boldsymbol{\Sigma}_t^*\,dt + \boldsymbol{\Sigma}_t^*\left[\frac{\partial \mathbf{a}_t(\mathbf{x}_t^*)}{\partial \mathbf{x}_t^*}\right]dt + \mathbf{B}_t\mathbf{B}_t'\,dt$$

$$+\boldsymbol{\Sigma}_t^*\left\{\sum_i\left[\frac{\partial^2 \ln\lambda_{t,\mathbf{U}_i}(\mathbf{x}_t^*)}{\partial \mathbf{x}_t^{*2}}\right]N(dt\times\{\mathbf{U}_i\})\right.$$

$$\left.-\sum_i\left[\frac{\partial^2 \lambda_{t,\mathbf{U}_i}(\mathbf{x}_t^*)}{\partial \mathbf{x}_t^{*2}}\right]dt\right\}\boldsymbol{\Sigma}_t^*, \tag{7.58}$$

$$\boldsymbol{\Sigma}_{t_0}^* = \operatorname{cov}(\mathbf{x}_0) = E(\mathbf{x}_0\mathbf{x}_0') - \quad E(\mathbf{x}_0)E(\mathbf{x}_0'),$$

where $\partial\mathbf{a}(\mathbf{x})/\partial\mathbf{x}$ denotes the Jacobian matrix having as its i-row, j-column element $\partial a_j(\mathbf{x})/\partial x_i$ and where $\partial^2\lambda(\mathbf{x})/\partial\mathbf{x}^2 = \partial[\partial\lambda(\mathbf{x})/\partial\mathbf{x}]\partial\mathbf{x}$. Here, $\mathbf{x}^*$ and $\boldsymbol{\Sigma}^*$ approximate $\hat{\mathbf{x}}$ and $\hat{\boldsymbol{\Sigma}}$ in the same sense as in Section 6.5.3. Thus, (7.57) and (7.58) generalize (6.164) and (6.165) to accomodate both marks and self-excitation in the observed point process. The approximation of $\mathbf{x}^*$ to $\mathbf{x}$ may be expected to be tight under conditions when the estimation error is small, but just what these conditions are is difficult to identify in general. Consequently, the usefulness of the suboptimal estimator defined by (7.57) and (7.58) must be investigated in individual applications.

Example 7.2.13 (Optical Position Sensing, continued). Consider Example 7.2.9 again. Assume that the photodetector is fixed so that the four mark intensities $\lambda_{t,Q_i}(\alpha_t,\beta_t)$, $i=1,2,3,4$, are given by (7.26). Also, assume that $\{\alpha_t; t\geq 0\}$ and $\{\beta_t; t\geq 0\}$ are independent, identically distributed Wiener processes such that $E(\alpha_t^2)=E(\beta_t^2)=t(2\tau_d)^{-1}$. The parameter τ_d characterizes the beam's stability; the beam undergoes a one unit (rms) displacement from the origin in τ_d units of time. We may then take the state vector $\mathbf{x}_t$ in (7.54) to be

$$\mathbf{x}_t = \begin{bmatrix}\alpha_t\\ \beta_t\end{bmatrix} = (2\tau_d)^{-1/2}\mathbf{w}_t,$$

and for this choice, $\mathbf{a}_t(\mathbf{x}_t)=\mathbf{0}$ and $\mathbf{B}_t=(2\tau_d)^{-1/2}\mathbf{I}$, where $\mathbf{I}$ is the identity matrix. Equation (7.57) becomes

$$d\begin{bmatrix} \alpha_t^* \\ \beta_t^* \end{bmatrix} = \Sigma_t^* \left\{ \sum_{i=1}^{4} \begin{bmatrix} \dfrac{\partial \lambda_{t,Q_i}(\alpha_t^*, \beta_t^*)}{\partial \alpha_t^*} \\ \dfrac{\partial \lambda_{t,Q_i}(\alpha_t^*, \beta_t^*)}{\partial \beta_t^*} \end{bmatrix} \lambda_{t,Q_i}^{-1}(\alpha_t^*, \beta_t^*)\Big[N(dt \times \{Q_i\}) \right.$$

$$\left. - \lambda_{t,Q_i}(\alpha_t^*, \beta_t^*)\, dt \Big] \right\}. \tag{7.59}$$

A schematic diagram for the position estimator specified by (7.59) is shown in Figure 7.2. The processes $\dot{N}([t_0,t)\times\{Q_i\})$, $i=1,2,3,4$, model the four pulse processes obtained from the four-quadrant photodetector; for example, an impulse in $\dot{N}([t_0,t)\times\{Q_1\})$ at time w corresponds to the pulse generated by a photoevent at time w in the first quadrant. The process Σ_t^* must be implemented according to (7.58).

Summary. We have seen in this section that marked point processes having a denumerable mark space subsume a wide variety of point and jump processes that have proven useful in practice. In particular, the following are all included as special cases:

1. the Poisson process of Chapter 2;
2. the compound Poisson process of Chapter 3 and its generalizations in Example 7.2.1, which include the compound renewal process;
3. the self-exciting point process of Chapter 5 and its multichannel extension in Example 7.2.6;
4. the generalized birth-death process of Example 7.2.3;
5. the denumerable-state Markov jump process of Example 7.2.5.

In addition, we have discussed doubly stochastic marked point processes in which an underlying information process influences the evolution of both marks and occurrence times. These processes include as special cases doubly-stochastic versions of (1–5) above.

For brevity, we have confined our investigation to certain aspects of the theory that relate to decision and estimation problems in which a marked point process models observed data. The principal results may be summarized as follows:

1. The sample-function density for a marked point process having a denumerable mark space is given in (7.21). In view of the greater

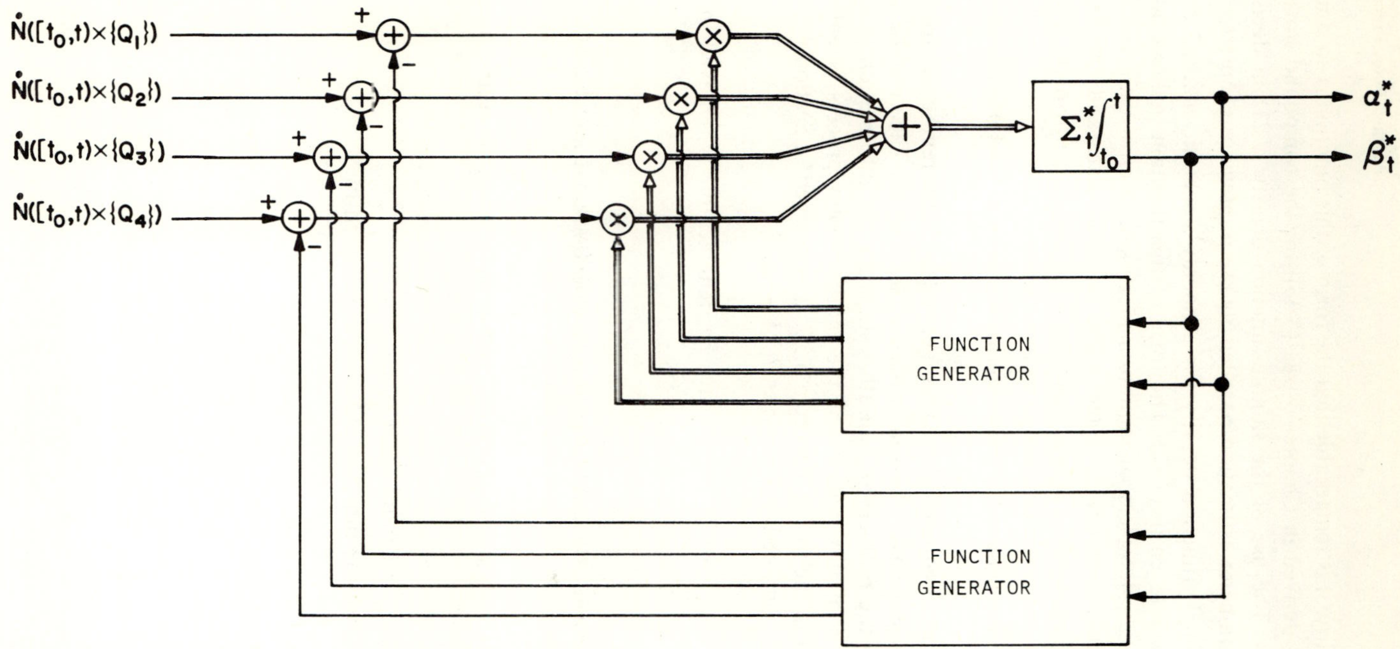

Figure 7.2. Position estimator for a four-quadrant photodetector.

generality, it is remarkable that the form of this function is so similar to that derived in Chapter 3 [see (3.36)] for the compound Poisson process. The use of the sample-function density (7.21) in decision and parameter estimation parallels its use in previous chapters.

2. The sample-function density for a doubly stochastic marked point process is given in (7.32). The form of this function makes evident the basic role of causal minimum mean square-error estimation. A separation theorem of detection is an immediate and important consequence of (7.32) that generalizes the separation theorem for the doubly stochastic Poisson process in Section 6.4.
3. A characterization for the posterior statistics of a Markovian information process that influences an observed marked point process is given in Theorem 7.2.1. This generalizes Theorem 6.5.2 for the doubly stochastic Poisson process. The role of Theorem 7.2.1 in solving the problem of causally estimating the information process in terms of observations of the marked point process is entirely parallel to the use of Theorem 6.5.2 as developed in Section 6.5 and is briefly outlined in Examples 7.2.11–7.2.13.

The restriction that the mark space be denumerable is removed in the next section.

7.3 MARKED POINT PROCESSES HAVING A NONDENUMERABLE MARK SPACE

Suppose now that the mark space $\mathfrak{U}$ is a region of a finite-dimensional Euclidean space. Marks then have a continuum of possible values. Assume that:

a. $\{N_t; t \geqslant t_0\}$ is conditionally orderly;
b. the limit of the function $a(\Delta t, \Delta \mathbf{U}, N_t)$ defined by

$$a(\Delta t, \Delta \mathbf{U}, N_t)$$

$$= \begin{cases} \left(\Delta t \prod_i \Delta U_i\right)^{-1} Pr[N([t,t+\Delta t)\times[\mathbf{U},\mathbf{U}+\Delta\mathbf{U}))=1|N_t], & N_t=0 \\ \left(\Delta t \prod_i \Delta U_i\right)^{-1} Pr[N([t,t+\Delta t)\times[\mathbf{U},\mathbf{U}+\Delta\mathbf{U}))=1|N_t; \\ \qquad w_1,\ldots,w_{N_t},\mathbf{u}_1,\ldots,\mathbf{u}_{N_t}], & N_t \geqslant 1 \end{cases} \tag{7.60}$$

exists as Δt and $|\Delta\mathbf{U}|$ tend to zero for almost every realization of the marked point process;

c. $\Pr[N_{t_0}=0]=1$.

The modification of Section 2 that accounts for a nondenumberable mark space is contained in assumption (b). Here, $[\mathbf{U},\mathbf{U}+\Delta\mathbf{U})$ is an elemental rectangle in $\mathfrak{U}$ of volume $\prod_i \Delta U_i$. The limit in (b) will be denoted by $\mu_{t,\mathbf{U}}(0)$ for $N_t=0$ and by $\mu_{t,\mathbf{U}}(N_t; w_1,\ldots,w_{N_t},\mathbf{u}_1,\ldots,\mathbf{u}_{N_t})$ for $N_t \geqslant 1$; thus,

$$\mu_{t,\mathbf{U}}(0)=\lim\left(\Delta t\prod_i \Delta U_i\right)^{-1} Pr\left[N([t+\Delta t)\times[\mathbf{U},\mathbf{U}+\Delta\mathbf{U})])=1|N_t\right] \quad (7.61)$$

for $N_t=0$, and

$$\mu_{t,\mathbf{U}}(N_t; w_1,\ldots,w_{N_t},\mathbf{u}_1,\ldots,\mathbf{u}_{N_t})$$
$$=\lim\left(\Delta t\prod_i \Delta U_i\right)^{-1} Pr\Big[N([t,t+\Delta t)\times[\mathbf{U},\mathbf{U}+\Delta\mathbf{U}))=1|N_t;$$
$$w_1,\ldots,w_{N_t},\mathbf{u}_1,\ldots,\mathbf{u}_{N_t}\Big] \quad (7.62)$$

for $N_t \geqslant 1$, where the limit is taken as Δt and $|\Delta\mathbf{U}|$ tend to zero. Define the intensity process $\{\lambda_{t,\mathbf{U}}; t \geqslant t_0, \mathbf{U}\in\mathfrak{U}\}$ by

$$\lambda_{t,\mathbf{U}}=\begin{cases}\mu_{t,\mathbf{U}}(0), & t_0 \leqslant t \leqslant w_1\\ \mu_{t,\mathbf{U}}(N_t; w_1,\ldots,w_{N_t},\mathbf{u}_1,\ldots,\mathbf{u}_{N_t}), & w_{N_t} < t \leqslant w_{N_t+1}.\end{cases} \quad (7.63)$$

This process is the counterpart of (7.4) for a nondenumerable mark space. It is a left-continuous function of t. The conditional probability of there being a point in $[t,t+\Delta t)$ with mark in $[\mathbf{U},\mathbf{U}+\Delta\mathbf{U})$ given the number of points in $[t_0,t)$, their occurrence times, and their marks is approximated by $\lambda_{t,\mathbf{U}}\Delta t\prod_i \Delta U_i$.

Here are some examples of processes included as special cases in this self-exciting, marked point process model.

Example 7.3.1 (Generalized Compound Poisson Process). Points occur in a compound Poisson process as a Poisson process and marks are assigned independently to each point. If v_t is the intensity of the Poisson process and if marks are continuous random variables with probability density function $p_{\mathbf{u}}(\mathbf{U})$, then $\lambda_{t,\mathbf{U}}$ has the form

$$\lambda_{t,\mathbf{U}}=p_{\mathbf{u}}(\mathbf{U})v_t.$$

In parallel with Example 7.2.1, there are various straightforward generalizations of this situation. For instance, if marks are assigned independently to points of a general self-exciting point process of Chapter 5, and if marks are continuous random variables with probability density function $p_{\mathbf{u}}(\mathbf{U})$, then $\lambda_{t,\mathbf{U}}$ has the form

$$\lambda_{t,\mathbf{U}} = \begin{cases} p_{\mathbf{u}}(\mathbf{U})\mu_t(0), & t_0 \leqslant t \leqslant w_1 \\ p_{\mathbf{u}}(\mathbf{U})\mu_t(N_t; w_1, \dots, w_{N_t}), & w_{N_t} < t \leqslant w_{N_t+1} \end{cases}$$

where the functions μ are defined in (5.2). A marked renewal process is a special case of interest in applications.

Example 7.3.2 (Nondenumerable-State, Markov Jump Process). Nondenumerable-state Markov jump processes are discussed by I. Gikhman and A. Skorokhod [6, pp. 311–319]. These may be imbedded in a marked point process in the manner of Example 7.2.4.

Example 7.3.3 (Randomly Sampled Random Process). Suppose $\{\mathbf{u}_t; t \geqslant t_0\}$ is a continuous random process with state space $\mathfrak{U}$ and that this process is sampled at event times of a self-exciting point process of the type studied in Chapter 5. The mark on a point occurring at time w_i is then $\mathbf{u}_{w_i}$. The function $\lambda_{t,\mathbf{U}}$ has the form

$$\lambda_{t,\mathbf{U}} = \begin{cases} p_{\mathbf{u}_t}(\mathbf{U})\mu_t(0), & t_0 \leqslant t \leqslant w_1 \\ p_{\mathbf{u}_t}(\mathbf{U}|\mathbf{u}_{w_1}, \dots, \mathbf{u}_{w_{N_t}})\mu_t(N_t; w_1, \dots, w_{N_t}), & w_{N_t} < t \leqslant w_{N_t+1} \end{cases}$$

where $p_{\mathbf{u}_t}(\mathbf{U})$ is the probability density function for $\mathbf{u}_t$ and $p_{\mathbf{u}_t}(\mathbf{U}|\mathbf{u}_{w_1}, \dots, \mathbf{u}_{w_n})$ is the conditional probability density function for $\mathbf{u}_t$ given $\mathbf{u}_{w_1}, \dots,$ and $\mathbf{u}_{w_n}$. We refer to F. Beutler and O. Leneman [1] and R. Kahn and B. Liu [7, 8] for details on the important practical problem of reconstructing $\{\mathbf{u}_t; t \geqslant t_0\}$ from its randomly timed samples.

Example 7.3.4 (Time-Space Point Processes). There are many and diverse applications in which points occur randomly in time and space. For instance, in nuclear medicine where the gamma camera (see Example 6.6.1) is used to monitor radioactivity in an organ under study, each detected gamma photon has both a temporal and spatial coordinate corresponding, respectively, to the time and position its absorption occurs in the crystal used to detect radiation. Seismic events occur randomly in time and in space on the earth's surface, as do lightning discharges, storms, and so forth. Photoevents in the two-dimensional photodetector of Example 3.3.6 have both a temporal and spatial coordinate. These time-space point processes are marked point processes in which the mark indicates the location of the point in the space $\mathfrak{U}$. The model we consider may, therefore, be viewed as a self-exciting time-space point process in which both the occurrence time and location of a point are influenced by the occurrence times and locations of all previous points.

All the results of Section 7.2 for marked point processes having a denumerable mark space have natural counterparts for a nondenumerable mark space. In fact, we shall see that the modification of the expressions for the sample function density in (7.21) and the conditional characteristic function in (7.39) is simply the replacement of the discrete sums over elements of the denumerable mark space by continuous sums (that is, integrals) over the elements of the nondenumerable mark space. The use of these quantities is precisely the same as in Section 7.2 and earlier chapters, so the discussion will be cursory.

Sample-Function Density: Nondenumerable Mark Space

In parallel with (3.45), the quantity $p[\{\mathbf{z}_\sigma; t_0 \leqslant \sigma < t\}]$ defined below is the sample-function density for a marked point process on $[t_0, t)$ with a nondenumerable mark space:

$$p[\{\mathbf{z}_\sigma; t_0 \leqslant \sigma < t\}] = \begin{cases} Pr(N_t = 0), & N_t = 0 \\ p_{\mathbf{w},\mathbf{u}_1,\ldots,\mathbf{u}_n}(\mathbf{W}, \mathbf{U}_1, \ldots, \mathbf{U}_n, \mathbf{N}_t = n), & N_t = n \geqslant 1 \end{cases} \tag{7.64}$$

where

$$p_{\mathbf{w},\mathbf{u}_1,\ldots,\mathbf{u}_n}(\mathbf{W}, \mathbf{U}_1, \ldots, \mathbf{U}_n) = \lim \left(\prod_{i=1}^{n} \Delta W_i |\Delta \mathbf{U}_i| \right)^{-1} Pr(E_n, N_t = n)$$

in which E_n denotes the event

$$\{N([t_0, W_1) \times \mathfrak{U}) = 0, N([W_1, W_1 + \Delta W_1) \times [\mathbf{U}_1, \mathbf{U}_1 + \Delta \mathbf{U}_1)) = 1\}$$

for $n = 1$ and the event

$$\begin{aligned} \{N([t_0, W_1) \times \mathfrak{U}) = 0, N([W_1, W_1 + \Delta W) \times [\mathbf{U}_1, \mathbf{U}_1 + \Delta \mathbf{U}_1)) = 1, \ldots \\ \ldots N([W_{n-1} + \Delta W_{n-1}, W_n) \times \mathfrak{U}) = 0, \\ N([W_n, W_n + \Delta W_n) \times [\mathbf{U}_n, \mathbf{U}_n + \Delta \mathbf{U}_n)) = 1\} \end{aligned}$$

for $n \geqslant 2$, and where the limit is taken as $\max_{1 \leqslant i \leqslant n}(\Delta W_i)$ and $\max_{1 \leqslant i \leqslant n} |\Delta \mathbf{U}_i|$ tend to zero. It is evident that $p[\{\mathbf{z}_\sigma; t_0 \leqslant \sigma < t\}]\Pi_i \Delta W_i |\Delta \mathbf{U}_i|$ may be interpreted roughly as the probability of obtaining a particular realization of the marked point process on the interval $[t_0, t)$ with $N_t = n$ points located at times $w_1 = W_1, \ldots, w_n = W_n$ and marks $\mathbf{u}_1 = \mathbf{U}_1, \ldots, \mathbf{u}_n = \mathbf{U}_n$.

Now, by paralleling the derivation of (7.21), we obtain the following expression for the sample-function density when the mark space is nondenumerable:

$$p[\{\mathbf{z}_\sigma; t_0 \leqslant \sigma < t\}] = \exp\left[-\int_{t_0}^{t} \int_{\mathfrak{U}} \lambda_{\sigma, \mathbf{U}} \, d\sigma \prod_i dU_i \right.$$

$$\left. + \int_{t_0}^{t} \int_{\mathfrak{U}} \ln \lambda_{\sigma, \mathbf{U}} N(d\sigma \times d\mathbf{U}) \right]. \tag{7.65}$$

From Example 7.3.1, the intensity function $\lambda_{t, \mathbf{U}}$ has the factored form $\lambda_{t, \mathbf{U}} = p_{\mathbf{u}}(\mathbf{U}) v_t$ for a compound Poisson process. In this case, (7.65) reduces to (3.48). More generally, the similarity in the forms of (7.65) and (3.48) is striking in view of the much greater generality of (7.65). Moreover, the use of (7.65) in estimation and decision problems involving observed marked point processes is precisely the same as the use of (3.48) for observed compound Poisson processes. This is outlined in the following example for parameter estimation.

Example 7.3.5 (Parameter Estimation for Marked Point Processes with a Nondenumerable Mark Space). This problem is motivated in Example 7.2.7 when the mark space is denumerable, and the situation is analogous here. Suppose the intensity process $\lambda_{t, \mathbf{U}}(\boldsymbol{\theta})$ is a function of unknown parameters $\boldsymbol{\theta}$ and that it is desired to estimate these parameters on the basis of observations of the marked point process on $[t_0, t)$. The maximum likelihood estimate $\hat{\boldsymbol{\theta}}_{\text{ML}}$ of $\boldsymbol{\theta}$ in terms of the data $N_t = n$, $w_1 = W_1, \ldots, w_n = W_n$, $\mathbf{u}_1 = \mathbf{U}_1, \ldots,$ and $\mathbf{u}_n = \mathbf{U}_n$ is the value of $\boldsymbol{\theta}$ maximizing the likelihood function

$$l(\boldsymbol{\theta}) = \ln p[\{\mathbf{z}_\sigma; t_0 \leqslant \sigma < t\} | \boldsymbol{\theta}]$$

$$= -\int_{t_0}^{t} \int_{\mathfrak{U}} \lambda_{\sigma, \mathbf{U}}(\boldsymbol{\theta}) \, d\sigma \prod_i dU_i + \int_{t_0}^{t} \int_{\mathfrak{U}} \ln \lambda_{\sigma, \mathbf{U}}(\boldsymbol{\theta}) N(d\sigma \times d\mathbf{U}), \tag{7.66}$$

where the right side is to be evaluated for the observed data. This maximization is subject to any constraints on the possible parameter values and will generally require numerical search.

As a particular example, consider the following marked point process obtained by randomly sampling a Wiener process at the times of an inhomogeneous Poisson process. Let $\{u_t; t \geqslant 0\}$ be a Wiener process with parameter θ. Thus, $u_0 = 0$, and u has independent, normally distributed increments; $u_t - u_s$ has zero mean and variance $\theta \cdot (t-s)$ for $0 \leqslant s \leqslant t$. Suppose u is observed at the event times of a Poisson process with intensity function v_t for $t \geqslant 0$, and on the basis of the observed values of u on the interval $[t_0, t)$, it is desired to estimate θ. From Example 7.3.3, we have

$$\lambda_{t,u}(\theta) = \frac{v_t}{\sqrt{2\pi\theta \cdot (t - w_{N_t})}} \exp\left[-\frac{(U - u_{w_{N_t}})^2}{2\theta \cdot (t - w_{N_t})} \right], \tag{7.67}$$

where we set $w_0 = 0$. Thus, from (7.66), the likelihood function for estimating θ is given by

$$l(\theta) = -\int_{t_0}^{t} v_\sigma \, d\sigma$$

for $N_t = 0$ and by

$$l(\theta) = -\int_{t_0}^{t} v_\sigma \, d\sigma + \sum_{i=1}^{N_t} \ln v_{w_i} - \frac{1}{2} \sum_{i=1}^{N_t} \ln [2\pi\theta \cdot (w_i - w_{i-1})]$$

$$- \frac{1}{2\theta} \sum_{i=1}^{N_t} \frac{(u_{w_i} - u_{w_{i-1}})^2}{(w_i - w_{i-1})}$$

for $N_t \geqslant 1$. We deduce from this that the maximum likelihood estimate $\hat{\theta}_{\mathrm{ML}}$ of θ in terms of the data $N_t = n \geqslant 1$, $w_1 = W_1, \ldots, w_n = W_n$, $u_{w_1} = U_1, \ldots$, and $u_{w_n} = U_n$ is given by

$$\hat{\theta}_{\mathrm{ML}} = \frac{1}{n} \sum_{i=1}^{n} \frac{(U_i - U_{i-1})^2}{(W_i - W_{i-1})}, \tag{7.68}$$

where $W_0 = U_0 = 0$.

Doubly Stochastic Marked Point Processes: Nondenumerable Mark Space

Suppose next that the intensity process $\{\lambda_{t,\mathbf{u}}(\mathbf{x}_t); t \geqslant t_0, \mathrm{U} \in \mathfrak{U}\}$ is a function of an information process $\{\mathbf{x}_t; t \geqslant t_0\}$. This extension is treated in a manner paralleling that in Section 7.2 for a denumerable mark space. The results are summarized as follows:

a. *Sample-Function Density.* The counterpart of (7.32) for a nondenumerable mark space is

$$p[\{\mathbf{z}_\sigma; t_0 \leqslant \sigma < t\}] = \exp\left[-\int_{t_0}^{t} \int_{\mathfrak{U}} \hat{\lambda}_{\sigma,\mathbf{U}}\, d\sigma \prod_i dU_i + \int_{t_0}^{t} \int_{\mathfrak{U}} \ln \hat{\lambda}_{\sigma,\mathbf{U}} N(d\sigma \times d\mathbf{U}) \right], \qquad (7.69)$$

where

$$\hat{\lambda}_{t,\mathbf{U}} = E[\lambda_{t,\mathbf{U}}(\mathbf{x}_t) | N_t; w_1, \ldots, w_{N_t}, \mathbf{u}_1, \ldots, \mathbf{u}_{N_t}]$$

is the minimum mean square-error estimate of $\lambda_{t,\mathbf{U}}(\mathbf{x}_t)$ in terms of the marked point process on $[t_0, t)$.

b. *Posterior Characteristic Function.* The counterpart of (7.39) for a nondenumerable mark space is

$$d\hat{M}_t(j\mathbf{v}) = \hat{E}\{\exp[j\langle \mathbf{v}, \mathbf{x}_t\rangle] \psi_t(\mathbf{v}|\mathbf{x}_t)\}\, dt + \int_{\mathfrak{U}} \hat{E}\left\{\exp[j\langle \mathbf{v}, \mathbf{x}_t\rangle](\lambda_{t,\mathbf{U}}(\mathbf{x}_t) - \hat{\lambda}_{t,\mathbf{U}})\right\} \times \hat{\lambda}_{t,\mathbf{U}}^{-1}\left[N(dt \times d\mathbf{U}) - \hat{\lambda}_{t,\mathbf{U}}\, dt \prod_i dU_i\right], \qquad (7.70)$$

$$\hat{M}_{t_0}(j\mathbf{v}) = E\{\exp[j\langle \mathbf{v}, \mathbf{x}_{t_0}\rangle]\},$$

where $(\hat{\cdot})$ and $\hat{E}\{(\cdot)\}$ denote the conditional expectation of $(\cdot)$ given $N_t, w_1, \ldots, w_{N_t}, \mathbf{u}_1, \ldots, \mathbf{u}_{N_t}$.

The proofs of (7.69) and (7.70) parallel the proofs of their counterparts (7.32) and (7.39) with only minor modifications. The use of these equations is the same as the use of (7.32) and (7.39) and, therefore, will not be explored in detail. We mention only that by starting with (7.70) and following the procedure used to derive (7.57), the equations we obtain for the suboptimal estimation of the Markov diffusion process (7.54) are

$$d\mathbf{x}_t^* = \mathbf{a}_t(\mathbf{x}_t^*)\,dt + \frac{1}{2}\sum_j \sum_k \sigma_{jk}^* \frac{\partial^2 \mathbf{a}_t(\mathbf{x}_t^*)}{\partial x_j^* \partial x_k^*}\,dt$$

$$+\boldsymbol{\Sigma}_t^* \int_{\mathfrak{U}} \left[\frac{\partial \lambda_{t,\mathbf{U}}(\mathbf{x}_t^*)}{\partial \mathbf{x}_t^*}\right] \lambda_{t,\mathbf{U}}^{-1}(\mathbf{x}_t^*) \left[N(dt \times d\mathbf{U}) - \lambda_{t,\mathbf{U}}(\mathbf{x}_t^*)\,dt \prod_i dU_i\right], \tag{7.71}$$

$\mathbf{x}_{t_0}^* = E(\mathbf{x}_0)$,

$$d\boldsymbol{\Sigma}_t^* = \left[\frac{\partial \mathbf{a}_t(\mathbf{x}_t^*)}{\partial \mathbf{x}_t^*}\right]' \boldsymbol{\Sigma}_t^*\,dt + \boldsymbol{\Sigma}_t^* \left[\frac{\partial \mathbf{a}_t(\mathbf{x}_t^*)}{\partial \mathbf{x}_t^*}\right] dt + \mathbf{B}_t \mathbf{B}_t'\,dt$$

$$+\boldsymbol{\Sigma}_t^* \left\{ \int_{\mathfrak{U}} \left[\frac{\partial^2 \ln \lambda_{t,\mathbf{U}}(\mathbf{x}_t^*)}{\partial \mathbf{x}_t^{*2}}\right] N(dt \times d\mathbf{U}) \right.$$

$$\left. - \int_{\mathfrak{U}} \left[\frac{\partial^2 \lambda_{t,\mathbf{U}}(\mathbf{x}_t^*)}{\partial \mathbf{x}_t^{*2}}\right] dt \prod_i dU_i \right\} \boldsymbol{\Sigma}_t^*, \tag{7.72}$$

$\boldsymbol{\Sigma}_{t_0}^* = E(\mathbf{x}_0 \mathbf{x}_0') - E(\mathbf{x}_0) E(\mathbf{x}_0')$,

where all terms are defined as in (7.57) and (7.58).

Example 7.3.6 (Optical Position Sensing, continued). Consider Example 7.2.13 again, but assume now that the photosensitive surface is not partitioned into four quadrants. Rather, as in Example 3.3.6, the number, time of occurrence, and position of each photoevent is measured during $[t_0, t)$ to determine the location of the wandering beam at time t. The mark space corresponds to positions in the photosensitive surface, $\mathfrak{U} = \{\mathbf{U} | -1 \leqslant U_1 \leqslant 1, -1 \leqslant U_2 \leqslant 1\}$. The intensity process has the form

$$\lambda_{t,\mathbf{U}}(\alpha_t, \beta_t) = I_0 \exp\left\{ -\frac{1}{2\rho^2}\left[(U_1 - \alpha_t)^2 + (U_2 - \beta_t)^2\right]\right\}, \tag{7.73}$$

where the processes $\{\alpha_t; t \geqslant 0\}$ and $\{\beta_t; t \geqslant 0\}$ defining the beam location are independent Wiener processes such that $E(\alpha_t^2) = E(\beta_t^2) = t(2\tau_d)^{-1}$. As in Example 7.2.13, we take $\mathbf{x}_t$ in (7.56) to be

$$\mathbf{x}_t = \begin{bmatrix} \alpha_t \\ \beta_t \end{bmatrix} = (2\tau_d)^{-1/2}\mathbf{w}_t,$$

in which case $\mathbf{a}_t(\mathbf{x}_t) = \mathbf{0}$ and $\mathbf{B}_t = (2\tau_d)^{-1/2}\mathbf{I}$.

Equations (7.71) and (7.72) do not simplify appreciably in general for the intensity process (7.73), although the estimator can be readily implemented. If the photosensitive surface is large compared to the beam size (that is, $\rho \ll 1$) and displacement (that is, $t \ll \tau_d$), then we may approximate the mark space by $\mathfrak{U} = \{\mathbf{U} | -\infty < U_1 < \infty, -\infty < U_2 < \infty\}$. With this approximation, (7.71) and (7.72) become

$$d\begin{bmatrix} \alpha_t^* \\ \beta_t^* \end{bmatrix} = \rho^{-2}\boldsymbol{\Sigma}_t^* \int_{-\infty}^{\infty}\int_{-\infty}^{\infty} \begin{bmatrix} U_1 - \alpha_t^* \\ U_2 - \beta_t^* \end{bmatrix}$$

$$\times N(dt \times dU_1 \times dU_2), \qquad \begin{bmatrix} \alpha_0^* \\ \beta_0^* \end{bmatrix} = \mathbf{0} \tag{7.74}$$

and

$$d\boldsymbol{\Sigma}_t^* = (2\tau_d)^{-1}\mathbf{I}\,dt - \rho^{-2}\boldsymbol{\Sigma}_t^*\boldsymbol{\Sigma}_t^* N(dt \times \mathfrak{U}), \qquad \boldsymbol{\Sigma}_t^* = \mathbf{0}. \tag{7.75}$$

Note from (7.75) that $\boldsymbol{\Sigma}_t^*$ is a diagonal matrix.

The evolution of the estimates, α_t^* and β_t^*, defined by (7.74) and (7.75) may be interpreted as follows. At $t = 0$, $\alpha_0^* = 0$ and $\beta_0^* = 0$ because the beam is known to be at the origin. The position estimate remains at the origin until the first photoevent is observed, say at time W_1 and location $\mathbf{U} = (a_1, b_1)$. During the interval $0 \leqslant t \leqslant W_1$, $\boldsymbol{\Sigma}_t^* = t(2\tau_d)^{-1}\mathbf{I}$, which is the uncertainty in (α_t, β_t). At time W_1+, the estimated position is

$$\begin{bmatrix} \alpha_{W_1+}^* \\ \beta_{W_1+}^* \end{bmatrix} = \rho^{-2}W_1(2\tau_d)^{-1}\begin{bmatrix} a_1 \\ b_1 \end{bmatrix}.$$

Thus, the estimate moves in the direction of the observed photoevent by an amount proportional to the uncertainty in (α_t, β_t), as represented by $W_1(2\tau_d)^{-1}$, and inversely proportional to the beam spread, as represented by ρ^{-2}. The position estimate remains fixed at $(\alpha_{W_1+}^*, \beta_{W_1+}^*)$ until the second photoevent is observed, say at time W_2 and location $\mathbf{U}=(a_2,b_2)$. During the interval $W_1<t\leqslant W_2$,

$$\Sigma_t^*=(t-W_1)(2\tau_d)^{-1}\mathbf{I}+W_1(2\tau_d)^{-1}\left[1-\rho^{-2}W_1(2\tau_d)^{-1}\right]\mathbf{I}.$$

At time W_2+, the estimated position is

$$\begin{bmatrix}\alpha_{W_2+}^*\\ \beta_{W_2+}^*\end{bmatrix}=\begin{bmatrix}\alpha_{W_1+}^*\\ \beta_{W_1+}^*\end{bmatrix}+\rho^{-2}\Sigma_{W_2}^*\begin{bmatrix}a_2-\alpha_{W_1+}^*\\ b_2-\beta_{W_1+}^*\end{bmatrix}.$$

Thus, the estimate moves in the direction of the observed photoevent by an amount proportional to the conditional uncertainty in (α_t,β_t) given the first photoevent, as represented by $\Sigma_{W_2}^*$, and inversely proportional to the beam spread. The estimated position (α_t^*,β_t^*) continues in this fashion, remaining fixed between photoevents and instantaneously moving in the direction of each photoevent as it occurs, the displacement being proportional to the conditional uncertainty in (α_t,β_t) and inversely proportional to the beam spread.

We mention that the information process $\mathbf{x}$ in this example can be extended to include position processes (α_t,β_t) that are general Markov diffusions, fading processes (Rayleigh, Ricean, and lognormal) caused by atmospheric turbulence, and analog modulation signals. The extended information process may be estimated in principle using (7.71) and (7.72). We mention that the position sensing problem considered here is solved without any approximation by D. Snyder and P. Fishman [20].

Summary. Marked point processes having a nondenumerable mark space have been outlined in this section. Expressions for the sample-function density and for the posterior statistics of an information process influencing the marked point process are given in (7.65), (7.69), and (7.70). The use of these expressions in decision and estimation problems in which a marked point process models observed data is entirely analogous to Section 7.2 and earlier chapters.

We refer to P. Fishman [4] and P. Fishman and D. Snyder [5] for a more rigorous treatment of marked point processes having a nondenumerable mark space and for further extensions. A particularly important extension from an applications viewpoint is the inclusion of an information process that depends on spatial as well as temporal coordinates; that is, an information process that is a random time-space field. A general theory for estimating a random time-space field from observations of a random time-space point process it influences is developed. A large number of applications seem to fit the model, including imaging in the photographic and optical sciences.

REFERENCES

1. F. J. Beutler and O. A. Leneman, "Random Sampling of Random Processes: Stationary Point Processes," *Information and Control*, **9**, 325 (1966).
2. A. T. Bharucha-Reid, *Elements of the Theory of Markov Processes and Their Applications*, McGraw-Hill, New York, 1960.
3. W. Feller, *An Introduction to Probability Theory and Its Applications: Volume II*, Wiley, New York, 1966.
4. P. M. Fishman, "Statistical Inference for Space-Time Point Processes," D. Sc. Thesis, Sever Institute of Technology, Washington University, St. Louis, Mo., June 1974.
5. P. M. Fishman and D. L. Snyder, "Estimation of a Random Space-Time Field that Influences an Observed Space-Time Point Process," Proceedings of the Eleventh Annual Allerton Conference on Circuit and System Theory, Univ. of Ill., Urbana, October 1973, pp. 597–604.
6. I. Gikhman and A. Skorokhod, *Introduction to the Theory of Random Processes*, Saunders, Philadelphia, 1969.
7. R. E. Kahn and B. Liu, "Sampling with Time Jitter," Princeton Univ. Communications Lab., Tech. Rpt. No. 8, June 1964.
8. R. E. Kahn and B. Liu, "Optimum Prefiltering in Sampled-Data Systems with Read-In Jitter," *IEEE International Convention Record*, Pt. 7, pp. 127–133, 1965.
9. A. J. Lawrance, "Selective Interaction of a Stationary Point Process and a Renewal Process," *J. Appl. Prob.*, **7**, 483–489 (1973).
10. A. J. Lawrance, "Selective Interaction of a Poisson and Renewal Process," *J. Appl. Prob.*, **8**, 170–183 (March 1971).
11. E. Parzen, *Stochastic Processes*, Holden-Day, San Francisco, 1962.
12. D. H. Perkel, G. L. Gerstein, and G. P. Moore, "Neuronal Spike Trains and Point Processes: Simultaneous Spike Trains," *Biophys. J.*, **7**, 419–440 (1967).
13. R. Pyke, "Markov Renewal Processes: Definitions and Preliminary Properties," *Ann. Math. Stat.*, **32**, 1231–1242 (1961).
14. R. Pyke, "Markov Renewal Processes with Finitely Many States," *Ann. Math. Stat.*, **32**, 1243–1249 (1961).
15. S. M. Ross, *Applied Probability Models with Optimization Applications*, Holden-Day, San Francisco, 1970.
16. I. Rubin, "Regular Jump Processes and Their Detection and Filtering," Tech. Rpt. UCLA-ENG-7306, Univ. Calif., Los Angeles, December 1972.
17. D. Snyder, "Information Processing for Observed Jump Processes," *Information and Control*, **22**, No. 1, 69–78 (Feb. 1973).
18. M. Ten-Hoppen and H. Reuver, "Recurrent Point Processes with Dependent Interference with Reference to Neuronal Spike Trains," *Math. Biosci.*, **2**, 1–10 (1968).
19. A. I. Yashin, "Filtering of Jump Processes," *Avtomat. i Telemeh.*, **5**, 52–58 (1970).
20. D. L. Snyder and P. M. Fishman, "How to Track a Swarm of Fireflies by Observing Their Flashes," *IEEE Trans. on Information Theory*, (Nov. 1975).

AUTHOR INDEX

Anderson, B.D.O., 308, 320, 403
Anger, H. O., 384, 403
Aoki, M., 106
Apostol, T. M., 44, 49, 106, 192, 222
Aroian, L., 106, 110

Baggeroer, A. B., 304, 308, 403
Barankin, E. W., 106, 121
Bar-David, I., 73, 106, 222
Beutler, F. J., 468, 476
Bharucha-Reid, A. T., 26, 34, 205, 209, 222, 237, 264, 272, 441, 476
Breiman, L., 140, 157
Brémaud, P. M., 383, 403
Brockett, R. W., 188, 214, 222
Bucy, R. S., 320, 403

Chernoff, H., 98, 106
Cinlar, E., 256, 272
Clark, J. R., 289, 290, 317, 403, 404, 409
Cox, D. R., 71, 106, 272, 284, 403
Cox, J. R., Jr., 77, 107
Curran, T. F., 97, 106

Daley, D. J., 42, 106, 237, 272
Davenport, W. B., 165, 222, 236, 272
Davis, D. O., 108, 405
DeSoer, C. A., 188, 223
Dollery, C. T., 108
Doob, J. L., 13, 29, 33, 34, 191, 209, 223
Dorogovcev, A. Ja., 200, 209, 223
Dunn, K. P., 383, 403

Eichling, J. O., 108, 405
Esposito, R., 314, 403
Evans, J., 3, 34, 166, 223, 269, 270, 272
Evans, R. D., 71, 107

Feller, W., 262, 264, 272, 337, 403, 476
Fishman, P. M., 13, 34, 200, 223, 272, 383, 404, 435, 475, 476
Fisz, M., 71, 107
Fluchel, R. B., 107, 123
Forrester, R. H., 190, 221, 223, 379, 404
Franken, P., 255, 272

Gagliardi, R. M., 72, 107, 190, 220, 223, 290, 404
Geiger, H., 108, 119
Gerstein, G. L., 442, 476
Gihman, I. I., 11, 35, 191, 197, 200, 209, 223
Gikhman, I. I., 13, 26, 31, 35, 43, 46, 47, 50, 107, 110, 143, 157, 209, 223, 264, 272, 468, 476
Goodman, J. W., 384, 404
Grandell, J., 296, 317, 404
Grigelionis, B., 255, 273

Haber, S., 342, 404
Haight, F., 56, 107, 109, 110
Halme, S. J., 288, 314, 404
Hance, H., 107, 123
Haney, M., 190, 220, 223
Harger, R. O., 288, 314, 404
Harris, C. C., 141, 157, 384, 405
Helstrom, C., 73, 106, 107
Herrmann, G. F., 107, 123
Hoversten, E. V., 73, 107, 174, 223, 288, 289, 314, 384, 403, 404
Huang, H., 144, 157
Hugh-Jones, P., 108

Itô, K., 191, 223

Jackson, J., 378, 405
Jacques, J. A., 72, 107
Jansson, B., 62, 107
Jazwinski, A. H., 11, 35, 320, 346, 353, 404
Joseph, P. D., 320, 403
Joy, M. L. G., 404

Kahn, R. E., 468, 476
Kailath, T., 304, 404
Karp, S., 72, 107, 290, 404, 409
Kennedy, R. S., 166, 223, 289, 404
Khinchin, A. Y., 40, 42, 43, 107
König, D., 128, 157
Kushner, H. J., 12, 35
Kuznetsov, P. I., 237, 273, 276

Lachs, B., 291, 404
Langevin, P., 186, 223
Lawrance, A. J., 442, 476
Lee, G. M., 107, 123
Leneman, O. A., 468, 476
Lewis, P. A. W., 71, 106
Lindsey, W. C., 190, 223
Littlefield, W. M., 74, 107
Liu, B., 468, 476
Liusternik, L., 208, 223
Lo, J. T., 403, 404
Loève, M., 13, 31, 35
Loo, S. G., 308, 320, 403
Lundberg, O., 404, 409

Macchi, O., 237, 273, 304, 317, 404
Mandel, L., 72, 107
Markham, J., 77, 107
Matthes, K., 128, 129, 157
McFadden, J. A., 237, 273
Mehta, C. L., 290, 404
Melsa, J. L., 11, 13, 35
Miller, H. D., 272
Moody, N. F., 384, 404
Moore, G. P., 403, 442, 476

Moore, J. B., 308, 320
Moyal, J. E., 13, 35, 237, 273
Mullani, N., 141, 157

Neveu, J., 13, 35

O'Neill, E. L., 72, 107, 290, 404
Ortega, J., 107

Paddon, E. A., 107, 123
Papoulis, A., 13, 35, 183, 184, 223
Parzen, E., 13, 35, 66, 107, 165, 185, 223, 272, 273, 441, 476
Paul, W., 404
Peierls, R., 107, 121
Perkel, D. H., 442, 476
Picinbono, B., 404
Pratt, W. K., 188, 223, 323, 404
Pyke, R., 441, 442, 476

Ralston, A., 62, 107
Ramakrishnan, A., 237, 273
Rao, C. R., 80, 107
Reuver, H., 442, 476
Rheinholdt, W., 107
Rhodes, I. B., 372, 383, 403, 405
Rice, S. O., 237, 273
Riordan, J., 108, 110
Robinson, L. M., 367, 404
Root, W. L., 165, 222, 236, 272
Rosenblatt, M., 290, 405
Ross, D. A., 141, 157, 384, 405
Ross, M., 97, 106
Ross, S. M., 441, 442, 476
Roydan, H. L., 33, 35
Rubin, I., 237, 257, 273, 435, 476
Rudemo, M., 405
Rutherford, E., 108, 119

Sage, A. P., 11, 13, 35
Saleh, B. E. A., 405, 415
Savage, L. J., 71, 108
Segall, A., 383, 405
Serfozo, R. F., 296, 405
Sheppard, C. W., 72, 108
Siebert, W. M., 73, 108
Simon, M. K., 190, 223
Skorokhod, A. V., 11, 13, 26, 31, 35, 43, 46, 47, 50, 107, 110, 143, 157, 191, 197, 198, 200, 205, 209, 223, 264, 272, 468, 476
Smith, W. L., 237, 272, 273
Snyder, D. L., 13, 35, 73, 77, 107, 190, 221, 223, 353, 372, 379, 381, 382, 383, 403, 404, 405, 435, 475, 476
Sobolev, V., 208, 223
Srinivasan, S. K., 236, 273
Stratonovich, R. L., 12, 35, 237, 273, 276

Ten-Hoppen, M., 442, 476
Ter-Pogossian, M. M., 74, 108, 384, 405
Titterton, P. J., 314, 405
Torretta, C., 107, 123

Van Schuppen, J. H., 383, 405
Van Trees, H. L., 71, 80, 98, 106, 108, 290, 304, 307, 308, 314, 320, 332, 374, 405, 409
Vasudevan, R., 236, 273
Vere-Jones, D., 7, 35, 42, 106, 237, 272
Viterbi, A. J., 190, 220, 223, 378, 405

Welch, M. J., 108, 405
West, J. B., 108
Wilf, H., 62, 107
Wilks, S. S., 108
Wold, H., 237, 273
Wong, E., 11, 30, 35, 191, 209, 223, 320, 334, 405

Yashin, A. I., 476
Yovits, M., 378, 405

Zadeh, L. A., 188, 223

SUBJECT INDEX

Aftereffects, evolution without, 11, 42
Almost surely, convergence, 14
Amplitude modulation, 322, 427
Anger scintillation camera, 141
Angle modulation, performance, 428
Astronomy, 4
Asymptotically efficient estimate, 84
Atmospheric noise, 3, 166, 269
Auditory electrophysiology, 2, 73, 78, 120
Automatic gain control, 430

Barankin bound, 121
Bayes' criterion, 89
Bayesian regression, 322
Bias, 79
Birth process, 111, 440
  linear, 111
Bounded convergence theorem, 33
Brownian motion, 186

Campbell's theorem, 172
Central limit theorem, 178
Characteristic form of differential generator, 332
Characteristic function, 19, 20, 21
  for compound Poisson process, 133
  for doubly stochastic Poisson process, 407
  for filtered Poisson process, 170, 227
  for Poisson process, 68
Characteristic functional, 67
  for compound Poisson process, 131
  for filtered Poisson process, 169, 227
  for Poisson process, 67
Chernoff bound, 98, 158
Compound Poisson process, 129-162, 284
  characteristic function, 132
  characteristic functional, 131
  first- and second-order statistics, 135
  sample-function density, 148
  statistical inference, 145
Conditional density, 22
  of filtering, 393
  of parameter estimation, 393, 394, 397
  of smoothing, 393, 401
Conditional expectation, 22-24
  optimality of, 24, 330
Conditional median estimate, 331
Conditional mode estimate, 332
Conditional orderliness, 50
  for self-exciting point processes, 238
Conditional probability, 22
Conditional survival probability, 245, 309
Convergence, almost everywhere, 31
  in distribution, 32
  everywhere, 31
  in mean square, 31
  in probability, 31
  in quadratic mean, 31
  in *rth* mean, 31
  with probability one, 31
Count-conditional intensity, 240
Counting integral, 64, 191-198
  definition, 195
  existence, 195
  properties, 195
Counting loss correction, 251
Counting process, 7
Counting statistics, 7
  for self-exciting point process, 240-244
  for doubly stochastic Poisson process, 286-300
Count-rate histogram, 74
Count-record data, 77
  Cramér-Rao bound, 80
  likelihood ratio, 93
Covariance matrix, 18
Cramér-Rao bound, 80, 161, 374
Cumulant generating function, 109
Cumulants, 109

Dark current, 177
Dead time, 246, 281
  correction for, 251
Differential generator, 213
Differential rule, 199-204, 385
  component product, 203
  exponential, 203
  generalized, 209
  logarithm, 204
  multichannel, 385
  power law, 202
Discharges, auditory nerve, 2, 73, 78
  lightning, 3, 43, 269
Doubly stochastic Poisson process, 283-432
  as a self-exciting point process, 292
  conditional survival probability, 309
  counting statistics, 286-300
  cumulants, 407
  filtered, 408
  filtering problem, 314-383
  hypothesis testing, 310
  interarrival-time distribution, 414
  joint-occurrence density, 309
  linear filtering, 317-329
  multichannel, 383
  nonlinear filtering, 330-352
  posterior statistics, 336
  sample-function density, 301, 310
  smoothing, 392
  suboptimal filtering, 352
  time statistics, 300-310

Efficient estimate, 84
Electron emissions, 2, 164, 173
Electron microscopy, 4
Emission, electron, 2, 164, 173

photoelectron, 4, 174
radioactive, 4, 141, 153, 363
Ensemble, 24
Equilibrium renewal process, 267
Estimation, maximum likelihood, 75
minimum mean square error, 317, 330
Events, measurable, 14
Evolution without aftereffects, 42
Expectation, 17
conditional, 22
iterated, 23
Explosion, 259

False alarm, 87
Filtered Poisson process, 163-233
characteristic function, 170
characteristic functional, 168
convergence to a Gaussian process, 178-184
covariance function, 172
cumulants, 171
Filtering performance, 371
multichannel, 391
Finite-dimensional covariance, 307
Fixed-point smoothing, 399
Fokker-Planck equation, 334
Forward Kolmogorov differential operator, 334
Forward Kolmogorov equation, diffusion process, 334
jump process, 336
Fredholm determinent, 303
Frequency modulation, 365
performance, 380

Gamma-ray camera, 384
Gaussian light, 288, 303
Generalized bounded convergence, 33
Generalized Poisson process, 134

Hazard function, 269
Histogram data, 74
Cramér-Rao bound, 80
likelihood ratio, 93
Hypothesis testing, 88-106
Bayes' criterion, 89
binary, 88
M-ary, 103
multichannel, 388
performance, 94

Index of dispersion, 407
Information process, 284
Integration by parts, 231
Intensity function, 54
Intensity interferometry, 384
Intensity process, 239, 437
for *ith* mark, 437
Intensity uncertainty, 422
Interarrival time, 56
distribution, 414
Iterated expectation, 23

Joint occurrence-density, 248, 309
Joint probability-density, 16
Joint probability-distribution, 16

Kalman-Bucy filtering, 320
Khinchin decomposition, 117

Langevin's equation, 186
Likelihood ratio, 91
compound Poisson process, 152
doubly-stochastic Poisson process, 311
marked point process, 454
Poisson process, 94
renewal process, 271
self-exciting point process, 253
Linear birth process, 111
Lightning discharges, 3, 269
Limit theorems, counting statistics, 295
filtered point process, 178
pooled point process, 256
Linear dynamical system, 187, 213
covariance for, 215
Lognormal fading, 371
automatic gain control, 430
Lognormal process, 426
Low-energy coherence, 307

Mark accumulator process, 129, 435
Mark space, 166
Marked point process, 128, 433-476
filtering, 455-466
hypothesis testing, 454
parameter estimation, 447, 470
posterior statistics, 456
Markov diffusion process, 333
conditional moments, 346-349
estimation of, 346
Markov jump processes, 335, 441, 468
estimation of, 349
estimation of transition rates, 447
Markov process, 25, 210
Poisson driven, 185-221
transition distribution, 211
Martingale, 194, 231
Maximum *a posteriori* probability estimate, 332
Maximum entropy counting process, 413
Maximum likelihood detection, 91
Maximum likelihood estimation, 75
Mean square-error matrix, 79
Mean square performance, 371
Mean value, 17
Medicine, nuclear, 4, 5, 71, 384
Microscopy, electron, 4
Miss, 88
Mixed Poisson process, 284
m-Memory point process, 257
Modified renewal process, 267
Modulation, amplitude, 322, 427
angle, 364
frequency, 365
on-off, 96, 324, 350, 425
phase, 365
pulse amplitude, 73
pulse frequency, 73
pulse position, 73, 78, 85, 100, 104
Moment factorization, 353

Moving average estimate, 317
Multichannel analysis, 141
Multichannel, differential rule, 385
  doubly stochastic Poisson process, 383
  filtering performance, 391
  hypothesis testing, 388
  linear filtering, 388
  nonlinear filtering, 390
  sample-function density, 386
  suboptimal filtering, 390
Multiple isotope studies, 384
Multiprobe detection system, 384
Mutually exciting point process, 442

Nearest neighbor distance, 113
Noise, atmospheric, 3, 166
  shot, 164, 173, 184
Nonlinear filtering, doubly stochastic Poisson process, 330-382
  marked point process, 455-464, 473-475
  multichannel, 390
Nuclear medicine, 4, 5, 71, 384

Occurrence time, 56
On-off light, 324, 350, 425
Optical detection, 72
Optical position sensing, 149, 156, 450, 463, 473
Optical radar, 312
Optical range finding, 73, 78, 85
Optical range-rate, 73
Optimum linear estimator, 317
Orderliness, 41
  conditional, 50
  uniform, 41
  Khinchin, 42
Order statistics, 65
Ordinary renewal process, 267
Output equation, 188
Oxygen-15, 363

Paralyzable counter, 281
Parameter estimation, accuracy, 79-86
  for compound Poisson process, 148, 156
  for marked point process, 447, 470
  for Poisson process, 70-87
  for self-exciting point process, 251
Performance, angle modulation, 428
  Chernoff bound, 98, 158
  Cramér-Rao bound, 80, 161, 374
  frequency modulation, 380
  nonlinear filtering, 371-383
  phase modulation, 377
Phase-lock loop, 190, 220
Phase modulation, 365, 377
Phase synchronization, 370, 378
Phase tracking, 188, 217, 285
Photo conductive detector, 165
Photocount statistics, Gaussian light, 288
Photoelectron emissions, 4
Photodetector models, 176
Pinned conditional filtering density, 393
Poisson driven Markov process, 185-221
  calculus for, 192
  estimation of, 343
  existence and uniqueness, 204
  statistics for, 210
Poisson process, characteristic function, 55, 67
  conditional occurrence times, 65
  conditions for, 43, 50
  exponential interarrivals, 60
  forward occurrence density, 59
  gamma occurrence times, 61
  generalized, 134
  homogeneous, 54
  inhomogeneous, 54
  intensity function, 40
  joint occurrence density, 57
  likelihood function, 75, 77
  multichannel, doubly stochastic, 383
  parameter function, 39
  sample function density, 62, 63
  simulation of, 62
  stochastic continuity, 55
Polya process, 288
Pooled point process, 253
Population explosion, 259
Posterior characteristic function, 472
Power-density spectrum, 29
Preemphasized frequency modulation, 366
Probability, density function, 15
  conditional, 22
  distribution function, 15
  measure, 14
Process, continuous time, 24
  counting, 7
  discrete time, 24
  Gaussian, 29
  independent increment, 26
  Markov, 25-26, 185-222, 332-336
  orthogomal increment, 26
  Poisson, 37
  white, 30
  wide sense stationary, 29
  Wiener, 27
Processor equation, 354
Product density, 275-278
Pseudo estimate, 312

Quantum limit, 97

Radar clutter, 166
Radioactive, emissions, 2, 4, 363
  decay, 71
Randomly sampled random process, 468
Random process, characterization of, 25
Random telegraph wave, 112
Random variable, Bose-Einstein, 20
  complex, 19
  continuous, 15, 21
  definition, 14
  discrete, 15, 20
  exponenetial, 21
  gamma, 21
  negative binomial, 20
  normal, 21
  Poisson, 20
  uniform, 21
Rayleigh fading, 314

Realization, 24
Renewal process, 267
  hypothesis testing for, 271
  parameter estimation for, 269
Riccati equation, 321
Ricean fading, 314

Sample Function Density, compound poisson process, 146, 156
  definition, 63
  doubly stochastic Poisson Process, 301, 310
  Gaussian light, 303
  marked point process, 443-448, 472
  moments of, 65
  multichannel, 386
  self-exciting point process, 443-449
Sample space, 13
Scatter communication channel, 166
Schottky's formula, 174
Scintigraphy, 4
Seismic events, 5
Self-exciting birth-death process, 440
Self-exciting point process, 235-282
  birth-death process, 440
  counting moments, 274
  hypothesis testing, 252
  joint-occurrence density, 248
  limited memory, 256-272
  parameter estimation, 251
  sample function density, 249
  survival probability, 245
  time statistics, 244-253
Semi-invariant, 109
Semi-Markov process, 441
Separable covariance, 307
Separation theorem of detection, 310-314, 388, 419, 452
Shot noise, 164, 173, 184
Sigma field, 13
Signal-to-noise ratio, 174
Smoothing, 392, 418, 432
  fixed-interval sweep, 392, 401
  fixed lag, 392, 402
  fixed point, 392, 399-401
Space, probability, 13
  sample, 13
Spatial diversity, 384
State equation, 188
State-transition matrix, 188
Stationary increments, 28
Statistics, counting, 7
  time, 7
Stochastic differential equation, 186
Subcarrier angle modulation, 364
Suboptimal nonlinear filtering, 352
  multichannel, 390
Sums of point processes, 253

Tanlock loop, 190, 221, 367, 368
Tchebycheff inequality, 32
Time-space point process, 468
Time statistics, 7
Transition characteristic function, 212

Union bound, 105
Uniform Lipschitz condition, 205
Updating algorithm, 342, 350, 354

Variance, 18
Variance equation, 354

Waiting time, 56
Weiner process, 27
White noise process, 30
Wide-sense stationary, 29
With probability one, convergence, 14

Yovits-Jackson equation, 378
Yule-Furry distribution, 112

EXAMPLES INDEX

Example		page
1.1.1	Electron Emissions	2
1.1.2	Electrical Activity in Nerves	2
1.1.3	Radioactivity	2
1.1.4	Lightning Discharges	3
1.1.5	Photoelectron Emissions	4
1.1.6	Astronomy	4
1.1.7	Scintigraphy in Nuclear Medicine	4
1.1.8	Electron Microscopy	4
1.1.9	Dynamic Studies in Nuclear Medicine	5
1.1.10	Seismic Events	5
2.4.1	Radioactive Decay	71
2.4.2	Nuclear Medicine	71
2.4.3	Optical Detection	72
2.4.4	Auditory Electrophysiology	73
2.4.5	Scaled Inhomogeneous Poisson Process	76
2.4.6	Pulse-Position Modulation, Optical Range Finding	78
2.4.7	Auditory Electrophysiology	78
2.4.8	Scaled Inhomogeneous Poisson Process	84
2.4.9	Pulse-Position Modulation, Optical Range Finding	85
2.5.1	Binary Hypothesis Testing for a Poisson Process: Histogram Data	94
2.5.2	Binary Hypothesis Testing for a Poisson Process: Count-Record Data	94
2.5.3	On-Off Modulation	96
2.5.4	Binary Pulse-Position Modulation	100
2.5.5	M-ary Pulse-Position Modulation	104
3.2.1	Characteristic Function	132
3.2.2	Independent Increments	133
3.2.3	Poisson Process with Nonunit Jumps	133
3.2.4	Denumerable Mark Space: Generalized Poisson Counting Process	134
3.3.1	Multichannel Analysis	141
3.3.2	Parameter Estimation for a Compound Poisson Process with a Denumerable Mark Space	148
3.3.3	Optical Position Sensing	149
3.3.4	Hypothesis Testing for a Compound Poisson Process with a Denumerable Mark Space	152
3.3.5	Distinguishing Between Two Radioactive Elements	153
3.3.6	Optical Position Sensing	156
4.1.1	Shot Noise	164
4.1.2	Busy Telephone Channels	165
4.1.3	Photoconductive Detector	165
4.1.4	ELF-VLF Atmospheric Noise	166
4.1.5	Radar Clutter, Scatter Communication Channels	166
4.1.6	Linear, Time-Invariant Filter	171
4.1.7	Campbell's Theorem	172
4.1.8	Shot Noise	173
4.1.9	Signal-to-Noise Ratio in Photodetection	174
4.1.10	Homogeneous Point Process	182
4.1.11	Bounded Response Function	183
4.1.12	Time-Invariant and Bandlimited Response Function	183
4.1.13	Shot Noise	184
4.2.1	Brownian Motion	186
4.2.2	Linear, Dynamical Systems	187
4.2.3	Phase-Tracking Loop Excited by Shot Noise	188

4.2.4	Evaluation of an Integral	198
4.2.5	Power Law	202
4.2.6	Component Product	203
4.2.7	Exponential	203
4.2.8	Logarithm	204
4.2.9	Poisson Counting Process	213
4.2.10	Linear Dynamical System	213
4.2.11	Covariance for a Linear Dynamical System	215
4.2.12	Phase-Tracking Loop Excited by Shot Noise	217
5.2.1	Counting Statistics for a Particular Self-Exciting Point Process	243
5.2.2	Dead-Time in Gamma-Photon Detectors	246
5.2.3	Parameter Estimation for Self-Exciting Point Processes	251
5.2.4	Correction for Counting Losses Due to Fixed Dead-Time	251
5.2.5	Hypothesis Testing for Self-Exciting Point Processes	252
5.2.6	Hypothesis Testing for a Particular Self-Exciting Point Process	253
5.3.1	A Population Explosion	259
6.1.1	Optical Detection	285
6.1.2	Optical Phase Tracking	285
6.2.1	Randomly Scaled Inhomogeneous Poisson Process	287
6.2.2	Photocount Statistics for Gaussian Light	288
6.2.3	Randomly Scaled Inhomogeneous Poisson Process	294
6.2.4	Poisson Process	298
6.2.5	Gamma Distributed Parameter	298
6.3.1	Randomly Scaled Inhomogeneous Poisson Process	301
6.3.2	Sample Function Density for Gaussian Light	303
6.4.1	Optical Radar	312
6.5.1	Bayesian Regression	322
6.5.2	Subcarrier Amplitude Modulation in Direct-Detection Optical Communication Systems	322
6.5.3	On-Off Modulated Light	324
6.5.4	Estimation of Random Variables	340
6.5.5	Estimation of a Poisson Driven Markov Process	343
6.5.6	Estimation of a Markov Diffusion-Process	346
6.5.7	Conditional Moments of a Markov Diffusion-Process	346
6.5.8	Estimation of a Markov Jump-Process	349
6.5.9	On-Off Modulated Light	350
6.5.10	One-Parameter Exponential Intensity	358
6.5.11	Two-Parameter Exponential Intensity	361
6.5.12	Subcarrier Angle Modulation	364
6.5.13	Binary Detection in the Presence of Lognormal Fading	371
6.5.14	Performance for Subcarrier Phase Modulation	377
6.5.15	Performance for Subcarrier Frequency Modulation	380
6.6.1	Nuclear Medicine	384
6.6.2	Direct-Detection Optical Communication Systems with Spatial Diversity	384
6.6.3	Intensity Interferometry	384
7.2.1	Generalized Compound Poisson Process	439
7.2.2	Self-Exciting Counting Process	440
7.2.3	Self-Exciting Birth-Death Process	440
7.2.4	Denumerable-State Markov Jump Processes	441
7.2.5	Semi-Markov Jump-Processes	441
7.2.6	Mutually Exciting Point Processes	442
7.2.7	Parameter Estimation for Marked Point Processes with a Denumerable Mark Space	447
7.2.8	Estimating the Transition Rates of a Markov Jump Process	447
7.2.9	Optical Position Sensing	450
7.2.10	Hypothesis Testing for Marked Point Processes	454
7.2.11	Estimation of a Markov Jump-Process	460
7.2.12	Estimation of a Markov Diffusion-Process that Influences an Observed Marked Point-Process	461
7.2.13	Optical Position Sensing	463
7.3.1	Generalized Compound Poisson Process	467

7.3.2	Nondenumerable-State, Markov Jump-Processes	468
7.3.3	Randomly Sampled Random Process	468
7.3.4	Time-Space Point Processes	468
7.3.5	Parameter Estimation for Marked Point Processes with a Non-denumerable Mark Space	470
7.3.6	Optical Position Sensing	473